VOLUME ONE HUNDRED AND NINETEEN

ADVANCES IN
CANCER RESEARCH

VOLUME ONE HUNDRED AND NINETEEN

ADVANCES IN
CANCER RESEARCH

Edited by

KENNETH D. TEW

Professor and Chairman,
Department of Cell and Molecular Pharmacology,
John C. West Chair of Cancer Research,
Medical University of South Carolina,
South Carolina, USA

PAUL B. FISHER

Professor and Chair,
Department of Human & Molecular Genetics,
Director, VCU Institute of Molecular Medicine,
Thelma Newmeyer Corman Chair in Cancer Research,
VCU Massey Cancer Center,
Virginia Commonwealth University, School of Medicine,
Richmond, Virginia, USA

AMSTERDAM • BOSTON • HEIDELBERG • LONDON
NEW YORK • OXFORD • PARIS • SAN DIEGO
SAN FRANCISCO • SINGAPORE • SYDNEY • TOKYO
Academic Press is an imprint of Elsevier

Academic Press is an imprint of Elsevier
525 B Street, Suite 1800, San Diego, CA 92101-4495, USA
225 Wyman Street, Waltham, MA 02451, USA
32 Jamestown Road, London, NW1 7BY, UK
The Boulevard, Langford Lane, Kidlington, Oxford, OX5 1GB, UK
Radarweg 29, PO Box 211, 1000 AE Amsterdam, The Netherlands

First edition 2013

Notice
No responsibility is assumed by the publisher for any injury and/or damage to persons or
property as a matter of products liability, negligence or otherwise, or from any use or
operation of any methods, products, instructions or ideas contained in the material herein.
Because of rapid advances in the medical sciences, in particular, independent verification of
diagnoses and drug dosages should be made.

ISBN: 978-0-12-407190-2
ISSN: 0065-230X

For information on all Academic Press publications
visit our website at store.elsevier.com

Printed and bound in USA

13   14   15   16   12   11   10   9   8   7   6   5   4   3   2   1

Working together
to grow libraries in
developing countries

www.elsevier.com • www.bookaid.org

CONTENTS

CONTRIBUTORS

Swadesh K. Das
Department of Human and Molecular Genetics, and VCU Institute of Molecular Medicine, Virginia Commonwealth University, School of Medicine, Richmond, Virginia, USA

Santanu Dasgupta
Department of Human and Molecular Genetics; VCU Institute of Molecular Medicine, and VCU Massey Cancer Center, Virginia Commonwealth University, School of Medicine, Richmond, Virginia, USA

Robert DeSalle
VCU Institute of Molecular Medicine, Virginia Commonwealth University, School of Medicine, Richmond, Virginia; Sackler Institute for Comparative Genomics, American Museum of Natural History, and New York University, New York, New York, USA

Luni Emdad
Department of Human and Molecular Genetics; VCU Institute of Molecular Medicine, and VCU Massey Cancer Center, Virginia Commonwealth University, School of Medicine, Richmond, Virginia, USA

Victoria J. Findlay
Department of Pathology and Laboratory Medicine, Hollings Cancer Center, Medical University of South Carolina, Charleston, South Carolina, USA

Paul B. Fisher
Department of Human and Molecular Genetics; VCU Institute of Molecular Medicine, and VCU Massey Cancer Center, Virginia Commonwealth University, School of Medicine, Richmond, Virginia, USA

Chunqing Guo
Department of Human and Molecular Genetics; VCU Institute of Molecular Medicine, and VCU Massey Cancer Center, Virginia Commonwealth University School of Medicine, Richmond, Virginia, USA

Agnieszka Jezierska-Drutel
Department of Cell and Molecular Pharmacology and Experimental Therapeutics, Medical University of South Carolina, Charleston, South Carolina, USA

Amanda C. LaRue
Department of Pathology and Laboratory Medicine, Hollings Cancer Center, Medical University of South Carolina, and Medical Research Service, Ralph H. Johnson Veterans Affairs Medical Center, Charleston, South Carolina, USA

Masoud H. Manjili
VCU Massey Cancer Center, and Department of Microbiology and Immunology, Virginia Commonwealth University School of Medicine, Richmond, Virginia, USA

Carlos T. Moraes
Graduate Program in Cancer Biology; Department of Neurology, and Department of Cell
Biology, University of Miami Miller School of Medicine, Miami, Florida, USA

Carola A. Neumann
Department of Cell and Molecular Pharmacology and Experimental Therapeutics,
Medical University of South Carolina, Charleston, South Carolina, and Department of
Pharmacology and Chemical Biology, University of Pittsburgh Cancer Institute,
Magee-Womens-Research Institute, Pittsburgh, Pennsylvania, USA

Susana Peralta
Department of Neurology, University of Miami Miller School of Medicine, Miami, Florida,
USA

Steven A. Rosenzweig
Department of Cell and Molecular Pharmacology and Experimental Therapeutics, Medical
University of South Carolina, Charleston, South Carolina, USA

Devanand Sarkar
Department of Human and Molecular Genetics; VCU Institute of Molecular Medicine, and
VCU Massey Cancer Center, Virginia Commonwealth University, School of Medicine,
Richmond, Virginia, USA

Rita Shiang
Department of Human and Molecular Genetics, Virginia Commonwealth University,
School of Medicine, Richmond, Virginia, USA

Upneet K. Sokhi
Department of Human and Molecular Genetics, Virginia Commonwealth University,
School of Medicine, Richmond, Virginia, USA

John R. Subjeck
Department of Cell Stress Biology, Roswell Park Cancer Institute, Buffalo, New York, USA

David P. Turner
Department of Pathology and Laboratory Medicine, Hollings Cancer Center, Medical
University of South Carolina, Charleston, South Carolina, USA

Xiang-Yang Wang
Department of Human and Molecular Genetics; VCU Institute of Molecular Medicine, and
VCU Massey Cancer Center, Virginia Commonwealth University School of Medicine,
Richmond, Virginia, USA

Xiao Wang
Graduate Program in Cancer Biology, University of Miami Miller School of Medicine,
Miami, Florida, USA

Dennis K. Watson
Department of Pathology and Laboratory Medicine, and Department of Biochemistry and
Molecular Biology, Hollings Cancer Center, Medical University of South Carolina,
Charleston, South Carolina, USA

Patricia M. Watson
Department of Medicine, Hollings Cancer Center, Medical University of South Carolina, Charleston, South Carolina, USA

Inken Wierstra
Wißmannstr. 17, D-30173 Hannover, Germany

Michael D. Wyatt
Department of Drug Discovery and Biomedical Sciences, South Carolina College of Pharmacy, University of South Carolina, Columbia, South Carolina, USA

CHAPTER ONE

Understanding the Role of ETS-Mediated Gene Regulation in Complex Biological Processes

Victoria J. Findlay[*], Amanda C. LaRue[*,†], David P. Turner[*],
Patricia M. Watson[‡], Dennis K. Watson[*,§,1]

[*]Department of Pathology and Laboratory Medicine, Hollings Cancer Center, Medical University of South Carolina, Charleston, South Carolina, USA
[†]Medical Research Service, Ralph H. Johnson Veterans Affairs Medical Center, Charleston, South Carolina, USA
[‡]Department of Medicine, Hollings Cancer Center, Medical University of South Carolina, Charleston, South Carolina, USA
[§]Department of Biochemistry and Molecular Biology, Hollings Cancer Center, Medical University of South Carolina, Charleston, South Carolina, USA
[1]Corresponding author: e-mail address: watsondk@musc.edu

Contents

Advances in Cancer Research, Volume 119
ISSN 0065-230X
http://dx.doi.org/10.1016/B978-0-12-407190-2.00001-0

Abstract

Ets factors are members of one of the largest families of evolutionarily conserved transcription factors, regulating critical functions in normal cell homeostasis, which when
perturbed contribute to tumor progression. The well-documented alterations in ETS factor expression and function during cancer progression result in pleiotropic effects
manifested by the downstream effect on their target genes. Multiple ETS factors bind
to the same regulatory sites present on target genes, suggesting redundant or competitive functions. The anti- and prometastatic signatures obtained by examining specific
ETS regulatory networks will significantly improve our ability to accurately predict tumor
progression and advance our understanding of gene regulation in cancer. Coordination
of multiple ETS gene functions also mediates interactions between tumor and stromal
cells and thus contributes to the cancer phenotype. As such, these new insights may
provide a novel view of the ETS gene family as well as a focal point for studying the
complex biological control involved in tumor progression. One of the goals of molecular
biology is to elucidate the mechanisms that contribute to the development and progression of cancer. Such an understanding of the molecular basis of cancer will provide
new possibilities for: (1) earlier detection, as well as better diagnosis and staging of disease; (2) detection of minimal residual disease recurrences and evaluation of response to
therapy; (3) prevention; and (4) novel treatment strategies. Increased understanding of
ETS-regulated biological pathways will directly impact these areas.

1. INTRODUCTION

1.1. The ETS gene family

The oncogene v–ets was discovered in 1983 as part of the transforming
fusion protein (p135, gag–myb–ets) of E26, a replication–defective avian retrovirus. Both v–ets and v–myb contribute to the transformation of different
lineages and cell types. The name ets is derived from *E26 transforming
sequence* or *E-twenty-six* specific sequence. The v–ets oncogene transforms
fibroblasts, myeloblasts, and erythroblasts *in vitro* and causes mixed
erythroid–myeloid and lymphoid leukemia *in vivo* (reviewed in Blair &

Athanasiou, 2000). Molecular comparisons with the predicted chicken c-Ets1 protein demonstrated that the v-ets contained three internal amino acid substitutions and unique carboxy terminal amino acids. This change resulted from the inversion of the 3′ sequences of the chicken gene during retroviral transduction (Lautenberger & Papas, 1993).

All ETS family members are defined by a conserved sequence that encodes the DNA-binding (ETS) domain (Fig. 1.1 and Table 1.1). Identification of v-ets-related genes from metazoan species has established ETS as one of the largest families of transcriptional regulators, consisting of 28 ETS genes in humans, 27 in mice, 11 in sea urchin, 10 in *Caenorhabditis elegans*, and 9 in *Drosophila* (for reviews, see Gutierrez-Hartmann, Duval, & Bradford, 2007; Hollenhorst, McIntosh, & Graves, 2011; Hsu, Trojanowska, & Watson, 2004; Seth & Watson, 2005; Turner & Watson, 2008; Watson, Turner, Scheiber, Findlay, & Watson, 2010 and references therein). The human ETS factors are classified into 12 subgroups based upon ETS domain sequence homology: ETS, ERG, PEA3, ETV, TCF, GABP, ELF1, SPI1, TEL, ERF, SPDEF, and ESE (Hollenhorst, McIntosh, et al., 2011; Seth & Watson, 2005; Watson et al., 2010; see Table 1.1 for subgroup members). In addition, a subset of four ETS family genes (ELF3, ELF5, EHF, SPDEF) has been placed in a unique subgroup based upon their restricted expression to tissues with high epithelial cell content (Feldman, Sementchenko, & Watson, 2003). In this review, the Unigene Names will be used (alternative nomenclatures are provided in Table 1.1).

1.2. ETS protein domains and DNA-binding specificity

The DNA-binding (ETS) and pointed (PNT) domains are the two most common domains present in ETS proteins and will be discussed briefly below. Other domains present in smaller subsets of ETS proteins have been described in previous reviews, and these include the OST GABPA and B-box (TCF subfamily; ELK1, ELK3, ELK4) domains (Hollenhorst, McIntosh, et al., 2011).

The Ets domain is an ~85-amino acid region that forms the winged helix-turn-helix (wHTH) DNA-binding domain composed of three alpha helices and a four-stranded antiparallel beta sheet that recognizes a core GGAA/T sequence (ETS binding site, EBS). The HTH motif is formed by helices H2 and H3. The third alpha helix makes major groove contacts with the DNA (GGAA/T core). Two invariant arginine residues present in helix H3 make contact with the two guanine residues of the EBS.

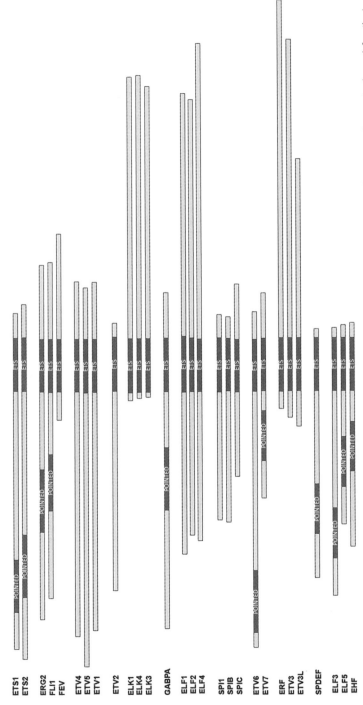

Figure 1.1 The human ETS family of transcription factors. The main structural organization of each human ETS protein by subfamily (see Table 1.1) is depicted. The ETS and Pointed domains are indicated. (For color version of this figure, the reader is referred to the online version of this chapter.)

Table 1.1 The human ETS gene family

Group	Name	Unigene name	Original name	Alternative names	Locus	Size	ETS domain	Pointed domain
1 ETS	ETS1	ETS1	v-ets erythroblastosis virus E26 oncogene homolog 1	EWSR1	11q23.3	441	331–416	54–135
2	ETS2	ETS2	v-ets erythroblastosis virus E26 oncogene homolog 2 (avian)		21q22.3	469	369–443	88–168
3 ERG	ERG2	ERG	v-ets erythroblastosis virus E26 oncogene-like (avian)	erg-3, p55	21q22.3	462	290–375	120–201
4	FLI1	FLI1	Friend leukemia integration 1 transcription factor	ERGB, EWSR2, SIC-1	11q24.1-q24.3	452	277–361	115–196
5	FEV	FEV	Fifth Ewing variant	PET-1, HSRNAFEV	2q23	238	43–126	None
6 PEA3	PEA3	ETV4	Polyoma enhancer binding	E1AF, PEAS3	17q21	462	315–399	None
7	ERM	ETV5	Ets-related molecule		3q28	510	368–449	None
8	ER81	ETV1	Ets variant gene 1		7p21.3	458	314–397	None
9 ETV	ER71	ETV2	Ets variant gene 2	ETSRP71	19q13.12	370	265–350	None
10 TCF	ELK1	ELK1			Xp11.2	428	7–92	None
11	SAP1	ELK4	SRF accessory protein 1A		1q32	431	4–89	None
12	NET	ELK3	Net transcription factor	SAP2, ERP	12q23	407	5–85	None
13 GABP	GABPα	GABPA	GA binding protein transcription factor A	E4TF1, NFT2, NRF2	21q21.3	454	318–400	171–249
14 ELF1	ELF1	ELF1	E74-like factor 1		13q13	619	207–289	None

Continued

Table 1.1 The human ETS gene family—cont'd

Group	Name	Unigene name	Original name	Alternative names	Locus	Size	ETS domain	Pointed domain
15	NERF	ELF2	E74-like factor 2	NERF1, NERF2, EU32	4q21	581	198–277	None
16	MEF	ELF4	E74-like factor 4	ELFR	Xq26	663	204–290	None
17 SPI1	SPI1	SPI1	Spleen focus forming virus (SFFV) proviral integration oncogene spi1	PU.1, SFPI1, SPIA	11p11.2	264	168–240	None
18	SPIB	SPIB	SpiB transcription factor		19q13.3–q13.4	262	169–251	None
19	SPIC	SPIC	SpiC transcription factor		12q23.2	248	111–193	None
20 TEL	TEL	ETV6	Ets variant gene 6		12p13	452	340–419	38–119
21	TEL2	ETV7	Ets variant gene 7	TEL-B, TREF	6p21	264	149–228	49–114
22 ERF	ERF	ERF	Ets2 repressor factor		19q13	548	26–106	None
23	ETV3	ETV3	Ets variant gene 3	METS, PE1, bA110J1.4	1q21–q23	512	34–118	None
24	ETV3L	ETV3L	Ets variant gene 3 like		1q23.1	361	38–121	None
25 PDEF	PDEF	SPDEF	SAM pointed domain containing Ets transcription factor	PSE	6p21.3	335	248–332	138–211
26 ESE	ESE1	ELF3	Epithelium-specific Ets transcription factor 1	ESX, JEN, ERT, EPR1	1q32.2	371	275–354	47–132
27	ESE2	ELF5	Epithelium-specific Ets transcription factor 2		11p13-p12	255	165–243	46–115
28	ESE3	EHF	Epithelium-specific Ets transcription factor 3	ESEJ	11p12	300	209–288	42–112

List of the known human ETS genes (grouped by subfamily), including gene names and alternative nomenclature, chromosomal location, size of protein (amino acids), approximate boundaries of the Ets domain and approximate boundaries of the Pointed domain (65–80 amino acids, if present).

Interestingly, crystallography data indicate that there are no direct contacts outside of the GGAA/T core. The DNA recognition sequence preference for several family members has been determined by *in vitro* selection of randomized oligonucleotides and indicates that target site recognition is dependent on sequences flanking the core motif, suggesting that DNA conformation may contribute to specificity for the flanking regions. A recent comprehensive genome-wide analysis of ETS factor binding specificities was conducted for 26 mouse and 27 human ETS genes using transcription factor DNA-binding specificity and protein-binding microarrays (Wei et al., 2010). These data support the model that ETS family DNA-binding specificities fall into four distinct classes, and identify key DNA-contact amino acids that contribute to class specificity based upon the published crystal structures for ETS1, GABPA, ELK1, ELF3, SPI1, and SPDEF. Class I contains 12 family members (ETS, ERG, PEA3, TCF, and ERF subfamilies) and is defined by an ACCGGAAGT consensus. Class II is composed of eight members (ELF, TEL, and ESE subfamilies) and the binding consensus differs in the first nucleotide, with a CCCGGAAGT sequence preference. Class III contains the three members of the SPI1 family, which bind to sites with an adenine-rich sequence 5′ to the core. Class IV contains a single family member, SPDEF, which has a GGAT core sequence rather than the GGAA. It is evident that ETS proteins often interact with EBS sequences that do not conform to the consensus binding site defined by *in vitro* selection experiments. Binding of ETS proteins to such subconsensus sequences is facilitated by the binding of other transacting factors to cis-elements in proximity to the EBS. Indeed, binding is often mediated by synergistic interaction with transcriptional partners on composite DNA elements. The most studied ETS synergistic interactions include those with AP1 (fos/jun), SRF, RUNX (AML), SP1, PAX5, and GATA1 (discussed further below).

DNA binding by multiple ETS factors is inhibited by two regions that flank the DNA-binding domain. Best exemplified by ETS1, this autoinhibition is stabilized by posttranslational modification (serine phosphorylation) on the region encoded by exon VII. Interestingly, exon VII undergoes alternative splicing, resulting in an isoform that binds DNA with higher affinity (Fisher et al., 1994). The alterations at the 3′ end of v-ets are functionally critical to the transforming properties of the virus, since the residues encoded by the 3′ region of c-ets have been shown to be capable of repressing the DNA-binding potential of c-ets; thus, the viral oncoprotein does not undergo autoinhibition and has higher DNA-binding activity.

ETS1 autoinhibition is also reduced by interaction with transcriptional cofactors, such as RUNX1 and PAX5.

The second conserved domain found in a subset of ETS genes is the pointed (PNT) domain. This 65–85 amino acid domain belongs to the sterile alpha motif (SAM) family and is found in 11 of 28 human ETS genes and has been shown to function in protein–protein interaction and homo- and heterooligomerization. The PNT domains of several ETS factors also provide the docking site for regulation of extracellular signaling pathways. For example, ERK phosphorylation of ETS1 and ETS2 on threonine phosphor-acceptor sites increases their resultant transcriptional activity through enhanced interaction with the histone acetylase CBP/p300 (Foulds, Nelson, Blaszczak, & Graves, 2004).

In summary, the DNA consensus sequences determined for the different ETS proteins are very similar, and thus specificity is dependent on other factors including interaction with other nuclear factors. Such a dependence of lower affinity ETS binding sequences on coexpression and binding of cofactors would be anticipated to provide greater biological specificity.

2. MODULATION OF ETS FUNCTION

ETS functional activity is modulated at multiple levels. As noted above, ETS factors are dependent on interaction with other factors for precise transcriptional regulation. Indeed, maximal transcriptional activation of multiple target genes is dependent on simultaneous expression of ETS and other transcription factors. Second, specific intracellular signaling pathways and posttranslational modifications directly affect the activity of several ETS proteins by regulating subcellular compartmentalization, DNA-binding activity, and transactivation potential or stability.

2.1. Regulation by protein–protein interactions

Transcriptional regulation is dependent upon the combinatorial interactions between multiple nuclear proteins. ETS proteins form complexes with many transcription factors and such interactions may strengthen the transcriptional activity and/or define target gene specificity. Tissue-specific combination of ETS with other cofactors also provides a mechanism for proper regulation of relevant target genes in a particular cell type. Many transcription factors have their DNA-binding sites adjacent to EBS (for reviews, see Li, Pei, & Watson, 2000; Verger & Duterque-Coquillaud, 2002). As mentioned above, well-studied ETS interactions with transcriptional cofactors include those with AP1 (fos/jun), SRF, RUNX (AML), PAX5, SP1, and GATA1. Depending on the precise sequence context, binding of an ETS protein near other transcription factors results in higher affinity interaction, synergistic activation, and/or repression of specific target genes.

Among the earliest characterized protein–protein interactions was that between ETS factors and the AP1 transcriptional complex. Interaction was shown to result in synergistic transcriptional activation of promoters containing composite AP1-EBS binding sites, including MMP1, uPA, GM-CSF, maspin, and TIMP-1. In contrast, MafB, an AP1-like protein, inhibits ETS1-mediated transactivation of the AP1-EBS sites (Sieweke, Tekotte, Frampton, & Graf, 1996). ETS/AP1-binding sequences are proto-typical RAS-responsive elements and oncogenic ETS factors (ETV1, ETV4, ETV5, and ERG) have been shown to activate a RAS/MAPK transcriptional program in prostate cells in the absence of MAPK activation (Hollenhorst, Ferris, et al., 2011).

Another well-characterized interaction involves SRF and ELK1 (or ELK3, ELK4, FLI1, EWS-FLI1) that together form a ternary complex with the SRE motif present in several genes, including c-fos, Egr-1, pip92 Mcl-1, and SRF(Buchwalter, Gross, & Wasylyk, 2004).

RUNX1 and ETS1 interaction counteracts autoinhibition of DNA-binding activity (Garvie, Pufall, Graves, & Wolberger, 2002) and homotypic ETS1 interaction enhances binding to palindromic EBS (Baillat, Begue, Stehelin, & Aumercier, 2002). Interaction with PAX5 allows ETS1, as well as other family members, to bind to a nonconsensus EBS present in the early B-cell-specific mb-1 promoter (Fitzsimmons, Lutz, Wheat, Chamberlin, & Hagman, 2001).

SPI1 family proteins can function as activators or repressors of transcription and have been shown to interact with ETS factors with cell- and promoter-specific consequences. For example, functional interaction of FLI1 with SP1 or SP3 is essential for the inhibitory function of Fli1 on the collagen A2 promoter (Czuwara-Ladykowska, Shirasaki, Jackers, Watson, & Trojanowska, 2001).

FLI1 and GATA-1 act synergistically to activate gene transcription of multiple megakaryocytic genes, including *gpIIb*, *gpVI*, *gpIX*, *gpIb*, and *c-mpl* (reviewed in Szalai, LaRue, & Watson, 2006). We and others have demonstrated that FLI1 and GATA1 co-occupy these promoters *in vivo* (Jackers, Szalai, Moussa, & Watson, 2004; Moussa et al., 2010; Pang et al., 2006).

Several proteins that modulate ETS function have been identified, including Daxx (EAP1 (ETS1-associated protein 1)), EAPII, and SP100 (Li, Pei, Watson, & Papas, 2000; Pei et al., 2003; Yordy et al., 2004). The notion that loss of corepressor protein expression is relevant to cancer was demonstrated using the NCoR corepressor protein and the coregulators SRC-1 and AIB1, all of which interact with both ETS1 and ETS2 (Myers et al., 2005). The strongest clinical association in breast cancer was for NCoR downregulation in more aggressive hormone-unresponsive tumors (Myers et al., 2005).

2.2. Regulation by posttranslational modification

A common feature of many tumors is deregulation of signal transduction pathways, resulting in constitutive and often ligand-independent activation. As end effectors of these pathways, ETS factor function is significantly altered in cancer. In addition to being downstream of many RTKs (e.g., HER2/neu), ETS factors regulate the expression of multiple receptors, including HER2/neu, M-CSF receptor, MET, c-kit, and VEGF receptor (Sementchenko & Watson, 2000).

ETS factor functions are controlled by phosphorylation, acetylation, sumoylation, ubiquitinylation, and glycosylation (for reviews, see Charlot, Dubois-Pot, Serchov, Tourrette, & Wasylyk, 2010; Tootle & Rebay, 2005; Yordy & Muise-Helmericks, 2000).

One of the best-studied posttranslational modifications is phosphorylation. Phosphorylation of ETS proteins mediates effects on DNA binding, protein–protein interaction, transcriptional activation, and subcellular localization. ERK, JNK, and p38 MAP kinases are downstream components of signaling cascades. ERKs are activated in response to mitogenic signals, while JNKs and p38/SAPKs respond to stress signals. Specific ETS factors, including ETS1, ETS2, ELK1, ELK3, ELK4, GABPA, SPIB, ETV1, ETV4, and ETV5, can be phosphorylated by MAPKs, resulting in increased transcriptional activation (Charlot et al., 2010).

As noted above, phosphorylation of a mitogen-activated protein kinase (ERK) site adjacent to the PNT domain has been shown to positively regulate transcriptional activities of ETS1 and ETS2. Although MAP kinase phosphorylation of ETS1 does not affect DNA binding, calcium-induced phosphorylation of ETS1 occurs at serine residues present adjacent to the DNA-binding domain and inhibits ETS1 DNA-binding activity without affecting nuclear localization. ETS1 and ETS2 activity may also be activated by PKC in invasive breast cancer cells (Lindemann, Braig, Ballschmieter, et al., 2003; Lindemann, Braig, Hauser, Nordheim, & Dittmer, 2003). In contrast, ETV6 activity is negatively regulated by MAPK phosphorylation, which results in its nuclear export and decreased DNA-binding activity. Processes that are reversibly controlled by protein phosphorylation require a balance between protein kinase and protein phosphatase activities. Thus, it is important to assess whether specific protein phosphatases are associated with de-phosphorylation of ETS proteins.

Often associated with phosphorylation, acetylation also regulates ETS gene function. Acetylation of ETV1 enhances its DNA-binding activity and ability to transcriptionally activate target genes (Goel & Janknecht, 2003). In response to TGFβ signaling, ETS1 is acetylated and dissociated from the CBP/p300 complexes (Czuwara-Ladykowska, Sementchenko, Watson, & Trojanowska, 2002). FLI1 activity is repressed through a series of sequential posttranslational modifications (Thr312 phosphorylation and acetylation by p300/CREB binding protein-associated factor), resulting in detachment from target gene (e.g., collagen) promoters in response to TGFβ (Asano et al., 2009; Asano & Trojanowska, 2009).

ETS factors undergo ubiquitination and subsequent proteosomal degradation. ETV1, ETV4, and ETV5 each contain three potential binding motifs for the ubiquitin ligase COP1. ETV1 is degraded after being ubiquitinated by COP1. Data support the notion that COP1 functions as a tumor suppressor mediated by its negative regulation of ETV1, ETV4, and ETV5. Indeed, COP1 deficiency in mouse prostate is correlated with elevated ETV1 and increased cell proliferation, hyperplasia, and early prostate intraepithelial neoplasia (Vitari et al., 2011).

Sumoylation has been shown to affect the stability, activity, and localization of its targets. SUMO modification has been found to alter the function of several transcription factors, including ETS family members. For example, FLK1 is modified by SUMO, and this modification is reversed by ERK–MAP kinase pathway activation. Mechanistically, it has been shown that sumoylation of ELK1 facilitates recruitment of histone deacetylase 2 activity to promoters. This recruitment leads to decreased histone acetylation and altered chromatin structure, resulting in transcriptional repression at ELK1 target genes (Yang & Sharrocks, 2004). In contrast, sumoylation within the pointed domain of ETV6 inhibits ETV6-mediated repression (Chakrabarti & Nucifora, 1999; Chakrabarti et al., 1999), associated with sequestering to subnuclear compartments. Mutation of SUMO acceptor site(s) results in increased transcriptional repression, presumably because of decreased nuclear export (Wood, Irvin, Nucifora, Luce, & Hiebert, 2003). Sumoylation of ETS1, ETV4, and ETV5 leads to reduced transcriptional activity.

Future studies will help elucidate the functional impact of specific post-translational modifications on the activity of ETS transcription factors. As specific antibodies are developed, it will be possible to determine the temporal relationships between specific posttranslational events. Through such analyses, it will also be possible to determine whether specific events work cooperatively or antagonistically.

3. DEFINING AND CHARACTERIZING ETS TARGET GENES

3.1. ETS target genes

The importance of the ETS family of transcription factors in various biological and pathological processes necessitates the identification of downstream cellular target genes of specific ETS proteins. Although some overlap in the biological function of different ETS proteins may exist, the emergence of a family of closely related transcription factors suggests that individual ETS members may have evolved unique roles, manifested through the control of specific target genes. Several key areas are critical for understanding what defines a functionally important ETS target gene: First, the functional importance of the EBS must be demonstrated by mutagenesis. Second, the specific ETS factor or factors responsible for transcriptional control of specific target genes need to be identified. While extensive publications have identified functionally important EBS and thus, ETS target genes (Sementchenko & Watson, 2000), fewer investigations have identified definitive target genes for a specific ETS factor.

ETS factors are known to act as positive or negative regulators of the expression of genes, including those that control response to various signaling cascades, cellular proliferation, differentiation, hematopoiesis, apoptosis, adhesion, migration, invasion and metastasis, tissue remodeling, ECM composition, and angiogenesis (Fig. 1.2). Our earlier literature survey enabled identification of over 200 ETS target genes (Sementchenko & Watson, 2000) and to date, over 700 ETS target genes have been defined, based upon the presence of functional EBS in their regulatory regions (Watson, D.K., unpublished). While most ETS factors were initially characterized as transcriptional activators or repressors, it has become evident that several ETS factors can function as either activators or repressors, depending upon the type of promoter and cellular context.

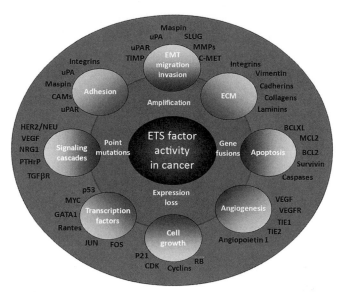

Figure 1.2 ETS factors regulate the expression of genes associated with cancer progression. Dysregulated ETS factor function leads to the altered expression of multiple target genes that are known to play critical roles in many of the processes required for cancer progression. While each of the target genes highlighted has functional EBS(s) in their regulatory regions, the role and relative affinities of specific ETS factors have only been examined in a limited subset. (For color version of this figure, the reader is referred to the online version of this chapter.)

During cancer progression, the oncogenic and tumor-suppressor activities of ETS factors are likely coordinated by their target genes. In the past few years, we and others have made significant strides in identifying and validating these target genes. Collectively, ETS genes have been shown to regulate the expression of genes that have important roles in malignant and metastatic processes (Fig. 1.2). Among these are those that function in control of cell proliferation (e.g., cyclins and cdks), motility (hepatocyte growth factor, HGF), invasion (uPA & uPAR, PAI, MMPs; TIMPs), extravasation (MMPs, Integrins), micro-metastasis (Osteopontin; BSP and Osteonectin), and establishment and maintenance of distant site metastasis and angiogenesis (Neovascularization and Neoangiogenesis (integrin β3, VEGF, Flt-1/KDR, Tie2; Sementchenko & Watson, 2000)). Aberrant expression of ETS factors results in the altered regulation of their target response genes. For example, upregulated ETS target genes include extracellular matrix (ECM)-degrading proteins (e.g., MMP1, MMP9, uPA), which are associated with clinical features such as lymph node status and prognosis in prostate cancer.

Significantly, altered ETS expression also provides a mechanism for the downregulation of response genes that include uPA and survivin. Recent analysis of gene expression signatures allowed correlation between expression of ETS factors, ETS target genes, and prostate cancer progression (Tomlins et al., 2007).

Several studies have demonstrated that a polymorphism that generates a functional EBS within the MMP1 promoter is a negative prognostic indicator (Benbow, Tower, Wyatt, Buttice, & Brinckerhoff, 2002).

Functional studies have demonstrated that SPDEF is a negative regulator of uPA and SLUG mRNA expression. Chromatin immunoprecipitation (ChIP, discussed further below) allows definition of direct target genes for specific ETS factors. However, ChIP alone does not indicate whether the interaction is functional (e.g., causing transcriptional activation or repression). Biological rescue experiments have been used to demonstrate the importance of specific target genes (e.g., SPDEF target genes, uPA, SLUG; Findlay et al., 2011; Turner, Findlay, Kirven, Moussa, & Watson, 2008). Correlation between ChIP and gene expression can further define functional ETS targets. For example, there is an inverse correlation between SPDEF and uPA in primary colon tumors (Moussa et al., 2009).

3.2. ETS gene coexpression

Initial expression analysis supported the notion that while some ETS factors showed rather ubiquitous expression (e.g., ETS2), others had more restricted expression in specific tissues or cells (e.g., ETS1). Subsequent studies have demonstrated the simultaneous expression of 14–25 ETS mRNAs in many human tissues and cell lines. For example, studies examining ETS factor expression profiles in normal and cancerous breast cells have demonstrated that a combination of up to 25 of 28 ETS family members examined is expressed at any one time in these cells (Galang, Muller, Foos, Oshima, & Hauser, 2004; Hollenhorst, Jones, & Graves, 2004). It should be noted that mRNA expression alone does not adequately define the ETS profile, as factors including, but not limited to, alternative splicing, mRNA translation, protein stability, posttranslational modifications, and protein localization ultimately contribute to define the level of functional proteins in a cell. Complete proteomic studies need to be performed to define the relative prevalence of ETS factors in specific tissues. ETS factor function is also highly dependent upon the presence and level of specific coregulatory proteins.

3.3. Whole genome analysis: Redundant and specific binding

Multiple ETS factors bind to the same regulatory sites present on target genes, suggesting redundant or competitive functions. Furthermore, additional events contribute to, or may be necessary for, target gene regulation. As technologies have advanced, it has become possible to identify the true regulatory targets of transcription factors. ChIP has become an established method for the analysis of protein–DNA (gene regulatory elements) interactions *in vivo*. Sequential ChIP is an extension of the ChIP protocol, in which the immunoprecipitated chromatin is subjected to sequential immunoprecipitations with antibodies of different specificity. This provides a method of examining co-occupancy of defined promoters by multiple regulatory proteins. Furthermore, sequential ChIP provides an experimental approach to simultaneously evaluate promoter occupancy and transcriptional status (e.g., histone H3 acetylation, phosphorylated RNAPII-CTD; Jackers et al., 2004). However, ChIP and sequential ChIP methods have been restricted to the analysis of small promoter regions, the boundaries defined by the sequences of the primers designed for the PCR amplification step.

To determine the more global location of *in vivo* promoter binding sites of a specific protein, ChIP protocols have been combined with whole genome analysis methods to produce "ChIP-on-chip" microarrays. ChIP products are amplified and hybridized to arrays consisting of promoter regions, limiting genome coverage. ChIP sequencing (ChIP-seq) is the next generation protocol for defining Protein–DNA transcriptomes (Farnham, 2009; Schmidt et al., 2009; Visel et al., 2009). It combines ChIP with new high throughput sequencing platforms, such as Genome Analyzer (Solexa/Illumina), generating significantly more informative data (Mardis, 2007).

In the context of an ETS transcription network, ChIP-Seq analysis can potentially identify the full transcriptome for each individual ETS family member in any given scenario. Furthermore, by comparing ChIP-Seq data with mRNA expression profiles obtained following modulation of ETS expression, direct and indirect targets for each ETS factor can be ascertained. Recent genome-wide analyses of ETS-factor occupancy have identified genomic regions, both promoters and enhancers, bound by individual ETS proteins in living cells. Nine ETS proteins have been assayed for genome-wide occupancy by either promoter microarrays (ETS1, GABPA, ELF1, ELK1, EWS-FLI1, and SPI1) or high throughput sequencing (GABPA, ETS1, ERG, FLI1, EWS-ERG, EWS-FLI1, SPI1, SPDEF, ETV1, and ELF1; Hollenhorst, McIntosh, et al., 2011).

Genome-wide occupancy data for several ETS proteins have been compared and found to have a high degree of similarity. Genomic targets of ETS transcription factors can be divided into two classes (Hollenhorst et al., 2009; Hollenhorst, McIntosh, et al., 2011; Hollenhorst, Shah, Hopkins, & Graves, 2007). Class 1: redundant binding sites found in the proximal promoters of housekeeping genes. Binding sites in this class are characterized by the consensus ETS sequence (CCGGAAGT) and have the potential to bind any ETS protein with relatively high affinity. DNA regions occupied by multiple ETS proteins are frequently found a short distance (~20–40 bp) upstream of transcription start sites. Class 2: specific binding sites that are found more often in enhancer regions associated with genes that mediate the specific biological functions of an ETS family member. Specific target sites are characterized by a lower-affinity ETS sequence and are sometimes flanked by binding sites for other transcription factors. Many predicted ETS sites are not occupied *in vivo* and conversely, many actual sites of genomic occupancy are not predicted.

4. ETS AND MicroRNA

MicroRNAs (miRNAs) are both upstream modulators and downstream effectors of ETS transcriptional factors. miRNAs are 19–25-nucleotide RNAs that have emerged as a novel class of small, evolutionarily conserved gene regulatory molecules involved in many critical developmental and cellular functions (Wiemer, 2007). miRNAs base-pair with target mRNA sequences primarily in their 3′ untranslated region. Through specific base pairing, miRNAs induce mRNA degradation, translational repression, or both depending upon the complementarity of the miRNA to its mRNA target. Each miRNA can target numerous mRNAs, often in combination with other miRNAs, therefore controlling complex regulatory networks. It is estimated that there are ~1000 miRNAs in mammalian cells, and that approximately one third of all genes are regulated by miRNAs (Rajewsky, 2006; Shilo, Roy, Khanna, & Sen, 2007). Over 3000 identified mature miRNAs exist in species ranging from plants to humans, suggesting that miRNAs are ancient players in gene regulation (Wang & Li, 2007). Their existence and conservation throughout species support the concept that they perform critical functions in gene regulation (Wang, Stricker, Gou, & Liu, 2007). Indeed, the conserved evolution of both miRNAs and transcription factors highlights their importance in and the complexity of gene regulation (Chen & Rajewsky, 2006). In fact, one of the most widely studied miRNAs is miR-34, which has been shown to be positively and negatively regulated by the transcription factors p53 and myc, respectively (Bui & Mendell, 2010; He, He, Lowe, & Hannon, 2007).

4.1. miRNAs targeting ETS factors

A summary of the published studies that identified miR-ETS interactions is provided in Table 1.2. The majority of the studies examining miRNA-regulated ETS factors are for ETS1. miRNA 125b has been shown to be dysregulated in cancer and can act as either a tumor suppressor or oncogene, depending on cellular context. This is true for many miRNAs and adds to the complexity of targeting miRNAs therapeutically. However, in invasive breast cancer, miR-125b is downregulated and predicts poor patient survival (Zhang, Yan, et al., 2011). miR-125b expression inhibits tumor growth *in vivo* and has ETS1 as one of its novel direct targets. Although ETS1 protein levels were decreased by miR-125b, ETS1 mRNA levels were unchanged, suggesting a translational repression mechanism of regulation. Like miR-125b, ETS1 overexpression in invasive breast cancer predicts poor patient prognosis. Another study in hepatocellular carcinoma (HCC) identified ETS1 as a direct target of miR-193b (Xu et al., 2010). miR-193b expression inhibits tumor growth *in vivo* and a negative correlation between miR-193b and ETS1 mRNA levels was defined in HCC tissue samples, suggesting that miR-193b regulation of ETS1 results in mRNA degradation as opposed to translation repression; however, this was not validated *in vitro*. Other studies that identified ETS1 as a direct target of miRs included roles in osteoblast differentiation, inflammation, migration, angiogenesis, and megakaryopoiesis (Table 1.2). Two miRNAs, miR-204 and miR-510, were defined as direct negative regulators of SPDEF by translational repression in breast and prostate cancer (Findlay et al., 2008; Turner et al., 2011). miR-204 and miR-510 expressions are both elevated in tumor samples compared to matched normal in breast cancer and explained the apparent discordance in the literature of reports of SPDEF being elevated or downregulated in breast cancer due to the fact that SPDEF mRNA levels can be elevated in the absence of protein. miR-145 was shown to directly target FLI1 in colon cancer and in pericytes, where it was shown to block migration in response to growth factor gradients (Larsson et al., 2009; Zhang, Guo, et al., 2011). This interaction was also observed in patients with the 5q syndrome, a subtype of myelodysplastic syndrome (MDS), in which the inhibition of FLI1 by miR-145 decreases the production of

Table 1.2 microRNAs targeting ETS factors

microRNA	ETS factor	Function/disease	References
569	SPI1	Systemic lupus erythematosis	Hikami et al. (2011)
510	SPDEF	Breast and prostate cancer	Findlay et al. (2008) and Turner et al. (2011)
204		Breast cancer	Findlay et al. (2008)
145, 214	ELK1	Smooth muscle cell proliferation	Park et al. (2011)
7	ERF	Lung cancer	Chou et al. (2010)
196a, 196b	ERG	Acute leukemia	Coskun et al. (2011)
145	FLI1	Megakaryocyte and erythroid differentiation	Kumar et al. (2011)
		Colon cancer	Zhang, Guo, et al. (2011)
		Migration of microvascular cells (pericytes)	Larsson et al. (2009)
125b	ETS1	Breast cancer	Zhang, Yan, et al. (2011)
193b		Hepatocellular carcinoma	Xu et al. (2010)
370		Osteoblast differentiation	Itoh, Ando, Tsukamasa, and Akao (2012)
155, 221/ 222		Inflammation, migration of endothelial cells	Wu et al. (2011) and Zhu et al. (2011)
200b		Angiogenesis	Chun et al. (2011)
208		Preosteoblast differentiation	Itoh, Takeda, and Akao (2010)
155		Megakaryopoiesis	Romania et al. (2008)
320	ETS2	Stroma breast cancer	Bronisz et al. (2012)
221		Endothelial cell motility	Wu et al. (2011)
378	GABPA	Metabolic shift breast cancer	Eichner et al. (2010)

megakaryocytic cells relative to erythroid cells contributing to the pheno-type of the human malignancy (Kumar et al., 2011). Finally, an elegant study by the Ostrowski group showed a link between miR-320 and ETS2 in the stromal fibroblasts (Bronisz et al., 2012). Although many studies have inves-tigated a role for miRNAs in the epithelial tumor cells, very few have focused on their regulation in the stromal compartment. This study showed that miR-320 is a critical target of PTEN in stromal fibroblasts and directly controls ETS2 expression and instructs the tumor microenvironment to sup-press many of the aggressive phenotypes associated with advanced stages of breast cancer, including tumor cell invasiveness and increased angiogenic networks.

4.2. ETS factors targeting miRNAs

Only a handful of studies are published on the role of ETS factor modulated miRNAs (Table 1.3); therefore, we expect many studies in the future in this underdeveloped area of research. In ovarian cancer, a study reported that EGFR signaling leads to transcriptional repression of miR-125a through the ETS family transcription factor ETV4 (Cowden Dahl et al., 2009). It is known that overexpression of EGFR in ovarian cancer correlates with poor disease outcome and induces epithelial–mesenchymal transition (EMT) in ovarian cancer cells (Cowden Dahl et al., 2008; Nicholson, Gee, & Harper, 2001). Overexpression of miR-125a induced a conversion of highly invasive ovarian cancer cells from a mesenchymal to an epithelial morphology, suggesting that miR-125a is a negative regulator of EMT. A study to distinguish serous ovarian cancer from normal ovarian tissue using miRNA profiling identified miR-125a as downregulated (Nam et al., 2008). This correlates well with the previous study; however, miR-21 was also identified as part of the signature in this study and was reported as being upregulated. This is interesting because miR-21 was also shown to be repressed by ETV4 in colon cancer (Kern et al., 2012); therefore, this exemplifies the importance of context when studying miRNAs. Some miRNAs are regulated by the same ETS factor as illustrated by miR-126 in endothelial cells (Harris et al., 2010). miR-126 is abundantly expressed in endothelial cells, and promoter analysis showed that multiple ETS factors led to increased expression, but ETS1 and ETS2 were the most robust.

Some studies have also focused on the role of ETS fusion proteins on miRNAs. In particular, one study performed a genome-wide analysis of miRNAs affected by RNAi-mediated silencing of EWS-FLI1 in Ewing's

Table 1.3 ETS factors targeting microRNAs

ETS factor	microRNA	Action	Function/disease	References
SPI1	29b	Activation	Neutrophil differentiation (PML)	Batliner et al. (2012)
ETV5	21	Activation	Spermatogonial stem cell self-renewal	Niu et al. (2011)
ETV4	125a	Repression	EMT in ovarian cancer	Cowden Dahl, Dahl, Kruichak, and Hudson (2009)
	21	Repression	Colorectal cancer	Kern et al. (2012)
ELK1	34a	Activation	Oncogene-induced senescence	Christoffersen et al. (2010)
ETS1	126	Activation	Endothelial cells	Harris et al. (2010)
ETS2	126	Activation	Endothelial cells	Harris et al. (2010)
	196b	Repression	Gastric cancer	Liao et al. (2012)
Fusions				
TEL/ AML	494, 320a	Repression	Leukemia	Diakos et al. (2010)
EWS/ FLI1	30a-5p	Activation	Ewings sarcoma	Franzetti et al. (2012)
	let-7a			De Vito et al. (2011)
	145, 100, 125b	Repression	Ewings sarcoma	Ban et al. (2011)
	22, 221/ 222			McKinsey et al. (2011)
	27a, 29a, 145			Riggi et al. (2010)

sarcoma cell lines and identified miR-145 as the top repressed miRNA (Ban et al., 2011), which is interesting as previously mentioned miR-145 is a negative regulator of FLI1 suggesting a possible feedback loop mechanism of regulation of this fusion gene.

5. ETS MOUSE KNOCKOUT AND MUTANT MODELS
5.1. Phenotypes of mice with genetically altered Ets

To date, 23 of the 27 murine Ets genes have been genetically altered (knockouts or mutant mice; Table 1.4). Diverse biological roles of individual ETS family members are supported by the wide range of phenotypes displayed in these models. Most of these models have specific phenotypes, with the exception of Elf1 and Elk1, demonstrating nonredundant functions for the majority of Ets factors. For these, subtle phenotypes have (Elk1, minor defects in neuronal gene activation, and, Elf1, reduced NK-T cell development and function) been identified. Complete or significant embryonic and/or postnatal lethality is observed for 11 family members. Consistent with their tissue expression profiles, the majority have phenotypes that demonstrate their important functions in hematopoiesis, either exclusively or in combination with other lineage defects. There is often a wide range of phenotypes observed even within an Ets subfamily. For example, in the Spi1 subfamily, phenotypes range from Spi1, a principal regulator of myelolymphopoiesis, to SpiB, which regulates the proper function of terminally differentiated lineages and SpiC that is necessary for the function of a subset of macrophages. Ets1 and Elf4 are important regulators of T cell (T, NK, and NKT) development. Ets family members such as Fli1, Etv2, and Etv6 display functions in hematopoiesis and/or vasculo/angiogenesis. Nonhematopoietic defects were observed for Ets2 which has phenotypes related to extraembryonic development. Etv1, Etv4, and Fev each have defective neurogenesis, Etv4 and Etv5 affect male fertility. Consistent with their restricted epithelial-specific expression, the Ese and Spdef subfamilies (Elf3, Ehf, and Spdef) show tissue-specific (e.g., intestine, mammary gland) phenotypes, albeit at significantly different severities.

These constitutive knockout models reveal only the earliest/most distinct functions of each of these Ets family members. A better understanding of the roles and hierarchies of Ets family members in cellular differentiation and function will come with the generation of new null alleles in untargeted family members, double knockouts, ES cell differentiation and chimera rescue experiments, and tissue-specific inducible knockouts.

Table 1.4 Mouse Ets gene knockout and mutant mice

Gene name	Severity of mutation	Phenotype	Gene targets identified
Ets1	Viable and fertile with 50% neonatal lethality at 4 weeks	Hematopoietic cell defects. Reduced number of B, T, and NK cells	MDM2; Cyclin G (Xu et al., 2002), Foxp3 (Mouly et al., 2010), T-bet (Nguyen et al., 2012), MMP2, 3, and 13, TIMP1, 2, and 3 (Hahne et al., 2006)
Ets2	Embryonic lethal (<E8.5)	Extraembryonic tissue defect. Rescued neomorph has hair follicle defect	MMP3, MMP9, MMP13 (Yamamoto et al., 1998)
Erg	Embryonic lethal at E11–12.5	Lack of definitive hematopoiesis; defective thrombopoiesis in heterozygotes	Igfr2, MMP3 (Loughran et al., 2008)
Fli1	Knockout is embryonic lethal at E12.5.	Thrombocytopenia, reduced erythroid progenitors, defective vascularization, and B-cell maturation	Spyropoulos et al. (2000) and Starck et al. (2010)
Fev	Viable with reduced numbers of homozygous knockouts	Defective development of 5-HT neurons; increased anxiety and aggression	TPH, SERT, AADC (Hendricks et al., 2003)
Etv4	Viable; males infertile	Male sexual defect; lack of dendrite patterning in motor neurons	
Etv5	Viable; males infertile	Defective sperm production	CXCL12, CXCL5, CCL7 (Chen et al., 2005)
Etv1	Viable	Lack of motor coordination, synaptic defect, limb ataxia	
Etv2	Embryonic lethal E8–9.5	Defects in hematopoiesis and vasculogenesis. Lack of endothelial and endocardial lineages.	Tie-2 (Ferdous et al., 2009), SOX9 (DiTacchio et al., 2012)

Elk1	Viable; no phenotype	Phenotypically normal, minor effects on neuronal gene activation	Cesari et al. (2004)
Elk4	Viable	Defects in thymocytes and peripheral T cells	Costello, Nicolas, Watanabe, Rosewell, and Treisman (2004)
Elk3	Hypomorphic mutant dies at birth	Vascular defect leading to respiratory failure	EGR-1 (Ayadi et al., 2001)
Gabpa	Lethal prior to implantation	Mitochondrial defects; T cells	AChRδ and AChRε; AChE; Utrophin (O'Leary et al., 2007); IL7Rα (Yu, Zhao, Jothi, & Xue, 2010)
Elf1	Minor defect	Reduced NK–T cell development and function	(Choi et al., 2011)
Elf4	Viable and fertile	Reduced NK and NK–T cell development and function	Perforin (Lacorazza et al., 2002)
Spi1	Embryonic lethal at E18.5	Lack of hematopoiesis; no B, T cells or monocytes	Colucci et al. (2001)
SpiB	Viable	Defective B-cell responses; reduced intestinal immunity	Schotte, Nagasawa, Weijer, Spits, and Blom (2004), de Lau et al. (2012), and Kanaya et al. (2012)
SpiC	Viable	Defect in red pulp macrophages and red blood cell recycling and iron homeostasis	Vcam1 (Kohyama et al., 2009)
Etv6	Embryonic lethal E11	Lack of embryonic angiogenesis; defective hematopoiesis	Wang et al. (1998) and Ciau-Uitz, Pinheiro, Gupta, Enver, and Patient (2010)

Continued

Table 1.4 Mouse Ets gene knockout and mutant mice—cont'd

Gene name	Severity of mutation	Phenotype	Gene targets identified
Erf	Embryonic lethal E10.5	Defective placental development	Papadaki et al. (2007)
Spdef	Viable; fertile	Defective mucosal development in the gastrointestinal tract and respiratory tract. Increased gastrointestinal malignancies	Muc6,Tff (Horst et al., 2010)
Elf3	30% Embryonic lethality E11.5	Abnormal morphogenesis and differentiation of intestinal epithelium	TGFβ (Ng et al., 2002)
Elf5	Embryonic lethal E7.5	Embryonic patterning defect; heterozygotes deficient in mammary gland development	WAP, casein (Zhou et al., 2005); Notch pathway (Chakrabarti, Wei, et al., 2012); Snai2 (Chakrabarti, Hwang, et al., 2012)
Ets1 −/−; Ets2A72/ A72	Embryonic lethal E11.5–15.5	Endothelial cell defects leading to failure of vascular branching	MMP9, BCL-xL, cIAP2 (Wei et al., 2009)

Analyses of ES cell differentiation and chimeric and mutant mice were used to evaluate postembryonic phenotypes observed in the constitutive Fli1 knockout mice. These studies demonstrated that Fli1 also plays an important role in multiple non-megakaryocytic hematopoietic lineages, including erythroid, granulocyte, monocyte, and lymphocyte lineages (Zhang et al., 2008). Mutant mice lacking one of two regulatory domains (Fli1ΔCTA) provide novel evidence for the importance of Fli1 in megakaryocytic differentiation and platelet function. These approaches have also established Fli1 as an important regulator of fibroblast functions (Asano et al., 2009; Asano & Trojanowska, 2009; Kubo et al., 2003).

Conditional knockouts have further allowed definition of additional phenotypes. Fli1 conditional knockout mice in combination with Tie2-Cre have shown that mice with reduced endothelial Fli1 expression have compromised vessel integrity, markedly increased vessel permeability, and impaired pericyte/vascular smooth muscle cells coverage (Asano, Stawski, et al., 2010). Conditional Fli1 deletion in the adult results in mild thrombocytopenia associated with a maturation defect of bone marrow megakaryocytes (Starck et al., 2010), as previously observed in fetal liver of constitutive Fli1 knockout mice. In addition to the decrease in megakaryocytic cells, analysis of these mice revealed increases in natural killer (NK) cells and erythrocytic cells and a decrease in granulocytic cells, in agreement with the studies with chimeric Fli1 mice.

Fewer studies have examined the phenotypes of mice with deletions or mutations of two members. Phenotypes support similar roles for ETS subgroup (Ets1 and Ets2) in endothelial cells (Wei et al., 2009), ERG subgroup (Erg and Fli1) in hematopoietic cells (Kruse et al., 2009), TCF subfamily (Elk1 and Elk4) in thymocyte development (Costello et al., 2010), SPI1 subfamily (Spi1 and SpiB) in B-cell function (Garrett-Sinha et al., 1999), and PEA3 subfamily (Etv1 and Etv5) in limb-bud development (Zhang, Verheyden, Hassell, & Sun, 2009).

5.2. Identification of ETS target genes using genetic models

One important experimental approach for identifying Ets targets is the creation of null (knockout) or mutant mice lacking the function of a single or multiple family members. Analysis of these mice provides another means for the identification of genes whose expression or repression is dependent upon an Ets family member. Specific *in vivo* targets for Ets genes have been identified based on the knockout mice (Table 1.4). For example, *c-mpl* (Kawada et al., 2001) and Tie2 (Hart et al., 2000) have reduced expression in knockout Fli1 mice, consistent with the megakaryocytic lineage and vascular defects observed.

Mutant mice lacking one of two regulatory domains (Fli1ΔCTA) are thrombocytopenic and show significantly reduced expression of multiple megakaryocytic genes, including *c-mpl*, *platelet glycoprotein IIb (gpIIb)*, *gpIX*, and *gpV*. These mice also show reduced expression of genes associated with terminal differentiation of megakaryocytes to platelets. As noted above, Fli1 and GATA1 synergistically regulate gene transcription of multiple megakaryocytic genes. Transient-transfection studies indicate that only wild-type (WT) Fli1 can synergize with GATA-1, increasing promoter activity. Consistent with the failure of Fli1ΔCTA and GATA-1 to synergistically activate the *c-mpl* promoter-luciferase reporters *in vitro*, ChIP studies demonstrate that Fli1ΔCTA is not able to efficiently recruit GATA-1 to specific (*c-mpl*, PF4, and *gpIX*) promoters *in vivo* and further define these as Fli1 direct target genes (Moussa et al., 2010).

Fli1 has been shown to repress collagen synthesis in cultured dermal fibroblasts and mouse embryo fibroblasts and Fli1 mutant mice show significant upregulation of fibrillar collagen mRNA and altered expression of matrix-related genes, including decorin, fibromodulin, lumican, procollagen lysine, 2-oxoglutarate 5-dioxgenase 2 (PLOD2), and lysyl oxidase (Asano et al., 2009; Kubo et al., 2003).

Conditional Fli1 knockout mice with Tie2-Cre endothelial cell-specific disruption support the notion that Fli1 may function in maintenance of vascular homeostasis by directly regulating VE-cadherin, PECAM1, Tie2, MMP9, PDGF B, and S1P1 receptor (Asano, Stawski, et al., 2010).

The impact of altered expression of specific Ets response genes can be assessed by performing genetic rescue experiments. This approach is nicely represented by studies of Elf3 knockout mice (Flentjar et al., 2007; Ng et al., 2002). Elf3 knockout mice show significant embryonic and postnatal lethality, due to aberrant morphogenesis and terminal differentiation of the small intestine. Elf3 knockout mice express significantly less TGF-βRII protein. To perform a rescue experiment, transgenic mice that express human TGF-βRII specifically in the intestinal epithelium were crossed to the knockout animals. Significantly, the TGF-βRII transgenic Elf3 −/− mice displayed normal small intestinal morphology.

6. ETS FACTORS AND CANCER

The hallmark features of a cancer cell consist of uncontrolled proliferation, loss of differentiation, sustained cell division, increased angiogenesis, loss of apoptosis, and a capacity to migrate and invade to other tissues and organs. All of these processes are driven by transient and/or permanent changes in gene expression profiles conferred through the activation or repression of cancer-associated genes. It is therefore clear that the role of transcriptional gene regulation in cancer progression cannot be understated and many transcription factors including ETS family members have been assigned as candidate oncogenes or tumor repressors. The importance of ETS genes in human carcinogenesis is supported by the observations that during cancer progression, ETS genes acquire point mutations (e.g., SPI1, ETS1), genomic amplification (ETS1, ETS2, ERG), increased (ETS1, ETS2, ERG) or decreased (e.g., SPDEF, EHF) expression, or rearrangements (ETV6, FLI1, ERG; Seth & Watson, 2005), resulting in altered ETS gene expression which disrupts the regulated control of many complex biological processes, promoting cellular proliferation and inhibiting apoptosis, enhancing cell migration, invasiveness, and metastasis as well as angiogenesis (Fig. 1.2).

6.1. ETS expression in cancer

Altered ETS gene expression levels are correlated with tumor progression in human neoplasias, including thyroid, pancreas, liver, prostate, colon, lung, and breast carcinomas and leukemias (Seth & Watson, 2005; Watson et al., 2010). Furthermore, in breast cancer, upregulation of multiple ETS factors, including ETS1 (Buggy et al., 2004), ETS2 (Buggy et al., 2006), ETV4 (Benz et al., 1997; Chang et al., 1997), ETV5 (Yarden & Sliwkowski, 2001), and ETV1 (Bosc, Goueli, & Janknecht, 2001), is associated with poor prognosis and metastasis. In contrast, other ETS factors including SPDEF (Doane et al., 2006; Feldman, Sementchenko, Gayed, Fraig, & Watson, 2003), ELF5 (Zhou et al., 1998), and EHF (Tugores et al., 2001) are down-regulated during breast cancer progression within the same context. The impact of multiple ETS factors (e.g., ETS1, ETS2, ETV4, ELF3, SPDEF, ERG) on phenotypes and molecular regulation in cancer cells has been demonstrated through *in vitro* gain-of-function as well as loss-of-function experiments.

In addition to ETS-mediated transcriptional activation of multiple genes associated with cancer progression, analysis of androgen receptor (AR) genomic targets demonstrated an enrichment of ETS transcription factor family members and, more specifically, an interaction between the AR and ETS1 at a subset of the AR promoter targets was found (Massie et al., 2007). These studies support the model that ETS proteins, including ETS1, regulate genes, including androgen response genes, which contribute to prostate cancer progression.

6.2. ETS translocations

Cancer involves many chromosomal aberrations, the most studied being nonrandom chromosomal translocations resulting in recombinant chromosomes. Tumor cell formation results from the translocation associated production of FLI1 chimeric proteins as has been shown for Ewing's sarcomas (EWS) and related primitive neuroectodermal tumors (PNET; reviewed in Arvand & Denny, 2001). In this instance, chimeric transcripts result from the fusion of the amino terminal region of the EWS gene with the carboxyl terminal DNA-binding domain of the FLI1 gene (Delattre et al., 1992; Zucman et al., 1992). The chimeric fusion protein lacks the putative RNA-binding domain of EWS and one of the transactivation domains of FLI1. It has also been shown that the EWS-FLI1 fusion is a more potent transcriptional activator than the FLI1 protein. In other Ewing's sarcoma and PNET tumors, translocations fuse the EWS gene to other members of the ETS family, including ERG, ETV1, ETV4, and FEV.

ETV6 was originally identified by its rearrangement in specific cases of chronic myelomonocytic leukemia (CMML) presenting a t(5,12)(q33; p13) chromosomal translocation (Golub, Barker, Lovett, & Gilliland, 1994). ETV6 is rearranged in CMML, acute myelogenous leukemia (AML), acute myeloblastic leukemia (AML-M2), MDS, and acute lymphoblastic leukemia. Either the PNT domain or the ETS domain or both domains of ETV6 have been identified in over 20 different translocations observed in human leukemia and more rarely solid tumors (reviewed in Mavrothalassitis & Ghysdael, 2000). Fusions involving the PNT domain of ETV6 often lead to oligomerization that is necessary for constitutive activation of kinase activity of receptor or protein tyrosine kinases. Fusions that retain the DNA-binding domain of ETV6 are expected to result in aberrant regulation of ETS target genes.

ERG is highly expressed in over 60% of prostate tumor cells relative to benign tissues. A molecular mechanism to account for ERG overexpression in prostate cancer was subsequently provided by the identification of chromosomal rearrangements that result in the fusion between the 5' end of the androgen-regulated, prostate-specific transmembrane serine protease TMPRSS2 gene to ERG (Soller et al., 2006; Tomlins et al., 2005). Collective studies show that the TMPRSS2-ERG fusion is present in 40–80% of prostate cancers (recently reviewed in Kumar-Sinha, Tomlins, & Chinnaiyan, 2008; Shah & Chinnaiyan, 2009; Shah & Small, 2010). TMPRSS2 gene fusions involving other ETS transcription factors ETV1, ETV4, or ETV5 have been identified in prostate cancer; however, TMPRSS2-ERG fusion and mRNA overexpression accounts for the majority of cases. Possible mechanistic insights are provided by observation that TMPRSS2-ERG fusion activates MYC and abrogates prostate epithelial cell differentiation. An 87-gene signature has been associated with TMPRSS2:ERG fusion tumors. Collective data suggest that the TMPRSS2-ERG fusions define a subset of prostate cancer and specific fusions predict poor prognosis and survival.

6.3. ETS target gene expression and function

Functional studies demonstrate the impact of such altered expression on the regulation of genes associated with proliferation, transformation, migration, invasion, anti-apoptosis, and angiogenesis (Seth & Watson, 2005) and include but are not exclusive to Her2/neu, uPA, MMPs, TIMPs, MET, Bcl2, maspin, and VEGFR (Sementchenko & Watson, 2000; Fig. 1.2).

Alterations in cell cycle control are a critical step in carcinogenesis. Cell cycle arrest at the G_1–S transition by upregulation of the cyclin-dependent kinase inhibitor p21 occurs in response to DNA damage or oncogenic insult. The elevated level of p21 is known to be mediated through p53 and we have demonstrated SPDEF-mediated regulation of p21 expression is associated with inhibition of growth *in vitro* (Feldman, Sementchenko, Gayed, et al., 2003) and *in vivo* (Schaefer et al., 2010). Indeed, SPDEF-mediated inhibition of breast cancer xenograft growth can be reversed by shRNA targeting of p21 (Schaefer et al., 2010). These observations combined with ChIP demonstrate that p21 is a key direct target of SPDEF used to control cellular growth. The increased expression of the p21-activated kinase (PAK1) has been shown to be correlated with more aggressive breast cancer (Salh, Marotta, Wagey, Sayed, & Pelech, 2002). Furthermore, studies have shown that PAK1 regulates the activity of ELF3 by phosphorylation (Manavathi, Rayala, & Kumar, 2007). This novel finding raises the possibility that using a specific inhibitor to the upstream effector of ELF3 (e.g., PAK1-specific inhibitor CEP-1347) may represent a novel approach for targeting a transcription factor in breast cancer.

Migration and invasion, critical steps in the metastatic process, requires changes in cell-to-cell adhesion as well as cell adhesion to the ECM. Migration and invasion are often associated with EMT and resultant downregulation of E-cadherin. Invasion is mediated in part by proteolytic degradation of the ECM by MMPs and uPA. Indeed, activation of the uPA system is associated with a poor prognosis in breast cancer.

Significantly, we and others have shown that ETS factors are critical regulators of EMT, protease expression, and ECM (discussed further below, microenvironment). For example, studies using breast, prostate, colon, or ovarian cancer cells have demonstrated the antimigratory and antiinvasive properties of SPDEF, by negative regulation of the EMT regulator SLUG and mesenchymal genes, proteases (uPA, MMPs).

6.4. ETS conversion

To date, ETS research has mainly focused on the molecular mechanisms and functions of individual transcription factors and has produced insights into ETS factor function in both normal and cancer cells. In many cells, multiple ETS factors with similar or opposite functions are present simultaneously and the cell's fate may depend ultimately on the balance between the activities of distinct ETS factors.

ETS factor dysregulation disturbs normal cellular homeostasis, increasing cancer growth, invasion, and metastasis. While some ETS factors are lost during cancer progression, others show increased expression: tumor suppressive and oncogenic ETS factors. We hypothesize that the balance of "tumor suppressor" and "oncogenic" ETS factors could be a marker for aggressive cancer. Taken together, accumulating evidence suggests that multiple ETS factors act in concert to positively and negatively regulate the pathways that control progression to metastatic cancer. This indicates a possible ETS conversion mechanism of gene regulation which provides the cell with an integrated mechanism by which to respond to a variety of intra- and extracellular signals efficiently (Hsu et al., 2004; Turner, Findlay, Moussa, & Watson, 2007; Watson et al., 2010). Several Ets factors are deregulated in the development of breast cancers. During cancer progression, the expression of some ETS factors (e.g., ETS1, ETS2, ETV4, ETV5, ELF3) is often increased, while the expression of other ETS factors (SPDEF, EHF) is reduced or lost (Turner, Findlay, et al., 2007; Watson et al., 2010). The ETS conversion model further hypothesizes that the change in expression pattern from what is observed in normal or benign tissues (e.g., SPDEF expression) to that observed in invasive cancer (e.g., elevated ETS1) is necessary for cancer progression to proceed.

Reciprocal ETS regulation of a metastasis-associated gene can be illustrated by the uPA promoter. ETS regulation of uPA has both positive and negative effects on cancer progression depending on the specific Ets factor expressed. ETS1 is overexpressed in invasive breast and aggressive prostate cancer and associated with increased uPA expression. In noninvasive

(ETS1 negative) breast cancer cells, reexpression of ETS1 increases uPA levels leading to more aggressive tumorigenic phenotypes, including increased cell growth, migration, and invasion. In contrast, the expression of another ETS family member, SPDEF is present in noninvasive, but lost in invasive, breast cancer cells. SPDEF reexpression in invasive cells represses endogenous uPA transcription leading to an inhibition of cell migration and invasion and an antimetastatic phenotype (Feldman, Sementchenko, Gayed, et al., 2003; Turner, Moussa, Sauane, Fisher, & Watson, 2007). Significantly, a statistically significant inverse correlation between SPDEF and uPA expression is observed in colon cancer clinical specimens (Moussa et al., 2009). Intriguingly, although several potential EBS are found in the uPA promoter, both ETS1 and SPDEF have been demonstrated to bind at the same consensus EBS *in vivo*.

Many ETS factors (including ETS1, ETS2, ETV4) transcriptionally activate multiple MMPs, most commonly in cooperation with AP1 complexes. In contrast, SPDEF is a repressor of MMP7 (Moussa et al., 2009) and MMP9 (Johnson et al., 2010)

Another example of reciprocal regulation is provided by the maspin promoter. Maspin is a type II tumor-suppressor gene that has been shown to have antimetastatic properties when expressed in invasive breast and prostate cancer cells (Zou et al., 1994). The maspin promoter has been shown to be regulated by SPDEF (Feldman, Sementchenko, Gayed, et al., 2003; Yamada, Tamai, Miyamoto, & Nozaki, 2000). Significantly, this activation appeared to be specific for SPDEF, since neither FLI1 nor ETS1 was able to activate this promoter. Indeed, ETS1 expression inhibited SPDEF-mediated transactivation of the maspin promoter.

6.5. ETS regulatory network

Taken as a whole, this evidence strongly suggests the existence of distinct ETS expression regulatory networks that act in concert to positively or negatively regulate cancer-associated genes. Significantly, each ETS network would result in distinct patterns of target gene expression, the elucidation of which may identify prometastatic and antimetastatic signatures of gene expression that may predict the aggressive behavior of cancer cells. The ETS Regulatory Network is comprised of the ETS factors themselves, their upstream modulators, their coregulatory proteins, and their target genes (Fig. 1.3). Inflammatory cells are recruited by tumors through their secretion of chemokines, cytokines, and growth factors (1). In response, the recruited

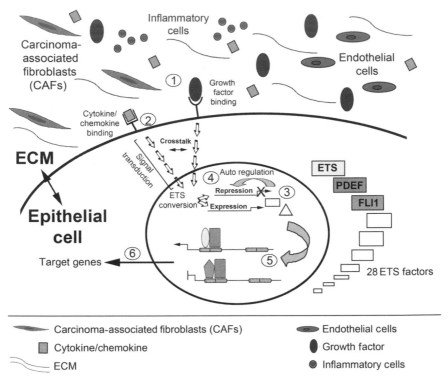

Figure 1.3 Hypothetical model of the ETS regulatory network in cancer. See text for details. (See Page 1 in Color Section at the back of the book.)

inflammatory cells and other cells of the microenvironment (e.g., fibroblasts promote tumor proliferation and progression through additional secretion of biologically active molecules). This in turn results in the activation of intra–cellular signaling cascades via ligand binding at the cell surface of epithelial cells (2). The activated cascades directly or indirectly (through crosstalk) result in the expression and repression of varying combinations of the 28 ETS family members (3). ETS factors can regulate their own expression and/or that of other family members (4). The composition of Ets factors defines the transcriptional regulation of their target genes, many known to be involved in cancer progression (5). The altered expression of these genes has profound consequences on many cancer-related pathways (6).

6.5.1 ETS-mediated anti- and prometastatic signatures

Gene expression signatures consist of sets of gene profiles that are known to be predictive of a disease state and/or patient response to treatment. The combined statistical analysis of multiple gene sets obtained from independent gene microarray studies has resulted in an increased number of putative and validated "metastatic signatures" that predict the outcome of disease in cancer. In addition, comparison of gene expression profiles from primary and metastatic tumors in multiple cancer types reveals highly specific signatures that allow discrimination between primary and metastatic tumors. Similarly, by elucidating the expression networks conferred by ETS family members that elicit a prometastatic response (ETS1, etc.) and an antimetastatic response (SPDEF, etc.), improved pro- and antimetastatic signatures may be isolated which predict the aggressive behavior of cancer cells. As such, these new insights may provide a novel view of the ETS gene family as well as a focal point for studying the complex biological control involved in tumor progression.

7. THE ROLE OF ETS FACTORS IN THE MICROENVIRONMENT

The majority of cancer-related deaths are due to tumor progression, whereby cells from the primary tumor migrate, invade, and reestablish at distant metastatic sites (Guarino, Rubino, & Ballabio, 2007; Turner, Moussa, et al., 2007). The progression of solid tumors corresponds with progressive alterations in the tumor microenvironment, suggesting crosstalk between epithelium and stroma. Increasing evidence suggests that these stromal–epithelial interactions play a critical role in regulating tumor growth and progression. However, this aspect of tumorigenesis remains little understood. Previous studies have indicated that members of the ETS transcription factor family are abnormally expressed in both tumor and stromal compartments. This aberrant expression of ETS factors has been associated with cancer progression and frequently correlates with poor prognosis. For example, ETS1 is frequently overexpressed in epithelial, endothelial, and stromal cells in various tumors (Behrens, Rothe, Florin, Wellmann, & Wernert, 2001; Behrens, Rothe, Wellmann, Krischler, & Wernert, 2001; Takai, Miyazaki, Nishida, Nasu, & Miyakawa, 2002; Trojanowska, 2000). Studies have demonstrated that ETS1 is a strong independent predictor of poor prognosis in breast cancer (Myers et al., 2005; Span et al., 2002). Further, drug-resistant breast cancer cells have been shown to overexpress ETS1 (Kars, Iseri, & Gunduz, 2010), suggesting that ETS factors may play a significant role in tumor aggressiveness and contribute to failed therapies. These studies highlight the importance of understanding the mechanisms by which ETS factors function in both the epithelium and microenvironment.

The stromal compartment consists of fibroblasts, endothelial cells, peri-vascular cells, blood-borne cells, nerves, and intervening ECM. The fibroblasts of the tumor microenvironment, termed carcinoma-associated fibroblasts (CAFs), are thought to promote tumor progression by establishing a reactive tumor stroma, stimulating growth, sustaining angiogenesis, inhibiting the immune response, promoting the malignant phenotype, and promoting inva-sion and metastasis (Hanahan & Coussens, 2012; Hanahan & Weinberg, 2000, 2011; Karnoub et al., 2007; Orimo & Weinberg, 2006). While these previous studies have identified important functions for CAFs, the factors regulating the functions of these cells are undefined.

Growing evidence suggests that ETS family members are critical regu-lators of stromal activation. The reactive tumor stroma is characterized by excessive remodeling of the ECM via CAF production of matrix molecules (i.e., collagen-1), matrix-degrading factors (i.e., matrix metalloproteinases, MMPs), and growth factors (i.e., TGFβ). Fli1 has been established as a reg-ulator of fibroblast function (Asano et al., 2009; Kubo et al., 2003; Truong & Ben-David, 2000; Watson et al., 1992). A hallmark of the CAF is the expres-sion of alpha-smooth muscle actin (αSMA) and studies have shown that Fli1 reduction leads to αSMA upregulation (Nakerakanti, Kapanadze, Yamasaki, Markiewicz, & Trojanowska, 2006). Fli1 has also shown to func-tion as a physiological transcription repressor of collagen type I gene *in vivo* (Asano, Bujor, & Trojanowska, 2010; Czuwara-Ladykowska et al., 2001). The absence of Fli1 correlates with elevated collagen synthesis (Kubo et al., 2003), a second hallmark of the activated tumor stroma. Fli1 has also been shown to regulate expression of tenascin-C, an additional ECM pro-tein associated with wound healing and tumor stroma activation (Shirasaki, Makhluf, LeRoy, Watson, & Trojanowska, 1999). At least part of the Ets functions in tumor stromal cells is the regulation of ECM-degrading enzymes including MMPs, uPA, and collagenases, molecules that are crucial for the establishment of a reactive stroma and onset of metastasis (Westermarck, Seth, & Kahari, 1997). ETS1 and FLI1 have been shown to modulate MMP1 expression (Gavrilov, Kenzior, Evans, Calaluce, & Folk, 2001; Nakerakanti et al., 2006). ETS1 and ETS2 have also been impli-cated in uPA and MMP9 activation (Watabe et al., 1998). A recent study identified novel ETS1 target genes by subtractive hybridization in stromal fibroblasts under bFGF stimulation (Hahne, Fuchs, et al., 2011; Hahne, Okuducu, Fuchs, Florin, & Wernert, 2011). MMP1, MMP3, PAI-1, and collagen Iα2 were confirmed as ETS1 target genes. Several additional targets were identified which may play a role in generation of the activated tumor stroma: cathepsin, a lysosomal proteinase whose elevated expression is associated with several cancers; lumican, a proteoglycan that binds collagen I and II to sequester growth factors in matrix; decorin, a proteoglycan that binds collagen I during matrix assembly and interacts with fibronectin, thrombospondin, epidermal growth factor receptor, and TGFβ to affect their functions; gremlin, a secreted antagonist of BMPs that promotes cancer cell survival and proliferation and is overexpressed in stroma of many cancers; HSP-90, a heat shock protein that acts to stabilize various growth factor receptors, is required for the induction of VEGF and nitric oxide synthase, and assists MMP2 to promote invasion/metastasis. While these studies did not experimentally demonstrate effects of ETS1 on the promoters of potential ETS1 target genes identified, promoter analysis showed the presence of potential EBS in the promoter regions of each gene identified.

ETS factors have also been shown to directly regulate the expression of cytokines as well as the response to specific growth factors and chemokines (Turner et al., 2008; Turner, Moussa, et al., 2007; Turner & Watson, 2008). For example, ETS1 is a downstream effector of the stroma-derived EMT-promoting HGF and an activator of its receptor, c-Met, thereby regulating a positive feedback loop whereby HGF/c-Met affects both tumor stroma and tumor cells (Hsu et al., 2004). HGF has also been shown to induce MMP1 protein expression in cultured human dermal fibroblasts. Studies showed that the balance of ETS1 and FLI1 binding to the EBS in the MMP1 promoter regulated the effects of HGF, with ETS1 binding leading to upregulation of MMP1 and FLI1 antagonizing this expression (Jinnin, Ihn, Mimura, et al., 2005). The activities of FLI1 and ETS1 toward the expression of Tenascin-C and connective tissue growth factor (CTGF/CCN2), novel Ets target genes (Jinnin et al., 2006; Nakerakanti et al., 2006; Shirasaki et al., 1999), are modulated by acetylation in a TGFβ-dependent manner (Asano, Czuwara, & Trojanowska, 2007; Asano et al., 2009; Asano & Trojanowska, 2009). These data suggest that ETS1 and FLI1 are the effectors of the TGFβ signaling pathway through novel, previously undescribed regulatory mechanisms. Elevated ETS1 has also been shown to be an antagonist of TGF-β functions in stromal cells (Czuwara-Ladykowska et al., 2002). Significantly, ETS1 and FLI1 are targets of the TGF-β signaling pathway, the primary regulator of fibroblast maturation, activation, and function. HGF-activated ETS1 has also been shown to regulate CXCL12/CXCR4-dependent promotion of tumor cell chemoinvasion (Maroni, Bendinelli, Matteucci, & Desiderio, 2007).

Several *in vivo* studies have demonstrated a correlation between stromal expression of ETS factors, dysregulation of matrix factors, and tumor progression. For example, stromal upregulation of ETS1, MMP1, and MMP9 has been observed in invasive ductal and lobular breast cancers (Behrens, Rothe, Wellmann, et al., 2001) and in invasive HNPCC and sporadic colon cancer (Behrens et al., 2003). Stromal cell expression of a specific ETS target gene (MMP9) has been shown to play a critical role in angiogenesis and growth of ovarian tumors in mice (Huang et al., 2002). Together, these studies suggest that targeting Ets factors in cells of the microenvironment may be an effective antitumor therapy. To demonstrate this potential, specific inactivation of Ets2 in the CAF population in a Pten murine mammary tumor model led to a reduction in tumor size (Li, Wallace, & Ostrowski, 2010). The absence of Ets2 in fibroblasts led to decreased epithelial cell proliferation and delayed tumor progress, illustrating the ability of ETS factors to regulate crosstalk between epithelial and stromal compartments and the importance of targeting this interaction.

In addition to their role in the fibroblastic and ECM components of tumor stroma, the altered expression of several ETS factors has been suggested to regulate angiogenesis, another key step in tumor progression and metastasis. FLI1 is normally expressed in vascular cells including hematopoietic cells, perivascular cells, and endothelial cells (Jinnin, Ihn, Yamane, et al., 2005; Kubo et al., 2003; Lelievre, Lionneton, & Soncin, 2001; Lelievre, Lionneton, Soncin, & Vandenbunder, 2001; Liu, Walmsley, Rodaway, & Patient, 2008; Pimanda et al., 2007; Spyropoulos et al., 2000; Truong & Ben-David, 2000; Watson et al., 1992). Loss of Fli1 results in embryonic lethality due in part to the absence of megakaryocytes, aberrant vasculogenesis, and disruption of tissue integrity (Kawada et al., 2001; Spyropoulos et al., 2000). FLI1 expression is reduced or lost in stromal cells in epithelial tumors, suggesting that this loss of FLI1 could have a direct effect on tumor vasculogenesis. Stromal-derived VEGF can induce ETS1 expression in endothelial cells (Lavenburg, Ivey, Hsu, & Muise-Helmericks, 2003) and activated transcription of VEGFR2/Flt-1 in concert with HIF-2α (Elvert et al., 2003). In addition, expression of ERG and FLI1 has been correlated with Tie2 gene expression, which is involved in the formation and remodeling of normal vascular networks (Mattot, Vercamer, Soncin, Fafeur, & Vandenbunder, 1999).

Evolving data indicate that Fli1 plays an important role in multiple hematopoietic lineages, including erythroid, granulocyte, monocyte, and lymphocyte lineages (Hart et al., 2000; Kawada et al., 2001; Masuya et al.,

2005; Nowling, Fulton, Chike-Harris, & Gilkeson, 2008; Spyropoulos et al., 2000; Zhang et al., 2008). Tumor-associated macrophages (TAMs), a cell of monocyte origin, have been implicated in tumor progression by mediating angiogenesis, invasion, and immunosuppression (Sica et al., 2008). Studies have demonstrated that ETS2 is an important downstream mediator of CSF1-R (colony stimulating factor-1, a growth factor that regulates macrophage survival, proliferation, and differentiation) signaling in TAMs. Macrophage-specific ablation of Ets2 in the *PyMT* tumor model resulted in significant decrease in mammary tumor metastasis to lung (Lin, Nguyen, Russell, & Pollard, 2001; Zabuawala et al., 2010). Gene expression profiling studies have demonstrated that Ets2 target genes are not only tumor specific, but compartment specific between CAFs and TAMs (reviewed in Li et al., 2010). These studies reinforce the idea that cellular context defines the direction and magnitude of response to ETS factors.

8. ETS FACTORS AND OTHER DISEASES

While less attention has been directed toward the elucidation of the roles for ETS transcription factors in diseases other than cancer, clear roles for ETS factors in autoimmune diseases have been defined; these and some other diseases will be briefly discussed below.

Transgenic mice overexpressing Fli1 develop a lupus-like disease (Zhang et al., 1995). It was also previously demonstrated that FLI1 is overexpressed in peripheral blood lymphocytes of systemic lupus erythematosis (SLE) patients compared to normal healthy controls and that NZB/NZW mice, a murine lupus model, have higher Fli1 mRNA expression in splenic lymphocytes than normal control mice (Georgiou et al., 1996). When Fli1 heterozygous mice were crossed with MRL/*lpr* mice, another model of SLE, Fli-1+/− MRL/*lpr* mice had significantly decreased serum levels of total IgG and anti-dsDNA antibodies as disease progressed. In addition, these mice had significantly increased splenic CD8+ and naive T cells and markedly decreased proteinuria and significantly lower pathologic renal scores compared to Fli1+/+ MRL/*lpr* mice. At 48 weeks of age, survival was significantly increased in the Fli1+/− MRL/*lpr* mice as 100% were alive, in contrast to only 27% of Fli1+/+ mice. Both *in vivo* and *in vitro* production of MCP-1 were significantly decreased in Fli1+/− MRL/*lpr* mice (Zhang et al., 2004). Similar findings were obtained in NZM2410 mice (derived from NZB X NZW F1 hybrids), where 93% of Fli1+/− NZM2410 mice survived to the age of 52 weeks compared to only 35% of WT NZM2410 mice (Mathenia et al., 2010).The primary endothelial cells isolated from the kidneys of Fli1 +/− NZM2410 mice produced significantly less MCP-1. ChIP analysis demonstrated that Fli1 directly binds to the promoter of the MCP-1 gene. These data indicate that Fli1 impacts glomerulonephritis development by regulating expression of inflammatory chemokine MCP-1 and inflammatory cell infiltration in the kidneys. Together, these findings indicate that FLI1 expression is important in lupus-like disease development. The length of a GA microsatellite in the FLI1 promoter has been shown to be inversely correlated to promoter activity and is associated with SLE patients without nephritis (Morris et al., 2010). Recent genome-wide association studies have identified genetic variants of ETS1 associated with SLE (Yang et al., 2010) and Ets1 knockout mice develop lupus-like disease (high IgM and IgG autoantibodies, glomerulonephritis, and local complement activation; Wang, John, et al., 2005). It has been recently suggested that some of these phenotypes could be related to ETS1 functions, including negative regulation of Th17 and B-cell differentiation (Pan, Leng, Tao, Li, & Ye, 2011).

Systemic sclerosis (SSc) or scleroderma is an autoimmune inflammatory disease characterized by fibrosis of the skin and internal organs as well as microvessel injury. The importance of reduced FLI1 expression in the pathogenesis of SSc has been reviewed recently (Asano, Bujor, et al., 2010). Although FLI1 expression in dermal fibroblasts is relatively low, studies have shown that FLI1 plays a critical role in the regulation of ECM genes, including type I collagen (Jinnin, Ihn, Yamane, et al., 2005) and the multifunctional matricellular factor CTGF/CCN2 (Nakerakanti et al., 2006). Importantly, FLI1 has been shown to be a potent inhibitor of collagen biosynthesis in dermal fibroblasts and its aberrant expression has been implicated in the pathogenesis of cutaneous fibrosis in SSc (Kubo et al., 2003; Wang, Fan, & Kahaleh, 2006). Interestingly, MCP-1 (regulated by FLI1 in SLE) also has been shown to play a role in SSc fibrosis (Artlett, 2010). In humans, FLI1 is expressed in the healthy skin microvasculature; however, its presence is greatly reduced in endothelial and periendothelial cells in SSc skin (Kubo et al., 2003). Conditional deletion of Fli1 in the endothelium of mice results in vascular defects observed in SSc vasculature (Asano, Stawski, et al., 2010).

Jacobsen syndrome (11q-) is a rare chromosomal disorder caused by deletions in distal 11q. Individuals have thrombocytopenia with a subpopulation of cells having enlarged α-granules. In addition to platelet effects, Jacobsen syndrome patients also present with a wide spectrum of the most common congenital heart defects, including an unprecedented high frequency of hypoplastic left heart syndrome (HLHS). Both of these conditions are associated with deletions on the long arm of chromosome 11, including 11q23, where ETS1 and FLI1 are located. Thus, these patients have only one copy of these ETS genes due to a heterozygous loss of regions in Chromosome 11. FLI1 monoallelic expression combined with its hemizygous loss underlies Jacobsen thrombocytopenia (Raslova et al., 2004). Significantly, overexpression of FLI1 in patient CD34(+) cells restores the megakaryopoiesis *in vitro*, indicating that FLI1 hemizygous deletion contributes to the hematopoietic defects (Raslova et al., 2004). Ets1 is expressed in the endocardium and neural crest during early mouse heart development. Ets1 knockout mice show large membranous ventricular septal defects and a bifid cardiac apex, and less frequently a nonapex-forming left ventricle (one of the hallmarks of HLHS). These results implicate an important role for ETS1 in mammalian heart development and some of the most common forms of congenital heart disease (Ye et al., 2010).

The functional polymorphism in the MMP1 promoter affecting ETS binding may contribute to the pathogenesis of osteomyelitis by increased MMP1 expression (Montes et al., 2010) as well as higher disease severity in recessive dystrophic epidermolysis bullosa (Titeux et al., 2008). It has also been hypothesized that the ETS2 activation of Bcl-xL may protect glia from constitutive oxidative stress that is believed to be a key mechanism for amyotropic lateral sclerosis, an adult-onset neurodegenerative disease (Lee, Kannagi, Ferrante, Kowall, & Ryu, 2009).

Phenotypes of several of the knockout, mutant, and transgenic mice support the notion that ETS factors have roles in several diseases. For example, the importance of Fli1 and Erg in megakaryopoiesis would support a possible role for these ETS factors in other diseases affecting megakaryopoiesis (thrombocytopenia, megakaryocytopenia) or other conditions associated with thrombocytopenia (e.g., chronic liver disease, acquired immunodeficiency syndrome). Many ETS factors would be expected to have a critical role in other hematopoietic, vascular, and respiratory (e.g., asthma, cystic fibrosis, chronic obstructive pulmonary disease) diseases. ETS1 has been shown to be a mediator of inflammation and neointima formation in a model of carotid artery balloon injury (endoluminal vascular injury; Feng et al., 2010). Spdef is required for differentiation of pulmonary goblet cell and regulates genes associated with mucus production, supporting the model that Spdef plays a critical role in regulating a transcriptional network mediating the goblet cell differentiation and mucus hyperproduction associated with chronic pulmonary disorders (Park et al., 2007).

9. TARGETING THE ETS NETWORK

9.1. Therapeutic targeting of ETS transcription factors

Targeting transcription factors for therapeutic gain is the focus of intense research as being able to manipulate transcriptional expression patterns would provide a novel approach for the treatment of many human diseases. The primary limitations to targeting transcription factors are the potential for off-target effects and insufficient delivery within the cell. Overwhelming evidence suggests that the number of transcription factors whose aberrant function supports tumorigenesis is limited (Darnell, 2002). Additionally, this limited number of transcription factors function at critical focal points controlling many of the genes involved in cancer-associated processes. Therefore, targeting transcription factors has great potential for therapeutic gain.

9.2. Targeting ETS factor biology

ETS factor family members are associated with the positive and negative regulation of gene expression profiles affecting all the classic hallmarks of cancer, including sustaining proliferative signaling, evading growth suppressors, resisting apoptosis, replicative immortality, activated angiogenesis, and induced invasion and metastasis (Turner & Watson, 2008). Critically, an ever growing body of evidence demonstrates many genetic and epigenetic alterations of ETS transcription factor function and activity in cancer. ETS factors, therefore, provide potential targets for cancer therapy. Pharmacological intervention may be used to inhibit the altered expression of oncogenic ETS factors such as ETS1, ETS2, ELF3, and/or activate the expression of tumor-suppressive members such as SPDEF and ETV6 (Turner & Watson, 2008). Several of the multifaceted aspects of ETS factor biology have been explored in order to assess the potential of therapeutically targeting these proteins. Strategies have included: directly inhibiting the promoter of oncogenic ETS factors (Carbone, McGuffie, Collier, & Catapano, 2003; Carbone et al., 2004; Miwa et al., 2005; Sahin et al., 2005) or directly targeting specific ETS factor mRNA (Dohjima, Lee, Li, Ohno, & Rossi, 2003; Hu-Lieskovan, Heidel, Bartlett, Davis, & Triche, 2005; Kitange et al., 1999; Tomlins et al., 2007; Wernert et al., 1999) to prevent the expression of their target genes; directly targeting the ETS protein itself (Turner & Watson, 2008) or indirectly targeting ETS responsive promoters of transcriptional target genes (Hewett et al., 2006; Pourtier-Manzanedo et al., 2003;

Sementchenko, Schweinfest, Papas, & Watson, 1998; summarized in Table 1.5; Fig. 1.4; and in Turner & Watson, 2008).

9.3. Therapeutic drugs to target ETS factor regulation

A growing body of evidence shows that many of the drugs being tested for cancer treatment alter the activity of ETS factors and provide further rationale for developing ETS targeting therapeutics. The polyphenolic flavone Luteolin demonstrates antitumor activity in several types of cancer. In prostate cancer, Luteolin treatment has been show to increase SPDEF expression while decreasing AR expression to lower the levels of prostate-specific antigen (Tsui, Chung, Feng, Chang, & Juang, 2012). Increased SPDEF expression in this model induced BTG2, NDRG1, and Maspin gene expression to inhibit proliferation and induce apoptosis. A recent study demonstrated that the therapeutic blockade of angiotensin II type1 receptor (AGTR1) inhibits CRPC through the inhibition of ETS1 (Kosaka et al., 2010). Knockdown of AGTR1 inhibits cell proliferation and influences AR expression levels in prostate cancer cells. Wang et al. has successfully used a nanoparticle drug delivery system to deliver the natural extract gambogic acid (GA) into pancreatic cancer cells (Wang, Zhang, Chen, Shi, & Chen, 2012). GA is a potent anticancer agent that inhibits cell growth and the ability for motile function in a wide variety of tumor cells. GA-treated cells show significantly decreased expression of ETS1 as well as its downstream target genes cyclin D1, uPA, and VEGF (Wang et al., 2012).

9.4. Translational ETS targeting

While these studies demonstrate the potential of targeting ETS factors for therapeutic gain, translating this success to the bedside has been limited by the challenges of directly targeting transcriptional factors in complex physiological systems. Such challenges include successfully targeting their nuclear localization and the successful design of small-molecule inhibitors given the large surface area of their DNA–protein and protein–protein interactions (Konstantinopoulos & Papavassiliou, 2011). However, the potential benefits of being able to target transcription factors have been demonstrated for the nuclear hormone-receptor family (e.g. estrogen receptor targeting by tamoxifen in breast cancer patients) as they can be targeted on the cell surface before translocation to the nucleus (Jordan, 2003). The notion that transcription factors represent undruggable targets is slowly being eroded. Technological advances in drug delivery systems and drug design provide insight

Table 1.5 Approaches used to investigate the multifaceted aspects of ETS factor biology in order to assess the potential of therapeutically targeting these proteins

Approach	Mode of action	Method	Molecular target	References
ETS promoter occupancy	Competition for binding to promoters	Decoy oligonucleotides	ETS1, ETS2	Miwa et al. (2005)
		Triplex-forming oligonucleotides		Carbone et al. (2003, 2004) and Sahin et al. (2005)
ETS mRNA	Prevention of expression	Antisense oligonucleotides, RNA interference	Multiple ETS factors, EWS–FLI1	Dohjima et al. (2003), Hu-Lieskovan et al. (2005), Kitange et al. (1999), Tomlins et al. (2007), and Wernert et al. (1999)
ETS protein	Functional inhibition	Small-molecule inhibitor	ESE1–SER2, ETS fusions, ERG	Barber-Rotenberg et al. (2012), Rahim et al. (2011), Stegmaier et al. (2007), Turner and Watson (2008), and Patent app # 20110207675
ETS responsive promoters	Prevention of ETS transcriptional regulation	Triplex-forming oligonucleotides, dominant negative	TIE1, FGF2	Hewett et al. (2006), Pourtier-Manzanedo et al. (2003), and Sementchenko et al. (1998)

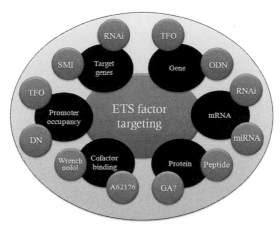

Figure 1.4 Therapeutic strategies for targeting ETS factor biology. Strategies have included directly inhibiting the promoter of oncogenic ETS factors; directly targeting specific ETS factor mRNA to prevent the expression of their target genes; directly targeting the ETS protein itself or indirectly targeting ETS responsive promoters of transcriptional target genes. See text for details. RNAi, RNA interference; miRNA, microRNA; GA, gambogic acid nanoparticles; ?, direct versus indirect effect; DN, dominant negative; ETS, E26 transforming sequence; ODN, decoy oligonucleotide; SMI, small-molecule inhibitor; TFO, triplex-forming oligonucleotide. (For color version of this figure, the reader is referred to the online version of this chapter.)

into how we may overcome these challenges (Konstantinopoulos & Papavassiliou, 2011). Recently, a cell-penetrating synthetic peptide has been developed to disrupt the ERG–DNA interaction in prostate cancer. The peptide is a potent inhibitor of ERG *in vitro* and *in vivo* to inhibit DNA damage, cell growth, invasion, and metastasis. Further structure/function studies are being performed to allow further optimization of the peptide (http://www.faqs.org/patents/app/20110207675). The transcriptional activity of ETS factors themselves is regulated by multiple ligands, coregulatory proteins, transcriptional cofactors, and chromatin remodeling components, which determine not only the expression status of a target gene but also the magnitude and duration of activation or repression. The binding sites of several other key transcription factors are found adjacent to ETS binding sites (e.g. ETS/AP-1, ETS/NFκB, ETS/AR) and such composite binding sites mediate the synergistic activation or repression of target genes (Turner & Watson, 2008). The ETS protein:cofactor binding interface is therefore crucial in the regulation of DNA binding, subcellular localization, target gene specificity, and transcriptional activity. The in-depth knowledge of both the overall structure and functional domains of specific ETS family members, including SPDEF (Y. Wang, Feng, et al., 2005) and ETS1 (G. M. Lee et al., 2005), makes targeting their interactions particularly attractive for drug discovery using small-molecule inhibitors. A series of small-molecule inhibitors have been shown to inhibit the interaction between the ETS factor ELF3 and its coactivator Sur2 (a RAS-linked subunit of the mediator complex). ELF3 with its coactivator Sur2 are required for the high HER2 promoter activation observed in cancer. The ELF3-Sur2 protein interaction is mediated by one face of an eight-amino acid alpha-helical region in the ELF3 activation domain (Shimogawa et al., 2004). Screening of an indole-mimicking π-electron-rich chemical library led to the development of wrenchnolol, a small inhibitor that mimics the alpha-helical region of the ELF3 activation domain (Shimogawa et al., 2004). *In vivo* studies using wrenchnolol in mice have been reported to be promising (Jung, Choi, & Uesugi, 2006). Secreted alkaline phosphatase screening has also identified a fluoroquinophenoxazine derivative that also disrupts the ELF3-Sur2 binding interface. A-62176 treatment arrests the cell cycle in the G1 phase via the downregulation of cyclin D1 and the upregulation of p27Kip1 in NCI-N87 gastric cancer cells (Kim et al., 2012). Disruption of the ELF3-Sur2 binding interface in cancer cells impairs the expression of HER2, inhibits HER2-mediated phosphorylation of MAPK/AKT, and restrains the activity of topoisomerase IIa (Kim et al., 2012).

There is a growing interest in developing miRNAs as therapeutic targets due to their ability to regulate multiple genes and networks of proteins. Multiple companies have been formed to exclusively pursue this research goal. One such company is Mirna Therapeutics, Inc., a biopharmaceutical company focused on miRNA-directed oncology therapies (http://mirnatherapeutics.com). Mirna Therapeutics currently has eight lead compounds in various phases of development with their lead compound, MRX34, an miR-34-mimic, due to be the first miRNA replacement therapy in Phase I clinic trials in 2013. This is an exciting area of research and one hopes to see great advances in the very near future.

9.5. Targeting ETS fusion proteins

ETS fusion proteins offer unique therapeutic targets as they are found only in certain cancer types and not in normal cells. Ewing's sarcoma ETS family fusion, EWS–FLI1, has been targeted using siRNA leading to more than 80% reduction in the EWS-FLI11 transcript and cell growth inhibition (Dohjima et al., 2003). Limitations to the use of siRNA approaches *in vivo* have been addressed by coating siRNA with a cyclodextrin-containing polycation to increase stability and by conjugating transferrin to siRNA to target to transferrin receptor-expressing tumor cells. Such modifications have successfully allowed the systemic delivery of EWS–FLI1 fusion targeting siRNA into a murine model of metastatic Ewing's sarcoma, inhibiting tumor growth with no observed side effects (Hu-Lieskovan et al., 2005). Recent research has used a gene expression approach to identify a 14-gene expression signature for the attenuation of EWS–FLI1 expression in Ewing's sarcoma (Stegmaier et al., 2007). The signature was then used to screen a small-molecule library highly enriched for FDA-approved drugs and identified cytosine arabinoside as a modulator of EWS–FLI1 protein expression and this compound reduced cell viability, transformation, and tumor growth in a xenograft model.

Oncogenic ETS fusion proteins are also therapeutic targets which have been targeted using small-molecule inhibitors. EWS-FLI1 fusions in Ewing's Sarcoma have been targeted using the compound YK-4-279 which inhibits EWS-FLI1 activity, induces apoptosis in cell lines and slows down tumor growth in mouse xenograft models (Barber-Rotenberg et al., 2012; Rahim et al., 2011). The potential of YK-4-279 has also been assessed against ETS fusions found in prostate cancer. It was found to inhibit the biological activity of both ERG and ETV1 in fusion positive cell lines to decrease migration and invasion (Rahim et al., 2011).

Given the successes in targeting ETS transcription factors for therapeutic gain, an intense examination of possible additional therapeutic approaches to target these factors is warranted.

10. CONCLUDING REMARKS

The ETS family is one of a limited number of fundamentally important gene families (Hsu et al., 2004; Turner, Findlay, et al., 2007; Watson, Ascione, & Papas, 1990; Watson et al., 2010). Although considerable work has been done on individual members of this family, little effort have been made in understanding the interrelationships between the family members and thus, why they exist as a family. Progress is being made toward addressing this fundamental question. We are just beginning to define unique versus redundant ETS functions. This will require understanding which proteins interact with each family member, which signal transduction pathways are ETS family members involved in, target genes of each family member, and the roles of ETS regulatory network in oncogenesis, tumor suppression, cell proliferation/death, and differentiation/development. During normal development, ETS factor expression is tightly controlled to regulate many biological processes including cell proliferation, differentiation, hematopoiesis, apoptosis, metastasis, tissue remodeling, angiogenesis, and transformation. In cancer, aberrant ETS factor expression results in the upregulation of genes known to drive cancer and the downregulation of genes known to suppress cancer. It is becoming increasingly evident that cellular context defines the direction and magnitude of response to ETS factors. In order to advance our understanding of the ETS-dependent regulation of cancer progression and metastasis, future studies should be directed toward elucidation of the effects of simultaneous expression of multiple transcription factors on the transcriptome of nonmetastatic and metastatic cancer. Collectively, we are beginning to define the molecular mechanisms that determine which ETS family member will regulate a particular target gene and are developing appropriate approaches to determine which target genes are necessary for ETS-dependent phenotypes.

In summary, while expression and promoter arrays will allow identification of new cancer-associated target genes that are regulated by ETS transcription factors, concomitant molecular studies will increase our understanding of the mechanisms by which ETS transcription factors act as oncogenes and tumor-suppressor genes. The holy grail of any therapeutic cancer regime is the reactivation of tumor-suppressor function and/or the inhibition of oncogene activation. Direct or indirect therapeutic intervention of ETS factor function or regulation offers intriguing possibilities in order to achieve this.

ACKNOWLEDGMENTS

We apologize to those researchers whose work could not be cited because of space limitations or was only cited indirectly by referring to reviews or more recent publications.

REFERENCES

Artlett, C. M. (2010). Animal models of scleroderma: Fresh insights. *Current Opinion in Rheumatology, 22*, 677–682.

Arvand, A., & Denny, C. T. (2001). Biology of EWS/ETS fusions in Ewing's family tumors. *Oncogene, 20*, 5747–5754.

Asano, Y., Bujor, A. M., & Trojanowska, M. (2010). The impact of Fli1 deficiency on the pathogenesis of systemic sclerosis. *Journal of Dermatological Science, 59*, 153–162.

Asano, Y., Czuwara, J., & Trojanowska, M. (2007). Transforming growth factor-beta regulates DNA binding activity of transcription factor Fli1 by p300/CREB-binding protein-associated factor-dependent acetylation. *Journal of Biological Chemistry, 282*, 34672–34683.

Asano, Y., Markiewicz, M., Kubo, M., Szalai, G., Watson, D. K., & Trojanowska, M. (2009). Transcription factor Fli1 regulates collagen fibrillogenesis in mouse skin. *Molecular and Cellular Biology*, *29*, 425–434.

Asano, Y., Stawski, L., Hant, F., Highland, K., Silver, R., Szalai, G., et al. (2010). Endothelial Fli1 deficiency impairs vascular homeostasis: A role in scleroderma vasculopathy. *The American Journal of Pathology*, *176*, 1983–1998.

Asano, Y., & Trojanowska, M. (2009). Phosphorylation of Fli1 at threonine 312 by protein kinase C delta promotes its interaction with p300/CREB-binding protein-associated factor and subsequent acetylation in response to transforming growth factor beta. *Molecular and Cellular Biology*, *29*, 1882–1894.

Ayadi, A., Zheng, H., Sobieszczuk, P., Buchwalter, G., Moerman, P., Alitalo, K., et al. (2001). Net-targeted mutant mice develop a vascular phenotype and up-regulate egr-1. *EMBO Journal*, *20*, 5139–5152.

Baillat, D., Begue, A., Stehelin, D., & Aumercier, M. (2002). ETS-1 transcription factor binds cooperatively to the palindromic head to head ETS-binding sites of the stromelysin-1 promoter by counteracting autoinhibition. *Journal of Biological Chemistry*, *277*, 29386–29398.

Ban, J., Jug, G., Mestdagh, P., Schwentner, R., Kauer, M., Aryee, D. N., et al. (2011). Hsa-mir-145 is the top EWS-FLI1-repressed microRNA involved in a positive feedback loop in Ewing's sarcoma. *Oncogene*, *30*, 2173–2180.

Barber-Rotenberg, J. S., Selvanathan, S. P., Kong, Y., Erkizan, H. V., Snyder, T. M., Hong, S. P., et al. (2012). Single enantiomer of YK-4-279 demonstrates specificity in targeting the oncogene EWS-FLI1. *Oncotarget*, *3*, 172–182.

Batliner, J., Buehrer, E., Federzoni, E. A., Jenal, M., Tobler, A., Torbett, B. E., et al. (2012). Transcriptional regulation of MIR29B by PU.1 (SPI1) and MYC during neutrophil differentiation of acute promyelocytic leukaemia cells. *British Journal of Haematology*, *157*, 270–274.

Behrens, P., Mathiak, M., Mangold, E., Kirdorf, S., Wellmann, A., Fogt, F., et al. (2003). Stromal expression of invasion-promoting, matrix-degrading proteases MMP-1 and -9 and the Ets 1 transcription factor in HNPCC carcinomas and sporadic colorectal cancers. *International Journal of Cancer*, *107*, 183–188.

Behrens, P., Rothe, M., Florin, A., Wellmann, A., & Wernert, N. (2001). Invasive properties of serous human epithelial ovarian tumors are related to Ets-1, MMP-1 and MMP-9 expression. *International Journal of Molecular Medicine*, *8*, 149–154.

Behrens, P., Rothe, M., Wellmann, A., Krischler, J., & Wernert, N. (2001). The Ets-1 transcription factor is up-regulated together with MMP 1 and MMP 9 in the stroma of pre-invasive breast cancer. *The Journal of Pathology*, *194*, 43–50.

Benbow, U., Tower, G. B., Wyatt, C. A., Buttice, G., & Brinckerhoff, C. E. (2002). High levels of MMP-1 expression in the absence of the 2G single nucleotide polymorphism is mediated by p38 and ERK1/2 mitogen-activated protein kinases in VMM5 melanoma cells. *Journal of Cellular Biochemistry*, *86*, 307–319.

Benz, C. C., O'Hagan, R. C., Richter, B., Scott, G. K., Chang, C. H., Xiong, X., et al. (1997). HER2/Neu and the Ets transcription activator PEA3 are coordinately upregulated in human breast cancer. *Oncogene*, *15*, 1513–1525.

Blair, D. G., & Athanasiou, M. (2000). Ets and retroviruses—Transduction and activation of members of the Ets oncogene family in viral oncogenesis. *Oncogene*, *19*, 6472–6481.

Bosc, D. G., Goueli, B. S., & Janknecht, R. (2001). HER2/Neu-mediated activation of the ETS transcription factor ER81 and its target gene MMP-1. *Oncogene*, *20*, 6215–6224.

Bronisz, A., Godlewski, J., Wallace, J. A., Merchant, A. S., Nowicki, M. O., Mathsyaraja, H., et al. (2012). Reprogramming of the tumour microenvironment by stromal PTEN-regulated miR-320. *Nature Cell Biology*, *14*, 159–167.

Buchwalter, G., Gross, C., & Wasylyk, B. (2004). Ets ternary complex transcription factors. *Gene, 324*, 1–14.

Buggy, Y., Maguire, T. M., McDermott, E., Hill, A. D., O'Higgins, N., & Duffy, M. J. (2006). Ets2 transcription factor in normal and neoplastic human breast tissue. *European Journal of Cancer, 42*, 485–491.

Buggy, Y., Maguire, T. M., McGreal, G., McDermott, E., Hill, A. D., O'Higgins, N., et al. (2004). Overexpression of the Ets-1 transcription factor in human breast cancer. *British Journal of Cancer, 91*, 1308–1315.

Bui, T. V., & Mendell, J. T. (2010). Myc: Maestro of microRNAs. *Genes & Cancer, 1*, 568–575.

Carbone, G. M., McGuffie, E. M., Collier, A., & Catapano, C. V. (2003). Selective inhibition of transcription of the Ets2 gene in prostate cancer cells by a triplex-forming oligonucleotide. *Nucleic Acids Research, 31*, 833–843.

Carbone, G. M., Napoli, S., Valentini, A., Cavalli, F., Watson, D. K., & Catapano, C. V. (2004). Triplex DNA-mediated downregulation of Ets2 expression results in growth inhibition and apoptosis in human prostate cancer cells. *Nucleic Acids Research, 32*, 4358–4367.

Cesari, F., Brecht, S., Vintersten, K., Vuong, L. G., Hofmann, M., Klingel, K., et al. (2004). Mice deficient for the ets transcription factor elk-1 show normal immune responses and mildly impaired neuronal gene activation. *Molecular and Cellular Biology, 24*, 294–305.

Chakrabarti, R., Hwang, J., Andres Blanco, M., Wei, Y., Lukacisin, M., Romano, R. A., et al. (2012). Elf5 inhibits the epithelial-mesenchymal transition in mammary gland development and breast cancer metastasis by transcriptionally repressing Snail2. *Nature Cell Biology, 14*, 1212–1222.

Chakrabarti, S. R., & Nucifora, G. (1999). The leukemia-associated gene TEL encodes a transcription repressor which associates with SMRT and mSin3A. *Biochemical and Biophysical Research Communications, 264*, 871–877.

Chakrabarti, S. R., Sood, R., Ganguly, S., Bohlander, S., Shen, Z., & Nucifora, G. (1999). Modulation of TEL transcription activity by interaction with the ubiquitin-conjugating enzyme UBC9. *Proceedings of the National Academy of Sciences of the United States of America, 96*, 7467–7472.

Chakrabarti, R., Wei, Y., Romano, R. A., DeCoste, C., Kang, Y., & Sinha, S. (2012). Elf5 regulates mammary gland stem/progenitor cell fate by influencing notch signaling. *Stem Cells, 30*, 1496–1508.

Chang, C. H., Scott, G. K., Kuo, W. L., Xiong, X., Suzdaltseva, Y., Park, J. W., et al. (1997). ESX: A structurally unique Ets overexpressed early during human breast tumorigenesis. *Oncogene, 14*, 1617–1622.

Charlot, C., Dubois-Pot, H., Serchov, T., Tourrette, Y., & Wasylyk, B. (2010). A review of post-translational modifications and subcellular localization of Ets transcription factors: Possible connection with cancer and involvement in the hypoxic response. *Methods in Molecular Biology, 647*, 3–30.

Chen, C., Ouyang, W., Grigura, V., Zhou, Q., Carnes, K., Lim, H., et al. (2005). ERM is required for transcriptional control of the spermatogonial stem cell niche. *Nature, 436*, 1030–1034.

Chen, K., & Rajewsky, N. (2006). Deep conservation of microRNA-target relationships and 3'UTR motifs in vertebrates, flies, and nematodes. *Cold Spring Harbor Symposia on Quantitative Biology, 71*, 149–156.

Choi, H. J., Geng, Y., Cho, H., Li, S., Giri, P. K., Felio, K., et al. (2011). Differential requirements for the Ets transcription factor Elf-1 in the development of NKT cells and NK cells. *Blood, 117*, 1880–1887.

Chou, Y. T., Lin, H. H., Lien, Y. C., Wang, Y. H., Hong, C. F., Kao, Y. R., et al. (2010). EGFR promotes lung tumorigenesis by activating miR-7 through a Ras/ERK/Myc pathway that targets the Ets2 transcriptional repressor ERF. *Cancer Research*, *70*, 8822–8831.

Christoffersen, N. R., Shalgi, R., Frankel, L. B., Leucci, E., Lees, M., Klausen, M., et al. (2010). p53-independent upregulation of miR-34a during oncogene-induced senescence represses MYC. *Cell Death and Differentiation*, *17*, 236–245.

Chun, C. Z., Remadevi, I., Schupp, M. O., Samant, G. V., Pramanik, K., Wilkinson, G. A., et al. (2011). Fli + etsrp + hemato-vascular progenitor cells proliferate at the lateral plate mesoderm during vasculogenesis in zebrafish. *PLoS One*, *6*, e14732.

Ciau-Uitz, A., Pinheiro, P., Gupta, R., Enver, T., & Patient, R. (2010). Tel1/ETV6 specifies blood stem cells through the agency of VEGF signaling. *Developmental Cell*, *18*, 569–578.

Colucci, F., Samson, S. I., DeKoter, R. P., Lantz, O., Singh, H., & Di Santo, J. P. (2001). Differential requirement for the transcription factor PU.1 in the generation of natural killer cells versus B and T cells. *Blood*, *97*, 2625–2632.

Coskun, E., von der Heide, E. K., Schlee, C., Kuhnl, A., Gokbuget, N., Hoelzer, D., et al. (2011). The role of microRNA-196a and microRNA-196b as ERG regulators in acute myeloid leukemia and acute T-lymphoblastic leukemia. *Leukemia Research*, *35*, 208–213.

Costello, P. S., Nicolas, R. H., Watanabe, Y., Rosewell, I., & Treisman, R. (2004). Ternary complex factor SAP-1 is required for Erk-mediated thymocyte positive selection. *Nature Immunology*, *5*, 289–298.

Costello, P., Nicolas, R., Willoughby, J., Wasylyk, B., Nordheim, A., & Treisman, R. (2010). Ternary complex factors SAP-1 and Elk-1, but not net, are functionally equivalent in thymocyte development. *Journal of Immunology*, *185*, 1082–1092.

Cowden Dahl, K. D., Dahl, R., Kruichak, J. N., & Hudson, L. G. (2009). The epidermal growth factor receptor responsive miR-125a represses mesenchymal morphology in ovarian cancer cells. *Neoplasia*, *11*, 1208–1215.

Cowden Dahl, K. D., Symowicz, J., Ning, Y., Gutierrez, E., Fishman, D. A., Adley, B. P., et al. (2008). Matrix metalloproteinase 9 is a mediator of epidermal growth factor-dependent e-cadherin loss in ovarian carcinoma cells. *Cancer Research*, *68*, 4606–4613.

Czuwara-Ladykowska, J., Sementchenko, V. I., Watson, D. K., & Trojanowska, M. (2002). Ets1 is an effector of the transforming growth factor Beta (tgf-Beta) signaling pathway and an antagonist of the profibrotic effects of tgf-Beta. *Journal of Biological Chemistry*, *277*, 20399–20408.

Czuwara-Ladykowska, J., Shirasaki, F., Jackers, P., Watson, D. K., & Trojanowska, M. (2001). Fli-1 inhibits collagen type I production in dermal fibroblasts via an Sp1-dependent pathway. *Journal of Biological Chemistry*, *276*, 20839–20848.

Darnell, J. E. (2002). Transcription factors as targets for cancer therapy. *Nature Reviews. Cancer*, *2*, 740–749.

Delattre, O., Zucman, J., Plougastel, B., Desmaze, C., Melot, T., Peter, M., et al. (1992). Gene fusion with an ETS DNA-binding domain caused by chromosome translocation in human tumours. *Nature*, *359*, 162–165.

de Lau, W., Kujala, P., Schneeberger, K., Middendorp, S., Li, V. S., Barker, N., et al. (2012). Peyer's patch M cells derived from Lgr5(+) stem cells require SpiB and are induced by RankL in cultured "miniguts" *Molecular and Cellular Biology*, *32*, 3639–3647.

De Vito, C., Riggi, N., Suva, M. L., Janiszewska, M., Horlbeck, J., Baumer, K., et al. (2011). Let-7a is a direct EWS-FLI-1 target implicated in Ewing's sarcoma development. *PLoS One*, *6*, e23592.

Diakos, C., Zhong, S., Xiao, Y., Zhou, M., Vasconcelos, G. M., Krapf, G., et al. (2010). TEL-AML1 regulation of survivin and apoptosis via miRNA-494 and miRNA-320a. *Blood*, *116*, 4885–4893.

DiTacchio, L., Bowles, J., Shin, S., Lim, D. S., Koopman, P., & Janknecht, R. (2012). Transcription factors ER71/ETV2 and SOX9 participate in a positive feedback loop in fetal and adult mouse testis. *Journal of Biological Chemistry*, *287*, 23657–23666.

Doane, A. S., Danso, M., Lal, P., Donaton, M., Zhang, L., Hudis, C., et al. (2006). An estrogen receptor-negative breast cancer subset characterized by a hormonally regulated transcriptional program and response to androgen. *Oncogene*, *25*, 3994–4008.

Dohjima, T., Lee, N. S., Li, H., Ohno, T., & Rossi, J. J. (2003). Small interfering RNAs expressed from a Pol III promoter suppress the EWS/Fli-1 transcript in an Ewing sarcoma cell line. *Molecular Therapy*, *7*, 811–816.

Eichner, L. J., Perry, M. C., Dufour, C. R., Bertos, N., Park, M., St-Pierre, J., et al. (2010). miR-378(*) mediates metabolic shift in breast cancer cells via the PGC-1beta/ERRgamma transcriptional pathway. *Cell Metabolism*, *12*, 352–361.

Elvert, G., Kappel, A., Heidenreich, R., Englmeier, U., Lanz, S., Acker, T., et al. (2003). Cooperative interaction of hypoxia-inducible factor-2alpha (HIF-2alpha) and Ets-1 in the transcriptional activation of vascular endothelial growth factor receptor-2 (Flk-1). *Journal of Biological Chemistry*, *278*, 7520–7530.

Farnham, P. J. (2009). Insights from genomic profiling of transcription factors. *Nature Reviews. Genetics*, *10*, 605–616.

Feldman, R. J., Sementchenko, V. I., Gayed, M., Fraig, M. M., & Watson, D. K. (2003). Pdef expression in human breast cancer is correlated with invasive potential and altered gene expression. *Cancer Research*, *63*, 4626–4631.

Feldman, R. J., Sementchenko, V. I., & Watson, D. K. (2003). The epithelial-specific Ets factors occupy a unique position in defining epithelial proliferation, differentiation and carcinogenesis. *Anticancer Research*, *23*, 2125–2131.

Feng, W., Xing, D., Hua, P., Zhang, Y., Chen, Y. F., Oparil, S., et al. (2010). The transcription factor ETS-1 mediates proinflammatory responses and neointima formation in carotid artery endoluminal vascular injury. *Hypertension*, *55*, 1381–1388.

Ferdous, A., Caprioli, A., Iacovino, M., Martin, C. M., Morris, J., Richardson, J. A., et al. (2009). Nkx2-5 transactivates the Ets-related protein 71 gene and specifies an endothelial/endocardial fate in the developing embryo. *Proceedings of the National Academy of Sciences of the United States of America*, *106*, 814–819.

Findlay, V. J., Turner, D. P., Moussa, O., & Watson, D. K. (2008). MicroRNA-mediated inhibition of prostate-derived Ets factor messenger RNA translation affects prostate-derived Ets factor regulatory networks in human breast cancer. *Cancer Research*, *68*, 8499–8506.

Findlay, V. J., Turner, D. P., Yordy, J. S., McCarragher, B., Shriver, M. R., Szalai, G., et al. (2011). Prostate-derived ETS factor regulates epithelial-to-mesenchymal transition through both slug-dependent and independent mechanisms. *Genes & Cancer*, *2*, 120–129.

Fisher, R. J., Fivash, M., Casas-Finet, J., Erickson, J. W., Kondoh, A., Bladen, S. V., et al. (1994). Real-time DNA binding measurements of the ETS1 recombinant oncoproteins reveal significant kinetic differences between the p42 and p51 isoforms. *Protein Science*, *3*, 257–266.

Fitzsimmons, D., Lutz, R., Wheat, W., Chamberlin, H. M., & Hagman, J. (2001). Highly conserved amino acids in Pax and Ets proteins are required for DNA binding and ternary complex assembly. *Nucleic Acids Research*, *29*, 4154–4165.

Flentjar, N., Chu, P. Y., Ng, A. Y., Johnstone, C. N., Heath, J. K., Ernst, M., et al. (2007). TGF-betaRII rescues development of small intestinal epithelial cells in Elf3-deficient mice. *Gastroenterology*, *132*, 1410–1419.

Foulds, C. E., Nelson, M. L., Blaszczak, A. G., & Graves, B. J. (2004). Ras/mitogen-activated protein kinase signaling activates Ets-1 and Ets-2 by CBP/p300 recruitment. *Molecular and Cellular Biology*, *24*, 10954–10964.

Franzetti, G. A., Laud-Duval, K., Bellanger, D., Stern, M. H., Sastre-Garau, X., & Delattre, O. (2012). MiR-30a-5p connects EWS-FLI1 and CD99, two major therapeutic targets in Ewing tumor. *Oncogene,* http://dx.doi.org/10.1038/onc.2012.403.

Galang, C. K., Muller, W. J., Foos, G., Oshima, R. G., & Hauser, C. A. (2004). Changes in the expression of many Ets family transcription factors and of potential target genes in normal mammary tissue and tumors. *Journal of Biological Chemistry, 279,* 11281–11292.

Garrett-Sinha, L. A., Su, G. H., Rao, S., Kabak, S., Hao, Z., Clark, M. R., et al. (1999). PU.1 and Spi-B are required for normal B cell receptor-mediated signal transduction. *Immunity, 10,* 399–408.

Garvie, C. W., Pufall, M. A., Graves, B. J., & Wolberger, C. (2002). Structural analysis of the autoinhibition of Ets-1 and its role in protein partnerships. *Journal of Biological Chemistry, 277,* 45529–45536.

Gavrilov, D., Kenzior, O., Evans, M., Calaluce, R., & Folk, W. R. (2001). Expression of urokinase plasminogen activator and receptor in conjunction with the ets family and AP-1 complex transcription factors in high grade prostate cancers. *European Journal of Cancer, 37,* 1033–1040.

Georgiou, P., Maroulakou, I. G., Green, J. E., Dantis, P., Romano-Spica, V., Kottaridis, S., et al. (1996). Expression of ets family of genes in systemic lupus erythematosus and Sjogren's syndrome. *International Journal of Oncology, 9,* 9–18.

Goel, A., & Janknecht, R. (2003). Acetylation-mediated transcriptional activation of the ETS protein ER81 by p300, P/CAF, and HER2/Neu. *Molecular and Cellular Biology, 23,* 6243–6254.

Golub, T. R., Barker, G. F., Lovett, M., & Gilliland, D. G. (1994). Fusion of PDGF receptor beta to a novel ets-like gene, tel, in chronic myelomonocytic leukemia with t(5; 12) chromosomal translocation. *Cell, 77,* 307–316.

Guarino, M., Rubino, B., & Ballabio, G. (2007). The role of epithelial-mesenchymal transition in cancer pathology. *Pathology, 39,* 305–318.

Gutierrez-Hartmann, A., Duval, D. L., & Bradford, A. P. (2007). ETS transcription factors in endocrine systems. *Trends in Endocrinology and Metabolism, 18,* 150–158.

Hahne, J. C., Fuchs, T., El Mustapha, H., Okuducu, A. F., Bories, J. C., & Wernert, N. (2006). Expression pattern of matrix metalloproteinase and TIMP genes in fibroblasts derived from Ets-1 knock-out mice compared to wild-type mouse fibroblasts. *International Journal of Molecular Medicine, 18,* 153–159.

Hahne, J. C., Fuchs, T., Florin, A., Edwards, D., Pourtier, A., Soncin, F., et al. (2011). Evaluation of effects caused by differentially spliced Ets-1 transcripts in fibroblasts. *International Journal of Oncology, 39,* 1073–1082.

Hahne, J. C., Okuducu, A. F., Fuchs, T., Florin, A., & Wernert, N. (2011). Identification of ETS-1 target genes in human fibroblasts. *International Journal of Oncology, 38,* 1645–1652.

Hanahan, D., & Coussens, L. M. (2012). Accessories to the crime: Functions of cells recruited to the tumor microenvironment. *Cancer Cell, 21,* 309–322.

Hanahan, D., & Weinberg, R. A. (2000). The hallmarks of cancer. *Cell, 100,* 57–70.

Hanahan, D., & Weinberg, R. A. (2011). Hallmarks of cancer: The next generation. *Cell, 144,* 646–674.

Harris, T. A., Yamakuchi, M., Kondo, M., Oettgen, P., & Lowenstein, C. J. (2010). Ets-1 and Ets-2 regulate the expression of microRNA-126 in endothelial cells. *Arteriosclerosis, Thrombosis, and Vascular Biology, 30,* 1990–1997.

Hart, A., Melet, F., Grossfeld, P., Chien, K., Jones, C., Tunnacliffe, A., et al. (2000). Fli-1 is required for murine vascular and megakaryocytic development and is hemizygously deleted in patients with thrombocytopenia. *Immunity, 13,* 167–177.

He, L., He, X., Lowe, S. W., & Hannon, G. J. (2007). microRNAs join the p53 network—Another piece in the tumour-suppression puzzle. *Nature Reviews. Cancer, 7,* 819–822.

Hendricks, T. J., Fyodorov, D. V., Wegman, L. J., Lelutiu, N. B., Pehek, E. A., Yamamoto, B., et al. (2003). Pet-1 ETS gene plays a critical role in 5-HT neuron development and is required for normal anxiety-like and aggressive behavior. *Neuron, 37*, 233–247.

Hewett, P. W., Daft, E. L., Laughton, C. A., Ahmad, S., Ahmed, A., & Murray, J. C. (2006). Selective inhibition of the human tie-1 promoter with triplex-forming oligonucleotides targeted to Ets binding sites. *Molecular Medicine, 12*, 8–16.

Hikami, K., Kawasaki, A., Ito, I., Koga, M., Ito, S., Hayashi, T., et al. (2011). Association of a functional polymorphism in the 3'-untranslated region of SPI1 with systemic lupus erythematosus. *Arthritis and Rheumatism, 63*, 755–763.

Hollenhorst, P. C., Chandler, K. J., Poulsen, R. L., Johnson, W. E., Speck, N. A., & Graves, B. J. (2009). DNA specificity determinants associate with distinct transcription factor functions. *PLoS Genetics, 5*, e1000778.

Hollenhorst, P. C., Ferris, M. W., Hull, M. A., Chae, H., Kim, S., & Graves, B. J. (2011). Oncogenic ETS proteins mimic activated RAS/MAPK signaling in prostate cells. *Genes & Development, 25*, 2147–2157.

Hollenhorst, P. C., Jones, D. A., & Graves, B. J. (2004). Expression profiles frame the promoter specificity dilemma of the ETS family of transcription factors. *Nucleic Acids Research, 32*, 5693–5702.

Hollenhorst, P. C., McIntosh, L. P., & Graves, B. J. (2011). Genomic and biochemical insights into the specificity of ETS transcription factors. *Annual Review of Biochemistry, 80*, 437–471.

Hollenhorst, P. C., Shah, A. A., Hopkins, C., & Graves, B. J. (2007). Genome-wide analyses reveal properties of redundant and specific promoter occupancy within the ETS gene family. *Genes & Development, 21*, 1882–1894.

Horst, D., Gu, X., Bhasin, M., Yang, Q., Verzi, M., Lin, D., et al. (2010). Requirement of the epithelium-specific Ets transcription factor Spdef for mucous gland cell function in the gastric antrum. *Journal of Biological Chemistry, 285*, 35047–35055.

Hsu, T., Trojanowska, M., & Watson, D. K. (2004). Ets proteins in biological control and cancer. *Journal of Cellular Biochemistry, 91*, 896–903.

Huang, S., Van Arsdall, M., Tedjarati, S., McCarty, M., Wu, W., Langley, R., et al. (2002). Contributions of stromal metalloproteinase-9 to angiogenesis and growth of human ovarian carcinoma in mice. *Journal of the National Cancer Institute, 94*, 1134–1142.

Hu-Lieskovan, S., Heidel, J. D., Bartlett, D. W., Davis, M. E., & Triche, T. J. (2005). Sequence-specific knockdown of EWS-FLI1 by targeted, nonviral delivery of small interfering RNA inhibits tumor growth in a murine model of metastatic Ewing's sarcoma. *Cancer Research, 65*, 8984–8992.

Itoh, T., Ando, M., Tsukamasa, Y., & Akao, Y. (2012). Expression of BMP-2 and Ets1 in BMP-2-stimulated mouse pre-osteoblast differentiation is regulated by microRNA-370. *FEBS Letters, 586*, 1693–1701.

Itoh, T., Takeda, S., & Akao, Y. (2010). MicroRNA-208 modulates BMP-2-stimulated mouse preosteoblast differentiation by directly targeting V-ets erythroblastosis virus E26 oncogene homolog 1. *Journal of Biological Chemistry, 285*, 27745–27752.

Jackers, P., Szalai, G., Moussa, O., & Watson, D. K. (2004). Ets-dependent regulation of target gene expression during megakaryopoiesis. *Journal of Biological Chemistry, 279*, 52183–52190.

Jinnin, M., Ihn, H., Asano, Y., Yamane, K., Trojanowska, M., & Tamaki, K. (2006). Upregulation of tenascin-C expression by IL-13 in human dermal fibroblasts via the phosphoinositide 3-kinase/Akt and the protein kinase C signaling pathways. *The Journal of Investigative Dermatology, 126*, 551–560.

Jinnin, M., Ihn, H., Mimura, Y., Asano, Y., Yamane, K., & Tamaki, K. (2005). Matrix metalloproteinase-1 up-regulation by hepatocyte growth factor in human dermal fibroblasts via ERK signaling pathway involves Ets1 and Fli1. *Nucleic Acids Research, 33,* 3540–3549.

Jinnin, M., Ihn, H., Yamane, K., Mimura, Y., Asano, Y., & Tamaki, K. (2005). Alpha2(I) collagen gene regulation by protein kinase C signaling in human dermal fibroblasts. *Nucleic Acids Research, 33,* 1337–1351.

Johnson, T. R., Koul, S., Kumar, B., Khandrika, L., Venezia, S., Maroni, P. D., et al. (2010). Loss of PDEF, a prostate-derived Ets factor is associated with aggressive phenotype of prostate cancer: Regulation of MMP 9 by PDEF. *Molecular Cancer, 9,* 148.

Jordan, V. C. (2003). Tamoxifen: A most unlikely pioneering medicine. *Nature Reviews. Drug Discovery, 2,* 205–213.

Jung, D., Choi, Y., & Uesugi, M. (2006). Small organic molecules that modulate gene transcription. *Drug Discovery Today, 11,* 452–457.

Kanaya, T., Hase, K., Takahashi, D., Fukuda, S., Hoshino, K., Sasaki, I., et al. (2012). The Ets transcription factor Spi-B is essential for the differentiation of intestinal microfold cells. *Nature Immunology, 13,* 729–736.

Karnoub, A. E., Dash, A. B., Vo, A. P., Sullivan, A., Brooks, M. W., Bell, G. W., et al. (2007). Mesenchymal stem cells within tumour stroma promote breast cancer metastasis. *Nature, 449,* 557–563.

Kars, M. D., Iseri, O. D., & Gunduz, U. (2010). Drug resistant breast cancer cells overexpress ETS1 gene. *Biomedicine & Pharmacotherapy, 64,* 458–462.

Kawada, H., Ito, T., Pharr, P. N., Spyropoulos, D. D., Watson, D. K., & Ogawa, M. (2001). Defective megakaryopoiesis and abnormal erythroid development in Fli-1 gene-targeted mice. *International Journal of Hematology, 73,* 463–468.

Kern, H. B., Niemeyer, B. F., Parrish, J. K., Kerr, C. A., Yaghi, N. K., Prescott, J. D., et al. (2012). Control of MicroRNA-21 expression in colorectal cancer cells by oncogenic epidermal growth factor/Ras signaling and Ets transcription factors. *DNA and Cell Biology, 31,* 1403–1411.

Kim, H. L., Jeon, K. H., Jun, K. Y., Choi, Y., Kim, D. K., Na, Y., et al. (2012). A-62176, a potent topoisomerase inhibitor, inhibits the expression of human epidermal growth factor receptor 2. *Cancer Letters, 325,* 72–79.

Kitange, G., Shibata, S., Tokunaga, Y., Yagi, N., Yasunaga, A., Kishikawa, M., et al. (1999). Ets-1 transcription factor-mediated urokinase-type plasminogen activator expression and invasion in glioma cells stimulated by serum and basic fibroblast growth factors. *Laboratory Investigation, 79,* 407–416.

Kohyama, M., Ise, W., Edelson, B. T., Wilker, P. R., Hildner, K., Mejia, C., et al. (2009). Role for Spi-C in the development of red pulp macrophages and splenic iron homeostasis. *Nature, 457,* 318–321.

Konstantinopoulos, P. A., & Papavassiliou, A. G. (2011). Seeing the future of cancer-associated transcription factor drug targets. *JAMA: The Journal of the American Medical Association, 305,* 2349–2350.

Kosaka, T., Miyajima, A., Shirotake, S., Kikuchi, E., Hasegawa, M., Mikami, S., et al. (2010). Ets-1 and hypoxia inducible factor-1alpha inhibition by angiotensin II type-1 receptor blockade in hormone-refractory prostate cancer. *The Prostate, 70,* 162–169.

Kruse, E. A., Loughran, S. J., Baldwin, T. M., Josefsson, E. C., Ellis, S., Watson, D. K., et al. (2009). Dual requirement for the ETS transcription factors Fli-1 and Erg in hematopoietic stem cells and the megakaryocyte lineage. *Proceedings of the National Academy of Sciences of the United States of America, 106,* 13814–13819.

Kubo, M., Czuwara-Ladykowska, J., Moussa, O., Markiewicz, M., Smith, E., Silver, R. M., et al. (2003). Persistent down-regulation of Fli1, a suppressor of collagen transcription, in fibrotic scleroderma skin. *The American Journal of Pathology, 163,* 571–581.

Kumar, M. S., Narla, A., Nonami, A., Mullally, A., Dimitrova, N., Ball, B., et al. (2011). Coordinate loss of a microRNA and protein-coding gene cooperate in the pathogenesis of 5q- syndrome. *Blood, 118*, 4666–4673.

Kumar-Sinha, C., Tomlins, S. A., & Chinnaiyan, A. M. (2008). Recurrent gene fusions in prostate cancer. *Nature Reviews. Cancer, 8*, 497–511.

Lacorazza, H. D., Miyazaki, Y., Di Cristofano, A., Deblasio, A., Hedvat, C., Zhang, J., et al. (2002). The ETS protein MEF plays a critical role in perforin gene expression and the development of natural killer and NK-T cells. *Immunity, 17*, 437–449.

Larsson, E., Fredlund Fuchs, P., Heldin, J., Barkefors, I., Bondjers, C., Genove, G., et al. (2009). Discovery of microvascular miRNAs using public gene expression data: miR-145 is expressed in pericytes and is a regulator of Fli1. *Genome Medicine, 1*, 108.

Lautenberger, J. A., & Papas, T. S. (1993). Inversion of a chicken ets-1 proto-oncogene segment in avian leukemia virus E26. *Journal of Virology, 67*, 610–612.

Lavenburg, K. R., Ivey, J., Hsu, T., & Muise-Helmericks, R. C. (2003). Coordinated functions of Akt/PKB and ETS1 in tubule formation. *The FASEB Journal, 17*, 2278–2280.

Lee, G. M., Donaldson, L. W., Pufall, M. A., Kang, H. S., Pot, I., Graves, B. J., et al. (2005). The structural and dynamic basis of Ets-1 DNA binding autoinhibition. *Journal of Biological Chemistry, 280*, 7088–7099.

Lee, J., Kannagi, M., Ferrante, R. J., Kowall, N. W., & Ryu, H. (2009). Activation of Ets-2 by oxidative stress induces Bcl-xL expression and accounts for glial survival in amyotrophic lateral sclerosis. *The FASEB Journal, 23*, 1739–1749.

Lelievre, E., Lionneton, F., & Soncin, F. (2001). Role of the ETS transcription factors in the control of endothelial-specific gene expression and in angiogenesis. *Bulletin du Cancer, 88*, 137–142.

Lelievre, E., Lionneton, F., Soncin, F., & Vandenbunder, B. (2001). The Ets family contains transcriptional activators and repressors involved in angiogenesis. *The International Journal of Biochemistry & Cell Biology, 33*, 391–407.

Li, R., Pei, H., & Watson, D. K. (2000). Regulation of Ets function by protein–protein interactions. *Oncogene, 19*, 6514–6523.

Li, R., Pei, H., Watson, D. K., & Papas, T. S. (2000). EAP1/Daxx interacts with ETS1 and represses transcriptional activation of ETS1 target genes. *Oncogene, 19*, 745–753.

Li, F., Wallace, J. A., & Ostrowski, M. C. (2010). ETS transcription factors in the tumor microenvironment. *Open Cancer Journal, 3*, 49–54.

Liao, Y. L., Hu, L. Y., Tsai, K. W., Wu, C. W., Chan, W. C., Li, S. C., et al. (2012). Transcriptional regulation of miR-196b by ETS2 in gastric cancer cells. *Carcinogenesis, 33*, 760–769.

Lin, E. Y., Nguyen, A. V., Russell, R. G., & Pollard, J. W. (2001). Colony-stimulating factor 1 promotes progression of mammary tumors to malignancy. *The Journal of Experimental Medicine, 193*, 727–740.

Lindemann, R. K., Braig, M., Ballschmieter, P., Guise, T. A., Nordheim, A., & Dittmer, J. (2003). Protein kinase Calpha regulates Ets1 transcriptional activity in invasive breast cancer cells. *International Journal of Oncology, 22*, 799–805.

Lindemann, R. K., Braig, M., Hauser, C. A., Nordheim, A., & Dittmer, J. (2003). Ets2 and protein kinase Cepsilon are important regulators of parathyroid hormone-related protein expression in MCF-7 breast cancer cells. *Biochemical Journal, 372*, 787–797.

Liu, F., Walmsley, M., Rodaway, A., & Patient, R. (2008). Fli1 acts at the top of the transcriptional network driving blood and endothelial development. *Current Biology, 18*, 1234–1240.

Loughran, S. J., Kruse, E. A., Hacking, D. F., de Graaf, C. A., Hyland, C. D., Willson, T. A., et al. (2008). The transcription factor Erg is essential for definitive hematopoiesis and the function of adult hematopoietic stem cells. *Nature Immunology, 9*, 810–819.

Manavathi, B., Rayala, S. K., & Kumar, R. (2007). Phosphorylation-dependent regulation of stability and transforming potential of ETS transcriptional factor ESE-1 by p21-activated kinase 1. *Journal of Biological Chemistry*, *282*, 19820–19830.

Mardis, E. R. (2007). ChIP-seq: Welcome to the new frontier. *Nature Methods*, *4*, 613–614.

Maroni, P., Bendinelli, P., Matteucci, E., & Desiderio, M. A. (2007). HGF induces CXCR4 and CXCL12-mediated tumor invasion through Ets1 and NF-kappaB. *Carcinogenesis*, *28*, 267–279.

Massie, C. E., Adryan, B., Barbosa-Morais, N. L., Lynch, A. G., Tran, M. G., Neal, D. E., et al. (2007). New androgen receptor genomic targets show an interaction with the ETS1 transcription factor. *EMBO Reports*, *8*, 871–878.

Masuya, M., Moussa, O., Abe, T., Deguchi, T., Higuchi, T., Ebihara, Y., et al. (2005). Dysregulation of granulocyte, erythrocyte, and NK cell lineages in Fli-1 gene-targeted mice. *Blood*, *105*, 95–102.

Mathenia, J., Reyes-Cortes, E., Williams, S., Molano, I., Ruiz, P., Watson, D. K., et al. (2010). Impact of Fli-1 transcription factor on autoantibody and lupus nephritis in NZM2410 mice. *Clinical and Experimental Immunology*, *162*, 362–371.

Mattot, V., Vercamer, C., Soncin, F., Fafeur, V., & Vandenbunder, B. (1999). Transcription factors of the Ets family and morphogenesis of the vascular tree. *Journal de la Societe de Biologie*, *193*, 147–153.

Mavrothalassitis, G., & Ghysdael, J. (2000). Proteins of the ETS family with transcriptional repressor activity. *Oncogene*, *19*, 6524–6532.

McKinsey, E. L., Parrish, J. K., Irwin, A. E., Niemeyer, B. F., Kern, H. B., Birks, D. K., et al. (2011). A novel oncogenic mechanism in Ewing sarcoma involving IGF pathway targeting by EWS/Fli1-regulated microRNAs. *Oncogene*, *30*, 4910–4920.

Miwa, K., Nakashima, H., Aoki, M., Miyake, T., Kawasaki, T., Iwai, M., et al. (2005). Inhibition of ets, an essential transcription factor for angiogenesis, to prevent the development of abdominal aortic aneurysm in a rat model. *Gene Therapy*, *12*, 1109–1118.

Montes, A. H., Valle-Garay, E., Alvarez, V., Pevida, M., Garcia Perez, E., Paz, J., et al. (2010). A functional polymorphism in MMP1 could influence osteomyelitis development. *Journal of Bone and Mineral Research*, *25*, 912–919.

Morris, E. E., Amria, M. Y., Kistner-Griffin, E., Svenson, J. L., Kamen, D. L., Gilkeson, G. S., et al. (2010). A GA microsatellite in the Fli1 promoter modulates gene expression and is associated with systemic lupus erythematosus patients without nephritis. *Arthritis Research & Therapy*, *12*, R212.

Mouly, E., Chemin, K., Nguyen, H. V., Chopin, M., Mesnard, L., Leite-de-Moraes, M., et al. (2010). The Ets-1 transcription factor controls the development and function of natural regulatory T cells. *The Journal of Experimental Medicine*, *207*, 2113–2125.

Moussa, O., LaRue, A. C., Abangan, R. S., Jr., Williams, C. R., Zhang, X. K., Masuya, M., et al. (2010). Thrombocytopenia in mice lacking the carboxy-terminal regulatory domain of the Ets transcription factor Fli1. *Molecular and Cellular Biology*, *30*, 5194–5206.

Moussa, O., Turner, D. P., Feldman, R. J., Sementchenko, V. I., McCarragher, B. D., Desouki, M. M., et al. (2009). PDEF is a negative regulator of colon cancer cell growth and migration. *Journal of Cellular Biochemistry*, *108*, 1389–1398.

Myers, E., Hill, A. D., Kelly, G., McDermott, E. W., O'Higgins, N. J., Buggy, Y., et al. (2005). Associations and interactions between Ets-1 and Ets-2 and coregulatory proteins, SRC-1, AIB1, and NCoR in breast cancer. *Clinical Cancer Research*, *11*, 2111–2122.

Nakerakanti, S. S., Kapanadze, B., Yamasaki, M., Markiewicz, M., & Trojanowska, M. (2006). Fli1 and Ets1 have distinct roles in connective tissue growth factor/CCN2 gene regulation and induction of the profibrotic gene program. *Journal of Biological Chemistry*, *281*, 25259–25269.

Nam, E. J., Yoon, H., Kim, S. W., Kim, H., Kim, Y. T., Kim, J. H., et al. (2008). MicroRNA expression profiles in serous ovarian carcinoma. *Clinical Cancer Research*, *14*, 2690–2695.

Ng, A. Y., Waring, P., Ristevski, S., Wang, C., Wilson, T., Pritchard, M., et al. (2002). Inactivation of the transcription factor Elf3 in mice results in dysmorphogenesis and altered differentiation of intestinal epithelium. *Gastroenterology, 122*, 1455–1466.

Nguyen, H. V., Mouly, E., Chemin, K., Luinaud, R., Despres, R., Fermand, J. P., et al. (2012). The Ets-1 transcription factor is required for Stat1-mediated T-bet expression and IgG2a class switching in mouse B cells. *Blood, 119*, 4174–4181.

Nicholson, R. I., Gee, J. M., & Harper, M. E. (2001). EGFR and cancer prognosis. *European Journal of Cancer, 37*(Suppl. 4), S9–S15.

Niu, Z., Goodyear, S. M., Rao, S., Wu, X., Tobias, J. W., Avarbock, M. R., et al. (2011). MicroRNA-21 regulates the self-renewal of mouse spermatogonial stem cells. *Proceedings of the National Academy of Sciences of the United States of America, 108*, 12740–12745.

Nowling, T. K., Fulton, J. D., Chike-Harris, K., & Gilkeson, G. S. (2008). Ets factors and a newly identified polymorphism regulate Fli1 promoter activity in lymphocytes. *Molecular Immunology, 45*, 1–12.

O'Leary, D. A., Noakes, P. G., Lavidis, N. A., Kola, I., Hertzog, P. J., & Ristevski, S. (2007). Targeting of the ETS factor GABPalpha disrupts neuromuscular junction synaptic function. *Molecular and Cellular Biology, 27*, 3470–3480.

Orimo, A., & Weinberg, R. A. (2006). Stromal fibroblasts in cancer: A novel tumor-promoting cell type. *Cell Cycle, 5*, 1597–1601.

Pan, H. F., Leng, R. X., Tao, J. H., Li, X. P., & Ye, D. Q. (2011). Ets-1: A new player in the pathogenesis of systemic lupus erythematosus? *Lupus, 20*, 227–230.

Pang, L., Xue, H. H., Szalai, G., Wang, X., Wang, Y., Watson, D. K., et al. (2006). Maturation stage-specific regulation of megakaryopoiesis by pointed-domain Ets proteins. *Blood, 108*, 2198–2206.

Papadaki, C., Alexiou, M., Cecena, G., Verykokakis, M., Bilitou, A., Cross, J. C., et al. (2007). Transcriptional repressor erf determines extraembryonic ectoderm differentiation. *Molecular and Cellular Biology, 27*, 5201–5213.

Park, C., Hennig, G. W., Sanders, K. M., Cho, J. H., Hatton, W. J., Redelman, D., et al. (2011). Serum response factor-dependent MicroRNAs regulate gastrointestinal smooth muscle cell phenotypes. *Gastroenterology, 141*, 164–175.

Park, K. S., Korfhagen, T. R., Bruno, M. D., Kitzmiller, J. A., Wan, H., Wert, S. E., et al. (2007). SPDEF regulates goblet cell hyperplasia in the airway epithelium. *The Journal of Clinical Investigation, 117*, 978–988.

Pei, H., Yordy, J. S., Leng, Q., Zhao, Q., Watson, D. K., & Li, R. (2003). EAPII interacts with ETS1 and modulates its transcriptional function. *Oncogene, 22*, 2699–2709.

Pimanda, J. E., Ottersbach, K., Knezevic, K., Kinston, S., Chan, W. Y., Wilson, N. K., et al. (2007). Gata2, Fli1, and Scl form a recursively wired gene-regulatory circuit during early hematopoietic development. *Proceedings of the National Academy of Sciences of the United States of America, 104*, 17692–17697.

Pourtier-Manzanedo, A., Vercamer, C., Van Belle, E., Mattot, V., Mouquet, F., & Vandenbunder, B. (2003). Expression of an Ets-1 dominant-negative mutant perturbs normal and tumor angiogenesis in a mouse ear model. *Oncogene, 22*, 1795–1806.

Rahim, S., Beauchamp, E. M., Kong, Y., Brown, M. L., Toretsky, J. A., & Uren, A. (2011). YK-4-279 inhibits ERG and ETV1 mediated prostate cancer cell invasion. *PLoS One, 6*, e19343.

Rajewsky, N. (2006). microRNA target predictions in animals. *Nature Genetics, 38*(Suppl.), S8–S13.

Raslova, H., Komura, E., Le Couedic, J. P., Larbret, F., Debili, N., Feunteun, J., et al. (2004). FLI1 monoallelic expression combined with its hemizygous loss underlies Paris-Trousseau/Jacobsen thrombopenia. *The Journal of Clinical Investigation, 114*, 77–84.

Riggi, N., Suva, M. L., De Vito, C., Provero, P., Stehle, J. C., Baumer, K., et al. (2010). EWS-FLI-1 modulates miRNA145 and SOX2 expression to initiate mesenchymal stem cell reprogramming toward Ewing sarcoma cancer stem cells. *Genes & Development, 24,* 916–932.

Romania, P., Lulli, V., Pelosi, E., Biffoni, M., Peschle, C., & Marziali, G. (2008). Micro-RNA 155 modulates megakaryopoiesis at progenitor and precursor level by targeting Ets-1 and Meis1 transcription factors. *British Journal of Haematology, 143,* 570–580.

Sahin, A., Velten, M., Pietsch, T., Knuefermann, P., Okuducu, A. F., Hahne, J. C., et al. (2005). Inactivation of Ets 1 transcription factor by a specific decoy strategy reduces rat C6 glioma cell proliferation and mmp-9 expression. *International Journal of Molecular Medicine, 15,* 771–776.

Salh, B., Marotta, A., Wagey, R., Sayed, M., & Pelech, S. (2002). Dysregulation of phosphatidylinositol 3-kinase and downstream effectors in human breast cancer. *International Journal of Cancer, 98,* 148–154.

Schaefer, J. S., Sabherwal, Y., Shi, H. Y., Sriraman, V., Richards, J., Minella, A., et al. (2010). Transcriptional regulation of p21/CIP1 cell cycle inhibitor by PDEF controls cell proliferation and mammary tumor progression. *Journal of Biological Chemistry, 285,* 11258–11269.

Schmidt, D., Wilson, M. D., Spyrou, C., Brown, G. D., Hadfield, J., & Odom, D. T. (2009). ChIP-seq: Using high-throughput sequencing to discover protein-DNA interactions. *Methods, 48,* 240–248.

Schotte, R., Nagasawa, M., Weijer, K., Spits, H., & Blom, B. (2004). The ETS transcription factor Spi-B is required for human plasmacytoid dendritic cell development. *The Journal of Experimental Medicine, 200,* 1503–1509.

Sementchenko, V. I., Schweinfest, C. W., Papas, T. S., & Watson, D. K. (1998). ETS2 function is required to maintain the transformed state of human prostate cancer cells. *Oncogene, 17,* 2883–2888.

Sementchenko, V. I., & Watson, D. K. (2000). Ets target genes: Past, present and future. *Oncogene, 19,* 6533–6548.

Seth, A., & Watson, D. K. (2005). ETS transcription factors and their emerging roles in human cancer. *European Journal of Cancer, 41,* 2462–2478.

Shah, R. B., & Chinnaiyan, A. M. (2009). The discovery of common recurrent transmembrane protease serine 2 (TMPRSS2)-erythroblastosis virus E26 transforming sequence (ETS) gene fusions in prostate cancer: Significance and clinical implications. *Advances in Anatomic Pathology, 16,* 145–153.

Shah, S., & Small, E. (2010). Emerging biological observations in prostate cancer. *Expert Review of Anticancer Therapy, 10,* 89–101.

Shilo, S., Roy, S., Khanna, S., & Sen, C. K. (2007). MicroRNA in cutaneous wound healing: A new paradigm. *DNA and Cell Biology, 26,* 227–237.

Shimogawa, H., Kwon, Y., Mao, Q., Kawazoe, Y., Choi, Y., Asada, S., et al. (2004). A wrench-shaped synthetic molecule that modulates a transcription factor-coactivator interaction. *Journal of the American Chemical Society, 126,* 3461–3471.

Shirasaki, F., Makhluf, H. A., LeRoy, C., Watson, D. K., & Trojanowska, M. (1999). Ets transcription factors cooperate with Sp1 to activate the human tenascin-C promoter. *Oncogene, 18,* 7755–7764.

Sica, A., Larghi, P., Mancino, A., Rubino, L., Porta, C., Totaro, M. G., et al. (2008). Macrophage polarization in tumour progression. *Seminars in Cancer Biology, 18,* 349–355.

Sieweke, M. H., Tekotte, H., Frampton, J., & Graf, T. (1996). MafB is an interaction partner and repressor of Ets-1 that inhibits erythroid differentiation. *Cell, 85,* 49–69.

Soller, M. J., Isaksson, M., Elfving, P., Soller, W., Lundgren, R., & Panagopoulos, I. (2006). Confirmation of the high frequency of the TMPRSS2/ERG fusion gene in prostate cancer. *Genes, Chromosomes & Cancer, 45,* 717–719.

Span, P. N., Manders, P., Heuvel, J. J., Thomas, C. M., Bosch, R. R., Beex, L. V., et al. (2002). Expression of the transcription factor Ets-1 is an independent prognostic marker for relapse-free survival in breast cancer. *Oncogene, 21,* 8506–8509.

Spyropoulos, D. D., Pharr, P. N., Lavenburg, K. R., Jackers, P., Papas, T. S., Ogawa, M., et al. (2000). Hemorrhage, impaired hematopoiesis, and lethality in mouse embryos carrying a targeted disruption of the Fli1 transcription factor. *Molecular and Cellular Biology*, *20*, 5643–5652.

Starck, J., Weiss-Gayet, M., Gonnet, C., Guyot, B., Vicat, J. M., & Morle, F. (2010). Inducible Fli-1 gene deletion in adult mice modifies several myeloid lineage commitment decisions and accelerates proliferation arrest and terminal erythrocytic differentiation. *Blood*, *116*, 4795–4805.

Stegmaier, K., Wong, J. S., Ross, K. N., Chow, K. T., Peck, D., Wright, R. D., et al. (2007). Signature-based small molecule screening identifies cytosine arabinoside as an EWS/FLI modulator in Ewing sarcoma. *PLoS Medicine*, *4*, e122.

Szalai, G., LaRue, A. C., & Watson, D. K. (2006). Molecular mechanisms of megakaryopoiesis. *Cellular and Molecular Life Sciences*, *63*, 2460–2476.

Takai, N., Miyazaki, T., Nishida, M., Nasu, K., & Miyakawa, I. (2002). c-Ets1 is a promising marker in epithelial ovarian cancer. *International Journal of Molecular Medicine*, *9*, 287–292.

Titeux, M., Pendaries, V., Tonasso, L., Decha, A., Bodemer, C., & Hovnanian, A. (2008). A frequent functional SNP in the MMP1 promoter is associated with higher disease severity in recessive dystrophic epidermolysis bullosa. *Human Mutation*, *29*, 267–276.

Tomlins, S. A., Laxman, B., Dhanasekaran, S. M., Helgeson, B. E., Cao, X., Morris, D. S., et al. (2007). Distinct classes of chromosomal rearrangements create oncogenic ETS gene fusions in prostate cancer. *Nature*, *448*, 595–599.

Tomlins, S. A., Rhodes, D. R., Perner, S., Dhanasekaran, S. M., Mehra, R., Sun, X. W., et al. (2005). Recurrent fusion of TMPRSS2 and ETS transcription factor genes in prostate cancer. *Science*, *310*, 644–648.

Tootle, T. L., & Rebay, I. (2005). Post-translational modifications influence transcription factor activity: A view from the ETS superfamily. *BioEssays*, *27*, 285–298.

Trojanowska, M. (2000). Ets factors and regulation of the extracellular matrix. *Oncogene*, *19*, 6464–6471.

Truong, A. H., & Ben-David, Y. (2000). The role of Fli-1 in normal cell function and malignant transformation. *Oncogene*, *19*, 6482–6489.

Tsui, K. H., Chung, L. C., Feng, T. H., Chang, P. L., & Juang, H. H. (2012). Upregulation of prostate-derived Ets factor by luteolin causes inhibition of cell proliferation and cell invasion in prostate carcinoma cells. *International Journal of Cancer*, *130*, 2812–2823.

Tugores, A., Le, J., Sorokina, I., Snijders, A. J., Duyao, M., Reddy, P. S., et al. (2001). The epithelium-specific ETS protein EHF/ESE-3 is a context-dependent transcriptional repressor downstream of MAPK signaling cascades. *Journal of Biological Chemistry*, *276*, 20397–20406.

Turner, D. P., Findlay, V. J., Kirven, A. D., Moussa, O., & Watson, D. K. (2008). Global gene expression analysis identifies PDEF transcriptional networks regulating cell migration during cancer progression. *Molecular Biology of the Cell*, *19*, 3745–3757.

Turner, D. P., Findlay, V. J., Moussa, O., Semenchenko, V. I., Watson, P. M., LaRue, A. C., et al. (2011). Mechanisms and functional consequences of PDEF protein expression loss during prostate cancer progression. *The Prostate*, *71*, 1723–1735.

Turner, D. P., Findlay, V. J., Moussa, O., & Watson, D. K. (2007). Defining ETS transcription regulatory networks and their contribution to breast cancer progression. *Journal of Cellular Biochemistry*, *102*, 549–559.

Turner, D. P., Moussa, O., Sauane, M., Fisher, P. B., & Watson, D. K. (2007). Prostate-derived ETS factor is a mediator of metastatic potential through the inhibition of migration and invasion in breast cancer. *Cancer Research*, *67*, 1618–1625.

Turner, D. P., & Watson, D. K. (2008). ETS transcription factors: Oncogenes and tumor suppressor genes as therapeutic targets for prostate cancer. *Expert Review of Anticancer Therapy*, *8*, 33–42.

Verger, A., & Duterque-Coquillaud, M. (2002). When Ets transcription factors meet their partners. *BioEssays*, *24*, 362–370.

Visel, A., Blow, M. J., Li, Z., Zhang, T., Akiyama, J. A., Holt, A., et al. (2009). ChIP-seq accurately predicts tissue-specific activity of enhancers. *Nature*, *457*, 854–858.

Vitari, A. C., Leong, K. G., Newton, K., Yee, C., O'Rourke, K., Liu, J., et al. (2011). COP1 is a tumour suppressor that causes degradation of ETS transcription factors. *Nature (London)*, *474*, 403–406.

Wang, Y., Fan, P. S., & Kahaleh, B. (2006). Association between enhanced type I collagen expression and epigenetic repression of the FLI1 gene in scleroderma fibroblasts. *Arthritis and Rheumatism*, *54*, 2271–2279.

Wang, Y., Feng, L., Said, M., Balderman, S., Fayazi, Z., Liu, Y., et al. (2005). Analysis of the 2.0 A crystal structure of the protein-DNA complex of the human PDEF Ets domain bound to the prostate specific antigen regulatory site. *Biochemistry*, *44*, 7095–7106.

Wang, D., John, S. A., Clements, J. L., Percy, D. H., Barton, K. P., & Garrett-Sinha, L. A. (2005). Ets-1 deficiency leads to altered B cell differentiation, hyperresponsiveness to TLR9 and autoimmune disease. *International Immunology*, *17*, 1179–1191.

Wang, C., & Li, Q. (2007). Identification of differentially expressed microRNAs during the development of Chinese murine mammary gland. *Journal of Genetics and Genomics*, *34*, 966–973.

Wang, Y., Stricker, H. M., Gou, D., & Liu, L. (2007). MicroRNA: Past and present. *Frontiers in Bioscience*, *12*, 2316–2329.

Wang, L. C., Swat, W., Fujiwara, Y., Davidson, L., Visvader, J., Kuo, F., et al. (1998). The TEL/ETV6 gene is required specifically for hematopoiesis in the bone marrow. *Genes & Development*, *12*, 2392–2402.

Wang, C., Zhang, H., Chen, Y., Shi, F., & Chen, B. (2012). Gambogic acid-loaded magnetic Fe(3)O(4) nanoparticles inhibit Panc-1 pancreatic cancer cell proliferation and migration by inactivating transcription factor ETS1. *International Journal of Nanomedicine*, *7*, 781–787.

Watabe, T., Yoshida, K., Shindoh, M., Kaya, M., Fujikawa, K., Sato, H., et al. (1998). The Ets-1 and Ets-2 transcription factors activate the promoters for invasion-associated urokinase and collagenase genes in response to epidermal growth factor. *International Journal of Cancer*, *77*, 128–137.

Watson, D. K., Ascione, R., & Papas, T. S. (1990). Molecular analysis of the ets genes and their products. *Critical Reviews in Oncogenesis*, *1*, 409–436.

Watson, D. K., Smyth, F. E., Thompson, D. M., Cheng, J. Q., Testa, J. R., Papas, T. S., et al. (1992). The ERGB/Fli-1 gene: Isolation and characterization of a new member of the family of human ETS transcription factors. *Cell Growth & Differentiation*, *3*, 705–713.

Watson, D. K., Turner, D. P., Scheiber, M. N., Findlay, V. J., & Watson, P. M. (2010). ETS transcription factor expression and conversion during prostate and breast cancer progression. *Open Cancer Journal*, *3*, 24–39.

Wei, G. H., Badis, G., Berger, M. F., Kivioja, T., Palin, K., Enge, M., et al. (2010). Genome-wide analysis of ETS-family DNA-binding in vitro and in vivo. *EMBO Journal*, *29*, 2147–2160.

Wei, G., Srinivasan, R., Cantemir-Stone, C. Z., Sharma, S. M., Santhanam, R., Weinstein, M., et al. (2009). Ets1 and Ets2 are required for endothelial cell survival during embryonic angiogenesis. *Blood*, *114*, 1123–1130.

Wernert, N., Stanjek, A., Kiriakidis, S., Hugel, A., Jha, H. C., Mazitschek, R., et al. (1999). Inhibition of angiogenesis in vivo by ets-1 antisense oligonucleotides-inhibition of Ets-1 transcription factor expression by the antibiotic fumagillin. *Angewandte Chemie International Edition*, *38*, 3228–3231 (in English).

Westermarck, J., Seth, A., & Kahari, V. M. (1997). Differential regulation of interstitial collagenase (MMP-1) gene expression by ETS transcription factors. *Oncogene*, *14*, 2651–2660.

Wiemer, E. A. (2007). The role of microRNAs in cancer: No small matter. *European Journal of Cancer*, *43*, 1529–1544.

Wood, L. D., Irvin, B. J., Nucifora, G., Luce, K. S., & Hiebert, S. W. (2003). Small ubiquitin-like modifier conjugation regulates nuclear export of TEL, a putative tumor suppressor. *Proceedings of the National Academy of Sciences of the United States of America*, *100*, 3257–3262.

Wu, Y. H., Hu, T. F., Chen, Y. C., Tsai, Y. N., Tsai, Y. H., Cheng, C. C., et al. (2011). The manipulation of miRNA-gene regulatory networks by KSHV induces endothelial cell motility. *Blood*, *118*, 2896–2905.

Xu, C., Liu, S., Fu, H., Li, S., Tie, Y., Zhu, J., et al. (2010). MicroRNA-193b regulates proliferation, migration and invasion in human hepatocellular carcinoma cells. *European Journal of Cancer*, *46*, 2828–2836.

Xu, D., Wilson, T. J., Chan, D., De Luca, E., Zhou, J., Hertzog, P. J., et al. (2002). Ets1 is required for p53 transcriptional activity in UV-induced apoptosis in embryonic stem cells. *EMBO Journal*, *21*, 4081–4093.

Yamada, N., Tamai, Y., Miyamoto, H., & Nozaki, M. (2000). Cloning and expression of the mouse Pse gene encoding a novel Ets family member. *Gene*, *241*, 267–274.

Yamamoto, H., Flannery, M. L., Kupriyanov, S., Pearce, J., McKercher, S. R., Henkel, G. W., et al. (1998). Defective trophoblast function in mice with a targeted mutation of Ets2. *Genes & Development*, *12*, 1315–1326.

Yang, S. H., & Sharrocks, A. D. (2004). SUMO promotes HDAC-mediated transcriptional repression. *Molecular Cell*, *13*, 611–617.

Yang, W., Shen, N., Ye, D. Q., Liu, Q., Zhang, Y., Qian, X. X., et al. (2010). Genome-wide association study in Asian populations identifies variants in ETS1 and WDFY4 associated with systemic lupus erythematosus. *PLoS Genetics*, *6*, e1000841.

Yarden, Y., & Sliwkowski, M. X. (2001). Untangling the ErbB signalling network. *Nature Reviews. Molecular Cell Biology*, *2*, 127–137.

Ye, M., Coldren, C., Liang, X., Mattina, T., Goldmuntz, E., Benson, D. W., et al. (2010). Deletion of ETS-1, a gene in the Jacobsen syndrome critical region, causes ventricular septal defects and abnormal ventricular morphology in mice. *Human Molecular Genetics*, *19*, 648–656.

Yordy, J. S., Li, R., Sementchenko, V. I., Pei, H., Muise-Helmericks, R. C., & Watson, D. K. (2004). SP100 expression modulates ETS1 transcriptional activity and inhibits cell invasion. *Oncogene*, *23*, 6654–6665.

Yordy, J. S., & Muise-Helmericks, R. C. (2000). Signal transduction and the Ets family of transcription factors. *Oncogene*, *19*, 6503–6513.

Yu, S., Zhao, D. M., Jothi, R., & Xue, H. H. (2010). Critical requirement of GABPalpha for normal T cell development. *Journal of Biological Chemistry*, *285*, 10179–10188.

Zabuawala, T., Taffany, D. A., Sharma, S. M., Merchant, A., Adair, B., Srinivasan, R., et al. (2010). An ets2-driven transcriptional program in tumor-associated macrophages promotes tumor metastasis. *Cancer Research*, *70*, 1323–1333.

Zhang, L., Eddy, A., Teng, Y.-T. , Fritzler, M., Kluppel, M., Melet, F., et al. (1995). An immunological renal disease in transgenic mice that overexpress *Fli-1*, a member of the *ets* family of transcription factor genes. *Molecular and Cellular Biology*, *15*, 6961–6970.

Zhang, X. K., Gallant, S., Molano, I., Moussa, O. M., Ruiz, P., Spyropoulos, D. D., et al. (2004). Decreased expression of the Ets family transcription factor Fli-1 markedly prolongs survival and significantly reduces renal disease in MRL/lpr mice. *Journal of Immunology*, *173*, 6481–6489.

Zhang, J., Guo, H., Zhang, H., Wang, H., Qian, G., Fan, X., et al. (2011). Putative tumor suppressor miR-145 inhibits colon cancer cell growth by targeting oncogene Friend leukemia virus integration 1 gene. *Cancer, 117,* 86–95.

Zhang, X. K., Moussa, O., LaRue, A., Bradshaw, S., Molano, I., Spyropoulos, D. D., et al. (2008). The transcription factor Fli-1 modulates marginal zone and follicular B cell development in mice. *Journal of Immunology, 181,* 1644–1654.

Zhang, Z., Verheyden, J. M., Hassell, J. A., & Sun, X. (2009). FGF-regulated Etv genes are essential for repressing Shh expression in mouse limb buds. *Developmental Cell, 16,* 607–613.

Zhang, Y., Yan, L. X., Wu, Q. N., Du, Z. M., Chen, J., Liao, D. Z., et al. (2011). miR-125b is methylated and functions as a tumor suppressor by regulating the ETS1 proto-oncogene in human invasive breast cancer. *Cancer Research, 71,* 3552–3562.

Zhou, J., Chehab, R., Tkalcevic, J., Naylor, M. J., Harris, J., Wilson, T. J., et al. (2005). Elf5 is essential for early embryogenesis and mammary gland development during pregnancy and lactation. *EMBO Journal, 24,* 635–644.

Zhou, J., Ng, A. Y., Tymms, M. J., Jermiin, L. S., Seth, A. K., Thomas, R. S., et al. (1998). A novel transcription factor, ELF5, belongs to the ELF subfamily of ETS genes and maps to human chromosome 11p13-15, a region subject to LOH and rearrangement in human carcinoma cell lines. *Oncogene, 17,* 2719–2732.

Zhu, N., Zhang, D., Chen, S., Liu, X., Lin, L., Huang, X., et al. (2011). Endothelial enriched microRNAs regulate angiotensin II-induced endothelial inflammation and migration. *Atherosclerosis, 215,* 286–293.

Zou, Z., Anisowicz, A., Hendrix, M. J., Thor, A., Neveu, M., Sheng, S., et al. (1994). Maspin, a serpin with tumor-suppressing activity in human mammary epithelial cells. *Science, 263,* 526–529.

Zucman, J., Delattre, O., Desmaze, C., Plougastel, B., Joubert, I., Melot, T., et al. (1992). Cloning and characterization of the Ewing's sarcoma and peripheral neuroepithelioma t(11; 22) translocation breakpoints. *Genes, Chromosomes & Cancer, 5,* 271–277.

CHAPTER TWO

Advances in Understanding the Coupling of DNA Base Modifying Enzymes to Processes Involving Base Excision Repair

Michael D. Wyatt[1]

Department of Drug Discovery and Biomedical Sciences, South Carolina College of Pharmacy, University of South Carolina, Columbia, South Carolina, USA
[1]Corresponding author: e-mail address: wyatt@sccp.sc.edu

Contents

Abstract

This chapter describes some of the recent, exciting developments that have characterized and connected processes that modify DNA bases with DNA repair pathways. It begins with AID/APOBEC or TET family members that covalently modify bases within DNA. The modified bases, such as uracil or 5-formylcytosine, are then excised by DNA glycosylases including UNG or TDG to initiate base excision repair (BER). BER is known to preserve genome integrity by removing damaged bases. The newer studies underscore the necessity of BER following enzymes that deliberately damage DNA. This includes the role of BER in antibody diversification and more recently, its requirement for

Advances in Cancer Research, Volume 119
ISSN 0065-230X
http://dx.doi.org/10.1016/B978-0-12-407190-2.00002-2

demethylation of 5-methylcytosine in mammalian cells. The recent advances have shed light on mechanisms of DNA demethylation, and have raised many more questions. The potential hazards of these processes have also been revealed. Dysregulation of the activity of base modifying enzymes, and resolution by unfaithful or corrupt means can be a driver of genome instability and tumorigenesis. The understanding of both DNA and histone methylation and demethylation is now revealing the true extent to which epigenetics influence normal development and cancer, an abnormal development.

ABBREVIATIONS

5-caC 5-carboxylcytosine
5-formylC 5-formylcytosine
5-hmC 5-hydroxymethylcytosine
5-hmU 5-hydroxymethyluracil
5-meC 5-methylcytosine
AID activation-induced cytosine deaminase
AP site abasic site in DNA
APE abasic site endonuclease
APOBEC apolipoprotein B mRNA-editing enzyme, catalytic polypeptide
BER base excision repair
CpG 5′-cytosine–guanine dinucleotide sequence
CSR class switch recombination
DNMT DNA methyltransferase
DSB DNA double-strand break
HR homologous recombination
IDH isocitrate dehydrogenase
MBD4 methyl-binding domain protein 4
MMR mismatch repair
NHEJ nonhomologous end joining
PARP poly(ADP-ribose) polymerase
Pol β DNA polymerase β
SAM S-adenosyl methionine
SHM somatic hypermutation
SMUG1 single-strand uracil DNA glycosylase
SSB DNA single-strand break
TDG thymine DNA glycosylase
TET ten-eleven translocation
UNG uracil DNA glycosylase
XRCC1 X-ray cross-complementation group 1

1. INTRODUCTION

1.1. Defining terms and participants

It is important to define terms used in the context of this review that are likely obvious to most readers but nevertheless can be misunderstood. Here,

DNA methylation refers to the addition of a methyl group to the 5-position of cytosine (5-meC) in DNA, a modification enzymatically added by DNA methyltransferases (DNMTs). It is not surprising that 5-meC is called the fifth base, given the profound role of 5-meC in the epigenetic regulation of gene expression through its occurrence in $5'$-cytosine–guanine dinucleotide sequences (CpGs) that occur nonrandomly in the genome. Moreover, methylation of cytosine in other sequence contexts is beginning to be appreciated (Jones, 2012). The connection between changes in methylation status and cancer has long been known but has also grown to a much deeper appreciation of the complexity. Global surveys of methylation status are providing a more comprehensive examination (Zhuang et al., 2012). In contrast to the process of methylation, elucidating the biochemical processes that actively demethylate CpGs has been a much longer and arduous journey (Ooi & Bestor, 2008). Recent exciting developments have advanced the concept that enzymatic modification of 5-meC followed by excision of the modified base via the base excision repair (BER) pathway is one means by which mammalian cells carry out demethylation. Along the way, this chapter examines other systems of base modifications coupled with DNA repair pathways to bring about genetic changes. These discoveries shed light on the importance of these processes during normal development as well as tumorigenesis, an abnormal development.

The methylation and demethylation of ε-amino groups of lysines on histones are certainly intimately related to DNA methylation, demethylation, and regulation of gene expression (Hashimoto, Vertino, & Cheng, 2010), but is not covered in detail and will be touched upon in the summary. Enzymatic methylation of other DNA base sites that occur in other organisms, for example, adenine N^6 methylation in *Escherichia coli*, will not be covered. It is also important to distinguish enzymatic methylation from reactions with electrophilic environmental carcinogens such as nitrosamines and nitrosoureas, methyl iodide, and azoxymethane, to name a few (Wyatt & Pittman, 2006). These reactions with sites in DNA bases such as guanine N7, guanine O^6, and purine-N1 in the bases of DNA are sources of mutagenic DNA damage and should not be confused with enzymatic methylation. In the long run, a better understanding of both chemical methylation (alkylation) and epigenetic methylation will provide powerful insights into the forces that drive multistage tumor progression (Grady & Ulrich, 2007).

The focus of the review is on mammalian players, with mention of examples from other organisms limited to those that offer an interesting comparison to how mammalian cells accomplish the tasks. Here, "demethylation" refers to restoring a cytosine in a DNA sequence once occupied by a 5-meC. The name "demethylase" should define in a strict biochemical sense an enzyme that removes a methyl group from a substrate. It is important to point out that there are not any known mammalian enzymes that break the carbon–carbon bond between the carbon at the 5-position of cytosine and the added methyl group. Enzymes can catalyze carbon–carbon bond breakage from pyrimidine rings; for example, an acid-catalyzed decarboxylation reaction in *de novo* pyrimidine synthesis converts orotidylate to uridylate (UMP; Berg, Tymoczko, & Stryer, 2007). An analogous decarboxylation reaction of 5-carboxylcytosine (5-caC) in DNA has also recently been reported (Schiesser et al., 2012). However, the leaving group in the decarboxylation reactions is carbon dioxide, which is quite different from considering a methyl species as a leaving group. There are proteins and enzymes that catalyze the removal of chemically induced methyl groups (and larger alkyl groups) from a few individual base positions mentioned previously (Wyatt & Pittman, 2006). For example, methylguanine methyltransferase (also known as alkylguanine transferase) is a well-characterized suicide protein that removes alkyl groups from the O^6-position of guanine via an active site cysteine (Pegg, 2000). Also, the AlkB family of iron (II), oxygen, and α-ketoglutarate-dependent dioxygenases catalyze the oxidative dealkylation of purines and pyrimidines modified at the N1-position and the N3-position, respectively (Yi, Yang, & He, 2009). In this reaction, a methyl group on these nitrogen positions is first oxidized to a hydroxymethyl group and then lost as formaldehyde. It is interesting to note that the AlkB family is considered to be a part of the superfamily of dioxygenases including the ten-eleven translocation (TET) and JBP enzymes, which will be discussed in Section 2. However, TET enzymes do not appear to be strict "demethylases" like the AlkB family members. What is becoming apparent is the enzymatic means to modify 5-meC that pave the way for subsequent restoration of a native cytosine via BER. As will be described in the following sections, BER has received attention as a potential means of demethylation and evidence exists in plants and chicken cells for direct excision of 5-meC by DNA glycosylases (Morales-Ruiz et al., 2006; Zhu, 2009; Zhu et al., 2000). Yet, the emerging consensus is that no mammalian DNA glycosylase directly removes 5-meC (Cortazar et al., 2011; Cortellino et al., 2011), and the story has turned out to be more complex.

1.2. 5-meC: The fifth base in DNA

This section attempts to put in context the stability of cytosine, 5-meC, and their derivatives (Fig. 2.1). DNA and RNA polymerases read the four-letter alphabet of DNA through the distinct chemical structures of each of the nitrogenous bases. The downside to the utilization of these bases is the inevitable chemical reactions to which the different functional groups are susceptible. Among, the simplest of the potential modifications are hydrolytic deamination reactions to the exocyclic amino groups of adenine, guanine, and cytosine, leading to hypoxanthine, xanthine, and uracil, respectively. In other words, the water in which DNA is bathed can drive base modifications with mutagenic consequences. The deamination of cytosine to uracil, which occurs at a biologically relevant rate (Lindahl & Nyberg, 1974), causes C:G to T:A transition mutations after replication. The ability to discriminate deaminated cytosine (uracil) as a damaged base is commonly presumed to be the evolutionary rationale for the utilization of thymine in DNA instead of uracil. DNA repair mechanisms are necessary to maintain DNA integrity. An early pioneer in the field was Lindahl (1974), who discovered enzymes

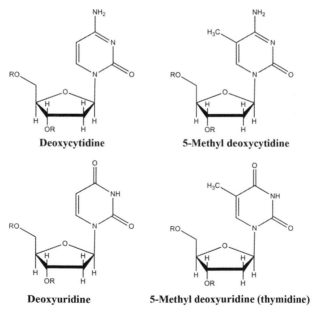

Figure 2.1 Structures of deoxycytidine, deoxyuridine, 5-methyl deoxycytidine, and 5-methyl deoxyuridine (thymidine).

capable of recognizing and removing uracil from DNA. It came to be appreciated that the restoration of DNA following excision of modified bases occurred in a multistep process known as BER, which will be described in Section 3. The deamination of cytosine in single-stranded DNA occurs at a much higher rate than in double-stranded DNA, which has interesting implications regarding to what extent transient denaturing of DNA during replication and transcription *in vivo* accelerates the deamination of cytosine (Friedberg et al., 2006).

5-MeC (Fig. 2.1) is a detectable base in DNA from a wide variety of prokaryotic and eukaryotic species (Ehrlich, Zhang, & Inamdar, 1990). Discoveries that 5-meC in CpGs of promoters silenced gene expression in mammals profoundly changed the appreciation of this base. Regarding its stability in DNA, 5-meC is more susceptible to deamination than native cytosine (Ehrlich et al., 1990). It has been calculated that 5-meC in double-stranded DNA deaminates at a rate 2- to 3-fold higher than cytosine (Shen, Rideout, & Jones, 1994), which is also true in single-stranded DNA (Wang, Kuo, Gehrke, Huang, & Ehrlich, 1982). As is apparent from the structures, deamination of 5-meC in DNA produces a T:G mismatch (Fig. 2.1). Unlike the recognition of an incorrect DNA base (uracil), correcting base:base mismatches requires presumptions about the origin of the mismatch to identify the incorrect base in the base pair. In a T:G mismatch generated by 5-meC deamination, it is always the thymine that is incorrect and therefore possible for the evolution of DNA glycosylases that excise thymine. Early efforts of Jiricny and colleagues uncovered an enzymatic activity in human cells that excises thymine from a T:G mispair (Wiebauer & Jiricny, 1989). In fact, two such DNA glycosylases with the ability to excise thymine from a T:G mispair were subsequently identified, commonly known as thymine DNA glycosylase (TDG) and methyl-binding domain protein 4 (MBD4), and will be discussed in Section 3.1. Note the significance of such an activity to excise a native undamaged base from a mispair by DNA glycosylases in BER as compared to mismatch recognition carried out by mismatch repair (MMR).

Single base mismatches generated by replication errors are repaired by postreplicative MMR, and although not a focus of this review, MMR will receive some mention because of interesting connections. In contrast to mismatch DNA glycosylases, MMR is only directed to remove the stretch of the daughter strand containing a mismatch irrespective of base identity, and can recognize U:G and T:G mismatches. It is worth noting the *dam* methylase system in *E. coli* as a means of discriminating the parental strand from the daughter strand. Dam methylase adds a methyl group to the adenine N6-position in GATC sequences of double-stranded DNA, and replicative synthesis with dATP creates a transiently "hemimethylated" state, a GATC sequence only methylated on the parental strand. Adenine-N6 methylation is the signal by which MMR discriminates parental from daughter strand in *E. coli* (Friedberg et al., 2006). However, the concept that the strand discrimination signal in mammalian systems might analogously proceed via recognition of hemimethylated CpGs turned out to be lacking in convincing evidence.

How frequently does deamination of cytosine and 5-meC occur compared to other types of endogenous base damage? Deamination of cytosine and 5-meC occurs less than a 1/10th as frequently as depurination (e.g., Friedberg et al., 2006). However, given the frequency with which C:G to T:A transition mutations occur in CpGs of genes associated with cancer, it would appear that 5-meC deamination is an important cause of mutation in human cancer. It is interesting to consider the concept that the fifth base in DNA can be converted to a damaged base (or a sixth base?) that can be converted to a substrate for a repair process.

1.3. 5-hmC: A sixth base, an intermediate, and a damaged base

5-Hydroxymethylcytosine (5-hmC) has until recently received some attention in two research areas (Fig. 2.2). First, 5-hmC was discovered as a constituent of bacteriophage DNA, in which cytosines are largely replaced with 5-hmC and is α-glucosylated (Wyatt & Cohen, 1953). This glucosylation occludes the major groove and protects phage DNA from host endonucleases. Note that trypanosomatids contain 5-hydroxymethyluracil (5-hmU) and glucosylated 5-hmU, which is called DNA base J (Gommers-Ampt, Teixeira, van de Werken, van Dijk, & Borst, 1993; Gommers-Ampt, Van Leeuwen, et al., 1993). The common enzymatic mechanism by which hydroxylated bases are produced is discussed in Section 2.3. Next, 5-hmC is also one possible product of many different base modifications to DNA caused by ionizing radiation (IR) and reactive oxygen species (Breen & Murphy, 1995; Nackerdien, Olinski, & Dizdaroglu, 1992). Note that because IR can also cause cytosine deamination, the reaction products are mixtures of oxidized thymine, cytosine, and uracil derivatives. The 5-hmU adduct is measurable in cancer tissues (Olinski et al., 1992), and in tissues from animals treated

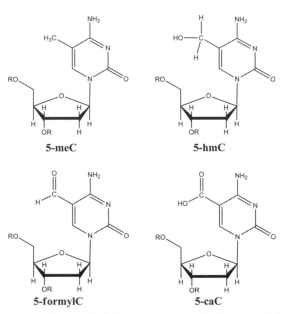

Figure 2.2 Structures of 5-methyl deoxycytidine, 5-hydroxymethyl deoxycytidine, 5-formyl deoxycytidine, and 5-carboxyl deoxycytidine.

with IR or oxidizing agents (Kasprzak et al., 1997; Mori, Hori, & Dizdaroglu, 1993; Toyokuni, Mori, & Dizdaroglu, 1994), although it would be debatable to what extent deamination of 5-hmC contributes to 5-hmU levels in these conditions *in vivo*. Early efforts identified a mammalian DNA glycosylase activity for 5-hmC, thus identifying it as a modified base that is actively removed from the genome (Boorstein, Chiu, & Teebor, 1989; Cannon-Carlson, Gokhale, & Teebor, 1989; Cannon, Cummings, & Teebor, 1988).

The related breakthroughs that have gained great attention were the discovery of 5-hmC in Purkinje neurons and granule cells in the brains of adult mice (Kriaucionis & Heintz, 2009) and the discovery of that the TET enzyme family oxidizes 5-meC to 5-hmC (Tahiliani et al., 2009), which will be discussed in Section 2.3. Interestingly, there were much earlier reports that genomic 5-hmC was detectable in brain tissue, although the levels in retrospect seem higher than more recent observations (Penn, 1976; Penn, Suwalski, O'Riley, Bojanowski, & Yura, 1972). Studies are rapidly appearing that measure the abundance of genomic 5-hmC in a number of contexts including in the brain (Chen, Dzitoyeva, & Manev, 2012; Chouliaras et al., 2012; Dzitoyeva, Chen, & Manev, 2012; Khare et al., 2012; Munzel et al., 2010), during early development (Booth et al., 2012), and in stem cells

(Pastor et al., 2011). For example, one report identifies the presence of 5-hmC and the absence of 5-meC in the paternal pronucleus, while the maternal appears to contain relatively abundant 5-meC and little 5-hmC (Iqbal, Jin, Pfeifer, & Szabo, 2011). Regarding cancer, reports show that 5-hmC levels are greatly reduced in brain cancer (Jin et al., 2011; Kraus et al., 2012), undetectable in HeLa cells (Kriaucionis & Heintz, 2009), and in other cancers (Haffner et al., 2011; Yang et al., 2013). In a global analysis of methylation status in glioma samples, higher levels of 5-meC were shown in the samples with reduced 5-hmC, providing suggestive evidence that 5-hmC is a metabolite of 5-meC (Xu et al., 2011).

5-HmC can be further oxidized by TETs to 5-formylcytosine (5-formylC) and 5-carboxylC (Fig. 2.2; He et al., 2011; Ito et al., 2011), and the presence of 5-formylC in embryonic stem cells has been reported (Pfaffeneder et al., 2011). The purpose of these additional oxidation events, as well as their regulation, is currently unclear. The potential for further oxidation also raises practical considerations about the measurement of the modified cytosines. For example, traditional bisulfite sequencing, which is by far the most common means of measuring 5-meC, cannot distinguish between 5-meC and 5-hmC (Huang et al., 2010; Jin, Kadam, & Pfeifer, 2010). In other words, bisulfite sequencing can detect the difference between unmodified and modified cytosine, but cannot discriminate among modified versions of 5-meC. There were much earlier references that modified cytosines behaved differently in bisulfite sequencing (Hayatsu & Shiragami, 1979; Ripley & Drake, 1984). However, a recent report identified a selective and sequential chemical oxidation of 5-hmC to 5-formylC, which can then be converted by bisulfite to uracil prior to sequencing (Booth et al., 2012). There are other methods for detecting 5-hmC, including the addition to 5-hmC of isotopically labeled glucosyl groups (Szwagierczak, Bultmann, Schmidt, Spada, & Leonhardt, 2010) or glucosyl groups subsequently labeled by biotinylation (Pastor, Huang, Henderson, Agarwal, & Rao, 2012), the use of novel restriction enzymes to detect genomic locations of 5-hmC (Wang et al., 2011), or conversion of 5-hmC by bisulfite to cytosine-5-methylenesulfonate followed by immunoprecipitation (Huang, Pastor, Zepeda-Martinez, & Rao, 2012).

The excitement surrounding the discovery of 5-hmC has already earned it a designation as the 6th base, but more work is required to fully understand its biological importance. If 5-hmC represents an intermediate state between 5-meC and native cytosine, its abundance in brain tissue suggests it must be a long-lived intermediate or there is extensive turnover in this tissue type. The protein motifs that recognize 5-meC in DNA are well characterized (Hashimoto et al., 2010). Evidence should be forthcoming that cellular components (besides repair machinery) can selectively distinguish 5-hmC from native cytosine and 5-meC. The important questions to be answered regard how modification of 5-meC to 5-hmC or other oxidation products affect such interactions, and the biological consequences of their recognition by as yet undiscovered protein motifs. Also, a better understanding of the biological significance of further oxidizing 5-hmC to 5-formylC and 5-caC is required.

2. ENZYMES THAT MODIFY BASES IN DNA

DNMTs are arguably the most studied enzyme family that modifies DNA bases. As they have been extensively reviewed elsewhere, only some salient points will be covered in Section 2.1. Two enzyme families with the ability to modify cytosine or 5-meC in DNA will also be discussed. The APOBEC/AID family of cytosine deaminases converts cytosine to uracil in DNA. The TET dioxygenases oxidize 5-meC to 5-hmC, 5-formylC, and 5-carboxyC in DNA. Enzymes that modify bases within native DNA have an interesting function because in a sense they can be seen to deliberately "damage" DNA. It is therefore essential for cells and organisms to carefully regulate this activity, as well as the downstream DNA repair activities, in order to appropriately manage this potentially renegade mutagenic activity (Di Noia & Neuberger, 2007; Pham, Bransteitter, & Goodman, 2005; Sousa, Krokan, & Slupphaug, 2007). Moreover, there is also a connection in some cases to downstream processes that involve a DNA double-strand break (DSB).

2.1. DNA methyltransferases

In a biochemical sense, methylation of cytosine is simplistically carried out with single enzymes, DNMTs. A few details will be covered here that are relevant for the rest of this narrative. First, DNMTs catalyze the methylation reaction via an active site cysteine thiolate carrying out a nucleophilic attack of the pyrimidine ring (Cheng & Roberts, 2001). Next, DNMTs utilize S-adenosyl methionine (SAM) as the cofactor for the methyl group, and studies have shown that limiting concentrations of SAM result in DNMTs increasing the deamination of substrate cytosines (Shen, Rideout, & Jones, 1992; Zingg, Shen, Yang, Rapoport, & Jones, 1996). Also, SAM is primarily derived from 5-methyl tetrahydrofolate. Consequently, there has been great interest in exploring the links between folate status, DNA methylation, and tumorigenesis. A review of this literature is beyond the scope of this chapter, but simply put, folate supplementation suppresses tumor initiation, yet also exacerbates progression of established tumors in humans (Cole et al., 2007; Ulrich & Potter, 2007). Interesting observations from select studies examining folate deficiency in relevant model systems will be touched upon throughout the chapter. Two other points regarding DNMT activity need to be raised. First, there is *de novo* methylation, that is, converting a CpG in double-stranded DNA by adding methyl groups to both cytosines in the CpG context. Second, maintenance methylation is necessary to maintain a proper methylation state in mitotically active cells. Semiconservative replication insures that daughter cells inherit half of the epigenetic marks of 5-meC, but a DNMT must add a methyl group to the newly synthesized DNA strand to restore the hemimethylated state to a fully methylated state. This points out a demethylation mechanism that can occur within a few cell divisions, a lack of maintenance methylation following replication. Such an obvious and passive mechanism did not leave much imagination as to the potential importance of active demethylation mechanisms, until developmental biology studies provided evidence of genome-wide demethylation events at precise times in development. The commonly held idea is that DNMT1 is the maintenance methyl transferase, whereas DNMT3a and DNMT3b are the *de novo* transferases, but both DNMT3 enzymes likely also contribute to maintenance methylation (Jones, 2012). How do cytosine base modifications such as deamination or oxidation affect the activity of DNMTs? There was a report that 5-hmC prevents DNMT1 binding to CpGs (Valinluck & Sowers, 2007). In other words, the oxidation of 5-meC on a parental strand followed by replication would prevent remethylation of the CpGs and cause the loss of epigenetic information. DNMT1 will not methylate cytosine in a CpGs containing a G:T mismatch, which implicates a similar loss of epigenetic information via deamination (Lao, Darwanto, & Sowers, 2010).

DNMTs actively maintain methylation of CpGs, and they play important but complex roles in tumorigenesis, as would be expected when considering the balance of methylation in oncogenes versus tumor suppressor genes. Knockout mice of each of the DNMTs are embryonic lethal, and hypomorphic expression of DNMT1 led to genome-wide hypomethylation and T-cell lymphomas, suggesting that even small changes in methylation have severe consequences (Gaudet et al., 2003). Depleting DNMT activity in HCT116 colorectal cancer cells ceased proliferation and caused mitotic catastrophe, suggesting that at least these microsatellite instability (MSI^+) cancer cells require methylation (Chen et al., 2007; Egger et al., 2006). In this vein, hypomorphic alleles of DNMTs reduce tumorigenesis in certain cancer predisposition models. For example, $Apc^{Min/+}$ mice harbor a mutation in the same gene mutated in familial adenomatous polyposis and are an extensively utilized model of intestinal tumorigenesis because they spontaneously develop intestinal tumors (Luongo, Moser, Gledhill, & Dove, 1994; Moser, Pitot, & Dove, 1990). Studies utilizing the $Apc^{Min/+}$ mouse model of intestinal tumorigenesis have found that dietary folate influences tumor number and size (Kadaveru, Protiva, Greenspan, Kim, & Rosenberg, 2012; Sibani et al., 2002; Song, Medline, Mason, Gallinger, & Kim, 2000; Trasler et al., 2003). Importantly, animal studies agree with the epidemiological studies cited earlier in suggesting that folate supplementation can have anti- and protumorigenic effects depending on the timing of supplementation. Furthermore, mouse strain differences and deficiencies in folate metabolizing enzymes can influence intestinal tumor susceptibility (Knock et al., 2008; Lawrance et al., 2007). Crosses of $Apc^{Min/+}$ mice with a DNMT1 hypomorphic allele, regardless of dietary folate status, decreased the number of polyps seen (Trasler et al., 2003). DNMT3b is overexpressed in human cancers and can methylate and silence tumor suppressor genes (Linhart et al., 2007; Nosho et al., 2009; Steine et al., 2011). $Apc^{Min/+}$ mice develop fewer intestinal tumors when crossed with mice harboring a tissue-specific deletion of DNMT3b (Lin et al., 2006). Other studies showed that reduced global hypomethylation increased the formation of smaller adenomas while also suppressing the formation of larger tumors (Yamada et al., 2005). The authors suggested that hypomethylation enhances tumorigenesis for those tumors with underlying chromosomal instability, whereas hypomethylation restrains the growth of tumors in which tumor suppressor genes are hypermethylated (Yamada et al., 2005).

2.2. APOBEC/AID cytosine deaminases

APOBEC stands for apolipoprotein B mRNA-editing enzyme, catalytic polypeptide. As the name suggests, this class of enzyme was originally identified as an enzyme that edits mRNA species by deaminating cytosine to uracil, which in this case produces a stop codon and truncated protein (Anant & Davidson, 2001). There are enzymes that perform analogous reactions, for example, an adenosine deaminase that acts on RNA (Rice et al., 2012). With the advent of next generation sequencing technologies and RNA sequencing, there should be interesting advances in understanding the biological importance of these editing processes (Peng et al., 2012; Vesely, Tauber, Sedlazeck, von Haeseler, & Jantsch, 2012). The APOBEC family has turned out to be large and diverse in their activities and biological functions, the extent of which is beyond the scope of this review.

Several of the APOBEC family members have been shown to act on cytosines in DNA. For example, APOBEC3 family members have been shown to "lethally edit" the reverse transcripts of retroviral DNA (Mangeat et al., 2003) and human papillomavirus DNA (Vartanian, Guetard, Henry, & Wain-Hobson, 2008). There has been much interest in determining the biochemical basis for how these enzymes achieve targeting (Holden et al., 2008; Kohli et al., 2010; Nowarski, Britan-Rosich, Shiloach, & Kotler, 2008; Rausch, Chelico, Goodman, & Le Grice, 2009). The other family member in this enzyme class is activation-induced cytosine deaminase (AID), which was originally discovered through its role in somatic hypermutation (SHM) and class switch recombination (CSR) in B cells (Muramatsu et al., 1999, 2000). Very briefly and simplistically, transient expression of AID in germinal center B cells drives genetic alterations, SHM and CSR, to increase the affinity of antibodies for their antigen and produce different Ig isoforms, respectively (Pham et al., 2005). Patients with inactivating mutations in *AID*, and $Aid^{-/-}$ mice, show a lack of SHM and CSR (Muramatsu et al., 2000; Revy et al., 2000), and the clinical manifestation in patients with defective AID is a hyper-IgM disorder with lymph node hyperplasia (Minegishi et al., 2000; Revy et al., 2000). Following the initial report that AID can act on DNA (Petersen-Mahrt, Harris, & Neuberger, 2002), there has been debate as to whether RNA or DNA is the ultimate target of AID in these processes, but a more recent study provided direct evidence of uracil residues in the DNA of variable and switch regions (Maul et al., 2011). One other important point is that AID acts on single-stranded DNA (Bransteitter, Pham, Scharff, & Goodman, 2003), which is proposed to be exposed during transcription (Yoshikawa et al., 2002). The fate of the uracil generated by AID will be discussed further in Section 4. It is not surprising that the process of SHM requires an enzyme that introduces mutagenic lesions in the DNA of B cells but also raises the possibility that dysregulation of this process can promote genome instability. Indeed, AID overexpression leads to large increases in point mutations and T-cell lymphomas (Okazaki et al., 2003), and AID has been associated with translocations in cancer (Lin et al., 2009; Robbiani et al., 2009).

How does 5-mC, 5-hmC, or any other modified base influence the ability of AID and APOBEC enzymes to remove the amino group of cytosine from these derivatives? It is known that 5-meC inhibits AID activity, suggesting that AID would not participate in an efficient demethylation process (Larijani et al., 2005). A recent study provided an extensive characterization of deaminase activity for three of the APOBEC family members and AID. The data showed that the deaminase activity of these enzymes for 5-mC is detectable but only occurs 10% as efficiently as the deamination of native cytosine, while no deamination activity was measurable for 5-hmC or other cytosine derivatives with increased steric bulk (Nabel et al., 2012). These are important factors to consider in the broader context of pathway connections and phenotypes of knockout mice, discussed in Section 4.

2.3. TET dioxygenases

TETs are interesting enzymes only recently appreciated for their significance. The name was derived from the first discovery of TET1 as the genetic loci fused to the MLL (mixed lineage leukemia) gene in some cases of acute myeloid leukemia (Lorsbach et al., 2003; Ono et al., 2002). MLL is notable as a histone methylase with multiple roles in chromatin regulation and gene expression, cell cycle progression, and development (Liu, Takeda, Cheng, & Hsieh, 2008). TET1 was also called LCX because the authors noted that the TET1 protein contained a zinc-binding CXXC domain (leukemia-associated CXXC protein; Ono et al., 2002). The cysteine-X-X-cysteine domain is notable as a methylated DNA binding domain that DNMTs and MLL also possess (Hashimoto et al., 2010). The TET1/MLL fusion protein retains the CXXC domain of MLL but has lost the TET1 CXXC domain (Lorsbach et al., 2003). Two other proteins with homology to TET1 were also identified as TET2 and TET3 (Lorsbach et al., 2003).

5-HmU and 5-hmC are intermediates in the synthesis of glucosylated thymidine and cytosine, respectively, and investigations into the enzymes that carry out these reactions provided a breakthrough in understanding the metabolism of 5-hmC in mammalian cells. 5-HmU was identified as an intermediate in the synthesis of a glucosylated thymidine called DNA base J, in trypanosomatids (Gommers-Ampt, Van Leeuwen, et al., 1993). Proteins from trypanosomatids that bind to base J possess were shown to possess thymidine hydroxylase activity, in other words the ability to oxidize thymidine to hydroxymethyl uracil in DNA (Cliffe et al., 2009; Iyer, Tahiliani, Rao, &

Aravind, 2009; Vainio, Genest, ter Riet, van Luenen, & Borst, 2009; Yu et al., 2007). Indeed, the JBP1 and TET enzymes are part of a family of iron (II), oxygen, and α-ketoglutarate-dependent dioxygenases (Iyer et al., 2009), and the TET enzymes were shown to convert 5-meC into 5-hmC (Tahiliani et al., 2009). Note that although the TET/JBP and AlkB enzymes mentioned previously are part of a superfamily of dioxygenases, they have distinct substrate preferences and products and should not be confused regarding their biological roles in DNA base modification for epigenetic control as opposed to repair of damaged bases that are mutagenic and replication blocking. One interesting note in common regards their use of α-ketoglutarate as the cofactor in the oxidation reaction. α-Ketoglutarate is synthesized from isocitrate by isocitrate dehydrogenase (IDH) as part of the citric acid cycle (Berg et al., 2007). Recent studies have identified mutations in the cytoplasmic (IDH1) or mitochondrial (IDH2) enzymes that gain an additional activity to convert α-ketoglutarate to 2-hydroxyglutarate (Dang et al., 2009; Gross et al., 2010; Ward et al., 2010; Xu et al., 2011). This dominant negative activity dramatically increases the 2-hydroxyglutarate levels in the glioma and leukemia cells examined. Moreover, elevation of this putative oncometabolite seems to have a direct consequence on TET activity, because a study of genome-wide methylation in glioma and leukemia revealed that tumors harboring the dominant-negative IDH1 or IDH2 mutations have distinct methylation profiles (Figueroa et al., 2010; Noushmehr et al., 2010). Biochemically, 2-hydroxyglutarate can act as a weak competitive inhibitor of TET2 activity, and to a lesser extent, TET1 activity (Xu et al., 2011). Gliomas with mutant IDH1 showed lower levels of genomic 5-hmC and higher levels of 5-mC, suggesting that TET activity is indeed modulated *in vivo* by the availability of α-ketoglutarate and IDH activity (Xu et al., 2011). Yet, more remains to be done in the field, because other studies that observed low to undetectable levels of 5-hmC did not see a correlation with IDH gain of function mutations (Jin et al., 2011).

Among the three TETs, TET2 has been most strongly implicated in cancer thus far. TET2 mutations have been identified in leukemias and myelodysplastic syndromes (Abdel-Wahab et al., 2009; Langemeijer et al., 2009; Perez et al., 2012). Lower 5-hmC levels are observed in leukemia patient samples with mutations in TET2, and TET2 activity is suggested to be important for myelopoiesis (Ko et al., 2010). Accordingly, $Tet2^{-/-}$ mice display disruptions in hematopoiesis and expanded proliferation that causes hematological malignancies and strongly suggest that TET2 mutations are causative in human cancer (Moran-Crusio et al., 2011; Quivoron et al., 2011).

TET1 is expressed in the adult brain of mice and has been proposed to promote demethylation via 5-hmC in neuronal tissues (Guo, Su, Zhong, Ming, & Song, 2011). TET3 is implicated in a role in zygotic programming of the male pronucleus (Gu et al., 2011; Inoue & Zhang, 2011; Iqbal et al., 2011). TET3 is reported to be highly expressed in oocytes and zygotes but rapidly declines by the two-cell stage. Interestingly, 5-hmC was also reported to persist through the first several divisions after fertilization and is gradually lost in preimplantation embryos, suggesting that 5-hmC is passively lost during DNA replication and is not removed *en masse* as part of a genome-wide reprogramming event early after fertilization (Inoue & Zhang, 2011; Iqbal et al., 2011). In this regard, there still remain important questions to be answered about TET activity to further oxidize 5-hmC to 5-formylC and 5-caC. If 5-hmC represents a sixth base while 5-formylC and 5-caC are the intermediates removed by BER, then this model requires a precise regulation of TET activity such that only the first oxidation occurs. The roles of TETs in cancer thus far are limited to TET2 in a few less frequent malignancies, but there is likely much more to be learned about this newly appreciated enzyme family and their potential broader roles in cancer.

3. THE REMOVAL OF ENZYMATICALLY MODIFIED BASES BY BER

As the name implies, the BER pathway can be initiated by the removal of a damaged base, although this is not the only means of engaging the remainder of the repair pathway. The pathway and many of its players have been characterized and reviewed elsewhere (e.g., Almeida & Sobol, 2007), so discussion here will be limited to the most pertinent points for this chapter. DNA glycosylases convert one form of DNA damage into another, which must be acted on by the rest of the pathway to achieve restoration of native DNA (Wyatt, Allan, Lau, Ellenberger, & Samson, 1999). Most DNA glycosylases, including the ones to be discussed below, are monofunctional, meaning the products are the liberated base and an abasic (AP) site remaining in DNA, which itself is a mutagenic and toxic DNA adduct that must be processed by the remainder of the pathway. Some DNA glycosylases have associated activities that catalyze deoxyribose cleavage via a β elimination reaction (Stivers & Jiang, 2003). The catalytic power of different DNA glycosylases varies but in most cases is quite modest. The trade-off for reduced catalytic capacity is presumed to be a reduction of the potential harm in an enzymatic activity that could aberrantly remove normal bases. Indeed, it has been shown with at least three different DNA glycosylases that single amino acid mutations can render the mutant enzymes capable of removing undamaged bases (Connor & Wyatt, 2002; Kavli et al., 1996; Maiti, Noon, Mackerell, Pozharski, & Drohat, 2012; O'Brien & Ellenberger, 2004). A second means of limiting the potential harm of a DNA glycosylase is achieved through a high binding affinity for the AP site product, and the implications of this will be discussed below.

The remainder of the BER pathway is highly coordinated by the physical association and structural complementarity of the proteins (Parikh, Mol, Hosfield, & Tainer, 1999; Wilson & Kunkel, 2000). This makes sense biologically, because the repair intermediates of DNA in BER are themselves DNA damage and, if left unattended and unprocessed, have more toxic consequences than the original damaged base removed by the DNA glycosylase. AP endonuclease (APE) activity produces a $3'$-OH and a $5'$-deoxyribose phosphate ($5'$-dRp) group, the remaining sugar group that must be removed. There are two subpathways by which this can occur, called short-patch BER and long-patch BER, of which the latter process is not discussed here. In SP-BER, single nucleotide gap-filling DNA synthesis and removal of the $5'$-dRp is performed by the bifunctional DNA polymerase β (Pol β). DNA ligase activity in coordination with an important scaffolding protein called XRCC1 (X-ray cross-complementing group 1) restores the intact DNA. Note also that the proteins downstream of DNA glycosylases can participate in a highly efficient repair mechanism for DNA single-strand breaks (SSBs). Defects in SSBR, which includes two proteins with DNA end-trimming activity, polynucleotide kinase $3'$-phosphatase and aprataxin, cause hereditary neurodegenerative diseases (Caldecott, 2008).

3.1. DNA glycosylases

The human genome contains four loci that code for distinct DNA glycosylases that can remove deaminated cytosine (uracil) or deaminated 5-meC (thymine), commonly called uracil DNA glycosylase (UNG), single-strand uracil DNA glycosylase (SMUG1), MBD4, and TDG. Numerous studies have elucidated distinct activities and roles for these enzymes (Cortazar, Kunz, Saito, Steinacher, & Schar, 2007; Krokan, Drablos, & Slupphaug, 2002). A few points regarding their activities relevant for considering downstream consequences are first covered. UNG removes uracil with no opposing base preference and can remove uracil from single-stranded or double-stranded DNA. UNG2 (the nuclear isoform) is the predominant cellular UDG activity, and the primary role of UNG2 seems to be counteracting uracil misincorporation during replication (Nilsen et al., 2000; Otterlei et al., 1999). Catalytically, UNG is the most proficient DNA glycosylase in single turnover measurements, between 100- and 10,000-fold better than TDG or SMUG1 (Stivers & Jiang, 2003). Also, UNG shows no product (AP site) inhibition (Pettersen et al., 2007), whereas SMUG1 and TDG show substantial affinities for the AP site that severely limit turnover (Hardeland, Bentele, Jiricny, & Schar, 2000; Pettersen et al., 2007; Waters, Gallinari, Jiricny, & Swann, 1999). Also, SMUG1 inhibits its cleavage by APE, whereas UNG was found to stimulate APE (Pettersen et al., 2007). In other words, there can be a difference how the downstream events of BER proceed depending on which DNA glycosylase activity has initiated base removal.

$Ung^{-/-}$ mice are viable without appreciable developmental defects during embryogenesis but are susceptible to developing B-lymphomas later in life (Nilsen et al., 2003). Evidence from $Ung^{-/-}$ mice and recessive mutations in patients with a hyper-IgM syndrome demonstrate the necessity of UNG in SHM and CSR (Imai et al., 2003; Rada et al., 2002). The model, in which enzymatic deamination of cytosines in DNA via AID is followed by UNG-initiated BER, will be discussed in greater detail in section 4. One other note, $Ung^{-/-}$ mice are also more susceptible to post-ischemic brain injury than wild-type mice (Endres et al., 2004). Going back for a moment to links with folate metabolism and methylation, there are a few other studies worth mentioning. First, $Ung^{-/-}$ mice on a folate deficient diet showed increased genomic uracil and hypomethylation in the colon epithelial cells and an increase in cytosine to thymine point mutations (Linhart et al., 2009). Although an increase in tumors was not observed, it was noted that the mouse strain used (50% C57BL/6J, 50% 129) is not sensitized to developing intestinal tumors, and also that the λCII assay from Big Blue™ mice would not detect chromosomal damage caused by strand break events (Linhart et al., 2009). Next, a few studies have examined the effect of folate deficiency on the neurocognitive function of $Ung^{-/-}$ mice compared to wild-type mice. It was concluded that the cognitive deficits and mood alterations caused by folate deficiency were exacerbated in the $Ung^{-/-}$ mice

(Kronenberg et al., 2008). It is speculative but intriguing to consider folate balance under conditions of stress in adult neuronal tissue, in which TET1 activity is prominent.

SMUG1 was originally identified in a screen with designed inhibitors of DNA glycosylases and named because of its activity for uracil in single-stranded DNA (Haushalter, Todd Stukenberg, Kirschner, & Verdine, 1999). Subsequent biochemical characterization confirmed activity for uracil in double-stranded DNA as well as 5-hmU (Boorstein et al., 2001). As mentioned in Section 1.3, 5-hmC and 5-hmU are products of IR damage, and SMUG1 knock down in $Ung^{-/-}$ MEFs caused sensitivity to IR (An, Robins, Lindahl, & Barnes, 2005). Also, an increase in C:G to T:A mutations was noted, which is consistent with a lack of repair of damaged cytosines. Deamination of 5-hmC produces 5-hmU (Cannon-Carlson et al., 1989), so it is tempting to speculate that SMUG1 might contribute to a demethylation process after enzymatic deamination of 5-hmC during development. Yet, $Smug1^{-/-}$ mice were very recently reported, and it is interesting that they are viable, fertile, and lacking an overtly disease-oriented phenotype identified as of yet (Kemmerich, Dingler, Rada, & Neuberger, 2012). Tissues from knockout mice lack detectable 5-hmU DNA glycosylase activity, implicating SMUG1 as the major DNA glycosylase for this adduct. $Ung^{-/-} Smug1^{-/-}$ double knockout mice were also generated and viable with no overt phenotype up to 1 year of age, which is striking considering that tissues from the double knockout mice lack any appreciable activity for uracil in DNA (Kemmerich et al., 2012). In other words, U:G mispairs resulting from cytosine deamination must be processed by other pathways, or that enzymatic and spontaneous deamination events do not occur at a rate that interferes with normal development. To test the involvement of MMR as a backup function in this regard, $Ung^{-/-} Smug1^{-/-}$ mice were crossed with $Msh2^{-/-}$ mice. Although the cohort number was small ($n = 6$), the triple knockouts suffered a substantially shorter lifespan (107 days) compared to the $Msh2^{-/-}$ mice, which themselves have a relatively short life span of ~6 months (de Wind, Dekker, Berns, Radman, & te Riele, 1995). The triple knockout mice were reported to succumb to mostly lymphoid tumors as do the $Msh2^{-/-}$ mice, suggesting that an inability to repair deaminated cytosine strongly predisposes to cancer (Kemmerich et al., 2012). Expanded analyses in larger cohorts of mice will allow firmer conclusions.

Two other DNA glycosylases with known activity for uracil are MBD4 (also known as MED1) and TDG. MBD4 removes uracil or thymine specifically in the CpG context (Hendrich, Hardeland, Ng, Jiricny, & Bird, 1999; Millar et al., 2002; Wong et al., 2002). There is reasonable evidence in zebrafish that AID deamination of 5-meC coupled to MBD4 activity against the T:G mismatch promotes demethylation (Rai et al., 2008), and that the loss of APC upregulates these demethylation components, leading to hypomethylation of genes involved in differentiation (Rai et al., 2010). However, the situation appears to be different in mammalian systems. $Mbd4^{-/-}$ mice are viable with no overt developmental defects or cancer predisposition reported thus far. $Mbd4^{-/-}$ mice show a 2- to 3-fold increased mutation frequency at CpG sites, which leads to the conclusion that the primary function of MBD4 is to protect against spontaneous muta-genesis at CpGs driven by hydrolytic deamination of 5-meC. The $Mbd4^{-/-}$ background accelerates the intestinal tumor predisposition of the $Apc^{Min/+}$ mice (Millar et al., 2002; Wong et al., 2002). Mutations in the $MBD4$ gene have been reported in human colorectal cancers with MMR defects that cause MSI^{+} (Bader et al., 1999; Riccio et al., 1999), cancers most commonly associated with defects in MMR. The MBD4 gene contains an exonic $(A)_{10}$ tract (Bader, Walker, & Harrison, 2000), and such polynucleotide runs are known sites of mutation in MSI^{+} tumors.

TDG was originally identified by its ability to recognize thymine or ura-cil mispaired opposite guanine (Neddermann et al., 1996; Neddermann & Jiricny, 1994), but TDG removes other damaged pyrimidines, including the products of bulkier modifications such as 5-halogenated uracils, 5-HmU, 5-hydroxyuracil, thymine glycol, and $3,N^{4}$-ethenocytosine (Cortazar et al., 2007). In most of these cases, the opposing base preference for guanine is retained. It was recently reported that TDG is also capable of removing 5-formylC and 5-caC, the oxidation products of 5-hmC, but sur-prisingly, not 5-hmC itself (He et al., 2011; Maiti & Drohat, 2011). Thus, sequential TET-catalyzed oxidations of 5-meC are explicitly linked to BER-mediated removal. As mentioned earlier, TDG binds with high affin-ity to the AP site product, but it was subsequently shown that modification of TDG by SUMOylation induces a conformational change that facilitates its release from the AP site (Hardeland, Steinacher, Jiricny, & Schar, 2002; Steinacher & Schar, 2005). The ability of TDG to remain bound to the AP site product, as well as a SUMOylation mechanism to dis-place TDG, suggests that the activity of this DNA glycosylase is highly reg-ulated. The other interesting aspect of TDG regards its reported interactions with DNMTs (Boland & Christman, 2008; Gallais et al., 2007; Li, Zhou,

Zheng, Walsh, & Xu, 2007), as well as transcription factors and histone methyltransferases (Chen et al., 2003; Tini et al., 2002; Um et al., 1998). In stark contrast to the murine knockouts of the other three UDGs, $Tdg^{-/-}$ mice are early embryonic lethal (Cortazar et al., 2011; Cortellino et al., 2011). Knockin of a catalytically inactive mutant of TDG failed to rescue the embryonic lethality, strongly suggesting that the DNA glycosylase activity of TDG was necessary (Cortellino et al., 2011). Interestingly, $Tdg^{-/-}$ MEFs are not sensitized to IR or hydrogen peroxide, suggesting that TDG activity is not important for removing the oxidized DNA base damage caused by these agents (Cortazar et al., 2011). Instead, the embryonic lethality of $Tdg^{-/-}$ mice appears to be due to a loss of epigenetic stability, as will be discussed in Section 4.

3.2. BER following DNA glycosylase activity

The downstream events and participants in BER are worth revisiting in light of the newly appreciated roles of BER and a few new studies. Mice lacking BER components downstream of DNA glycosylase activity are embryonic lethal or die immediately after birth, underscoring the overall importance of completing the repair process initiated by DNA glycosylases or for SSBR. The APE1 protein was independently identified as a nuclear redox factor that stimulates transcription factor activity (Xanthoudakis, Miao, Wang, Pan, & Curran, 1992). $Ape1^{-/-}$ mice are early embryonic lethal (Xanthoudakis, Smeyne, Wallace, & Curran, 1996), and cells depleted of APE cease proliferating within 24 h, underscoring the essentiality of this multifunctional enzyme (Fung & Demple, 2005; Izumi et al., 2005). $Ape1^{+/-}$ mice also display sensitivity to oxidative stress and heightened apoptotic response (Unnikrishnan et al., 2009). The severity of the phenotypes associated with APE1 deficiency not only underscores its necessity but also poses challenges in separating the several functions of the APE1 protein in addition to its role in BER. There is also an APE2 protein that possesses APE activity, but clearly does not act as a backup for APE1, as evidenced by the phenotypes of APE1-deficient cells and mice. $Ape2^{-/-}$ mice are viable but show growth retardation and defects in normal B-cell development (Guikema et al., 2011; Ide et al., 2004).

In SP-BER, single nucleotide, gap-filling DNA synthesis and end-trimming activity are performed by the bifunctional enzyme Pol β which also contains a 5′-dRp lyase domain (Matsumoto & Kim, 1995; Sobol et al., 2000).

Pol β^{−/−} mice are not viable (Gu, Marth, Orban, Mossmann, & Rajewsky, 1994), show growth retardation, defects in neurogenesis, and die immediately after birth (Sugo, Aratani, Nagashima, Kubota, & Koyama, 2000). Extensive neuronal death was observed in *Pol* β^{−/−} mice, which the authors speculated was a result of an inability to repair SSBs in postmitotic cells (Sugo et al., 2000). A *Pol* β knockin mice was recently reported, which expressed a variant with wild-type lyase activity but an inefficient polymerase domain (Y265C; Senejani et al., 2012). Mice expressing this mutant also show growth retardation, although in this cohort, 40% of the mice apparently survived after birth, and no characterization of neurodegenerative defects in the dead or surviving mice were reported. Pol β heterozygous mice have an increased likelihood of developing lymphoid hyperplasia and adenocarcinomas in several tissue sites (Cabelof et al., 2006).

Returning for a moment to the folate connection, combining folate deficiency and Pol β haploinsufficiency increased unrepaired BER intermediates (Cabelof et al., 2004). The authors were surprised to find that Pol β was not upregulated as a result of folate deficiency, when the prediction was that increased genomic uracil would generally upregulate BER in response to damage. It was subsequently reported that folate deficiency in Pol β heterozygous mice in fact attenuated colon tumorigenesis induced by the chemical methylating carcinogen *N,N*-dimethylhydrazine, which produces damage that BER repairs (Ventrella-Lucente et al., 2010). In other words, one can hypothesize that a reduced ability to complete BER might drive a proapoptotic phenotype that suppresses tumorigenesis.

In the final step of SP-BER, the long-held paradigm was that the ligation step is carried out by DNA ligase III in partnership with XRCC1. The solid biochemical evidence of this partnership includes the physical association and intracellular stability dependent on the interaction between ligase III and XRCC1. However, recent impressive mouse genetic experiments have suggested that ligase III is in fact essential for mitochondrial DNA integrity (Gao et al., 2011; Simsek et al., 2011). The authors propose that DNA ligase I acts as the ligase in nuclear BER, which is substantial shift in paradigm and should be further explored. *Xrcc1*^{−/−} mice are embryonic lethal (Tebbs et al., 1999), and neural specific knockouts of XRCC1 reveal the essentiality of this protein in normal neuronal development (Lee et al., 2009). In particular, XRCC1-mediated repair of SSBs resulting from DNA glycosylase activity and/or direct sources of strand breaks is essential for maintaining genome stability even in nondividing neuronal cells (Kulkarni, McNeill, Gleichmann, Mattson, & Wilson, 2008). The common model invokes the inability to properly repair endogenous oxidative damage and strand breaks as the driver of neuronal damage and loss. However, this phenotype becomes more interesting to consider in light of the expression of TET1 in adult brain tissue. TET activity to modify bases in these tissues and TDG-mediated removal of the modified bases is a potentially intriguing source of SSBs that must be resolved to preserve neuronal integrity.

Poly(ADP-ribose) polymerases (PARPs) are nearly impossible to ignore in broader discussions of DNA repair, even though their activities and functions are not easily described and categorized. The PARP family is up to 17 members with multiple and potentially overlapping interactions and functions, but only PARPs 1–3 thus far are directly implicated as DNA damage dependent (De Vos, Schreiber, & Dantzer, 2012). Even in the case of PARP1, the most studied member of the family, its precise functions in DNA repair responses remain enigmatic and misunderstood in the context of the great interest in PARP inhibitors in cancer chemotherapy (Helleday, 2011). Discussion in this chapter will be restricted to just a few relevant points. Single knockouts of PARP1 or PARP2 viable but sensitive to IR, while $Parp1^{-/-}$ $Parp2^{-/-}$ mice are early embryonic lethal, which underscores the necessity and redundancy of PARP1/2 activity (Menissier de Murcia et al., 2003). $Parp1^{-/-}$ mice have an altered immunoglobulin isoform expression and a reduced T-cell response (Ambrose et al., 2009). Moreover, it was also shown that although PARP1 and PARP2 are not required for CSR, PARP1 influences isoform switching and PARP2 suppresses translocations (Robert, Dantzer, & Reina-San-Martin, 2009). Biochemically, PARP1/2 can bind to DNA SSBs and DSBs and can also interact with and poly-ribosylate itself and a number of repair proteins. This includes not just proteins from SSBR/BER but also proteins from the two major DSB repair pathways, homologous recombination (HR) and nonhomologous end joining (NHEJ; De Vos et al., 2012). Recent characterization of PARP3 also suggests a direct role in DNA damage responses (Boehler et al., 2011). The older categorization of PARP1 is evolving with the realization that its role is quite distinct from the BER/SSBR proteins mentioned above. It is tempting to speculate that PARP1/2 and possibly PARP3 ribosylation of target proteins play roles in orchestrating the choreography of DSB repair.

4. RELATIONSHIPS BETWEEN THE ACTIVITY OF BASE MODIFYING ENZYMES AND DNA REPAIR

Section 4.1 covers connections between AID enzymes and downstream repair components. The relevance of AID in demethylation processes during embryogenesis or in differentiated tissue remains debatable, but AID has been implicated in genome instability leading to tumorigenesis. As mentioned earlier, AID is required in B cells for both SHM and CSR during antibody diversification, processes that require changes in the DNA structure. It is therefore not surprising that dysregulation of AID targeting leads to genome instability in B cells (Liu, Duke, et al., 2008; Okazaki et al., 2003). Indeed, this is not limited to B-cell-associated neoplasms (Robbiani et al., 2009), because AID was implicated in a mechanism of translocation formation in prostate cancer (Lin et al., 2009). Section 4.2 covers a combination of connections between DNMTs, TETs, and DNA repair, with a focus on TDG-initiated BER. Some of the challenges to be overcome in fully understanding these connections involve thinking about the participants in unanticipated ways, for example, considering the participation of some repair proteins in unfaithful repair processes.

4.1. AID, BER, and MMR

The U:G mispair in DNA generated by AID can proceed through one of three fates. First, replication across the uracil produces a C:G to T:A transition mutation on that strand. Next, excision of the uracil by UNG and error-prone DNA synthesis across the AP site can produce transition or transversion mutations (Kavli, Otterlei, Slupphaug, & Krokan, 2007). The third fate of the U:G mismatch is recognition by the MSH2-MSH6 heterodimer and recruitment of the error-prone polymerase η to generate mutations at A:T base pairs (Saribasak, Rajagopal, Maul, & Gearhart, 2009). At this point, a brief segue is necessary to describe the parts of mammalian MMR involved in antibody diversification. The repair of single base mismatches in canonical postreplicative MMR is initiated by the MSH2–MSH6 heterodimer, followed by the recruitment of the MLH1–PMS2 heterodimer and exonuclease I, which is required to digest the newly synthesized strand containing the misinserted base (Jiricny, 2006). Studies of knockout mice implicate MSH2–MSH6 and exonuclease 1 in SHM, while it is interesting to note that MLH1 and PMS2 do not seem to be required for SHM (Saribasak et al., 2009). Note, when it comes to the recognition of the U:G mispair, these BER and MMR proteins are participating in ways that at least some of the time deviate from the canonical steps of these repair pathways. Here, the invocation of UNG-mediated BER is specialized in that it requires at least a certain percentage of events be resolved unfaithfully to generate the mutations necessary for affinity maturation. SMUG1 is poorly expressed in B cells, and genomic uracil accumulates in B cells from patients with mutations in *UNG* (Kavli et al., 2005). Further evidence of this specialization came from a study showing that SMUG1 overexpression could only partially rescue antibody diversification in $Ung^{-/-}$ $Msh2^{-/-}$ B cells (Di Noia, Rada, & Neuberger, 2006). The results provided evidence that SMUG1 can access the AID-generated uracil to stimulate repair. However, SMUG1 excision activity appeared to bias events toward faithful repair of the original U:G mismatch, as opposed to unfaithful resolution that drives antibody diversification (Di Noia et al., 2006). It is tempting to speculate that the differences in AP site binding between UNG and SMUG1 described earlier might contribute, although specific protein interactions are undoubtedly required. Also, expression of TDG, which excises uracil from double-stranded DNA, cannot substitute at all, providing more evidence that the DNA intermediate is indeed single stranded (Di Noia et al., 2007). A recent report provides evidence that UNG can promote both unfaithful and faithful resolution of the repair process (Perez-Duran et al., 2012).

APE1 and APE2 activity are also required, which suggests AP site intermediates are involved and also strongly supports the idea that APE activity follows the DNA glycosylase activity of UNG (Guikema et al., 2011, 2007; Ide et al., 2004). At this point, error-prone DNA polymerases are proposed to be responsible for inducing mutations, reviewed elsewhere (Saribasak et al., 2009). In other words, faithful repair by canonical BER would prevent SHM from occurring. Therefore, the model also predicts a balance between faithful and nonfaithful repair that controls SHM. Indeed, studies in Pol β-deficient mice and cells confirm an increased mutation and CSR in the absence of faithful BER (Wu & Stavnezer, 2007). In other words, Pol β activity acts to suppress excessive SHM and CSR. Recent examination of $Xrcc1^{+/-}$ mice also revealed an increase in SHM with decreased XRCC1 (Saribasak et al., 2011). The balance also extends to the participation of MMR components, because it has been shown that MSH2–MSH6 can promote faithful repair of U:G mismatches in addition to its role in promoting mutation at A:T base pairs (Roa et al., 2010).

AID activity is required for CSR and its activity induces a response nearly identical to that seen for direct DSBs (Petersen et al., 2001). Also, AID does not just initiate the process, because studies have implicated AID as required for the resolution (Ranjit et al., 2011; Zan & Casali, 2008). The combined activities of AID, UNG, and APE on closely spaced cytosines on opposite strands can readily cause a DSB, yet note that the above-mentioned MMR components are also required for CSR (Schrader, Guikema, Linehan, Selsing, & Stavnezer, 2007). A more recent study proposed that direct inter-actions of the C terminus of AID with UNG and MSH2-MSH6 are impor-tant for resolution of the DSB (Ranjit et al., 2011). Very briefly, there are two major pathways for repairing DSBs, HR, and NHEJ. The requirement of NHEJ for V(D)J recombination is well established, but interestingly CSR occurs in NHEJ defective cells, suggesting an alternative means of repairing DSBs (Han & Yu, 2008; Soulas-Sprauel et al., 2007; Yan et al., 2007). AID-dependent activity in cells defective in XRCC2, a RAD51 paralog, caused extensive strand breaks and genome instability (Hasham et al., 2010), which suggests that AID activity coupled with defects in HR repair processes would promote genome instability. As the role of dysregulated AID activity in the generation of chromosomal damage such as translocations is consid-ered, it should be remembered that downstream processing by repair path-way proteins must participate in some manner to convert the genomic uracil to a DSB intermediate before a resolution by an unfaithful or corrupted means.

AID has also been implicated in mechanisms of demethylation in other systems. For example, AID was reported to be required for the repro-gramming of somatic cells toward pluripotency (Bhutani et al., 2010). How-ever, the system was a heterokaryon fusion of murine ES cells with human fibroblasts, and it is unclear how the reported cooperation among AID, UNG, and MSH2–MSH6 functions in this context. AID was also impli-cated during genome-wide erasure of methylation in primordial germ cells (Popp et al., 2010). However, the only resulting physiological phenotype in this cohort of $Aid^{-/-}$ mice was reported to be an altered regulation of litter size, which suggests a more subtle effect of altered epigenetics than might be expected, based on the phenotypes of DNMT, TET, or TDG knockout mice reported thus far.

4.2. DNMTs, TETs, and BER

Recall that $Tdg^{-/-}$ mice are embryonic lethal and show a host of phenotypes related to loss of epigenetic stability (Cortazar et al., 2011; Cortellino et al., 2011). In several genetic loci, the absence of TDG results in hypermethylation (Cortellino et al., 2011). Also, mutations were not observed in the promoter regions of the $Tdg^{-/-}$ cells, suggesting that the base modifying events and DNA glycosylase activity are coordinated. Cortazar et al. showed that APE1 and XRCC1 repair foci were visible in wild-type cells, while the $Tdg^{-/-}$ cells showed no foci for these proteins, which strongly implicates that the process also requires the downstream BER machinery to complete the repair and demethylation. It is intriguing to consider the potential influence of TET-mediated base modifications and TDG activity as an endogenous source of DNA damage contributing to the described phenotypes of genome instability in neuronal cells and neurodegeneration associated with defects downstream in the SSBR pathway. The link to hypermethylation of tumor suppressor genes is also apparent, and one example showed that oncogene overexpression inhibits active demethylation in part by preventing the recruitment of TDG (Thillainadesan et al., 2012).

The discovery that TET is capable of catalyzing the iterative oxidation of 5meC to 5-hmC, 5-formylC, and 5-caC raises interesting questions as to which base(s) is/are the most relevant substrate for TDG to initiate BER and complete demethylation. Some of the studies have also invoked AID/ABOBEC activity on these oxidized derivatives to produce the corresponding uracil derivatives (Cortellino et al., 2011; Guo et al., 2011). In this case, AID-catalyzed deamination of 5-hmC would create 5-hmU, which is a major substrate for SMUG1. Yet, the recent reports describing the inability of AID and three APOBECs to deaminate 5-hmC *in vitro* (Nabel et al., 2012), and the viability of SMUG1-deficient mice (Kemmerich et al., 2012) suggests that an AID/SMUG1 pathway is not required for demethylation processes at least during embryonic development. Studies of neurocognitive function in $Smug1^{-/-}$ mice could prove informative.

A pair of interesting papers provided evidence of fast, cyclical methylation, and demethylation in gene transcription in cancer cells (Kangaspeska et al., 2008; Metivier et al., 2008). TDG, APE1, Pol β, and Ligase I were shown by chromatin immunoprecipitation to associate at the promoter sites studied, implicating BER in this demethylation process (Metivier et al., 2008). One of the more remarkable findings was that DNMT3a and 3b also promote deamination of 5-meC (Metivier et al., 2008). The idea that a DNMT catalyzes deamination of 5-meC was quite unexpected. Recall earlier mention that DNMTs can catalyze increased rates of deamination of cytosine in the absence of the SAM cofactor. However, this and other studies noted that DNMT-catalyzed deamination of the product 5-meC occurs at a much lower rate than that for cytosine (Zingg et al., 1996), and also that DNMT binding to its product is weak (Cheng & Roberts, 2001). In short, more work is needed to reconcile these observations.

A few studies connecting PARP activity to DNA and histone demethylation are also briefly discussed here. First, deficiencies in poly(ADP) ribosylation caused by overexpression of the PAR-glycohydrolase have been reported to prevent DNA methylation via regulation of DNMT (Zampieri et al., 2009, 2012). PARP1 was also implicated with TET2 in somatic cell reprogramming toward pluripotency (Doege et al., 2012), and poly(ADP) ribosylation has been implicated in genome-wide erasure of methylation in primordial germ cells (Ciccarone et al., 2012). Future studies in these areas will have to tackle the challenge of dissecting the multiple and partially overlapping roles of PARP family members to better understand these connections.

5. SUMMARY

There has been an explosion in the number of studies exploring demethylation in a variety of contexts that implicate base modifying enzymes in processes that couple with DNA repair processes. The knowledge gained in the past 3 years alone has been astounding, with discovery of new enzyme classes, new roles for known enzymes, and an appreciation of the cooperation and competition among enzymes that modify DNA bases and pathways that remove modified bases. It is intriguing to consider that not all repair initiations meet with the same fate. For example, uracil in DNA generated by AID in B cells can result in faithful repair, transition or transversion mutations, or even translocations that generate isotype switching. Uncovering the regulation of this fate selection will be a great advance. The understanding of DNA repair coordination by ubiquitin and sumo-modification is rapidly growing (Bergink & Jentsch, 2009), and will no doubt also influence DNA repair invoked by base modifying enzymes. As one example, a ubiquitin ligase called RNF4 has been identified to participate in demethylation via TDG and APE1 interactions (Hu et al., 2010). It is highly likely that repair choice, even unfaithful repair, is coordinated by analogous ubiquitin and SUMO-regulated processes.

The exciting pace of discovery has to be tempered somewhat by the realization of the risk in drawing overly broad conclusions from the zygote all the way through to differentiated adult brain tissue. There are a number of unresolved issues requiring more study and ever more complex models to elucidate functions and connections. Regarding the impacts of these discoveries on our understanding of cancer, connections are already being drawn and much more is surely to come. For example, full elucidation of the APOBEC family of deaminases might uncover connections with their dysregulation and cancer in other contexts. New understandings of the TET family regulation and dysregulation should also be quite fruitful (Ko et al., 2013).

As these studies bring into focus mechanisms of DNA demethylation, it will be important to integrate this knowledge into the broader context of the multiple pathways that regulate epigenetics. As the connections between DNA methylation and demethylation are explored, a note of caution is urged in considering the experimental system to be used. As one simple example, most mammalian cell culture experiments are carried out in commercially available cell culture media containing supraphysiological concentrations of folic acid instead of more physiological ranges (Crott, Liu, Choi, & Mason, 2007, Crott et al., 2008; Hayashi, Sohn, Stempak, Croxford, & Kim, 2007; Mashiyama et al., 2004). Moreover, studies in mice show strain differences in the effect of folates on tumor progression (Kadaveru et al., 2012; Knock et al., 2008; Sibani et al., 2002; Song et al., 2000; Trasler et al., 2003), and it seems likely that epigenetics, the very thing being studied, is interacting with the environment factors to modify the outcomes.

Last but certainly not least, methylation and demethylation of lysine residues on histones is an equally important process in the regulation of gene expression (Greer & Shi, 2012). One of the more intriguing interconnections regards the use of α-ketoglutarate as the cofactor in the oxidation reactions carried out by TET enzymes. This relates not just to DNA demethylation, because the JmjC family of histone lysine demethylases also use α-ketoglutarate as a cofactor (Klose, Kallin, & Zhang, 2006). In other words, the potential depletion of α-ketoglutarate driven by oncogenic gain of function IDH enzymes can affect gene silencing via two related demethylation processes. These connections deserve closer attention in future studies. In summary, the importance of the influence of epigenetics in cancer has only grown in more recent times with the appreciation of cancer stem cells and dedifferentiation. Recent discoveries that directly connect base modifying enzymes with repair processes now provide mechanisms by which an important component, DNA demethylation, is carried out.

ACKNOWLEDGMENTS

Covering all aspects of these interactions in biochemical and cellular studies was overly ambitious and likely to have missed references worthy of mention. Apologies are extended to those whose work is not cited. M. D. W. was supported in part by a grant from the NIH/NCI CA 135136.

REFERENCES

Abdel-Wahab, O., Mullally, A., Hedvat, C., Garcia-Manero, G., Patel, J., Wadleigh, M., et al. (2009). Genetic characterization of TET1, TET2, and TET3 alterations in myeloid malignancies. *Blood, 114*, 144–147.

Almeida, K. H., & Sobol, R. W. (2007). A unified view of base excision repair: Lesion-dependent protein complexes regulated by post-translational modification. *DNA Repair (Amsterdam), 6*, 695–711.

Ambrose, H. E., Willimott, S., Beswick, R. W., Dantzer, F., de Murcia, J. M., Yelamos, J., et al. (2009). Poly(ADP-ribose) polymerase-1 (Parp-1)-deficient mice demonstrate abnormal antibody responses. *Immunology, 127*, 178–186.

An, Q., Robins, P., Lindahl, T., & Barnes, D. E. (2005). C → T mutagenesis and gamma-radiation sensitivity due to deficiency in the Smug1 and Ung DNA glycosylases. *The EMBO Journal, 24*, 2205–2213.

Anant, S., & Davidson, N. O. (2001). Molecular mechanisms of apolipoprotein B mRNA editing. *Current Opinion in Lipidology, 12*, 159–165.

Bader, S., Walker, M., & Harrison, D. (2000). Most microsatellite unstable sporadic colorectal carcinomas carry MBD4 mutations. *British Journal of Cancer, 83*, 1646–1649.

Bader, S., Walker, M., Hendrich, B., Bird, A., Bird, C., Hooper, M., et al. (1999). Somatic frameshift mutations in the MBD4 gene of sporadic colon cancers with mismatch repair deficiency. *Oncogene, 18*, 8044–8047.

Berg, J. M., Tymoczko, J. L., & Stryer, L. (2007). *Biochemistry*. New York: W. H. Freeman.

Bergink, S., & Jentsch, S. (2009). Principles of ubiquitin and SUMO modifications in DNA repair. *Nature, 458*, 461–467.

Bhutani, N., Brady, J. J., Damian, M., Sacco, A., Corbel, S. Y., & Blau, H. M. (2010). Reprogramming towards pluripotency requires AID-dependent DNA demethylation. *Nature, 463*, 1042–1047.

Boehler, C., Gauthier, L. R., Mortusewicz, O., Biard, D. S., Saliou, J. M., Bresson, A., et al. (2011). Poly(ADP-ribose) polymerase 3 (PARP3), a newcomer in cellular response to DNA damage and mitotic progression. *Proceedings of the National Academy of Sciences of the United States of America, 108*, 2783–2788.

Boland, M. J., & Christman, J. K. (2008). Characterization of Dnmt3b: Thymine-DNA glycosylase interaction and stimulation of thymine glycosylase-mediated methyltransferase(s) repair by DNA and RNA. *Journal of Molecular Biology, 379*, 492–504.

Boorstein, R. J., Chiu, L. N., & Teebor, G. W. (1989). Phylogenetic evidence of a role for 5-hydroxymethyluracil-DNA glycosylase in the maintenance of 5-methylcytosine in DNA. *Nucleic Acids Research, 17*, 7653–7661.

Boorstein, R. J., Cummings, A., Jr., Marenstein, D. R., Chan, M. K., Ma, Y., Neubert, T. A., et al. (2001). Definitive identification of mammalian 5-hydroxymethyluracil DNA N-glycosylase activity as SMUG1. *The Journal of Biological Chemistry, 276*, 41991–41997.

Booth, M. J., Branco, M. R., Ficz, G., Oxley, D., Krueger, F., Reik, W., et al. (2012). Quantitative sequencing of 5-methylcytosine and 5-hydroxymethylcytosine at single-base resolution. *Science, 336*, 934–937.

Bransteitter, R., Pham, P., Scharff, M. D., & Goodman, M. F. (2003). Activation-induced cytidine deaminase deaminates deoxycytidine on single-stranded DNA but requires the action of RNase. *Proceedings of the National Academy of Sciences of the United States of America, 100,* 4102–4107.

Breen, A. P., & Murphy, J. A. (1995). Reactions of oxyl radicals with DNA. *Free Radical Biology & Medicine, 18,* 1033–1077.

Cabelof, D. C., Ikeno, Y., Nyska, A., Busuttil, R. A., Anyangwe, N., Vijg, J., et al. (2006). Haploinsufficiency in DNA polymerase beta increases cancer risk with age and alters mortality rate. *Cancer Research, 66,* 7460–7465.

Cabelof, D. C., Raffoul, J. J., Nakamura, J., Kapoor, D., Abdalla, H., & Heydari, A. R. (2004). Imbalanced base excision repair in response to folate deficiency is accelerated by polymerase beta haploinsufficiency. *The Journal of Biological Chemistry, 279,* 36504–36513.

Caldecott, K. W. (2008). Single-strand break repair and genetic disease. *Nature Reviews. Genetics, 9,* 619–631.

Cannon, S. V., Cummings, A., & Teebor, G. W. (1988). 5-Hydroxymethylcytosine DNA glycosylase activity in mammalian tissue. *Biochemical and Biophysical Research Communications, 151,* 1173–1179.

Cannon-Carlson, S. V., Gokhale, H., & Teebor, G. W. (1989). Purification and characterization of 5-hydroxymethyluracil-DNA glycosylase from calf thymus. Its possible role in the maintenance of methylated cytosine residues. *The Journal of Biological Chemistry, 264,* 13306–13312.

Chen, H., Dzitoyeva, S., & Manev, H. (2012). Effect of aging on 5-hydroxymethylcytosine in the mouse hippocampus. *Restorative Neurology and Neuroscience, 30,* 237–245.

Chen, T., Hevi, S., Gay, F., Tsujimoto, N., He, T., Zhang, B., et al. (2007). Complete inactivation of DNMT1 leads to mitotic catastrophe in human cancer cells. *Nature Genetics, 39,* 391–396.

Chen, D., Lucey, M. J., Phoenix, F., Lopez-Garcia, J., Hart, S. M., Losson, R., et al. (2003). T:G mismatch-specific thymine-DNA glycosylase potentiates transcription of estrogen-regulated genes through direct interaction with estrogen receptor alpha. *The Journal of Biological Chemistry, 278,* 38586–38592.

Cheng, X., & Roberts, R. J. (2001). AdoMet-dependent methylation, DNA methyltransferases and base flipping. *Nucleic Acids Research, 29,* 3784–3795.

Chouliaras, L., van den Hove, D. L., Kenis, G., Keitel, S., Hof, P. R., van Os, J., et al. (2012). Age-related increase in levels of 5-hydroxymethylcytosine in mouse hippocampus is prevented by caloric restriction. *Current Alzheimer Research, 9,* 536–544.

Ciccarone, F., Klinger, F. G., Catizone, A., Calabrese, R., Zampieri, M., Bacalini, M. G., et al. (2012). Poly(ADP-ribosyl)ation acts in the DNA demethylation of mouse primordial germ cells also with DNA damage-independent roles. *PLoS One, 7,* e46927.

Cliffe, L. J., Kieft, R., Southern, T., Birkeland, S. R., Marshall, M., Sweeney, K., et al. (2009). JBP1 and JBP2 are two distinct thymidine hydroxylases involved in J biosynthesis in genomic DNA of African trypanosomes. *Nucleic Acids Research, 37,* 1452–1462.

Cole, B. F., Baron, J. A., Sandler, R. S., Haile, R. W., Ahnen, D. J., Bresalier, R. S., et al. (2007). Folic acid for the prevention of colorectal adenomas: A randomized clinical trial. *JAMA: The Journal of the American Medical Association, 297,* 2351–2359.

Connor, E. E., & Wyatt, M. D. (2002). Active-site clashes prevent the human 3-methyladenine DNA glycosylase from improperly removing bases. *Chemistry & Biology, 9,* 1033–1041.

Cortazar, D., Kunz, C., Saito, Y., Steinacher, R., & Schar, P. (2007). The enigmatic thymine DNA glycosylase. *DNA Repair (Amsterdam), 6,* 489–504.

Cortazar, D., Kunz, C., Selfridge, J., Lettieri, T., Saito, Y., MacDougall, E., et al. (2011). Embryonic lethal phenotype reveals a function of TDG in maintaining epigenetic stability. *Nature, 470,* 419–423.

Cortellino, S., Xu, J., Sannai, M., Moore, R., Caretti, E., Cigliano, A., et al. (2011). Thymine DNA glycosylase is essential for active DNA demethylation by linked deamination-base excision repair. *Cell, 146,* 67–79.

Crott, J. W., Liu, Z., Choi, S. W., & Mason, J. B. (2007). Folate depletion in human lymphocytes up-regulates p53 expression despite marked induction of strand breaks in exons 5–8 of the gene. *Mutation Research, 626,* 171–179.

Crott, J. W., Liu, Z., Keyes, M. K., Choi, S. W., Jang, H., Moyer, M. P., et al. (2008). Moderate folate depletion modulates the expression of selected genes involved in cell cycle, intracellular signaling and folate uptake in human colonic epithelial cell lines. *The Journal of Nutritional Biochemistry, 19,* 328–335.

Dang, L., White, D. W., Gross, S., Bennett, B. D., Bittinger, M. A., Driggers, E. M., et al. (2009). Cancer-associated IDH1 mutations produce 2-hydroxyglutarate. *Nature, 462,* 739–744.

De Vos, M., Schreiber, V., & Dantzer, F. (2012). The diverse roles and clinical relevance of PARPs in DNA damage repair: Current state of the art. *Biochemical Pharmacology, 84,* 137–146.

de Wind, N., Dekker, M., Berns, A., Radman, M., & te Riele, H. (1995). Inactivation of the mouse Msh2 gene results in mismatch repair deficiency, methylation tolerance, hyperrecombination, and predisposition to cancer. *Cell, 82,* 321–330.

Di Noia, J. M., & Neuberger, M. S. (2007). Molecular mechanisms of antibody somatic hypermutation. *Annual Review of Biochemistry, 76,* 1–22.

Di Noia, J. M., Rada, C., & Neuberger, M. S. (2006). SMUG1 is able to excise uracil from immunoglobulin genes: Insight into mutation versus repair. *The EMBO Journal, 25,* 585–595.

Di Noia, J. M., Williams, G. T., Chan, D. T., Buerstedde, J. M., Baldwin, G. S., & Neuberger, M. S. (2007). Dependence of antibody gene diversification on uracil excision. *The Journal of Experimental Medicine, 204,* 3209–3219.

Doege, C. A., Inoue, K., Yamashita, T., Rhee, D. B., Travis, S., Fujita, R., et al. (2012). Early-stage epigenetic modification during somatic cell reprogramming by Parp1 and Tet2. *Nature, 488,* 652–655.

Dzitoyeva, S., Chen, H., & Manev, H. (2012). Effect of aging on 5-hydroxymethylcytosine in brain mitochondria. *Neurobiology of Aging, 33,* 2881–2891.

Egger, G., Jeong, S., Escobar, S. G., Cortez, C. C., Li, T. W., Saito, Y., et al. (2006). Identification of DNMT1 (DNA methyltransferase 1) hypomorphs in somatic knockouts suggests an essential role for DNMT1 in cell survival. *Proceedings of the National Academy of Sciences of the United States of America, 103,* 14080–14085.

Ehrlich, M., Zhang, X. Y., & Inamdar, N. M. (1990). Spontaneous deamination of cytosine and 5-methylcytosine residues in DNA and replacement of 5-methylcytosine residues with cytosine residues. *Mutation Research, 238,* 277–286.

Endres, M., Biniszkiewicz, D., Sobol, R. W., Harms, C., Ahmadi, M., Lipski, A., et al. (2004). Increased postischemic brain injury in mice deficient in uracil-DNA glycosylase. *The Journal of Clinical Investigation, 113,* 1711–1721.

Figueroa, M. E., Abdel-Wahab, O., Lu, C., Ward, P. S., Patel, J., Shih, A., et al. (2010). Leukemic IDH1 and IDH2 mutations result in a hypermethylation phenotype, disrupt TET2 function, and impair hematopoietic differentiation. *Cancer Cell, 18,* 553–567.

Friedberg, E. C., Walker, G. C., Siede, W., Wood, R. D., Schultz, R. A., & Ellenberger, T. (2006). *DNA repair and mutagenesis.* Washington, D.C.: ASM Press.

Fung, H., & Demple, B. (2005). A vital role for Ape1/Ref1 protein in repairing spontaneous DNA damage in human cells. *Molecular Cell, 17,* 463–470.

Gallais, R., Demay, F., Barath, P., Finot, L., Jurkowska, R., Le Guevel, R., et al. (2007). Deoxyribonucleic acid methyl transferases 3a and 3b associate with the nuclear orphan receptor COUP-TFI during gene activation. *Molecular Endocrinology*, *21*, 2085–2098.

Gao, Y., Katyal, S., Lee, Y., Zhao, J., Rehg, J. E., Russell, H. R., et al. (2011). DNA ligase III is critical for mtDNA integrity but not Xrcc1-mediated nuclear DNA repair. *Nature*, *471*, 240–244.

Gaudet, F., Hodgson, J. G., Eden, A., Jackson-Grusby, L., Dausman, J., Gray, J. W., et al. (2003). Induction of tumors in mice by genomic hypomethylation. *Science*, *300*, 489–492.

Gommers-Ampt, J. H., Teixeira, A. J., van de Werken, G., van Dijk, W. J., & Borst, P. (1993). The identification of hydroxymethyluracil in DNA of Trypanosoma brucei. *Nucleic Acids Research*, *21*, 2039–2043.

Gommers-Ampt, J. H., Van Leeuwen, F., de Beer, A. L., Vliegenthart, J. F., Dizdaroglu, M., Kowalak, J. A., et al. (1993). beta-D-glucosyl-hydroxymethyluracil: A novel modified base present in the DNA of the parasitic protozoan T. brucei. *Cell*, *75*, 1129–1136.

Grady, W. M., & Ulrich, C. M. (2007). DNA alkylation and DNA methylation: Cooperating mechanisms driving the formation of colorectal adenomas and adenocarcinomas? *Gut*, *56*, 318–320.

Greer, E. L., & Shi, Y. (2012). Histone methylation: A dynamic mark in health, disease and inheritance. *Nature Reviews. Genetics*, *13*, 343–357.

Gross, S., Cairns, R. A., Minden, M. D., Driggers, E. M., Bittinger, M. A., Jang, H. G., et al. (2010). Cancer-associated metabolite 2-hydroxyglutarate accumulates in acute myelogenous leukemia with isocitrate dehydrogenase 1 and 2 mutations. *The Journal of Experimental Medicine*, *207*, 339–344.

Gu, T. P., Guo, F., Yang, H., Wu, H. P., Xu, G. F., Liu, W., et al. (2011). The role of Tet3 DNA dioxygenase in epigenetic reprogramming by oocytes. *Nature*, *477*, 606–610.

Gu, H., Marth, J. D., Orban, P. C., Mossmann, H., & Rajewsky, K. (1994). Deletion of a DNA polymerase β gene segment in T cells using cell type-specific gene targeting. *Science*, *265*, 103–106.

Guikema, J. E., Gerstein, R. M., Linehan, E. K., Cloherty, E. K., Evan-Browning, E., Tsuchimoto, D., et al. (2011). Apurinic/apyrimidinic endonuclease 2 is necessary for normal B cell development and recovery of lymphoid progenitors after chemotherapeutic challenge. *Journal of Immunology*, *186*, 1943–1950.

Guikema, J. E., Linehan, E. K., Tsuchimoto, D., Nakabeppu, Y., Strauss, P. R., Stavnezer, J., et al. (2007). APE1- and APE2-dependent DNA breaks in immunoglobulin class switch recombination. *The Journal of Experimental Medicine*, *204*, 3017–3026.

Guo, J. U., Su, Y., Zhong, C., Ming, G. L., & Song, H. (2011). Hydroxylation of 5-methylcytosine by TET1 promotes active DNA demethylation in the adult brain. *Cell*, *145*, 423–434.

Haffner, M. C., Chaux, A., Meeker, A. K., Esopi, D. M., Gerber, J., Pellakuru, L. G., et al. (2011). Global 5-hydroxymethylcytosine content is significantly reduced in tissue stem/progenitor cell compartments and in human cancers. *Oncotarget*, *2*, 627–637.

Han, L., & Yu, K. (2008). Altered kinetics of nonhomologous end joining and class switch recombination in ligase IV-deficient B cells. *The Journal of Experimental Medicine*, *205*, 2745–2753.

Hardeland, U., Bentele, M., Jiricny, J., & Schar, P. (2000). Separating substrate recognition from base hydrolysis in human thymine DNA glycosylase by mutational analysis. *The Journal of Biological Chemistry*, *275*, 33449–33456.

Hardeland, U., Steinacher, R., Jiricny, J., & Schar, P. (2002). Modification of the human thymine-DNA glycosylase by ubiquitin-like proteins facilitates enzymatic turnover. *The EMBO Journal*, *21*, 1456–1464.

Hasham, M. G., Donghia, N. M., Coffey, E., Maynard, J., Snow, K. J., Ames, J., et al. (2010). Widespread genomic breaks generated by activation-induced cytidine deaminase are prevented by homologous recombination. *Nature Immunology*, *11*, 820–826.

Hashimoto, H., Vertino, P. M., & Cheng, X. (2010). Molecular coupling of DNA methylation and histone methylation. *Epigenomics*, *2*, 657–669.

Haushalter, K. A., Todd Stukenberg, M. W., Kirschner, M. W., & Verdine, G. L. (1999). Identification of a new uracil-DNA glycosylase family by expression cloning using synthetic inhibitors. *Current Biology*, *9*, 174–185.

Hayashi, I., Sohn, K. J., Stempak, J. M., Croxford, R., & Kim, Y. I. (2007). Folate deficiency induces cell-specific changes in the steady-state transcript levels of genes involved in folate metabolism and 1-carbon transfer reactions in human colonic epithelial cells. *The Journal of Nutrition*, *137*, 607–613.

Hayatsu, H., & Shiragami, M. (1979). Reaction of bisulfite with the 5-hydroxymethyl group in pyrimidines and in phage DNAs. *Biochemistry*, *18*, 632–637.

He, Y. F., Li, B. Z., Li, Z., Liu, P., Wang, Y., Tang, Q., et al. (2011). Tet-mediated formation of 5-carboxylcytosine and its excision by TDG in mammalian DNA. *Science*, *333*, 1303–1307.

Helleday, T. (2011). The underlying mechanism for the PARP and BRCA synthetic lethality: Clearing up the misunderstandings. *Molecular Oncology*, *5*, 387–393.

Hendrich, B., Hardeland, U., Ng, H. H., Jiricny, J., & Bird, A. (1999). The thymine glycosylase MBD4 can bind to the product of deamination at methylated CpG sites. *Nature*, *401*, 301–304.

Holden, L. G., Prochnow, C., Chang, Y. P., Bransteitter, R., Chelico, L., Sen, U., et al. (2008). Crystal structure of the anti-viral APOBEC3G catalytic domain and functional implications. *Nature*, *456*, 121–124.

Hu, X. V., Rodrigues, T. M., Tao, H., Baker, R. K., Miraglia, L., Orth, A. P., et al. (2010). Identification of RING finger protein 4 (RNF4) as a modulator of DNA demethylation through a functional genomics screen. *Proceedings of the National Academy of Sciences of the United States of America*, *107*, 15087–15092.

Huang, Y., Pastor, W. A., Shen, Y., Tahiliani, M., Liu, D. R., & Rao, A. (2010). The behaviour of 5-hydroxymethylcytosine in bisulfite sequencing. *PLoS One*, *5*, e8888.

Huang, Y., Pastor, W. A., Zepeda-Martinez, J. A., & Rao, A. (2012). The anti-CMS technique for genome-wide mapping of 5-hydroxymethylcytosine. *Nature Protocols*, *7*, 1897–1908.

Ide, Y., Tsuchimoto, D., Tominaga, Y., Nakashima, M., Watanabe, T., Sakumi, K., et al. (2004). Growth retardation and dyslymphopoiesis accompanied by G2/M arrest in APEX2-null mice. *Blood*, *104*, 4097–4103.

Imai, K., Slupphaug, G., Lee, W. I., Revy, P., Nonoyama, S., Catalan, N., et al. (2003). Human uracil-DNA glycosylase deficiency associated with profoundly impaired immunoglobulin class-switch recombination. *Nature Immunology*, *4*, 1023–1028.

Inoue, A., & Zhang, Y. (2011). Replication-dependent loss of 5-hydroxymethylcytosine in mouse preimplantation embryos. *Science*, *334*, 194.

Iqbal, K., Jin, S. G., Pfeifer, G. P., & Szabo, P. E. (2011). Reprogramming of the paternal genome upon fertilization involves genome-wide oxidation of 5-methylcytosine. *Proceedings of the National Academy of Sciences of the United States of America*, *108*, 3642–3647.

Ito, S., Shen, L., Dai, Q., Wu, S. C., Collins, L. B., Swenberg, J. A., et al. (2011). Tet proteins can convert 5-methylcytosine to 5-formylcytosine and 5-carboxylcytosine. *Science*, *333*, 1300–1303.

Iyer, L. M., Tahiliani, M., Rao, A., & Aravind, L. (2009). Prediction of novel families of enzymes involved in oxidative and other complex modifications of bases in nucleic acids. *Cell Cycle*, *8*, 1698–1710.

Izumi, T., Brown, D. B., Naidu, C. V., Bhakat, K. K., Macinnes, M. A., Saito, H., et al. (2005). Two essential but distinct functions of the mammalian abasic endonuclease. *Proceedings of the National Academy of Sciences of the United States of America, 102,* 5739–5743.

Jin, S. G., Jiang, Y., Qiu, R., Rauch, T. A., Wang, Y., Schackert, G., et al. (2011). 5-Hydroxymethylcytosine is strongly depleted in human cancers but its levels do not correlate with IDH1 mutations. *Cancer Research, 71,* 7360–7365.

Jin, S.-G. , Kadam, S., & Pfeifer, G. P. (2010). Examination of the specificity of DNA methylation profiling techniques towards 5-methylcytosine and 5-hydroxymethylcytosine. *Nucleic Acids Research, 38,* e125.

Jiricny, J. (2006). The multifaceted mismatch-repair system. *Nature Reviews. Molecular Cell Biology, 7,* 335–346.

Jones, P. A. (2012). Functions of DNA methylation: Islands, start sites, gene bodies and beyond. *Nature Reviews. Genetics, 13,* 484–492.

Kadaveru, K., Protiva, P., Greenspan, E. J., Kim, Y. I., & Rosenberg, D. W. (2012). Dietary methyl donor depletion protects against intestinal tumorigenesis in Apc(Min/+) mice. *Cancer Prevention Research (Philadelphia, PA), 5,* 911–920.

Kangaspeska, S., Stride, B., Metivier, R., Polycarpou-Schwarz, M., Ibberson, D., Carmouche, R. P., et al. (2008). Transient cyclical methylation of promoter DNA. *Nature, 452,* 112–115.

Kasprzak, K. S., Jaruga, P., Zastawny, T. H., North, S. L., Riggs, C. W., Olinski, R., et al. (1997). Oxidative DNA base damage and its repair in kidneys and livers of nickel(II)-treated male F344 rats. *Carcinogenesis, 18,* 271–277.

Kavli, B., Andersen, S., Otterlei, M., Liabakk, N. B., Imai, K., Fischer, A., et al. (2005). B cells from hyper-IgM patients carrying UNG mutations lack ability to remove uracil from ssDNA and have elevated genomic uracil. *The Journal of Experimental Medicine, 201,* 2011–2021.

Kavli, B., Otterlei, M., Slupphaug, G., & Krokan, H. E. (2007). Uracil in DNA—General mutagen, but normal intermediate in acquired immunity. *DNA Repair (Amsterdam), 6,* 505–516.

Kavli, B., Slupphaug, G., Mol, C. D., Arvai, A. S., Peterson, S. B., Tainer, J. A., et al. (1996). Excision of cytosine and thymine from DNA by mutants of human uracil-DNA glycosylase. *The EMBO Journal, 15,* 3442–3447.

Kemmerich, K., Dingler, F. A., Rada, C., & Neuberger, M. S. (2012). Germline ablation of SMUG1 DNA glycosylase causes loss of 5-hydroxymethyluracil- and UNG-backup uracil-excision activities and increases cancer predisposition of Ung−/−Msh2−/− mice. *Nucleic Acids Research, 40,* 6016–6025.

Khare, T., Pai, S., Koncevicius, K., Pal, M., Kriukiene, E., Liutkeviciute, Z., et al. (2012). 5-hmC in the brain is abundant in synaptic genes and shows differences at the exon-intron boundary. *Nature Structural & Molecular Biology, 19,* 1037–1043.

Klose, R. J., Kallin, E. M., & Zhang, Y. (2006). JmjC-domain-containing proteins and histone demethylation. *Nature Reviews. Genetics, 7,* 715–727.

Knock, E., Deng, L., Wu, Q., Lawrance, A. K., Wang, X. L., & Rozen, R. (2008). Strain differences in mice highlight the role of DNA damage in neoplasia induced by low dietary folate. *The Journal of Nutrition, 138,* 653–658.

Ko, M., Huang, Y., Jankowska, A. M., Pape, U. J., Tahiliani, M., Bandukwala, H. S., et al. (2010). Impaired hydroxylation of 5-methylcytosine in myeloid cancers with mutant TET2. *Nature, 468,* 839–843.

Ko, M., An, J., Bandukwala, H. S., Chavez, L., Aijö, T., Pastor, W. A., et al. (2013). Modulation of TET2 expression and 5-methylcytosine oxidation by the CXXC domain protein IDAX. *Nature, 497,* 122–126.

Kohli, R. M., Maul, R. W., Guminski, A. F., McClure, R. L., Gajula, K. S., Saribasak, H., et al. (2010). Local sequence targeting in the AID/APOBEC family differentially impacts retroviral restriction and antibody diversification. *The Journal of Biological Chemistry, 285*, 40956–40964.

Kraus, T. F., Globisch, D., Wagner, M., Eigenbrod, S., Widmann, D., Munzel, M., et al. (2012). Low values of 5-hydroxymethylcytosine (5hmC), the "sixth base," are associated with anaplasia in human brain tumors. *International Journal of Cancer, 131*, 1577–1590.

Kriaucionis, S., & Heintz, N. (2009). The nuclear DNA base 5-hydroxymethylcytosine is present in Purkinje neurons and the brain. *Science, 324*, 929–930.

Krokan, H. E., Drablos, F., & Slupphaug, G. (2002). Uracil in DNA—Occurrence, consequences and repair. *Oncogene, 21*, 8935–8948.

Kronenberg, G., Harms, C., Sobol, R. W., Cardozo-Pelaez, F., Linhart, H., Winter, B., et al. (2008). Folate deficiency induces neurodegeneration and brain dysfunction in mice lacking uracil DNA glycosylase. *The Journal of Neuroscience, 28*, 7219–7230.

Kulkarni, A., McNeill, D. R., Gleichmann, M., Mattson, M. P., & Wilson, D. M., 3rd. (2008). XRCC1 protects against the lethality of induced oxidative DNA damage in nondividing neural cells. *Nucleic Acids Research, 36*, 5111–5121.

Langemeijer, S. M., Kuiper, R. P., Berends, M., Knops, R., Aslanyan, M. G., Massop, M., et al. (2009). Acquired mutations in TET2 are common in myelodysplastic syndromes. *Nature Genetics, 41*, 838–842.

Lao, V. V., Darwanto, A., & Sowers, L. C. (2010). Impact of base analogues within a CpG dinucleotide on the binding of DNA by the methyl-binding domain of MeCP2 and methylation by DNMT1. *Biochemistry, 49*, 10228–10236.

Larijani, M., Frieder, D., Sonbuchner, T. M., Bransteitter, R., Goodman, M. F., Bouhassira, E. E., et al. (2005). Methylation protects cytidines from AID-mediated deamination. *Molecular Immunology, 42*, 599–604.

Lawrance, A. K., Deng, L., Brody, L. C., Finnell, R. H., Shane, B., & Rozen, R. (2007). Genetic and nutritional deficiencies in folate metabolism influence tumorigenicity in Apcmin/+ mice. *The Journal of Nutritional Biochemistry, 18*, 305–312.

Lee, Y., Katyal, S., Li, Y., El-Khamisy, S. F., Russell, H. R., Caldecott, K. W., et al. (2009). The genesis of cerebellar interneurons and the prevention of neural DNA damage require XRCC1. *Nature Neuroscience, 12*, 973–980.

Li, Y. Q., Zhou, P. Z., Zheng, X. D., Walsh, C. P., & Xu, G. L. (2007). Association of Dnmt3a and thymine DNA glycosylase links DNA methylation with base-excision repair. *Nucleic Acids Research, 35*, 390–400.

Lin, H., Yamada, Y., Nguyen, S., Linhart, H., Jackson-Grusby, L., Meissner, A., et al. (2006). Suppression of intestinal neoplasia by deletion of Dnmt3b. *Molecular and Cellular Biology, 26*, 2976–2983.

Lin, C., Yang, L., Tanasa, B., Hutt, K., Ju, B. G., Ohgi, K., et al. (2009). Nuclear receptor-induced chromosomal proximity and DNA breaks underlie specific translocations in cancer. *Cell, 139*, 1069–1083.

Lindahl, T. (1974). An N-glycosidase from *Escherichia coli* that releases free uracil from DNA containing deaminated cytosine residues. *Proceedings of the National Academy of Sciences of the United States of America, 71*, 3649–3653.

Lindahl, T., & Nyberg, B. (1974). Heat-induced deamination of cytosine residues in deoxyribonucleic acid. *Biochemistry, 13*, 3405–3410.

Linhart, H. G., Lin, H., Yamada, Y., Moran, E., Steine, E. J., Gokhale, S., et al. (2007). Dnmt3b promotes tumorigenesis in vivo by gene-specific de novo methylation and transcriptional silencing. *Genes & Development, 21*, 3110–3122.

Linhart, H. G., Troen, A., Bell, G. W., Cantu, E., Chao, W. H., Moran, E., et al. (2009). Folate deficiency induces genomic uracil misincorporation and hypomethylation but does not increase DNA point mutations. *Gastroenterology, 136*, 227–235.

Liu, M., Duke, J. L., Richter, D. J., Vinuesa, C. G., Goodnow, C. C., Kleinstein, S. H., et al. (2008). Two levels of protection for the B cell genome during somatic hypermutation. *Nature*, *451*, 841–845.

Liu, H., Takeda, S., Cheng, E. H., & Hsieh, J. J. (2008). Biphasic MLL takes helm at cell cycle control: Implications in human mixed lineage leukemia. *Cell Cycle*, *7*, 428–435.

Lorsbach, R. B., Moore, J., Mathew, S., Raimondi, S. C., Mukatira, S. T., & Downing, J. R. (2003). TET1, a member of a novel protein family, is fused to MLL in acute myeloid leukemia containing the t(10;11)(q22;q23). *Leukemia*, *17*, 637–641.

Luongo, C., Moser, A. R., Gledhill, S., & Dove, W. F. (1994). Loss of Apc + in intestinal adenomas from Min mice. *Cancer Research*, *54*, 5947–5952.

Maiti, A., & Drohat, A. C. (2011). Thymine DNA glycosylase can rapidly excise 5-formylcytosine and 5-carboxylcytosine potential implications for active demethylation of CpG sites. *The Journal of Biological Chemistry*, *286*, 35334–35338.

Maiti, A., Noon, M. S., Mackerell, A. D., Jr., Pozharski, E., & Drohat, A. C. (2012). Lesion processing by a repair enzyme is severely curtailed by residues needed to prevent aberrant activity on undamaged DNA. *Proceedings of the National Academy of Sciences of the United States of America*, *109*, 8091–8096.

Mangeat, B., Turelli, P., Caron, G., Friedli, M., Perrin, L., & Trono, D. (2003). Broad anti-retroviral defence by human APOBEC3G through lethal editing of nascent reverse transcripts. *Nature*, *424*, 99–103.

Mashiyama, S. T., Courtemanche, C., Elson-Schwab, I., Crott, J., Lee, B. L., Ong, C. N., et al. (2004). Uracil in DNA, determined by an improved assay, is increased when deoxynucleosides are added to folate-deficient cultured human lymphocytes. *Analytical Biochemistry*, *330*, 58–69.

Matsumoto, Y., & Kim, K. (1995). Excision of deoxyribose phosphate residues by DNA polymerase β during DNA repair. *Science*, *269*, 699–702.

Maul, R. W., Saribasak, H., Martomo, S. A., McClure, R. L., Yang, W., Vaisman, A., et al. (2011). Uracil residues dependent on the deaminase AID in immunoglobulin gene variable and switch regions. *Nature Immunology*, *12*, 70–76.

Menissier de Murcia, J., Ricoul, M., Tartier, L., Niedergang, C., Huber, A., Dantzer, F., et al. (2003). Functional interaction between PARP-1 and PARP-2 in chromosome stability and embryonic development in mouse. *The EMBO Journal*, *22*, 2255–2263.

Metivier, R., Gallais, R., Tiffoche, C., Le Peron, C., Jurkowska, R. Z., Carmouche, R. P., et al. (2008). Cyclical DNA methylation of a transcriptionally active promoter. *Nature*, *452*, 45–50.

Millar, C. B., Guy, J., Sansom, O. J., Selfridge, J., MacDougall, E., Hendrich, B., et al. (2002). Enhanced CpG mutability and tumorigenesis in MBD4-deficient mice. *Science*, *297*, 403–405.

Minegishi, Y., Lavoie, A., Cunningham-Rundles, C., Bedard, P. M., Hebert, J., Cote, L., et al. (2000). Mutations in activation-induced cytidine deaminase in patients with hyper IgM syndrome. *Clinical Immunology*, *97*, 203–210.

Morales-Ruiz, T., Ortega-Galisteo, A. P., Ponferrada-Marin, M. I., Martinez-Macias, M. I., Ariza, R. R., & Roldan-Arjona, T. (2006). Demeter and repressor of silencing 1 encode 5-methylcytosine DNA glycosylases. *Proceedings of the National Academy of Sciences of the United States of America*, *103*, 6853–6858.

Moran-Crusio, K., Reavie, L., Shih, A., Abdel-Wahab, O., Ndiaye-Lobry, D., Lobry, C., et al. (2011). Tet2 loss leads to increased hematopoietic stem cell self-renewal and myeloid transformation. *Cancer Cell*, *20*, 11–24.

Mori, T., Hori, Y., & Dizdaroglu, M. (1993). DNA base damage generated in vivo in hepatic chromatin of mice upon whole body gamma-irradiation. *International Journal of Radiation Biology, 64,* 645–650.

Moser, A. R., Pitot, H. C., & Dove, W. F. (1990). A dominant mutation that predisposes to multiple intestinal neoplasia in the mouse. *Science, 247,* 322–324.

Munzel, M., Globisch, D., Bruckl, T., Wagner, M., Welzmiller, V., Michalakis, S., et al. (2010). Quantification of the sixth DNA base hydroxymethylcytosine in the brain. *Angewandte Chemie (International Ed. in English), 49,* 5375–5377.

Muramatsu, M., Kinoshita, K., Fagarasan, S., Yamada, S., Shinkai, Y., & Honjo, T. (2000). Class switch recombination and hypermutation require activation-induced cytidine deaminase (AID), a potential RNA editing enzyme. *Cell, 102,* 553–563.

Muramatsu, M., Sankaranand, V. S., Anant, S., Sugai, M., Kinoshita, K., Davidson, N. O., et al. (1999). Specific expression of activation-induced cytidine deaminase (AID), a novel member of the RNA-editing deaminase family in germinal center B cells. *The Journal of Biological Chemistry, 274,* 18470–18476.

Nabel, C. S., Jia, H., Ye, Y., Shen, L., Goldschmidt, H. L., Stivers, J. T., et al. (2012). AID/APOBEC deaminases disfavor modified cytosines implicated in DNA demethylation. *Nature Chemical Biology, 8,* 751–758.

Nackerdien, Z., Olinski, R., & Dizdaroglu, M. (1992). DNA base damage in chromatin of gamma-irradiated cultured human cells. *Free Radical Research Communications, 16,* 259–273.

Neddermann, P., Gallinari, P., Lettieri, T., Schmid, D., Truong, O., Hsuan, J. J., et al. (1996). Cloning and expression of human G/T mismatch-specific thymine-DNA glycosylase. *The Journal of Biological Chemistry, 271,* 12767–12774.

Neddermann, P., & Jiricny, J. (1994). Efficient removal of uracil from G.U mispairs by the mismatch-specific thymine DNA glycosylase from HeLa cells. *Proceedings of the National Academy of Sciences of the United States of America, 91,* 1642–1646.

Nilsen, H., Rosewell, I., Robins, P., Skjelbred, C. F., Andersen, S., Slupphaug, G., et al. (2000). Uracil-DNA glycosylase (UNG)-deficient mice reveal a primary role of the enzyme during DNA replication. *Molecular Cell, 5,* 1059–1065.

Nilsen, H., Stamp, G., Andersen, S., Hrivnak, G., Krokan, H. E., Lindahl, T., et al. (2003). Gene-targeted mice lacking the Ung uracil-DNA glycosylase develop B-cell lymphomas. *Oncogene, 22,* 5381–5386.

Nosho, K., Shima, K., Irahara, N., Kure, S., Baba, Y., Kirkner, G. J., et al. (2009). DNMT3B expression might contribute to CpG island methylator phenotype in colorectal cancer. *Clinical Cancer Research, 15,* 3663–3671.

Noushmehr, H., Weisenberger, D. J., Diefes, K., Phillips, H. S., Pujara, K., Berman, B. P., et al. (2010). Identification of a CpG island methylator phenotype that defines a distinct subgroup of glioma. *Cancer Cell, 17,* 510–522.

Nowarski, R., Britan-Rosich, E., Shiloach, T., & Kotler, M. (2008). Hypermutation by intersegmental transfer of APOBEC3G cytidine deaminase. *Nature Structural & Molecular Biology, 15,* 1059–1066.

O'Brien, P. J., & Ellenberger, T. (2004). Dissecting the broad substrate specificity of human 3-methyladenine-DNA glycosylase. *The Journal of Biological Chemistry, 279,* 9750–9757.

Okazaki, I. M., Hiai, H., Kakazu, N., Yamada, S., Muramatsu, M., Kinoshita, K., et al. (2003). Constitutive expression of AID leads to tumorigenesis. *The Journal of Experimental Medicine, 197,* 1173–1181.

Olinski, R., Zastawny, T., Budzbon, J., Skokowski, J., Zegarski, W., & Dizdaroglu, M. (1992). DNA base modifications in chromatin of human cancerous tissues. *FEBS Letters, 309,* 193–198.

Ono, R., Taki, T., Taketani, T., Taniwaki, M., Kobayashi, H., & Hayashi, Y. (2002). LCX, leukemia-associated protein with a CXXC domain, is fused to MLL in acute myeloid leukemia with trilineage dysplasia having t(10;11)(q22;q23). *Cancer Research, 62,* 4075–4080.

Ooi, S. K., & Bestor, T. H. (2008). The colorful history of active DNA demethylation. *Cell, 133,* 1145–1148.

Otterlei, M., Warbrick, E., Nagelhus, T. A., Haug, T., Slupphaug, G., Akbari, M., et al. (1999). Post-replicative base excision repair in replication foci. *The EMBO Journal, 18,* 3834–3844.

Parikh, S. S., Mol, C. D., Hosfield, D. J., & Tainer, J. A. (1999). Envisioning the molecular choreography of DNA base excision repair. *Current Opinion in Structural Biology, 9,* 37–47.

Pastor, W. A., Huang, Y., Henderson, H. R., Agarwal, S., & Rao, A. (2012). The GLIB technique for genome-wide mapping of 5-hydroxymethylcytosine. *Nature Protocols, 7,* 1909–1917.

Pastor, W. A., Pape, U. J., Huang, Y., Henderson, H. R., Lister, R., Ko, M., et al. (2011). Genome-wide mapping of 5-hydroxymethylcytosine in embryonic stem cells. *Nature, 473,* 394–397.

Pegg, A. E. (2000). Repair of O(6)-alkylguanine by alkyltransferases. *Mutation Research, 462,* 83–100.

Peng, Z., Cheng, Y., Tan, B. C., Kang, L., Tian, Z., Zhu, Y., et al. (2012). Comprehensive analysis of RNA-Seq data reveals extensive RNA editing in a human transcriptome. *Nature Biotechnology, 30,* 253–260.

Penn, N. W. (1976). Modification of brain deoxyribonucleic acid base content with maturation in normal and malnourished rats. *The Biochemical Journal, 155,* 709–712.

Penn, N. W., Suwalski, R., O'Riley, C., Bojanowski, K., & Yura, R. (1972). The presence of 5-hydroxymethylcytosine in animal deoxyribonucleic acid. *The Biochemical Journal, 126,* 781–790.

Perez, C., Martinez-Calle, N., Martin-Subero, J. I., Segura, V., Delabesse, E., Fernandez-Mercado, M., et al. (2012). TET2 mutations are associated with specific 5-methylcytosine and 5-hydroxymethylcytosine profiles in patients with chronic myelomonocytic leukemia. *PLoS One, 7,* e31605.

Perez-Duran, P., Belver, L., de Yebenes, V. G., Delgado, P., Pisano, D. G., & Ramiro, A. R. (2012). UNG shapes the specificity of AID-induced somatic hypermutation. *The Journal of Experimental Medicine, 209,* 1379–1389.

Petersen, S., Casellas, R., Reina-San-Martin, B., Chen, H. T., Difilippantonio, M. J., Wilson, P. C., et al. (2001). AID is required to initiate Nbs1/gamma-H2AX focus formation and mutations at sites of class switching. *Nature, 414,* 660–665.

Petersen-Mahrt, S. K., Harris, R. S., & Neuberger, M. S. (2002). AID mutates *E. coli* suggesting a DNA deamination mechanism for antibody diversification. *Nature, 418,* 99–103.

Pettersen, H. S., Sundheim, O., Gilljam, K. M., Slupphaug, G., Krokan, H. E., & Kavli, B. (2007). Uracil-DNA glycosylases SMUG1 and UNG2 coordinate the initial steps of base excision repair by distinct mechanisms. *Nucleic Acids Research, 35,* 3879–3892.

Pfaffeneder, T., Hackner, B., Truss, M., Munzel, M., Muller, M., Deiml, C. A., et al. (2011). The discovery of 5-formylcytosine in embryonic stem cell DNA. *Angewandte Chemie (International Ed. in English), 50,* 7008–7012.

Pham, P., Bransteitter, R., & Goodman, M. F. (2005). Reward versus risk: DNA cytidine deaminases triggering immunity and disease. *Biochemistry, 44,* 2703–2715.

Popp, C., Dean, W., Feng, S., Cokus, S. J., Andrews, S., Pellegrini, M., et al. (2010). Genome-wide erasure of DNA methylation in mouse primordial germ cells is affected by AID deficiency. *Nature, 463,* 1101–1105.

Quivoron, C., Couronne, L., Della Valle, V., Lopez, C. K., Plo, I., Wagner-Ballon, O., et al. (2011). TET2 inactivation results in pleiotropic hematopoietic abnormalities in mouse and is a recurrent event during human lymphomagenesis. *Cancer Cell, 20,* 25–38.

Rada, C., Williams, G. T., Nilsen, H., Barnes, D. E., Lindahl, T., & Neuberger, M. S. (2002). Immunoglobulin isotype switching is inhibited and somatic hypermutation perturbed in UNG-deficient mice. *Current Biology, 12,* 1748–1755.

Rai, K., Huggins, I. J., James, S. R., Karpf, A. R., Jones, D. A., & Cairns, B. R. (2008). DNA demethylation in zebrafish involves the coupling of a deaminase, a glycosylase, and gadd45. *Cell, 135,* 1201–1212.

Rai, K., Sarkar, S., Broadbent, T. J., Voas, M., Grossmann, K. F., Nadauld, L. D., et al. (2010). DNA demethylase activity maintains intestinal cells in an undifferentiated state following loss of APC. *Cell, 142,* 930–942.

Ranjit, S., Khair, L., Linehan, E. K., Ucher, A. J., Chakrabarti, M., Schrader, C. E., et al. (2011). AID binds cooperatively with UNG and Msh2-Msh6 to Ig switch regions dependent upon the AID C terminus. *Journal of Immunology, 187,* 2464–2475.

Rausch, J. W., Chelico, L., Goodman, M. F., & Le Grice, S. F. (2009). Dissecting APOBEC3G substrate specificity by nucleoside analog interference. *The Journal of Biological Chemistry, 284,* 7047–7058.

Revy, P., Muto, T., Levy, Y., Geissmann, F., Plebani, A., Sanal, O., et al. (2000). Activation-induced cytidine deaminase (AID) deficiency causes the autosomal recessive form of the Hyper-IgM syndrome (HIGM2). *Cell, 102,* 565–575.

Riccio, A., Aaltonen, L. A., Godwin, A. K., Loukola, A., Percesepe, A., Salovaara, R., et al. (1999). The DNA repair gene MBD4 (MED1) is mutated in human carcinomas with microsatellite instability. *Nature Genetics, 23,* 266–268.

Rice, G. I., Kasher, P. R., Forte, G. M., Mannion, N. M., Greenwood, S. M., Szynkiewicz, M., et al. (2012). Mutations in ADAR1 cause Aicardi-Goutieres syndrome associated with a type I interferon signature. *Nature Genetics, 44,* 1243–1248.

Ripley, L. S., & Drake, J. W. (1984). Bacteriophage T4 particles are refractory to bisulfite mutagenesis. *Mutation Research, 129,* 149–152.

Roa, S., Li, Z., Peled, J. U., Zhao, C., Edelmann, W., & Scharff, M. D. (2010). MSH2/MSH6 complex promotes error-free repair of AID-induced dU:G mispairs as well as error-prone hypermutation of A:T sites. *PLoS One, 5,* e11182.

Robbiani, D. F., Bunting, S., Feldhahn, N., Bothmer, A., Camps, J., Deroubaix, S., et al. (2009). AID produces DNA double-strand breaks in non-Ig genes and mature B cell lymphomas with reciprocal chromosome translocations. *Molecular Cell, 36,* 631–641.

Robert, I., Dantzer, F., & Reina-San-Martin, B. (2009). Parp1 facilitates alternative NHEJ, whereas Parp2 suppresses IgH/c-myc translocations during immunoglobulin class switch recombination. *The Journal of Experimental Medicine, 206,* 1047–1056.

Saribasak, H., Maul, R. W., Cao, Z., McClure, R. L., Yang, W., McNeill, D. R., et al. (2011). XRCC1 suppresses somatic hypermutation and promotes alternative non-homologous end joining in Igh genes. *The Journal of Experimental Medicine, 208,* 2209–2216.

Saribasak, H., Rajagopal, D., Maul, R. W., & Gearhart, P. J. (2009). Hijacked DNA repair proteins and unchained DNA polymerases. *Philosophical Transactions of the Royal Society of London. Series B, Biological Sciences, 364,* 605–611.

Schiesser, S., Hackner, B., Pfaffeneder, T., Muller, M., Hagemeier, C., Truss, M., et al. (2012). Mechanism and stem-cell activity of 5-carboxycytosine decarboxylation determined by isotope tracing. *Angewandte Chemie (International Ed. in English), 51,* 6516–6520.

Schrader, C. E., Guikema, J. E., Linehan, E. K., Selsing, E., & Stavnezer, J. (2007). Activation-induced cytidine deaminase-dependent DNA breaks in class switch recombination occur during G1 phase of the cell cycle and depend upon mismatch repair. *Journal of Immunology, 179,* 6064–6071.

Senejani, A. G., Dalal, S., Liu, Y. F., Nottoli, T. P., McGrath, J. M., Clairmont, C. S., et al. (2012). Y265C DNA polymerase beta knockin mice survive past birth and accumulate base excision repair intermediate substrates. *Proceedings of the National Academy of Sciences of the United States of America, 109,* 6632–6637.

Shen, J. C., Rideout, W. M., 3rd., & Jones, P. A. (1992). High frequency mutagenesis by a DNA methyltransferase. *Cell, 71,* 1073–1080.

Shen, J. C., Rideout, W. M., 3rd., & Jones, P. A. (1994). The rate of hydrolytic deamination of 5-methylcytosine in double-stranded DNA. *Nucleic Acids Research, 22,* 972–976.

Sibani, S., Melnyk, S., Pogribny, I. P., Wang, W., Hiou-Tim, F., Deng, L., et al. (2002). Studies of methionine cycle intermediates (SAM, SAH), DNA methylation and the impact of folate deficiency on tumor numbers in Min mice. *Carcinogenesis, 23,* 61–65.

Simsek, D., Furda, A., Gao, Y., Artus, J., Brunet, E., Hadjantonakis, A. K., et al. (2011). Crucial role for DNA ligase III in mitochondria but not in Xrcc1-dependent repair. *Nature, 471,* 245–248.

Sobol, R. W., Prasad, R., Evenski, A., Baker, A., Yang, X. P., Horton, J. K., et al. (2000). The lyase activity of the DNA repair protein β-polymerase protects from DNA-damage-induced cytotoxicity. *Nature, 405,* 807–810.

Song, J., Medline, A., Mason, J. B., Gallinger, S., & Kim, Y. I. (2000). Effects of dietary folate on intestinal tumorigenesis in the apcMin mouse. *Cancer Research, 60,* 5434–5440.

Soulas-Sprauel, P., Le Guyader, G., Rivera-Munoz, P., Abramowski, V., Olivier-Martin, C., Goujet-Zalc, C., et al. (2007). Role for DNA repair factor XRCC4 in immunoglobulin class switch recombination. *The Journal of Experimental Medicine, 204,* 1717–1727.

Sousa, M. M., Krokan, H. E., & Slupphaug, G. (2007). DNA-uracil and human pathology. *Molecular Aspects of Medicine, 28,* 276–306.

Steinacher, R., & Schar, P. (2005). Functionality of human thymine DNA glycosylase requires SUMO-regulated changes in protein conformation. *Current Biology, 15,* 616–623.

Steine, E. J., Ehrich, M., Bell, G. W., Raj, A., Reddy, S., van Oudenaarden, A., et al. (2011). Genes methylated by DNA methyltransferase 3b are similar in mouse intestine and human colon cancer. *The Journal of Clinical Investigation, 121,* 1748–1752.

Stivers, J. T., & Jiang, Y. L. (2003). A mechanistic perspective on the chemistry of DNA repair glycosylases. *Chemical Reviews, 103,* 2729–2759.

Sugo, N., Aratani, Y., Nagashima, Y., Kubota, Y., & Koyama, H. (2000). Neonatal lethality with abnormal neurogenesis in mice deficient in DNA polymerase beta. *The EMBO Journal, 19,* 1397–1404.

Szwagierczak, A., Bultmann, S., Schmidt, C. S., Spada, F., & Leonhardt, H. (2010). Sensitive enzymatic quantification of 5-hydroxymethylcytosine in genomic DNA. *Nucleic Acids Research, 38,* e181.

Tahiliani, M., Koh, K. P., Shen, Y., Pastor, W. A., Bandukwala, H., Brudno, Y., et al. (2009). Conversion of 5-methylcytosine to 5-hydroxymethylcytosine in mammalian DNA by MLL partner TET1. *Science, 324,* 930–935.

Tebbs, R. S., Flannery, M. L., Meneses, J. J., Hartmann, A., Tucker, J. D., Thompson, L. H., et al. (1999). Requirement for the Xrcc1 DNA base excision repair gene during early mouse development. *Developmental Biology, 208,* 513–529.

Thillainadesan, G., Chitilian, J. M., Isovic, M., Ablack, J. N., Mymryk, J. S., Tini, M., et al. (2012). TGF-beta-dependent active demethylation and expression of the p15(ink4b) tumor suppressor are impaired by the ZNF217/CoREST complex. *Molecular Cell, 46,* 636–649.

Tini, M., Benecke, A., Um, S. J., Torchia, J., Evans, R. M., & Chambon, P. (2002). Association of CBP/p300 acetylase and thymine DNA glycosylase links DNA repair and transcription. *Molecular Cell, 9*, 265–277.

Toyokuni, S., Mori, T., & Dizdaroglu, M. (1994). DNA base modifications in renal chromatin of Wistar rats treated with a renal carcinogen, ferric nitrilotriacetate. *International Journal of Cancer, 57*, 123–128.

Trasler, J., Deng, L., Melnyk, S., Pogribny, I., Hiou-Tim, F., Sibani, S., et al. (2003). Impact of Dnmt1 deficiency, with and without low folate diets, on tumor numbers and DNA methylation in Min mice. *Carcinogenesis, 24*, 39–45.

Ulrich, C. M., & Potter, J. D. (2007). Folate and cancer—Timing is everything. *JAMA: The Journal of the American Medical Association, 297*, 2408–2409.

Um, S., Harbers, M., Benecke, A., Pierrat, B., Losson, R., & Chambon, P. (1998). Retinoic acid receptors interact physically and functionally with the T:G mismatch-specific thymine-DNA glycosylase. *The Journal of Biological Chemistry, 273*, 20728–20736.

Unnikrishnan, A., Raffoul, J. J., Patel, H. V., Prychitko, T. M., Anyangwe, N., Meira, L. B., et al. (2009). Oxidative stress alters base excision repair pathway and increases apoptotic response in apurinic/apyrimidinic endonuclease 1/redox factor-1 haploinsufficient mice. *Free Radical Biology & Medicine, 46*, 1488–1499.

Vainio, S., Genest, P. A., ter Riet, B., van Luenen, H., & Borst, P. (2009). Evidence that J-binding protein 2 is a thymidine hydroxylase catalyzing the first step in the biosynthesis of DNA base J. *Molecular and Biochemical Parasitology, 164*, 157–161.

Valinluck, V., & Sowers, L. C. (2007). Endogenous cytosine damage products alter the site selectivity of human DNA maintenance methyltransferase DNMT1. *Cancer Research, 67*, 946–950.

Vartanian, J. P., Guetard, D., Henry, M., & Wain-Hobson, S. (2008). Evidence for editing of human papillomavirus DNA by APOBEC3 in benign and precancerous lesions. *Science, 320*, 230–233.

Ventrella-Lucente, L. F., Unnikrishnan, A., Pilling, A. B., Patel, H. V., Kushwaha, D., Dombkowski, A. A., et al. (2010). Folate deficiency provides protection against colon carcinogenesis in DNA polymerase beta haploinsufficient mice. *The Journal of Biological Chemistry, 285*, 19246–19258.

Vesely, C., Tauber, S., Sedlazeck, F. J., von Haeseler, A., & Jantsch, M. F. (2012). Adenosine deaminases that act on RNA induce reproducible changes in abundance and sequence of embryonic miRNAs. *Genome Research, 22*, 1468–1476.

Wang, H., Guan, S., Quimby, A., Cohen-Karni, D., Pradhan, S., Wilson, G., et al. (2011). Comparative characterization of the PvuRts1I family of restriction enzymes and their application in mapping genomic 5-hydroxymethylcytosine. *Nucleic Acids Research, 39*, 9294–9305.

Wang, R. Y., Kuo, K. C., Gehrke, C. W., Huang, L. H., & Ehrlich, M. (1982). Heat- and alkali-induced deamination of 5-methylcytosine and cytosine residues in DNA. *Biochimica et Biophysica Acta, 697*, 371–377.

Ward, P. S., Patel, J., Wise, D. R., Abdel-Wahab, O., Bennett, B. D., Coller, H. A., et al. (2010). The common feature of leukemia-associated IDH1 and IDH2 mutations is a neomorphic enzyme activity converting alpha-ketoglutarate to 2-hydroxyglutarate. *Cancer Cell, 17*, 225–234.

Waters, T. R., Gallinari, P., Jiricny, J., & Swann, P. F. (1999). Human thymine DNA glycosylase binds to apurinic sites in DNA but is displaced by human apurinic endonuclease 1. *The Journal of Biological Chemistry, 274*, 67–74.

Wiebauer, K., & Jiricny, J. (1989). In vitro correction of G.T mispairs to G.C pairs in nuclear extracts from human cells. *Nature, 339*, 234–236.

Wilson, S. H., & Kunkel, T. A. (2000). Passing the baton in base excision repair. *Nature Structural Biology*, 7, 176–178.

Wong, E., Yang, K., Kuraguchi, M., Werling, U., Avdievich, E., Fan, K., et al. (2002). Mbd4 inactivation increases Cright-arrowT transition mutations and promotes gastrointestinal tumor formation. *Proceedings of the National Academy of Sciences of the United States of America*, 99, 14937–14942.

Wu, X., & Stavnezer, J. (2007). DNA polymerase beta is able to repair breaks in switch regions and plays an inhibitory role during immunoglobulin class switch recombination. *The Journal of Experimental Medicine*, 204, 1677–1689.

Wyatt, M. D., Allan, J. M., Lau, A. Y., Ellenberger, T. E., & Samson, L. D. (1999). 3-methyladenine DNA glycosylases: Structure, function, and biological importance. *Bio Essays*, 21, 668–676.

Wyatt, G. R., & Cohen, S. S. (1953). The bases of the nucleic acids of some bacterial and animal viruses: The occurrence of 5-hydroxymethylcytosine. *The Biochemical Journal*, 55, 774–782.

Wyatt, M. D., & Pittman, D. L. (2006). Methylating agents and DNA repair responses: Methylated bases and sources of strand breaks. *Chemical Research in Toxicology*, 19, 1580–1594.

Xanthoudakis, S., Miao, G., Wang, F., Pan, Y. C., & Curran, T. (1992). Redox activation of Fos-Jun DNA binding activity is mediated by a DNA repair enzyme. *The EMBO Journal*, 11, 3323–3335.

Xanthoudakis, S., Smeyne, R. J., Wallace, J. D., & Curran, T. (1996). The redox/DNA repair protein, Ref-1, is essential for early embryonic development in mice. *Proceedings of the National Academy of Sciences of the United States of America*, 93, 8919–8923.

Xu, W., Yang, H., Liu, Y., Yang, Y., Wang, P., Kim, S. H., et al. (2011). Oncometabolite 2-hydroxyglutarate is a competitive inhibitor of alpha-ketoglutarate-dependent dioxygenases. *Cancer Cell*, 19, 17–30.

Yamada, Y., Jackson-Grusby, L., Linhart, H., Meissner, A., Eden, A., Lin, H. J., et al. (2005). Opposing effects of DNA hypomethylation on intestinal and liver carcinogenesis. *Proceedings of the National Academy of Sciences of the United States of America*, 102, 13580–13585.

Yan, C. T., Boboila, C., Souza, E. K., Franco, S., Hickernell, T. R., Murphy, M., et al. (2007). IgH class switching and translocations use a robust non-classical end-joining pathway. *Nature*, 449, 478–482.

Yang, H., Liu, Y., Bai, F., Zhang, J. Y., Ma, S. H., Liu, J., et al. (2013). Tumor development is associated with decrease of TET gene expression and 5-methylcytosine hydroxylation. *Oncogene*, 32, 663–669.

Yi, C., Yang, C. G., & He, C. (2009). A non-heme iron-mediated chemical demethylation in DNA and RNA. *Accounts of Chemical Research*, 42, 519–529.

Yoshikawa, K., Okazaki, I. M., Eto, T., Kinoshita, K., Muramatsu, M., Nagaoka, H., et al. (2002). AID enzyme-induced hypermutation in an actively transcribed gene in fibroblasts. *Science*, 296, 2033–2036.

Yu, Z., Genest, P. A., ter Riet, B., Sweeney, K., DiPaolo, C., Kieft, R., et al. (2007). The protein that binds to DNA base J in trypanosomatids has features of a thymidine hydroxylase. *Nucleic Acids Research*, 35, 2107–2115.

Zampieri, M., Guastafierro, T., Calabrese, R., Ciccarone, F., Bacalini, M. G., Reale, A., et al. (2012). ADP-ribose polymers localized on Ctcf-Parp1-Dnmt1 complex prevent methylation of Ctcf target sites. *The Biochemical Journal*, 441, 645–652.

Zampieri, M., Passananti, C., Calabrese, R., Perilli, M., Corbi, N., De Cave, F., et al. (2009). Parp1 localizes within the Dnmt1 promoter and protects its unmethylated state by its enzymatic activity. *PLoS One*, 4, e4717.

Zan, H., & Casali, P. (2008). AID- and Ung-dependent generation of staggered double-strand DNA breaks in immunoglobulin class switch DNA recombination: A post-cleavage role for AID. *Molecular Immunology*, *46*, 45–61.

Zhu, J. K. (2009). Active DNA demethylation mediated by DNA glycosylases. *Annual Review of Genetics*, *43*, 143–166.

Zhu, B., Zheng, Y., Hess, D., Angliker, H., Schwarz, S., Siegmann, M., et al. (2000). 5-methylcytosine-DNA glycosylase activity is present in a cloned G/T mismatch DNA glycosylase associated with the chicken embryo DNA demethylation complex. *Proceedings of the National Academy of Sciences of the United States of America*, *97*, 5135–5139.

Zhuang, J., Jones, A., Lee, S. H., Ng, E., Fiegl, H., Zikan, M., et al. (2012). The dynamics and prognostic potential of DNA methylation changes at stem cell gene loci in women's cancer. *PLoS Genetics*, *8*, e1002517.

Zingg, J. M., Shen, J. C., Yang, A. S., Rapoport, H., & Jones, P. A. (1996). Methylation inhibitors can increase the rate of cytosine deamination by (cytosine-5)-DNA methyltransferase. *Nucleic Acids Research*, *24*, 3267–3275.

CHAPTER THREE

Role of Oxidative Stress and the Microenvironment in Breast Cancer Development and Progression

Agnieszka Jezierska-Drutel*, Steven A. Rosenzweig*, Carola A. Neumann*,†,1

*Department of Cell and Molecular Pharmacology and Experimental Therapeutics, Medical University of South Carolina, Charleston, South Carolina, USA
†Department of Pharmacology and Chemical Biology, University of Pittsburgh Cancer Institute, Magee-Womens-Research Institute, Pittsburgh, Pennsylvania, USA
¹Corresponding author: e-mail address: neumannc@upmc.edu

Contents

Abstract

Breast cancer is a highly complex tissue composed of neoplastic and stromal cells. Carcinoma-associated fibroblasts (CAFs) are commonly found in the cancer stroma, where they promote tumor growth and enhance vascularity in the microenvironment. Upon exposure to oxidative stress, fibroblasts undergo activation to become myofibroblasts. These cells are highly mobile and contractile and often express numerous mesenchymal markers. CAF activation is irreversible, making them incapable of being removed by nemosis. In breast cancer, almost 80% of stromal fibroblasts acquire an activated phenotype that manifests by secretion of elevated levels of growth factors, cytokines, and metalloproteinases. They also produce hydrogen peroxide, which induces the generation of subsequent sets of activated fibroblasts and tumorigenic alterations in epithelial cells. While under oxidative stress, the tumor stroma releases high energy nutrients that fuel cancer cells and facilitate their growth and survival. This review describes how breast cancer progression is dependent upon oxidative stress activated stroma and proposes potential new therapeutic avenues.

Advances in Cancer Research, Volume 119
ISSN 0065-230X
http://dx.doi.org/10.1016/B978-0-12-407190-2.00003-4

1. OXIDATIVE STRESS CHANGES THE BREAST TUMOR MICROENVIRONMENT

Cancer is one of the most common diseases in the United States (Siegel, Naishadham, & Jemal, 2012). In 2012, breast cancer alone was expected to account for 29% (226,870) of all new cancer cases among women in the United States. It is known that one in eight women will develop breast cancer during their lives, and almost 40,000 will die because of breast cancer metastasis. Men are also at risk of breast cancer and it is estimated that in 2012, almost 2190 men will be diagnosed with breast cancer, and 410 will die due to this disease. Breast cancer is not only an isolated group of mutated somatic cells, but it is also a microenvironment system comprising breast cancer cells, fibroblasts, adipocytes, immune, and endothelial cells (Fig. 3.1). Close interactions between cancer cells and stroma are known to regulate breast cancer pathways. Among the many factors influencing development, progression, and metastasis, oxidative stress has an important role in the initiation and preservation of breast cancer progression. Oxidative stress results from an imbalance between unstable reactive species lacking one or more unpaired electrons (superoxide anion, hydrogen peroxide, hydroxyl radical, reactive nitrogen species) and antioxidants. Oxidative stress can be generated by ultraviolet light exposure, ionizing radiation, or carcinogen exposure. One way in which cellular reactive oxygen species (ROS) impact cell signaling is through their localized accumulation, if we consider ROS byproducts of the electron transport chain in mitochondria or activation of the NADPH oxidases (NOX). Induced pathways producing ROS function in all cell types, including both breast carcinoma and cancer stroma (Hecker et al., 2009; Sampson et al., 2011; Tobar, Guerrero, Smith, & Martinez, 2010). Stromal-derived NOX4 ROS are able to stimulate migration of MCF-7 cells in a paracrine manner (Tobar et al., 2010). Overexpression of NOX4 in normal breast epithelial cells results in cellular senescence, resistance to apoptosis, and tumorigenic transformation, as well as increased aggressiveness of breast cancer cells (Graham et al., 2010). Interestingly, several recent studies have confirmed that ROS, including hydrogen peroxide, alone are able to drive the differentiation of normal fibroblasts into myofibroblasts, which are able to generate high amounts of hydrogen peroxide themselves, increasing oxidative stress in the microenvironment (Comito et al., 2012; Taddei et al., 2012; Toullec et al., 2010; Waghray et al., 2005) (Fig. 3.2). Almost 80% of fibroblasts acquire an

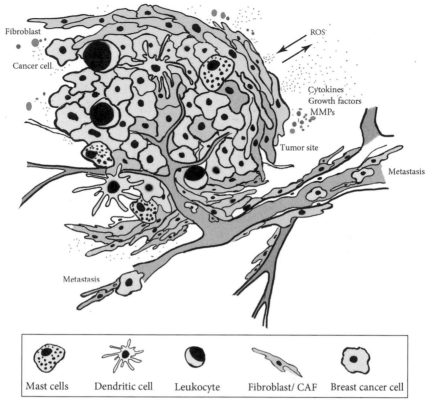

Figure 3.1 Breast tumor microenvironment. The tumor microenvironment consists of cancer, stromal, and nonmalignant cells supporting tumor growth and vascularization. Under oxidative stress, activated stromal cells generate tumor enhancing signals. Activated fibroblasts acquire a contractile phenotype and mobility affecting cancer cell migration and invasion. Immune cells infiltrate the tumor tissue and secrete cytokines supporting further microenvironmental stress. (See Page 2 in Color Section at the back of the book.)

activated phenotype in breast cancer (Kalluri & Zeisberg, 2006; Sappino, Skalli, Jackson, Schurch, & Gabbiani, 1988). Myofibroblasts are highly mobile and contractile cells that typically express mesenchymal markers such as α-smooth muscle actin (α-SMA), calponin, and vimentin (Kalluri & Zeisberg, 2006). Additionally, type I collagen secreted by myofibroblasts causes higher breast density, which contributes to mammary tumor formation and metastasis (Provenzano et al., 2008). It has also been shown that type I collagen contributes to decreased chemotherapeutic agent uptake and altered tumor cell sensitivity to a variety of chemotherapies. Not only collagen deposition but also collagen degradation is required for the physiologic remodeling of connective tissue during breast cancer growth,

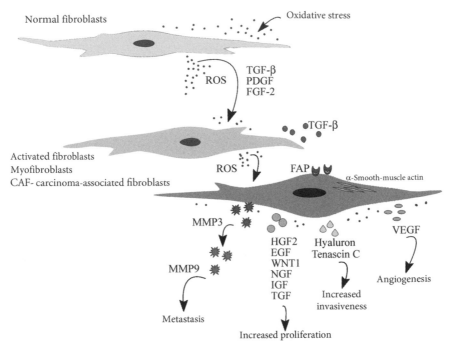

Figure 3.2 Oxidative stress activates stroma fibroblasts. In breast cancer, under oxidative stress, stroma-associated fibroblasts acquire an activated state and can be identified by numerous markers including α-smooth muscle actin (α-SMA), fibroblast specific protein (FSP), fibroblast activation protein (FAP), PDGF receptors-β, desmin, endosialin, tenascin-c, palladin, vimentin, pro-collagen, stromelysin-3, and cadherin-11. Oxidative stress can cause cell senescence, locking activated fibroblast in a nonproliferative state and creating a senescence-associated secretory phenotype, which manifests in releasing factors that affect neighboring cells. Activation of fibroblasts, followed by secretion of numerous factors including metalloproteinases (MMP-2, MMP-3, MMP9), growth factors (HGF-2, EGF, IGF, TGFβ, VEGF), the polysaccharide hyaluronan and tenascin-c significantly promotes proliferation and growth of premalignant epithelial cells and cancer cells, invasiveness, metastasis, and angiogenesis in breast cancer. (See Page 3 in Color Section at the back of the book.)

development, and cancer cell invasion. Secreted metalloproteinases (MMP) such as MMP-2, MMP-3, and MMP-9 increase extracellular matrix turnover and are themselves activated by oxidative stress (Fu, Kassim, Parks, & Heinecke, 2001; Koch et al., 2009; Provenzano et al., 2008; Radisky et al., 2005). Additionally secreted transforming growth factor β (TGFβ), insulin-like growth factor (IGF), platelet-derived growth factor (PDGF), fibroblast growth factor 2, and stromal-derived factor 1 (SDF1) are able to activate fibroblasts and increase cancer cell proliferation (Sappino et al., 1988). Growth factors such as TGFβ, IGF, TNFα, or PDGF stimulate ROS production through NOX (Basuroy, Bhattacharya, Leffler, & Parfenova, 2009;

Edderkaoui et al., 2011; Marumo, Schini-Kerth, Fisslthaler, & Busse, 1997; Meng, Lv, & Fang, 2008). Once activated, the tumor microenvironment network produces large amounts of ROS, initiating tumor growth. It would be clinically advantageous to detect this large ROS activity early enough to prevent further cancer progression. There is only one clinical study showing that such a possibility exists. The research involving an equal number of breast cancer patients and healthy controls showed increased levels of hydrogen peroxide in exhaled breath condensate from patients with localized breast malignancy, associated with increased clinical severity (Stolarek et al., 2010).

2. OXIDATIVE STRESS REGULATE CAVEOLIN-1 SIGNALING AND ENERGY METABOLISM IN THE TUMOR MICROENVIRONMENT

Since 1920, the "Warburg Effect" has been the leading principle in understanding breast cancer development resulting from dysfunctional mitochondrial oxidative phosphorylation (Koppenol, Bounds, & Dang, 2011). This mechanism was shown to be incomplete. Specifically based on the "Warburg Effect," the recently proposed "Reverse Warburg Effect" underscores an important and supportive role for oxidative stress and the microenvironment in cancer progression (Balliet et al., 2011). It was shown that aerobic glycolysis occurs in the adjacent stroma, rather than in cancer cells. Stromal autophagy drives cancer cell growth and progression by providing fatty acids, nucleotides, and free amino acids that are ready for use by cancer cells. Moreover, catabolites produced by stroma–associated fibroblasts not only stimulate mitochondrial biogenesis in breast cancer cells, they also protect them from autophagy and chemotherapy-induced apoptosis (Martinez-Outschoorn, Pavlides, et al., 2010; Martinez-Outschoorn, Goldberg, et al., 2011; Martinez-Outschoorn, Lin, et al., 2011). Comparing mitochondrial metabolic activity revealed a difference between stroma and epithelial cells (Balliet et al., 2011). While breast cancer epithelial cells appear to have increased mitochondrial oxidative activity, as shown by their elevated levels of NADH dehydrogenase, saccharopine dehydrogenase (SDH), and cyclooxygenase (Whitaker-Menezes, Martinez-Outschoorn, Flomenberg, et al., 2011). Activated stroma fibroblasts, on the other hand, exhibit a low mitochondrial oxidative capacity, low translocase of outer mitochondrial membrane 20, yet high expression of monocarboxylate transporter 4 (MCT4) and the lysosomal protease—cathepsin B. Not surprisingly, breast cancer tumors that exhibit upregulation of enzymes involved in oxidative phosphorylation are prone to metastasize to the brain, which itself is a lactate-rich environment (Sonveaux et al., 2008; Sotgia,

Martinez-Outschoorn, & Lisanti, 2011; Sotgia et al., 2012). This creates an ideal condition for tumor progression. A recent study has shown that cancer cells are able to induce drivers of oxidative stress, autophagy and mitophagy: HIF-1α and NFκB in surrounding stroma fibroblasts (Martinez-Outschoorn, Lin, Trimmer, et al., 2011). This process can lead to loss of stromal caveolin-1 (Cav-1) (Martinez-Outschoorn, Trimmer, et al., 2010). Cav-1 is highly expressed in differentiated fibroblasts, adipocytes, and endothelial cells (Lisanti et al., 1994). Loss of Cav-1 in CAFs is considered to be a "lethal" breast tumor microenvironment, as it is linked with tamoxifen resistance, increased lymph node metastasis, early tumor recurrence, and poor overall outcome (Martinez-Outschoorn, Balliet, et al., 2010; Witkiewicz et al., 2011). In triple negative breast cancer stroma, the loss of Cav-1 in $\sim$10% of cases was linked with a $\leq$5-year survival rate compared to approximately 75% of cases where Cav-1 stroma-positive patients had a 12 year survival rate (Witkiewicz et al., 2010). Interestingly, oxidative stress is sufficient to induce the loss of Cav-1 in fibroblasts due to lysosomal degradation of Cav-1 (Martinez-Outschoorn, Pavlides, et al., 2010; Martinez-Outschoorn, Balliet, et al., 2010). In support of this, knockdown of Cav-1 in immortalized human fibroblasts (hTERT-BJ1) results in increased hydrogen peroxide generation, which can further induce Cav-1 loss in neighboring fibroblasts (Trimmer et al., 2011). Studies show that loss of Cav-1 in adjacent breast cancer stroma fibroblasts can be prevented by treatment with N-acetyl cysteine, quercetin, or metformin (Martinez-Outschoorn, Balliet, et al., 2010; Witkiewicz et al., 2011). It was reported that overexpression of recombinant (SOD2) (Trimmer et al., 2011) or injection of SOD, catalase, or their pegylated counterparts can block recurrence and metastasis in mice (Goh et al., 2011; Hyoudou et al., 2008; Nishikawa et al., 2005). Of note, mammary stromal cells from Cav-1 ($-/-$) null mice have upregulated myofibroblast markers (vimentin, calponin2, tropomyosin, gelsolin, and prolyl 4-hydroxylase alpha), EF-1-δ (elongation factor 1-delta: which can drive cell transformation and tumorigenesis), signaling molecules (annexin A1, annexin A2, and RhoGDI), glycolytic enzymes (M2-isoform of pyruvate kinase-PKM2 and LDHA), secreted proteins (type I collagen and SPARC), and two peroxidases: catalase and peroxiredoxin-1 (Prdx1) (Pavlides et al., 2009). Consistent with this premise, a study examining mammary stroma from human breast cancer specimens using laser-capture microdissection and molecular profiling revealed that patients with Cav-1 negative breast cancer stroma had higher expression of genes associated with oxidative stress, hypoxia, redox signaling, apoptosis, autophagy, lysosomal degradation, glycolysis aging, "stemness," inflammation, DNA damage, and myofibroblast differentiation (Witkiewicz et al., 2011). The reduction of Cav-1 occurs in both subtypes of breast cancers: estrogen receptor positive ER(+), and estrogen receptor negative ER(−). However, it was shown to be significantly associated with ER(−) cancers, increased recurrence, and decreased overall survival (Witkiewicz et al., 2011), suggesting perhaps that hormone receptor negative breast cancers are linked to higher levels of ROS. Along those lines, when exposed to estrogenic polychlorinated biphenyls, the triple negative breast cancer cell line MDA-MB-231 produced significantly higher amounts of ROS than the poorly metastatic T47D cell line (Lin & Lin, 2006). Moreover, treatment of breast cancer cells with physiologic levels of 17 beta estradiol induces ROS production and activates redox sensitive signaling pathways in neighboring cells, leading to angiogenesis, tumor growth, and invasiveness (Felty, 2011). Lowered expression of Cav-1 not only leads to myofibroblast conversion and inflammation but also seems to impact aerobic glycolysis, leading to secretion of high energy metabolites such as pyruvate and lactate that drive mitochondrial oxidative phosphorylation in cancer cells (Balliet et al., 2011). Pyruvate dehydrogenase, the enzyme that converts pyruvate into acetyl-coA, is decreased in stromal fibroblasts due to Cav-1 loss (Martinez-Outschoorn, Balliet, et al., 2010). Depletion of Cav-1 activates HIF-1α and creates a lactate-rich environment due to the forced conversion of aerobic pyruvate into lactate (Sonveaux et al., 2008; Sotgia et al., 2011b). It is widely held that HIF-1α function is dependent upon its location within the tumor microenvironment. It acts as a tumor promoter in CAFs and as a tumor suppressor in cancer cells (Chiavarina et al., 2010). Oxidative stress generated by breast cancer cells activates HIF-1α and NFκB in fibroblasts, leading to autophagy and lysosomal degradation of Cav-1 (Martinez-Outschoorn, Balliet, et al., 2010). Thus, the level of Cav-1 in the microenvironment serves as an ideal marker for oxidative stress, hypoxia, and autophagy. The close relationship between oxidative stress, Cav-1, and energy metabolism is further illustrated by the finding that loss of Cav-1 results in the expression of two monocarboxylate transporters: MCT1 and MCT4. MCT4 expression is only increased in breast cancer-associated fibroblasts, whereas MCT1 expression appears to be restricted to breast cancer cells. As this allows for proper import of lactate from fibroblasts into cancer cells (Sonveaux et al., 2008; Whitaker-Menezes, Martinez-Outschoorn, Lin, et al., 2011), such findings again underscore the intimate relationship between these two compartments. In support of this, higher levels of MCT4 in the tumor stroma are associated with poor outcome in triple negative breast cancer (Sonveaux et al., 2008; Witkiewicz et al., 2012), while MCT4 expression in breast cancer cells had no significant predictive value (Sonveaux et al., 2008; Whitaker-Menezes, Martinez-Outschoorn, Lin, et al., 2011). In contrast, MCT1 expression in breast cancer cells was linked to p53 mutations, p53 loss, and poor outcome (Boidot et al., 2012). Given that the prognostic value of MCT4 is restricted to breast cancer stroma and MCT1 to breast cancer epithelial cells, MCT4, and MCT1 in combination with Cav-1, may be functional markers of oxidative stress and aerobic glycolysis, exclusively in tumor stroma.

3. OXIDATIVE STRESS CAUSES FIBROBLASTS TO UNDERGO SENESCENCE AND INDUCES MITOCHONDRIAL DYSFUNCTION

By the age of 40, breast cancer risk in women is 1 in 203, but it rapidly increases: for women over 40 years of age, it is 1 in 28 and by age 70, 1 in 15 females will develop breast cancer. Over the years, somatic mutations accumulate and increase the risk of developing breast cancer. Phenotypically, senescent cells partially resemble aging cells. For example, senescent fibroblasts are locked in a nonproliferative state, suffer multiple phenotypic changes and are able to secrete factors influencing endothelial cells, pushing them toward a carcinogenic pathway (Parrinello, Coppe, Krtolica, & Campisi, 2005). Besides other stressors, including ionizing radiation, oncogenes, and environmental factors, hydrogen peroxide is one of the main factors that can push fibroblasts and cancer cells into senescence (Capparelli, Chiavarina, et al., 2012). It was reported that hydrogen peroxide- induced senescence and autophagy are closely linked processes that may be mutually induced (Capparelli, Chiavarina, et al., 2012). Cyclin-dependent kinase (CDK) is a crucial factor in initiating DNA synthesis during the mammalian cell division cycle. CDKs are regulated by CDK inhibitors such as p16 and p19, which also function as tumor suppressors (Bardeesy et al., 2006; Capparelli, Chiavarina, et al., 2012; Sharpless, Ramsey, Balasubramanian, Castrillon, & DePinho, 2004; Weitzman, 2001). In CAFs, the CDK inhibitors p19 and smARF drive senescence and autophagy, resulting in increased angiogenesis and promotion of tumor growth. Fibroblasts expressing cell cycle inhibitors, such as p19 and smARF, are predisposed to hydrogen peroxide initiated senescence. In human ductal breast carcinoma *in situ* (DCIS), increases in p16 are linked with a nine-fold increased risk of recurrence (Capparelli, Chiavarina, et al., 2012). Recent data have revealed that cell cycle arrest, senescence, and autophagy are closely related biological processes, which can be induced by the same stimuli and can arise simultaneously in a given cell. Senescence can induce autophagy in CAFs and they undergo senescence–autophagy transition (Capparelli, Chiavarina, et al., 2012). It was shown that overexpression of autophagy inducers in fibroblasts leads to autophagy. Moreover, these fibroblasts express the active senescence markers β-galactosidase and p21 and they gain a senescence-associated secretory phenotype resulting in secretion of numerous growth factors, cytokines, and proteases that promote migration and invasion of cancer cells, independent of angiogenesis (Capparelli, Guido, et al., 2012; Capparelli, Whitaker-Menezes, et al., 2012; Coppe et al., 2010). Of note, selective induction of autophagy only in breast cancer cells inhibits tumor growth. Fibroblast senescence is able to attenuate the functional differentiation of cells and increases branching morphogenesis of normal mammary epithelial cells (Krtolica, Parrinello, Lockett, Desprez, & Campisi, 2001; Parrinello et al., 2005). Branching is a physiological process in which secretion of soluble factors such as hepatocyte growth factor (HGF), MMP2, and MMP3 by fibroblasts, induces epithelial cells to invade and migrate through collagen in a controlled manner. Therefore, during carcinogenesis, fibroblasts support branching but control of this process is lost (Parrinello et al., 2005). Moreover, in aging mammary glands, the number of senescent fibroblasts along with IL-6 and -8 secretion increases. This in turn promotes malignant processes such as epithelial–mesenchymal transition (EMT) in premalignant epithelial cells (Campisi, 2013).

Recent studies show that in the breast cancer microenvironment, oxidative stress causes mitochondrial dysfunction, which manifests itself by the upregulation of numerous factors, such as the nuclear respiratory factor 1 (NRF1), which was shown to be upregulated in cancer stroma due to oxidative stress (Witkiewicz et al., 2011). NRF1 is a transcription factor that activates the expression of genes regulating cellular growth, respiration, and mitochondrial DNA (mtDNA) transcription and replication (Witkiewicz et al., 2011). Hydrogen peroxide was shown to increase levels of mtDNA absence sensitive factor—GOLPH$_3$/MIDAS. This oncogene modulates mTOR signaling and is able to increase mitochondrial mass by regulating mitochondrial lipids in response to mitochondrial dysfunction (Witkiewicz et al., 2011). mtDNA is extremely sensitive to oxidative damage due to the lack of protective histones and efficient repair mechanisms located in the nucleus. This is the mechanism by which excess mitochondrial ROS promotes malignant cell transformation. It was recently demonstrated that transgenic mice expressing the human catalase gene (mCAT) in mammary tumors had reduced primary tumor invasiveness and decreased pulmonary metastatic tumor incidence compared to mice that lacked mCAT (Goh et al., 2011). The presence of mCAT in fibroblasts leads to altered phosphorylation of p38MAPK (downstream ROS signaling pathway) in response to hydrogen peroxide (Goh et al., 2011). This lowered oxidative stress and selectin expression and decreased the colonization ability and metastatic potential of cancer cells. Mitochondrial malfunction leads to oversaturation with ROS in cancer cells and thereby cancer risk. This detrimental process is comprised of numerous factors. For example, loss of mitochondrial transcription factor A results in compromised transcription of mtDNA (Balliet et al., 2011). Persistent exposure to oxidative stressors or other DNA damaging agents eventually leads to accumulation of DNA mutations in every tissue. Therefore, increased cancer risk with aging can be attributed to somatic mutations or alterations of gene copy number in the adjacent stroma. Particularly in breast cancer, somatic mutations in *TP53* and PTEN and gene copy number alterations at other loci in the adjacent stroma have been described (Haviv, Polyak, Qiu, Hu, & Campbell, 2009). PTEN, a protein tyrosine and lipid phosphatase, is a crucial tumor suppressor in most cancers, including breast. PTEN belongs to a class of phosphatases known to be prone to inactivation induced by oxidative stress. Supporting this premise, we have recently shown that Prdx1 prevents PTEN oxidation-induced inactivation preventing H-Ras and human epidermal growth factor 2 (ErbB-2)-driven transformations (Cao et al., 2009). Over the last few years, an impressive body of work describing PTEN's tumor suppressive role in epithelial cells, including breast, has accumulated (Gonzalez-Angulo et al., 2011; Heering, Erlmann, & Olayioye, 2009; Jensen et al., 2012; Noh et al., 2011; Russillo et al., 2011; Suda et al., 2012; Tanic et al., 2012). However, a very recent elegant study defined PTEN's tumor suppressive role in mammary stroma-associated fibroblasts (Trimboli et al., 2009), enforcing the importance of oxidative stress promoting cancer through regulating the tumor microenvironment. mtDNA polymorphisms and oxidative stress sensitivity may be underappreciated factors in breast carcinogenesis. The mtDNA 10398A allele was reported to increase breast cancer susceptibility in African-American women, whereas the mitochondrial NADH dehydrogenase subunit 3 (ND3) polymorphism (A10398G) is linked to sporadic breast cancer in Poland (Canter, Kallianpur, Parl, & Millikan, 2005; Czarnecka et al., 2010). It was also shown that mtDNA 10398G allele carriers have increased breast cancer risk due to alcohol consumption, perhaps as a result of the alcohol metabolite, acetaldehyde, which increases oxidative stress-induced damage of mtDNA (Pezzotti et al., 2009; Rohan, Wong, Wang, Haines, & Kabat, 2010). Smoking not only increases oxidative stress, but also augments mtDNA copy number, which correlates with an increased risk for breast cancer (Rohan et al., 2010; Shen, Platek, Mahasneh, Ambrosone, & Zhao, 2010).

It was reported that oxidative stress leads to HIF-1α accumulation and *junD* gene inactivation in fibroblasts. This event likely activates fibroblasts that express elevated levels of stromal cell-derived factor (SDF1) and have expression profiles similar to carcinoma-associated fibroblast profiles (Toullec et al., 2010). Not surprisingly, accumulation of SDF1 in ErbB-2 positive breast cancer is linked with increased numbers of activated fibroblasts and increased lymph node metastasis (Toullec et al., 2010). In addition, HIF-1α is known to regulate mitochondrial sirtuin NAD-dependent protein deacetylases 3 (SIRT3), a mitochondrial tumor suppressor, which maintains mitochondrial integrity and guards proper oxidative metabolism. Given the cancer preventive role of healthy mitochondria, it is not unexpected that loss of SIRT3 in fibroblasts would result in a tumorigenic phenotype and enhance breast cancer development (Kim et al., 2010). Healthy mitochondria are essential for maintaining energy, redox, and calcium balance. Impairments in mitochondrial function, and/or mechanisms controlling autophagy removal of damaged mitochondria lead to altered mitochondrial metabolisms and can promote cancer development. Another example of how oxidative stress may alter mitochondrial function to promote tumorigenesis through the microenvironment is via mammalian stanniocalcin 1 (STC1). STC1 has been found to be massively upregulated in mouse fibroblasts due to ROS accumulation (Nguyen, Chang, & Reddel, 2009). Examination of STC-1 mRNA levels in multiple tumor samples showed that this glycoprotein was upregulated in many cancer types and might serve as a marker of micrometastasis (Fujiwara et al., 2000).

4. CONCLUSIONS AND FUTURE PERSPECTIVES— SHOULD WE TARGET BREAST CANCER STROMA?

Martinez-Outschoorn et al. described the tight connection between breast cancer cells and the stroma, using an MCF7-fibroblast co-culture system (Martinez-Outschoorn, Lin, Trimmer, et al., 2011). In this environment, the cancer cells produce hydrogen peroxide and by driving the "Reverse Warburg Effect" initiate oxidative stress in fibroblasts. As a result of this process, fibroblasts exhibited reduced mitochondrial activity, increased glucose uptake, ROS, and metabolite production. That obstructing oxidative stress in the tumor microenvironment can lead to mitophagy and promote breast cancer shutdown is a promising discovery for the development of future therapeutic interventions. In healthy organisms, hydrogen peroxide produced during the wound healing process by a normal epithelium is sufficient to convert normal fibroblasts to activated myofibroblasts that go on to produce hydrogen peroxide themselves, enhancing the inflammatory signal. Activation of the cancer microenvironment activation is constant, and CAFs never undergo nemosis; this is in contrast to the normal wound healing process (Kankuri, Cholujova, Comajova, Vaheri, & Bizik, 2005; Kankuri et al., 2008). Nemosis is a type of nonapoptotic cell death characterized by a necrotic morphology in fibroblasts. Some investigators have promoted the idea of targeting CAFs, activated fibroblast, or fibroblasts in the senescence secretory phenotype state or the autophagy transition phase by cutting off all the "breast cancer fuel," L-lactate, ketone bodies, glutamine, and free fatty acids, thereby preventing breast cancer progression. Removing oxidative stress from the microenvironment by antioxidant treatment is a very tempting strategy. However, diets rich in antioxidants have fallen short in sufficiently preventing cancer. This clearly emphasizes the diverse roles of oxidative stress in a healthy organism and the need for a more detailed understanding of how oxidative stress specifically promotes cancer. It is clear that a healthy life style in combination with a healthy diet rich in fruits and vegetables, low in saturated fatty acids and free of alcohol and cigarette smoking is beneficial in supporting the prevention of breast cancer. Considerable data exists to support the premise that a lack of antioxidant enzymes tips the balance toward cancer initiation. For example, mice deficient in the peroxidase, Prdx1, die prematurely of oxidation-induced anemia and cancer (Neumann et al., 2003), suggesting a tumor preventive role for Prdx1. Fibroblasts isolated from mice lacking Prdx1 are characterized by hyperactive Akt, elevated cellular ROS and DNA damage. Prdx-1 null mice expressing oncogenic H-Ras in the mammary gland have a higher incidence of breast cancer compared to wild-type mice expressing oncogenic (Cao et al., 2009; Neumann, Cao, & Manevich, 2009). Prdx1 is a safeguard of the PI3K/Akt signaling pathway as a result of its ability to protect the tumor suppressor PTEN from oxidation-induced inactivation. Knockdown PTEN in mouse mammary stromal fibroblasts increases microenvironment remodeling, collagen deposition, angiogenesis, immune cell recruitment, and increased malignancy of mammary epithelial tumor (Trimboli et al., 2009), supporting the premise that oxidative stress may promote breast cancer in part, by inactivating PTEN. Along these lines, drugs targeting oxidative mitochondrial oxidative phosphorylation have recently been developed for support of breast cancer treatment (Whitaker-Menezes, Martinez-Outschoorn, Flomenberg, et al., 2011). Antioxidants like N-acetyl cysteine, hydroxy-chloroquine, and metformin prevent oxidative stress in

CAFs and ketones, lactate, and glutamine from being transported into breast cancer cells (Sonveaux et al., 2008; Whitaker-Menezes, Martinez-Outschoorn, Lin, et al., 2011). In conclusion, regulating the crosstalk between tumor and stroma (Fig. 3.3) or specifically targeting CAFs, which are genetically more stable than tumor cells may provide a viable option in the future

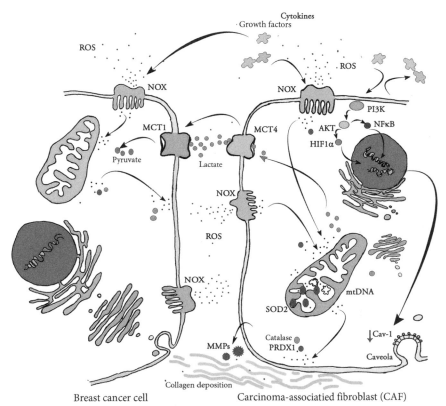

Figure 3.3 Interactions between CAFs and breast cancer cells in the tumor microenvironment. In the tumor microenvironment close interactive pathways are established, cross-linking breast cancer cells and stromal cells. Upregulated NADPH oxidase complex (NOX) bound in cell membranes is a source of large amounts of ROS released from these cells. Cancer cells secrete ROS, cytokines, and growth factors, prompting a reactive response in the stroma. Activated stromal cells influence cancer cell malignancy by increasing oxidative stress, collagen deposition, MMP, and growth factor secretion. Oxidative stress-induced loss of expression of membrane-bound scaffolding protein, caveolin-1 (Cav-1), in cancer-associated fibroblasts via autophagy also results in an activated tumor microenvironment. The uncontrolled expansion of cancer cells depends upon high ATP production and high glucose availability. CAFs release lactate *via* monocarboxylate transporter 4 (MCT4) and monocarboxylate transporter 1 (MCT1) on cancer cells facilitating lactate uptake. Thus, by providing high energy metabolites activated fibroblasts maintain continuous cancer progression. (See Page 3 in Color Section at the back of the book.)

for both the prevention and treatment of breast cancer (Hiscox, Barrett-Lee, & Nicholson, 2011; Loeffler, Kruger, Niethammer, & Reisfeld, 2006). However, to pursue this goal successfully, additional research is needed to define the molecular details of how oxidative stress promotes cancer through the tumor microenvironment.

REFERENCES

Balliet, R. M., Capparelli, C., Guido, C., Pestell, T. G., Martinez-Outschoorn, U. E., Lin, Z., et al. (2011). Mitochondrial oxidative stress in cancer-associated fibroblasts drives lactate production, promoting breast cancer tumor growth: Understanding the aging and cancer connection. *Cell Cycle, 10*, 4065–4073.

Bardeesy, N., Aguirre, A. J., Chu, G. C., Cheng, K. H., Lopez, L. V., Hezel, A. F., et al. (2006). Both p16(Ink4a) and the p19(Arf)-p53 pathway constrain progression of pancreatic adenocarcinoma in the mouse. *Proceedings of the National Academy of Sciences of the United States of America, 103*, 5947–5952.

Basuroy, S., Bhattacharya, S., Leffler, C. W., & Parfenova, H. (2009). Nox4 NADPH oxidase mediates oxidative stress and apoptosis caused by TNF-alpha in cerebral vascular endothelial cells. *American Journal of Physiology. Cell Physiology, 296*, C422–C432.

Boidot, R., Vegran, F., Meulle, A., Le Breton, A., Dessy, C., Sonveaux, P., et al. (2012). Regulation of monocarboxylate transporter MCT1 expression by p53 mediates inward and outward lactate fluxes in tumors. *Cancer Research, 72*, 939–948.

Campisi, J. (2013). Aging, cellular senescence, and cancer. *Annual Review of Physiology, 75*, 685–705.

Canter, J. A., Kallianpur, A. R., Parl, F. F., & Millikan, R. C. (2005). Mitochondrial DNA G10398A polymorphism and invasive breast cancer in African-American women. *Cancer Research, 65*, 8028–8033.

Cao, J., Schulte, J., Knight, A., Leslie, N. R., Zagozdzon, A., Bronson, R., et al. (2009). Prdx1 inhibits tumorigenesis via regulating PTEN/AKT activity. *The EMBO Journal, 28*, 1505–1517.

Capparelli, C., Chiavarina, B., Whitaker-Menezes, D., Pestell, T. G., Pestell, R. G., Hulit, J., et al. (2012). CDK inhibitors (p16/p19/p21) induce senescence and autophagy in cancer-associated fibroblasts, "fueling" tumor growth via paracrine interactions, without an increase in neo-angiogenesis. *Cell Cycle, 11*, 3599–3610.

Capparelli, C., Guido, C., Whitaker-Menezes, D., Bonuccelli, G., Balliet, R., Pestell, T. G., et al. (2012). Autophagy and senescence in cancer-associated fibroblasts metabolically supports tumor growth and metastasis via glycolysis and ketone production. *Cell Cycle, 11*, 2285–2302.

Capparelli, C., Whitaker-Menezes, D., Guido, C., Balliet, R., Pestell, T. G., Howell, A., et al. (2012). CTGF drives autophagy, glycolysis and senescence in cancer-associated fibroblasts via HIF1 activation, metabolically promoting tumor growth. *Cell Cycle, 11*, 2272–2284.

Chiavarina, B., Whitaker-Menezes, D., Migneco, G., Martinez-Outschoorn, U. E., Pavlides, S., Howell, A., et al. (2010). HIF1-alpha functions as a tumor promoter in cancer associated fibroblasts, and as a tumor suppressor in breast cancer cells: Autophagy drives compartment-specific oncogenesis. *Cell Cycle, 9*, 3534–3551.

Comito, G., Giannoni, E., Di Gennaro, P., Segura, C. P., Gerlini, G., & Chiarugi, P. (2012). Stromal fibroblasts synergize with hypoxic oxidative stress to enhance melanoma aggressiveness. *Cancer Letters, 324*, 31–41.

Coppe, J. P., Patil, C. K., Rodier, F., Krtolica, A., Beausejour, C. M., Parrinello, S., et al. (2010). A human-like senescence-associated secretory phenotype is conserved in mouse cells dependent on physiological oxygen. *PLoS One*, *5*, e9188.

Czarnecka, A. M., Krawczyk, T., Zdrozny, M., Lubinski, J., Arnold, R. S., Kukwa, W., et al. (2010). Mitochondrial NADH-dehydrogenase subunit 3 (ND3) polymorphism (A10398G) and sporadic breast cancer in Poland. *Breast Cancer Research and Treatment*, *121*, 511–518.

Edderkaoui, M., Nitsche, C., Zheng, L., Pandol, S. J., Gukovsky, I., & Gukovskaya, A. S. (2011). NADPH oxidase activation in pancreatic cancer cells is mediated through Akt-dependent up-regulation of p22phox. *The Journal of Biological Chemistry*, *286*, 7779–7787.

Felty, Q. (2011). Redox sensitive Pyk2 as a target for therapeutics in breast cancer. *Frontiers in Bioscience*, *16*, 568–577.

Fu, X., Kassim, S. Y., Parks, W. C., & Heinecke, J. W. (2001). Hypochlorous acid oxygenates the cysteine switch domain of pro-matrilysin (MMP-7). A mechanism for matrix metalloproteinase activation and atherosclerotic plaque rupture by myeloperoxidase. *The Journal of Biological Chemistry*, *276*, 41279–41287.

Fujiwara, Y., Sugita, Y., Nakamori, S., Miyamoto, A., Shiozaki, K., Nagano, H., et al. (2000). Assessment of Stanniocalcin-1 mRNA as a molecular marker for micrometastases of various human cancers. *International Journal of Oncology*, *16*, 799–804.

Goh, J., Enns, L., Fatemie, S., Hopkins, H., Morton, J., Pettan-Brewer, C., et al. (2011). Mitochondrial targeted catalase suppresses invasive breast cancer in mice. *BMC Cancer*, *11*, 191.

Gonzalez-Angulo, A. M., Ferrer-Lozano, J., Stemke-Hale, K., Sahin, A., Liu, S., Barrera, J. A., et al. (2011). PI3K pathway mutations and PTEN levels in primary and metastatic breast cancer. *Molecular Cancer Therapeutics*, *10*, 1093–1101.

Graham, K. A., Kulawiec, M., Owens, K. M., Li, X., Desouki, M. M., Chandra, D., et al. (2010). NADPH oxidase 4 is an oncoprotein localized to mitochondria. *Cancer Biology & Therapy*, *10*, 223–231.

Haviv, I., Polyak, K., Qiu, W., Hu, M., & Campbell, I. (2009). Origin of carcinoma associated fibroblasts. *Cell Cycle*, *8*, 589–595.

Hecker, L., Vittal, R., Jones, T., Jagirdar, R., Luckhardt, T. R., Horowitz, J. C., et al. (2009). NADPH oxidase-4 mediates myofibroblast activation and fibrogenic responses to lung injury. *Nature Medicine*, *15*, 1077–1081.

Heering, J., Erlmann, P., & Olayioye, M. A. (2009). Simultaneous loss of the DLC1 and PTEN tumor suppressors enhances breast cancer cell migration. *Experimental Cell Research*, *315*, 2505–2514.

Hiscox, S., Barrett-Lee, P., & Nicholson, R. I. (2011). Therapeutic targeting of tumor-stroma interactions. *Expert Opinion on Therapeutic Targets*, *15*, 609–621.

Hyoudou, K., Nishikawa, M., Kobayashi, Y., Ikemura, M., Yamashita, F., & Hashida, M. (2008). SOD derivatives prevent metastatic tumor growth aggravated by tumor removal. *Clinical & Experimental Metastasis*, *25*, 531–536.

Jensen, J. D., Knoop, A., Laenkholm, A. V., Grauslund, M., Jensen, M. B., Santoni-Rugiu, E., et al. (2012). PIK3CA mutations, PTEN, and pHER2 expression and impact on outcome in HER2-positive early-stage breast cancer patients treated with adjuvant chemotherapy and trastuzumab. *Annals of Oncology*, *23*, 2034–2042.

Kalluri, R., & Zeisberg, M. (2006). Fibroblasts in cancer. *Nature Reviews. Cancer*, *6*, 392–401.

Kankuri, E., Babusikova, O., Hlubinova, K., Salmenpera, P., Boccaccio, C., Lubitz, W., et al. (2008). Fibroblast nemosis arrests growth and induces differentiation of human leukemia cells. *International Journal of Cancer*, *122*, 1243–1252.

Kankuri, E., Cholujova, D., Comajova, M., Vaheri, A., & Bizik, J. (2005). Induction of hepatocyte growth factor/scatter factor by fibroblast clustering directly promotes tumor cell invasiveness. *Cancer Research*, *65*, 9914–9922.

Kim, H. S., Patel, K., Muldoon-Jacobs, K., Bisht, K. S., Aykin-Burns, N., Pennington, J. D., et al. (2010). SIRT3 is a mitochondria-localized tumor suppressor required for maintenance of mitochondrial integrity and metabolism during stress. *Cancer Cell*, *17*, 41–52.

Koch, S., Volkmar, C. M., Kolb-Bachofen, V., Korth, H. G., Kirsch, M., Horn, A. H., et al. (2009). A new redox-dependent mechanism of MMP-1 activity control comprising reduced low-molecular-weight thiols and oxidizing radicals. *Journal of Molecular Medicine*, *87*, 261–272.

Koppenol, W. H., Bounds, P. L., & Dang, C. V. (2011). Otto Warburg's contributions to current concepts of cancer metabolism. *Nature Reviews. Cancer*, *11*, 325–337.

Krtolica, A., Parrinello, S., Lockett, S., Desprez, P. Y., & Campisi, J. (2001). Senescent fibroblasts promote epithelial cell growth and tumorigenesis: A link between cancer and aging. *Proceedings of the National Academy of Sciences of the United States of America*, *98*, 12072–12077.

Lin, C. H., & Lin, P. H. (2006). Induction of ROS formation, poly(ADP-ribose) polymerase-1 activation, and cell death by PCB126 and PCB153 in human T47D and MDA-MB-231 breast cancer cells. *Chemico-Biological Interactions*, *162*, 181–194.

Lisanti, M. P., Scherer, P. E., Vidugiriene, J., Tang, Z., Hermanowski-Vosatka, A., Tu, Y. H., et al. (1994). Characterization of caveolin-rich membrane domains isolated from an endothelial-rich source: Implications for human disease. *The Journal of Cell Biology*, *126*, 111–126.

Loeffler, M., Kruger, J. A., Niethammer, A. G., & Reisfeld, R. A. (2006). Targeting tumor-associated fibroblasts improves cancer chemotherapy by increasing intratumoral drug uptake. *The Journal of Clinical Investigation*, *116*, 1955–1962.

Martinez-Outschoorn, U. E., Balliet, R. M., Rivadeneira, D. B., Chiavarina, B., Pavlides, S., Wang, C., et al. (2010). Oxidative stress in cancer associated fibroblasts drives tumor-stroma co-evolution: A new paradigm for understanding tumor metabolism, the field effect and genomic instability in cancer cells. *Cell Cycle*, *9*, 3256–3276.

Martinez-Outschoorn, U. E., Goldberg, A., Lin, Z., Ko, Y. H., Flomenberg, N., Wang, C., et al. (2011). Anti-estrogen resistance in breast cancer is induced by the tumor microenvironment and can be overcome by inhibiting mitochondrial function in epithelial cancer cells. *Cancer Biology & Therapy*, *12*, 924–938.

Martinez-Outschoorn, U. E., Lin, Z., Ko, Y. H., Goldberg, A. F., Flomenberg, N., Wang, C., et al. (2011). Understanding the metabolic basis of drug resistance: Therapeutic induction of the Warburg effect kills cancer cells. *Cell Cycle*, *10*, 2521–2528.

Martinez-Outschoorn, U. E., Lin, Z., Trimmer, C., Flomenberg, N., Wang, C., Pavlides, S., et al. (2011). Cancer cells metabolically "fertilize" the tumor microenvironment with hydrogen peroxide, driving the Warburg effect: Implications for PET imaging of human tumors. *Cell Cycle*, *10*, 2504–2520.

Martinez-Outschoorn, U. E., Pavlides, S., Whitaker-Menezes, D., Daumer, K. M., Milliman, J. N., Chiavarina, B., et al. (2010). Tumor cells induce the cancer associated fibroblast phenotype via caveolin-1 degradation: Implications for breast cancer and DCIS therapy with autophagy inhibitors. *Cell Cycle*, *9*, 2423–2433.

Martinez-Outschoorn, U. E., Trimmer, C., Lin, Z., Whitaker-Menezes, D., Chiavarina, B., Zhou, J., et al. (2010). Autophagy in cancer associated fibroblasts promotes tumor cell survival: Role of hypoxia, HIF1 induction and NFkappaB activation in the tumor stromal microenvironment. *Cell Cycle*, *9*, 3515–3533.

Marumo, T., Schini-Kerth, V. B., Fisslthaler, B., & Busse, R. (1997). Platelet-derived growth factor-stimulated superoxide anion production modulates activation of transcription factor NF-kappaB and expression of monocyte chemoattractant protein 1 in human aortic smooth muscle cells. *Circulation*, *96*, 2361–2367.

Meng, D., Lv, D. D., & Fang, J. (2008). Insulin-like growth factor-I induces reactive oxygen species production and cell migration through Nox4 and Rac1 in vascular smooth muscle cells. *Cardiovascular Research*, *80*, 299–308.

Neumann, C. A., Cao, J., & Manevich, Y. (2009). Peroxiredoxin 1 and its role in cell signaling. *Cell Cycle, 8*, 4072–4078.

Neumann, C. A., Krause, D. S., Carman, C. V., Das, S., Dubey, D. P., Abraham, J. L., et al. (2003). Essential role for the peroxiredoxin Prdx1 in erythrocyte antioxidant defence and tumour suppression. *Nature, 424*, 561–565.

Nguyen, A., Chang, A. C., & Reddel, R. R. (2009). Stanniocalcin-1 acts in a negative feedback loop in the prosurvival ERK1/2 signaling pathway during oxidative stress. *Oncogene, 28*, 1982–1992.

Nishikawa, M., Hyoudou, K., Kobayashi, Y., Umeyama, Y., Takakura, Y., & Hashida, M. (2005). Inhibition of metastatic tumor growth by targeted delivery of antioxidant enzymes. *Journal of Controlled Release, 109*, 101–107.

Noh, E. M., Lee, Y. R., Chay, K. O., Chung, E. Y., Jung, S. H., Kim, J. S., et al. (2011). Estrogen receptor alpha induces down-regulation of PTEN through PI3-kinase activation in breast cancer cells. *Molecular Medicine Reports, 4*, 215–219.

Parrinello, S., Coppe, J. P., Krtolica, A., & Campisi, J. (2005). Stromal-epithelial interactions in aging and cancer: Senescent fibroblasts alter epithelial cell differentiation. *Journal of Cell Science, 118*, 485–496.

Pavlides, S., Whitaker-Menezes, D., Castello-Cros, R., Flomenberg, N., Witkiewicz, A. K., Frank, P. G., et al. (2009). The reverse Warburg effect: Aerobic glycolysis in cancer associated fibroblasts and the tumor stroma. *Cell Cycle, 8*, 3984–4001.

Pezzotti, A., Kraft, P., Hankinson, S. E., Hunter, D. J., Buring, J., & Cox, D. G. (2009). The mitochondrial A10398G polymorphism, interaction with alcohol consumption, and breast cancer risk. *PLoS One, 4*, e5356.

Provenzano, P. P., Inman, D. R., Eliceiri, K. W., Knittel, J. G., Yan, L., Rueden, C. T., et al. (2008). Collagen density promotes mammary tumor initiation and progression. *BMC Medicine, 6*, 11.

Radisky, D. C., Levy, D. D., Littlepage, L. E., Liu, H., Nelson, C. M., Fata, J. E., et al. (2005). Rac1b and reactive oxygen species mediate MMP-3-induced EMT and genomic instability. *Nature, 436*, 123–127.

Rohan, T. E., Wong, L. J., Wang, T., Haines, J., & Kabat, G. C. (2010). Do alterations in mitochondrial DNA play a role in breast carcinogenesis? *Journal of Oncology*, 604304.

Russillo, M., Di Benedetto, A., Metro, G., Ferretti, G., Papaldo, P., Cognetti, F., et al. (2011). Assessment of PTEN and PI3K status in primary breast cancer and corresponding metastases: Is it worthwhile? *Journal of Clinical Oncology, 29*, 2834–2835 author reply 2835.

Sampson, N., Koziel, R., Zenzmaier, C., Bubendorf, L., Plas, E., Jansen-Durr, P., et al. (2011). ROS signaling by NOX4 drives fibroblast-to-myofibroblast differentiation in the diseased prostatic stroma. *Molecular Endocrinology, 25*, 503–515.

Sappino, A. P., Skalli, O., Jackson, B., Schurch, W., & Gabbiani, G. (1988). Smooth-muscle differentiation in stromal cells of malignant and non-malignant breast tissues. *International Journal of Cancer, 41*, 707–712.

Sharpless, N. E., Ramsey, M. R., Balasubramanian, P., Castrillon, D. H., & DePinho, R. A. (2004). The differential impact of p16(INK4a) or p19(ARF) deficiency on cell growth and tumorigenesis. *Oncogene, 23*, 379–385.

Shen, J., Platek, M., Mahasneh, A., Ambrosone, C. B., & Zhao, H. (2010). Mitochondrial copy number and risk of breast cancer: A pilot study. *Mitochondrion, 10*, 62–68.

Siegel, R., Naishadham, D., & Jemal, A. (2012). Cancer statistics, 2012. *CA: A Cancer Journal for Clinicians, 62*, 10–29.

Sonveaux, P., Vegran, F., Schroeder, T., Wergin, M. C., Verrax, J., Rabbani, Z. N., et al. (2008). Targeting lactate-fueled respiration selectively kills hypoxic tumor cells in mice. *The Journal of Clinical Investigation, 118*, 3930–3942.

Sotgia, F., Martinez-Outschoorn, U. E., Howell, A., Pestell, R. G., Pavlides, S., & Lisanti, M. P. (2012). Caveolin-1 and cancer metabolism in the tumor microenvironment: Markers, models, and mechanisms. *Annual Review of Pathology*, *7*, 423–467.

Sotgia, F., Martinez-Outschoorn, U. E., & Lisanti, M. P. (2011). Mitochondrial oxidative stress drives tumor progression and metastasis: Should we use antioxidants as a key component of cancer treatment and prevention? *BMC Medicine*, *9*, 62.

Stolarek, R. A., Potargowicz, E., Seklewska, E., Jakubik, J., Lewandowski, M., Jeziorski, A., et al. (2010). Increased H_2O_2 level in exhaled breath condensate in primary breast cancer patients. *Journal of Cancer Research and Clinical Oncology*, *136*, 923–930.

Suda, T., Oba, H., Takei, H., Kurosumi, M., Hayashi, S., & Yamaguchi, Y. (2012). ER-activating ability of breast cancer stromal fibroblasts is regulated independently of alteration of TP53 and PTEN tumor suppressor genes. *Biochemical and Biophysical Research Communications*, *428*, 259–263.

Taddei, M. L., Giannoni, E., Raugei, G., Scacco, S., Sardanelli, A. M., Papa, S., et al. (2012). Mitochondrial oxidative stress due to complex I dysfunction promotes fibroblast activation and melanoma cell invasiveness. *Journal of Signal Transduction*, *2012*, 684592.

Tanic, N., Milovanovic, Z., Dzodic, R., Juranic, Z., Susnjar, S., Plesinac-Karapandzic, V., et al. (2012). The impact of PTEN tumor suppressor gene on acquiring resistance to tamoxifen treatment in breast cancer patients. *Cancer Biology & Therapy*, *13*, 1165–1174.

Tobar, N., Guerrero, J., Smith, P. C., & Martinez, J. (2010). NOX4-dependent ROS production by stromal mammary cells modulates epithelial MCF-7 cell migration. *British Journal of Cancer*, *103*, 1040–1047.

Toullec, A., Gerald, D., Despouy, G., Bourachot, B., Cardon, M., Lefort, S., et al. (2010). Oxidative stress promotes myofibroblast differentiation and tumour spreading. *EMBO Molecular Medicine*, *2*, 211–230.

Trimboli, A. J., Cantemir-Stone, C. Z., Li, F., Wallace, J. A., Merchant, A., Creasap, N., et al. (2009). Pten in stromal fibroblasts suppresses mammary epithelial tumours. *Nature*, *461*, 1084–1091.

Trimmer, C., Sotgia, F., Whitaker-Menezes, D., Balliet, R. M., Eaton, G., Martinez-Outschoorn, U. E., et al. (2011). Caveolin-1 and mitochondrial SOD2 (MnSOD) function as tumor suppressors in the stromal microenvironment: A new genetically tractable model for human cancer associated fibroblasts. *Cancer Biology & Therapy*, *11*, 383–394.

Waghray, M., Cui, Z., Horowitz, J. C., Subramanian, I. M., Martinez, F. J., Toews, G. B., et al. (2005). Hydrogen peroxide is a diffusible paracrine signal for the induction of epithelial cell death by activated myofibroblasts. *The FASEB Journal*, *19*, 854–856.

Weitzman, J. B. (2001). p16(Ink4a) and p19(Arf): Terrible twins. *Trends in Molecular Medicine*, *7*, 489.

Whitaker-Menezes, D., Martinez-Outschoorn, U. E., Flomenberg, N., Birbe, R. C., Witkiewicz, A. K., Howell, A., et al. (2011). Hyperactivation of oxidative mitochondrial metabolism in epithelial cancer cells in situ: Visualizing the therapeutic effects of metformin in tumor tissue. *Cell Cycle*, *10*, 4047–4064.

Whitaker-Menezes, D., Martinez-Outschoorn, U. E., Lin, Z., Ertel, A., Flomenberg, N., Witkiewicz, A. K., et al. (2011). Evidence for a stromal-epithelial "lactate shuttle" in human tumors: MCT4 is a marker of oxidative stress in cancer-associated fibroblasts. *Cell Cycle*, *10*, 1772–1783.

Witkiewicz, A. K., Dasgupta, A., Sammons, S., Er, O., Potoczek, M. B., Guiles, F., et al. (2010). Loss of stromal caveolin-1 expression predicts poor clinical outcome in triple negative and basal-like breast cancers. *Cancer Biology & Therapy*, *10*, 135–143.

Witkiewicz, A. K., Kline, J., Queenan, M., Brody, J. R., Tsirigos, A., Bilal, E., et al. (2011). Molecular profiling of a lethal tumor microenvironment, as defined by stromal caveolin-1 status in breast cancers. *Cell Cycle, 10*, 1794–1809.

Witkiewicz, A. K., Whitaker-Menezes, D., Dasgupta, A., Philp, N. J., Lin, Z., Gandara, R., et al. (2012). Using the "reverse Warburg effect" to identify high-risk breast cancer patients: Stromal MCT4 predicts poor clinical outcome in triple-negative breast cancers. *Cell Cycle, 11*, 1108–1117.

CHAPTER FOUR

Mitochondrial Alterations During Carcinogenesis: A Review of Metabolic Transformation and Targets for Anticancer Treatments

Xiao Wang[*], Susana Peralta[†], Carlos T. Moraes[*,†,‡,1]
[*]Graduate Program in Cancer Biology, University of Miami Miller School of Medicine, Miami, Florida, USA
[†]Department of Neurology, University of Miami Miller School of Medicine, Miami, Florida, USA
[‡]Department of Cell Biology, University of Miami Miller School of Medicine, Miami, Florida, USA
[1]Corresponding author: e-mail address: cmoraes@med.miami.edu

Contents

Advances in Cancer Research, Volume 119
ISSN 0065-230X
http://dx.doi.org/10.1016/B978-0-12-407190-2.00004-6

Abstract

Mitochondria play important roles in multiple cellular processes including energy metabolism, cell death, and aging. Regulated energy production and utilization are critical in maintaining energy homeostasis in normal cells and functional organs. However, mitochondria go through a series of morphological and functional alterations during carcinogenesis. The metabolic profile in transformed cells is altered to accommodate their fast proliferation, confer resistance to cell death, or facilitate metastasis. These transformations also provide targets for anticancer treatment at different levels. In this review, we discuss the major modifications in cell metabolism during carcinogenesis, including energy metabolism, apoptotic and autophagic cell death, adaptation of tumor microenvironment, and metastasis. We also summarize some of the main metabolic targets for treatments.

1. AN OVERVIEW OF THE ROLE OF MITOCHONDRIA IN CANCER

After decades of extensive research and conceptual progress in cancer cell biology, malignant transformation has been summarized into eight essential alteration hallmarks in cell physiology: self-sufficient growth signaling, evasion from growth inhibition, invasion and metastasis, unlimited replication capacity, angiogenesis signaling, resistance to cell death, shifting in cellular energetics, and avoiding detection by immune surveillance (Hanahan & Weinberg, 2011). Genome instability and mutations are underlying characteristics that enable these transformations during multiple stages of carcinogenesis. In addition, the presence of tumor-associated inflammatory response in the tumor microenvironment, consisting of cancer cells as well as normal cells, has been shown to enhance tumorigenesis and progression by facilitating incipient neoplasias to acquire characteristic transformations (Hanahan & Weinberg, 2011).

Mitochondria are the power plant of the cell. Regulated energy production and utilization are crucial for maintaining energy homeostasis in normal cells and functional organisms. In the 1920s, Warburg described the phenomenon that transformed cells produce their energy predominantly by a high rate of glycolysis followed by lactic acid fermentation in the cytosol (Warburg, Wind, & Negelein, 1927). This is in contrast to "normal" cells that have a relatively lower rate of glycolysis followed by oxidation of the end product, pyruvate, in mitochondria. However, the Warburg effect does not represent the fundamental difference between normal and cancer cells; it is rather the consequence of a metabolic shift in dividing cells. In quiescent state, cells maintain a basal rate of glycolysis, converting glucose to pyruvate, which is then oxidized to carbon dioxide through tricarboxylic acid (TCA) cycle in the mitochondria. As a result, the majority of ATP is generated by oxidative phosphorylation that requires an aerobic environment. On the contrary, in proliferating cells, glycolytic flux is largely increased to produce ATP rapidly in the cytoplasm. The resulting pyruvate is either converted into lactate or used for biosynthesis of macromolecules such as nucleic acids, fatty acids, and amino acids (DeBerardinis, Lum, Hatzivassiliou, & Thompson, 2008a).

The Warburg effect initially led to the hypothesis that impaired mito-chondrial metabolism could be the cause of cancer. During several decades of research afterward, the concept that cancer cells switch to fermentation instead of oxidative respiration has been widely accepted, even though it is no longer deemed as the pivotal cause of cancer. It is now believed that mutations in oncogenes and tumor suppressor genes are responsible for the malignant transformations (Bertram, 2000). It has also been established that cancer cell can reprogram their energy production, by limiting their energy metabolism largely to glycolysis. This switch from oxidative respiration to glycolysis is beneficial to cancer cells in that it accommodates energy and substrates required for the fast proliferation.

Besides its role in ATP production, mitochondria are also the major source of reactive oxygen species (ROS) and crucial for intracellular Ca^{2+} homeostasis, both important for tumor cell physiology, cell growth, and survival (Gogvadze, Orrenius, & Zhivotovsky, 2008).

Mitochondria are also key organelles involved in apoptotic and auto-phagic cell death. Deregulation of apoptosis and/or autophagy has been implicated in various pathological conditions including cancer. Stimuli trig-gering apoptosis converge on mitochondria, where signals are coordinated, resulting in the release of cytochrome c (Cyt c) from the mitochondrial intermembrane space (IMS) into cytoplasm and activation of downstream effector caspases (Desagher & Martinou, 2000). Cancer cells evade apoptosis by gain-of-function mutation of oncogenes and loss-of-function mutation of tumor suppressor genes. These modifications modulate the permeability of mitochondrial membrane and control the release of Cyt c (Kilbride & Prehn, 2012).

Autophagy is a lysosome-based cellular catabolic process where the cell degrades unnecessary or dysfunctional components, which is essential for cell survival during starvation and organelle quality control (Mizushima,

2007). Selective elimination of mitochondria, or mitophagy, is an important quality control that eliminates aged and damaged mitochondria, as well as mitochondria removal during erythrocyte maturation (Youle & Narendra, 2011). It has been shown that oncogene-mediated mitophagy promoted cell survival during early tumorigenesis before the formation of hypoxia, as a key strategy for cancer cells to overcome energy deficit due to insufficient glucose and to protect from the release of proapoptotic proteins (Kim et al., 2011). However, the role of autophagy in tumor progression is differential and probably depends on tumor types and stages. Suppressing autophagy in apoptosis-defective tumor cells could lead to necrotic cell death associated with inflammation and accelerated tumor growth (Degenhardt et al., 2006). Currently, cancer treatments targeting mitophagy are still under extensive studies.

As mentioned earlier, alterations in mitochondrial function have a clear impact on the physiology of tumor cells. In this review, we focus on the mitochondrial function in cancer cells, in particular on its role in the reprogramming of energy metabolism and during cell survival versus cell death signaling.

2. ALTERATIONS IN ENERGY METABOLISM IN CANCER CELLS

Most tumor cells show high rates of glycolysis and lactate production, a phenomenon known as "Warburg effect." Why does the Warburg effect occur? It could be simply resulted from an adaptation of the fast proliferating cancer cells to the low-oxygen environment (hypoxia) within a solid tumor, or mutated oncogenes (*Ras*, *c-Myc*)/tumor suppressors (*p53*) repressing mitochondrial function (DeBerardinis, Lum, Hatzivassiliou, & Thompson, 2008b; Jones & Thompson, 2009). The shift toward glycolysis is achieved by the increased expression of glycolytic enzymes and glucose transporters (GLUTs) and by inhibition of mitochondrial metabolism. It is known that the yield of ATP per glucose consumed by glycolysis is lower than the one produced from oxidative phosphorylation. However, tumor cells compensate for this low efficiency in energy production by increasing glucose import and glycolytic enzymes (Guppy, Greiner, & Brand, 1993; Warburg, 1956). Moreover, the glycolytic metabolism, found in proliferating cells besides tumor cells, provides cells with intermediates needed for biosynthetic pathways. Glucose degradation provides ribose sugars for nucleotides; glycerol and citrate for lipids and nonessential amino acids; and through the oxidative pentose phosphate pathway, nicotinamide adenine dinucleotide phosphate (NADPH).

In tumor cells, most of the glycolysis end product, pyruvate, is converted to lactate through the activity of the enzyme lactate dehydrogenase A (LDH-A) and is subsequently secreted (DeBerardinis et al., 2008b). By converting pyruvate to lactate, LDH-A recovers the $NAD^+/NADH$ ratio needed to maintain glycolysis. This step has been shown to be critical for tumor proliferation *in vivo* (Fantin, St-Pierre, & Leder, 2006). Oxidation of pyruvate in the mitochondria requires its import followed by activity of highly regulated enzymes such as the pyruvate dehydrogenase (PDH) complex. Because glycolytic rate is so high in tumor cells, it has been postulated that it could outpace the capacity of pyruvate oxidation in the mitochondria. Tumor cells also increase lactate production, which is able to be excreted, to avoid pyruvate accumulation inside the cells. Accordingly, LDH-A is induced by oncogenes (*c-Myc*, *HER2/neu*, and others) (Fantin et al., 2006; Shim et al., 1997), and by the hypoxia-inducible factor 1, HIF-1 (Semenza et al., 1996).

Pyruvate inside the mitochondria is oxidized in the TCA cycle, which connects glucose metabolism in the cytosol to oxidative phosphorylation in the mitochondria. Genes encoding for enzymes of the TCA cycle have been described as tumor suppressors and found mutated in different types of cancer (Gottlieb & Tomlinson, 2005). These genes include the succinate dehydrogenase (SDH) genes, *SDHB*, *SDHC*, and *SDHD*, and the fumarate hydratase (*FH*) gene. The accumulation of succinate and fumarate due to SDH and FH deficiency, respectively, was shown to inhibit HIF-1α prolyl hydroxylases (PHDs) in the cytosol, leading to stabilization and activation of HIF-1α (Selak et al., 2005). These result in the transcription of a number of genes involved in enhanced glycolysis and angiogenesis. Thus, when mitochondrial respiration in tumor cells is downregulated, accumulation of the tricarboxylic cycle substrates might serve as a signal for increase of glycolysis promoting the Warburg effect. Moreover, mutations in the TCA enzymes isocitrate dehydrogenase 1 and 2 have been reported in glioblastoma and other human tumors (Yen, Bittinger, Su, & Fantin, 2010).

Currently, it is still under debate whether these metabolic changes are a consequence or a cause of tumorigenesis. However, the fact that metabolic genes such as *SDHs* and *FH* act as tumor suppressors in human strongly suggests that metabolic deregulation can initiate cancer processes.

2.1. Hypoxia and HIF-1-modulated alterations of cellular energy metabolism

In the case of inadequate oxygen supply, cancer cells survive by lowering the expression or mutating the tumor suppressor p53 (Moll & Schramm, 1998). Meanwhile, HIF-1 is stabilized and stimulates key steps of glycolysis to supply the cells with enough ATP (Wang, Jiang, Rue, & Semenza, 1995; Wang & Semenza, 1993). GLUTs are regulated in an isoenzyme-specific manner in response to HIF-1 induction. The high-affinity/low-capacity isoforms GLUT1 and GLUT3 are upregulated, while the low-affinity/high-capacity isoform GLUT2 is downregulated under hypoxic conditions (Ebert et al., 1996). Transcriptional upregulation by hypoxia is also observed for genes encoding other glycolytic enzymes, including phosphofructokinase-1 (PFK-1), aldolase A, pyruvate kinase muscle isozyme (PKM), triose-phosphate isomerase, and LDH-A (Semenza, Roth, Fang, & Wang, 1994).

Not only does HIF-1 stimulate glycolysis, but it also actively represses oxidative metabolism by directly inducing pyruvate dehydrogenase kinase 1 (PDK1), which phosphorylates and inhibits PDH from using pyruvate to fuel the mitochondrial TCA cycle (Kim, Tchernyshyov, Semenza, & Dang, 2006b; Papandreou, Cairns, Fontana, Lim, & Denko, 2006a). Besides, HIF-1 is shown to negatively regulate mitochondrial biogenesis and oxygen consumption by decreasing the expression of the transcriptional coactivator, peroxisome proliferator-activated receptor gamma coactivator-1beta (PGC-1β), which induces mitochondrial genes, in renal carcinoma cells lacking the von Hippel–Lindau (VHL) tumor suppressor (Zhang et al., 2007) (Fig. 4.1).

The level of HIF-1α protein is regulated through hydroxylation at the oxygen-dependent degradation (ODD) domain (Huang, Gu, Schau, & Bunn, 1998a). Two independent groups have demonstrated that the enzyme activity of HIF-1α PHDs is oxygen dependent and that HIF-1α PHDs serve as the oxygen sensors of the cell (Ivan et al., 2001; Jaakkola et al., 2001). Thus, PHDs activate HIF-1 degradation under normoxia but release HIF-1 complex in poorly oxygenated environment. The role of HIF-1α PHDs in tumor growth is complex because the outcome is mainly dependent on cancer-type-specific expression and the availability of PHD inhibitors secreted by the tumor (Jokilehto & Jaakkola, 2010). Besides, PHDs are also involved in signaling pathways other than HIF-1α regulation (Ameln et al., 2011). Clinically, overexpression of PHDs has been associated with

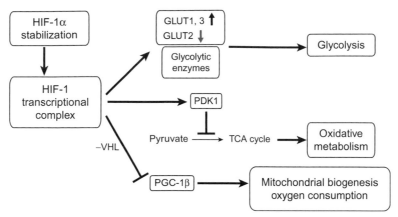

Figure 4.1 Roles of HIF-1 transcriptional complex in regulating mitochondrial function during carcinogenesis. Activated HIF-1 transcriptional complex increases glycolysis by upregulating glucose transportation and the expression of glycolytic enzymes. HIF-1 could prevent pyruvate from entering the TCA cycle for oxidative metabolism. Besides, HIF-1 also inhibits the expression of PGC-1β in the absence of tumor suppressor VHL, resulting in decreased mitochondrial biogenesis and oxygen consumption.

poor prognostics in non-small-cell lung carcinoma (NSCLC) patients (Andersen et al., 2011).

In addition to PHDs, mitochondria *per se* are also implicated as potential oxygen sensors by ROS production (Agani, Pichiule, Chavez, & LaManna, 2000). Whether ROS is involved in HIF-1 activation is equivocal. Some studies show that ROS generated at mitochondrial Complex III is regulated by oxygen tension that it stabilizes HIF-1α and increases HIF-1-dependent transcription. This regulation of HIF-1 is not observed in rho0 (ρ0) cells where mtDNA is depleted, since there is no functional electron transport chain (ETC) to generate ROS (Chandel et al., 1998, 2000). However, there is also a growing body of evidence showing that ρ0 cells are still able to stabilize HIF-1α protein under hypoxia (Enomoto, Koshikawa, Gassmann, Hayashi, & Takenaga, 2002), and that HIF-1α accumulation in low-oxygen conditions is independent of mitochondrial ROS production (Chua et al., 2010). In either case, activation of HIF-1 transcription complex is implicated in multiple carcinogenic events such as cell proliferation, metastasis, and neovascularization.

2.2. The mitochondrial respiratory chain complexes in cancer

The 16-kb mitochondrial genome only encodes for 13 mitochondrial proteins, which compose the catalytic cores of the oxidative phosphorylation complexes, while the other subunits are encoded by the nuclear DNA. Complex I (NADH-ubiquinone oxidoreductase) has 39 nuclear-encoded subunits and 7 mitochondrial-encoded subunits. A point mutation T14634C in *ND6* gene was identified in human glioma cell lines and predicted to alter the structure and orientation of the transmembrane helices of the ND6 protein, causing deregulation of proton flux and contributing to resistance to chemotherapy that requires redox cycling for activation (DeHaan et al., 2004). Besides, Complex I deficiency was linked with tumorigenesis in salivary gland, thyroid, and renal oncocytomas. Studies have revealed several types of mtDNA mutations that would disrupt the expression of genes encoding Complex I subunits (Gasparre et al., 2007; Mayr et al., 2008). As a result, oncocytomas usually present with densely accumulated mitochondria due to compensating the loss of Complex I activity and assembly (Simonnet et al., 2003). Moreover, Complex I dysfunction can also result in ROS production. In fibroblasts carrying Complex I mutations, HIF-1α is stabilized and a panel of downstream targets including VEGF-A (vascular endothelial growth factor-A) and HGF (hepatocyte growth factor) are elevated, leading to enhanced mobility and invasiveness (Taddei et al., 2012).

All four subunits of Complex II (succinate-ubiquinone oxidoreductase) are encoded by nuclear genes. Mutations in genes coding for subunits of mitochondrial Complex II have been identified in hereditary and sporadic paraganglioma (Baysal et al., 2000). The mutations could abolish Complex II activity and activate the hypoxia pathway (Dekker et al., 2003; Gimenez-Roqueplo et al., 2001). As discussed earlier, SDH defects lead to stabilization and activation of HIF-1α, prompting glycolysis and the Warburg effect (Selak et al., 2005).

Complex III (UQCR, ubiquinol-cytochrome *c* oxidoreductase) contains 10 nuclear-encoded subunits and 1 mitochondrial-encoded protein, cytochrome *b*. The function of Complex III, besides transferring electrons to Cyt *c*, has been shown to play a role in hypoxia-induced ROS production and oxygen sensing. Terpestacin, a small molecule that binds to and inactivates Complex III subunit UQCR, was shown to reduce ROS production and destabilize HIF-1α, providing a novel venue to selectively target angiogenesis and tumor progression (Jung et al., 2010). *In vitro* and *in vivo* studies have identified impaired Complex III activity associated with highly metastatic breast cancer (Owens, Kulawiec, Desouki, Vanniarajan, & Singh, 2011).

Complex IV (COX, cytochrome c oxidase) is composed of 10 nuclear-encoded subunits and 3 mitochondrial-encoded subunits: Cox1, Cox2, and Cox3. This is the last enzyme, and proposed to be the rate-limiting step, in the respiratory chain that pumps proton from mitochondrial matrix to the IMS, building up an electrochemical potential used for ATP synthesis (Herrmann et al., 2003). It is worth pointing out that COX composition and thus activity are regulated by oxygen levels and oxidative stress as well. This regulation is achieved through allosteric regulation, isoform expression, and regulation through cell signaling pathways. Multiple cell signaling pathways have been implicated in this process and manifested differently in the stress response of normal cells and carcinogenic transformation (Lee, Greeley, & Englander, 2008). It has been noted that an increase in the ratio of nuclear-encoded Cyt c oxidase subunits versus mitochondrial-encoded subunits commensurate with prostate cancer progression. This shift was not limited to prostate cancer but also found in ovarian, colon, and breast carcinomas, suggesting a potential new therapeutic target (Herrmann et al., 2003). More recently, COX subunit Va (COXVa) has been shown to be highly expressed in tumors and associated with distant metastasis in non–small-cell lung carcinoma (NSCLC) patients (Chen et al., 2012).

2.3. Suppression of mitochondrial respiration in cancer cells

The synthesis of ATP in the mitochondria is coupled with respiration. It is driven by the ATP synthase or Complex V, using as driving force the proton gradient generated by the electron respiratory chain complexes. The ATP synthase is a reversible engine that besides synthesizing ATP can hydrolyze ATP upon changes in cellular physiology. In cancer cells, one of the catalytic subunit of the mitochondrial ATP synthase (β-F1 ATPase) is significantly diminished and provides a bioenergetic signature of disease progression and of the response to chemotherapy (Cuezva et al., 2009). It has been shown that this repression is mediated by the ATPase inhibitory factor 1 (IF1), a physiological inhibitor of the ATP synthase complex. IF1 is expressed in mitochondria of normal human tissues at basal levels. In contrast, the mitochondria of prevalent human carcinomas have enormous amount of IF1 (Sanchez-Cenizo et al., 2010). The signaling pathway/mechanism whereby IF1 is upregulated in cancer cells remains unknown. However, clear evidence is showing that overexpression of IF1 in normal cells results in the inhibition of the ATP synthase and the switch to an increased aerobic glycolysis phenotype (Formentini, Sanchez-Arago, Sanchez-Cenizo, & Cuezva, 2012; Sanchez-Cenizo et al., 2010).

IF1-mediated inhibition of the ATP synthase indeed results in mitochondria hyperpolarization and the subsequent production of reactive oxygen species (ROS). ROS-mediated response in some cancer cells signals to the nucleus via NFκB and results in enhanced proliferation, invasion, and cell survival (Formentini et al., 2012). Therefore, in addition to the role of IF1 in the metabolic adaptation, the overexpression of IF1 in human cancer cells also triggers a retrograde signal to the nucleus to establish the appropriate adaptive cellular program needed for tumorigenesis and tumor progression.

Suppression of mitochondrial function in tumor cells can also be mediated by HIF-1 (Yen et al., 2010). The activity of the PDH enzyme, which is required for the metabolism of pyruvate inside the mitochondria, is controlled by PDK1. HIF-1 was shown to induce PDK1 and thereby inactivates PDH and mitochondrial respiration (Kim, Tchernyshyov, Semenza, & Dang, 2006a; Papandreou, Cairns, Fontana, Lim, & Denko, 2006b). HIF-1 was also shown to stimulate the expression of LDH-A, which facilitates conversion of pyruvate into lactate and therefore suppresses the mitochondrial respiration (Semenza et al., 1996).

In addition to what has been discussed earlier, mitochondrial respiration is also suppressed due to gain-of-function mutations of proto-oncogenes and loss-of-function modification of tumor suppressor genes in cancer cells. These changes collectively contribute to the process of carcinogenic transformation.

3. ONCOGENES AND TUMOR SUPPRESSORS INVOLVED IN MITOCHONDRIAL FUNCTION

Oncogenetic transformations involve a nonrandom set of gene deletions, amplifications, and mutations, and many oncogenes and tumor suppressor genes cluster along the signaling pathways that regulate c-Myc, HIF-1, and p53 (Yeung, Pan, & Lee, 2008). Although these transformations might be originally identified independently from mitochondrial dysfunction, accumulating evidence has suggested that the associated oncogenes are always, to certain extent, implicated in regulating mitochondrial metabolism and oxidative phosphorylation.

3.1. HIF-1α and the tumor suppressor VHL

HIF-1α is essential for the survival of transformed cells and plays a critical role in the regulation of glycolysis under hypoxia; it was also shown that HIF-1 transcription complex could also be stabilized and transactivate downstream gene expression in the presence of oxygen (Giaccia, Siim, & Johnson, 2003). The tumor suppressor *VHL* gene encodes an E3 ubiquitin–protein ligase. Mutations of the *VHL* gene are associated with VHL syndrome characterized by dominantly inherited hereditary predisposition to a variety of malignant and benign tumors in the nervous system, eye, kidney, and pancreas (Bader & Hsu, 2012). Under normal physiological conditions, the VHL protein binds to HIF-1α through the ODD domain in the presence of oxygen and targets HIF-1α to proteasome degradation. Loss of VHL tumor suppressor or deletion of the ODD domain can lead to insensitivity of HIF-1α to oxygen and stabilization under normoxia conditions (Huang, Gu, Schau, & Bunn, 1998b; Maxwell et al., 1999). As a result, the HIF-1 transcription complex is activated, and a group of downstream genes including *VEGF, PDGFB* (platelet-derived growth factor B), and genes involved in glucose uptake and metabolism are induced (Bader & Hsu, 2012). Besides VHL, HIF-1α could also be regulated through other mechanisms. It has been shown that lactate and pyruvate, the main metabolic products from aerobic glycolysis, regulated hypoxia-inducible gene expression under normoxia by stabilizing HIF-1α, activating the transcriptional complex HIF-1 DNA binding capacity, and enhancing the expression of several HIF-1 target genes (Lu, Forbes, & Verma, 2002).

3.2. p53

Many types of cancer have been characterized with mutation or loss of function of the p53 tumor suppressor. p53 was shown to repress the transcription of GLUTs, GLUT1 and GLUT4, and increase the transcription of TIGAR (Tumor protein p53-induced glycolysis and apoptosis regulator) to inhibit glycolysis and overall ROS production (Bensaad et al., 2006). The p53 protein was reported to have a direct effect on mitochondrial respiration by enhancing the transcription of SCO2 (synthesis of Cyt *c* oxidase 2) and increase the assembly of oxidative phosphorylation Complex IV (Matoba et al., 2006). Mutations of p53 in tumor cells lead to compromised oxidative phosphorylation due to Complex IV deficiency and a shift of cellular energy metabolism toward glycolysis. Recently, p53 was shown to be involved in the regulation of mitochondrial energy metabolism by the nuclear factor kappaB (NF-κB) transcriptional factor family. In the presence of p53, the interaction between NF-κB family member RelA and mitochondrial heat–shock protein (Hsp) mortalin/mtHsp70/Grp75 is blocked and RelA translocation to mitochondria is abolished. However, when p53 is absent, RelA enters mitochondria and represses mitochondrial gene expression, oxygen consumption, and cellular ATP synthesis (Johnson, Witzel, & Perkins, 2011).

In addition, the function of p53 could also be modulated through the transcriptional coactivator, PGC-1α. PGC-1α was first identified as the master regulator of mitochondrial biogenesis which coordinates and induces the expression of mitochondrial proteins (Puigserver et al., 1998). PGC-1α directly binds to p53 and enhances p53 transcriptional activity. Sen et al. showed that PGC-1α determines the specificity of p53-mediated transactivation of downstream genes, resulting in cell cycle arrest and responses upon metabolic stresses (Sen, Satija, & Das, 2011). p53 activation in turn binds and represses *PPARGC1A* and *PPARGC1B* (encoding *PGC-1α* and *PGC-1β*, respectively) promoters, which serves as a feedback mechanism to limit the p53 response, and may contribute to mitochondrial defects and ATP deficit in many pathological conditions including cancer (Sahin et al., 2011; Villeneuve et al., 2012).

3.3. c-Myc

c-Myc was found to be constitutively active in many types of cancer. It serves as a classical transcription factor as well as a global regulator of chromatin structure through histone acetylation (Cotterman et al., 2008). Li et al. demonstrated that c-Myc directly regulates the expression of mitochondrial transcription factor-A (TFAM), whose product is essential for mitochondrial DNA replication and gene transcription (Li et al., 2005). They found a tight association between c-Myc expression and mitochondrial load in the cell, which led to the hypothesis that c-Myc-induced genomic instability could result from the production of ROS (Vafa et al., 2002). ROS levels could be increased because of imbalanced expression of nuclear and mitochondrial-encoded proteins making defective respiratory chain complexes, or as a consequence of c-Myc stimulating abnormal mitochondria in transformed cells under local hypoxic conditions (KC, Carcamo, & Golde, 2006). In any case, clear evidence indicates oxidative damage in genomic DNA in response to c-Myc expression could be mediated by alterations in the mitochondria (Dang, Li, & Lee, 2005).

Deregulation of oncogene c-Myc was also shown to induce angiogenesis and promote metabolic switch in collaboration with HIF-1 (Kim, Gao, Liu,

Semenza, & Dang, 2007). Besides, c-Myc enhances the transcription of glutaminase-1 and upregulates glutamate production. It also stimulates transcription of ribosomal RNA and other genes, increasing the rate of protein synthesis of a cell (Dang, 1999; DeBerardinis et al., 2007). c-Myc also regulates glutaminolysis at the microRNA level by transcriptionally repressing miR-23a and miR-23b, which stabilize their target mitochondrial glutaminase-1 and upregulate glutamine catabolism and maintain homeostasis of reactive oxygen species (Gao et al., 2009). Interestingly, p53 was found to transcribe glutaminase-2, which enhances the rate of the TCA cycle and oxidative phosphorylation (Hu et al., 2010). Thus, the orchestration between these two glutaminases, regulated by c-Myc (glutaminase-1) and p53 (glutaminase-2), decides if a cell utilizes oxidative phosphorylation or glutamine metabolism, which controls the metabolic switch in transformed cells.

3.4. Oncogenic Ras

Ras represents a family of small GTPase proteins that orchestrate cell growth, differentiation, and survival. As a result, mutations causing constitutively active Ras can lead to malignant transformation. Oncogene *Ras* includes *H-Ras*, *N-Ras*, and *K-Ras*, and it is the most common oncogene in human cancer (Downward, 2003). It was reported in cultured human bronchial epithelial cells that introduction of active H-Ras promotes cell growth and increases TCA cycle reactions and oxygen consumption, along with a global upregulation of ETC activity (Telang, Lane, Nelson, Arumugam, & Chesney, 2007). Utilizing transcriptional profiling, Gaglio et al. detected an elevated expression of genes associated with glycolysis, glutamine metabolism, and nucleotide biosynthesis upon K-Ras-induced transformation (Gaglio et al., 2011). These results suggest a decoupling of glycolysis and TCA metabolism, with glutamine fueling the TCA cycle in cancer cells.

More recently, Guo et al. reported that an upregulation of basal autophagy is necessary for oncogenes *H-Ras-* and *K-Ras*-induced cell survival and tumor development *in vivo*. They found reduced oxidative metabolism and energy depletion in autophagy-deficient cells expressing Ras, due to accumulation of abnormal mitochondria. In addition, the downregulation of expression of essential autophagy proteins in cell lines with highly active *Ras* mutations impaired cell growth (Guo et al., 2011). Cancers with *Ras* mutations usually have poor prognosis. The reliance on autophagy in these types of cancer suggests that targeting autophagy and mitochondrial metabolism could become valuable new approaches to treat these malignant cancers.

3.5. Other oncogenes

Recent studies have also revealed other oncogenes with an effect on mitochondrial metabolism that are involved in carcinogenic transformation. The glycolytic enzyme, pyruvate kinase (PK), is exclusively expressed in the M2 isoform (PKM2) in self-renewing stem cells and most tumor cells (Mazurek, Boschek, Hugo, & Eigenbrodt, 2005). PKM2 inhibits PK reaction while diverting glycolytic substrates into alternative biosynthetic and reduced NADPH-generating pathways (Heiden, Cantley, & Thompson, 2009). It has been shown that expression of PKM2 is essential for aerobic glycolysis and provides selective growth advantage for tumorigenesis *in vivo* (Christofk et al., 2008). Meanwhile, *PKM2* gene transcription can be activated by HIF-1. The hydroxylation of PKM2 by the enzyme PHD3 further enhances its binding with HIF-1α and promotes transactivation of HIF-1-target genes, thus reprogramming glucose metabolism in carcinogenesis (Luo et al., 2011).

STAT3 (signal transducers and activators of transcription-3) is a pro-oncogenic transcription factor and is constitutively activated in a wide variety of tumors that often develop addiction to its activity. It was shown that STAT3 acts as a master regulator shifting cell metabolism toward aerobic glycolysis in a HIF-1α-dependent manner. They also proposed this metabolic transformation as the cause for tumor cells developing STAT3 addiction (Demaria et al., 2010).

Lee and colleagues reported a new function for Wnt/Snail signaling in tumor growth and progression. Wnt suppresses mitochondrial respiration and COX activity by inhibiting the expression of three COX subunits, namely, COXVIc, COXVIIa, and COXVIIc, whereas the expression of pyruvate carboxylase (PC), a key enzyme of anaplerosis, is induced. In addition, this group showed that the Wnt-induced mitochondrial repression and glycolytic switching occurred through the canonical β-catenin/T-cell factor 4/Snail pathway. These findings provided a new function for Wnt/Snail signaling in the regulation of mitochondrial respiration (via COX gene expression) and glucose metabolism (via PC gene expression) (Lee et al., 2012).

3.6. Mitochondrial tumor suppressors: SDH and FH

Both SDH and FH are TCA cycle enzymes that catalyze the reaction converting succinate to fumarate and fumarate to malate, successively. Mutations in *SDHB*, *SDHC*, or *SDHD*, and *FH* are associated with several types of hereditary and sporadic cancers: paraganglioma, pheochromocytoma, leiomyoma, and leiomyosarcoma, and renal cell carcinoma (Gottlieb & Tomlinson, 2005). In either case, loss of enzyme activity results in accumulation of metabolic intermediates succinate or fumarate, which could be transported to the cytosol and eventually stabilize HIF-1α even under normoxic conditions (Selak et al., 2005).

This "pseudohypoxia" phenotype has been described in several studies. Tumors with either SDH subunit or FH mutations showed an activation of the HIF-1 pathway and high vascularity (Gimenez-Roqueplo et al., 2001; Pollard et al., 2005). Recently, more detailed mechanistic studies by Tong and colleagues revealed a metabolic shift toward aerobic glycolysis in FH-deficient cells from patient with hereditary leiomyomatosis and renal cell carcinoma. It was found that FH inactivation caused decreases in AMP-activated kinase (AMPK) and tumor suppressor p53, and activation of ribosomal protein S6 and HIF-1α (Tong et al., 2011). These results suggest a strong link between mitochondrial function and carcinogenesis.

4. MITOCHONDRIA AS THE KEY ORGANELLES INVOLVED IN CELL DEATH AND SURVIVAL

Early biochemical studies have long forged the concept of mitochondria as the organelles responsible for cellular energy production. In the past two decades, a new field of research raised renewed interest in mitochondria because of their key role in apoptotic cell death. The process of mitophagy, on the other hand, has been shown to promote survival.

4.1. Mitochondria and apoptotic cell death

Apoptosis is a highly regulated process of cell suicide. It is an essential physiological process that is required for the normal development and maintenance of tissue homeostasis (Vaux & Korsmeyer, 1999). Deregulation of this process has been implicated in various pathological conditions including autoimmune disease and cancer. Stimuli triggering apoptosis converge on mitochondria, which coordinate the signals and release Cyt *c* from the

IMS into cytoplasm and activate downstream effector caspases (Desagher & Martinou, 2000).

A number of death signals are mediated through members of the Bcl-2 (B-cell lymphoma 2) protein family called "BH3-only" proteins that are closely associated with mitochondria. Bcl-2-associated X, or Bax, is a proapoptotic protein that goes through conformational shifts and inserts into mitochondrial outer membrane upon activation (Wolter et al., 1997). Membrane-bound Bax is shown to accelerate the opening of porin channel (voltage-dependent anion channel, VDAC) (Shimizu, Narita, & Tsujimoto, 1999) or forms mitochondrial apoptosis-inducing channel (MAC) with another proapoptotic Bcl-2 family member, Bak (Narita et al., 1998). As a result, Cyt c and other prodeath factors are released into cytoplasm and trigger the apoptotic cascade.

It is worth noting that the *BAX* gene itself is directly upregulated by tumor suppressor p53 (Narita et al., 1998). Besides, p53 also promotes Bax activation and its integration into mitochondrial membrane (Deng, Gao, Flagg, Anderson, & May, 2006). In addition, many cancerous cells display abnormal levels of Bcl-2 protein family member expression and/or relative quantity. Together, these modifications modulate the permeability of mitochondrial membrane and control the release of proapoptotic factors such as Cyt c (Kilbride & Prehn, 2012).

4.2. Mitophagy and cell survival

Mitochondria are also involved in cellular catabolic processes where the cell degrades unnecessary or dysfunctional components through the lysosomal machinery. This mechanism, autophagy, is essential for cell survival during starvation and organelle quality control (Mizushima, 2007). Mitochondria structural proteins, VDAC isoforms, have been found to be essential for the recruitment of E3 ubiquitin ligase Parkin and to promote mitochondrial autophagy (Sun, Vashisht, Tchieu, Wohlschlegel, & Dreier, 2012). This selective elimination of mitochondria, or mitophagy, is an important quality control that eliminates damaged mitochondria, as well as mitochondria removal during erythrocytes maturation (Youle & Narendra, 2011). It has been shown that oncogene-mediated mitophagy promoted cell survival during early tumorigenesis before the formation of hypoxia, as a key strategy for cancer cells to overcome energy deficit due to insufficient glucose (Kim et al., 2011). Cancer cells overexpressing oncogenic Ras have higher levels of basal autophagy, while defects in autophagy cause accumulation of abnormal mitochondria and further deplete TCA cycle metabolite and cellular ATP (Guo et al., 2011). However, the role of autophagy in tumor progression is differential and probably depends on tumor types and stages. In coculture models, increased mitophagy was observed in tumor stromal cells, which was mediated by oxidative stress and accumulated substrates/nutrients feeding cancer cells (Pavlides et al., 2010), whereas suppressing autophagy in apoptosis–defective tumor cells led to necrotic cell death associated with inflammation and accelerated tumor growth (Degenhardt et al., 2006). Treatments targeting mitophagy are still under extensive studies. Kim et al. showed that selenite induced mitophagic cell death in glioma cells via superoxide (Kim & Choi, 2008). A recent study utilizing linamarase/linamarin/glucose oxidase system based on the combination of cyanide and oxidative stress induced massive mitophagy, which effectively abrogated tumor growth *in vivo* and potentiated antiapoptosis treatment (Gargini, Garcia-Escudero, & Izquierdo, 2011).

5. CANCER METASTASIS: IMPLICATION OF MITOCHONDRIAL METABOLISM IN THE ADAPTATION TO MICROENVIRONMENT

Cancer metastasis is defined as the complex process by which certain cancer cells at the primary tumor location acquire the ability to penetrate and infiltrate surrounding normal tissues and/or lymphatic or blood vessels, migrate to a different tissue, take root, and proliferate to form a secondary tumor (Chaffer & Weinberg, 2011). Metastasis is one of the hallmarks of malignant cancers. Patients with metastatic diseases usually present with poor prognosis and decrease in life expectancy (Hanahan & Weinberg, 2000).

5.1. Mitochondria and metastasis

Mitochondrial function has been intimately tied with metastatic features of cancer. Several studies have demonstrated the relationship between mitochondria-generated ROS and cancer metastasis. Inhibition of mitochondrial function enhances gastric cancer metastasis through ROS-mediated induction of β5-integrin, an adhesion protein, and enhanced αVβ5-integrin expression on the cell surface (Hung et al., 2012). In addition, microarray analysis showed a distinctive expression profile of genes involved in extracellular matrix remodeling between OXPHOS-competent and OXPHOS-deficient cells. It was reported that MMP-1 (matrix metalloproteinase-1) and uPA (urokinase plasminogen activator) were significantly increased, while TIMPs (tissue inhibitor of metalloproteinases) were significantly decreased in mtDNA-null (ρ0) 143B cells compared to WT 143B cells. These modifications could be replicated through drug-induced respiratory inhibition and result in a more invasive phenotype in OXPHOS-deficient cells (van Waveren, Sun, Cheung, & Moraes, 2006). In human bladder cancer, increased ROS production, due to elevated SOD2 and diminished catalase levels, promotes the expression of prometastatic genes such as *VEGFA* and *MMP-9* (Hempel, Ye, Abessi, Mian, & Melendez, 2009). VEGF is a vascular endothelial growth factor which stimulates angiogenesis, and MMP-9 is the metalloproteinase involved in degradation of extracellular components, therefore assisting tumor cells for migration and metastasis (Kargiotis et al., 2008). These findings suggest that mitochondrial ROS is critically involved in cancer metastasis.

The emergence of mtDNA transfer technology provided a useful tool to study the role of mtDNA mutations in cancer cell metastasis. Ishikawa and colleagues demonstrated that a *ND6* mutation in the mtDNA caused deficiency in respiratory Complex I and overproduction of ROS, which conferred a high metastatic potential to the recipient cell line that was originally poorly metastatic (Ishikawa et al., 2008b). However, mtDNA mutation-induced enhancement in glycolysis, without the increase in ROS production, is not sufficient to induce metastasis (Ishikawa et al., 2008a). Further analysis showed this metastatic phenotype was mediated through an activation of HIF-1α transcription via the PI3K-Akt/PKC/HDAC signaling pathways (Koshikawa, Hayashi, Nakagawara, & Takenaga, 2009). Interestingly, mtDNA transfer in the highly metastatic MDA-MB-231 cells suggested defects in mitochondrial respiration were responsible for this phenotype independent of ROS-mediated pathways, indicating other oncogenic changes might account for the upregulation of metastasis pathway (Imanishi et al., 2011).

Clinically, solid tumors usually present pronounced heterogeneity of both neoplastic and stromal cells on the histological, genetic, and metabolic levels. It is widely accepted that tumor heterogeneity is triggered by local hypoxia due to defective microcirculation as a result of the fast and uncontrolled proliferation of cancer cells (Hockel & Vaupel, 2001). Because of the stabilization of HIF-1 and subsequent metabolic remodeling, sustained hypoxia in a developing tumor may cause cellular changes that can result in a more clinically aggressive phenotype (Walenta et al., 2000). As the hypoxia–driven malignance progresses, tumors may develop an increased potential for local invasive growth, perifocal tumor cell spreading, and regional and distant tumor cell spreading (Graham, Forsdike, Fitzgerald, & Macdonald-Goodfellow, 1999; Hockel, Schlenger, Hockel, & Vaupel, 1999; Jang & Hill, 1997). Likewise, enhanced resistance to radiation and other cancer treatments are also documented in association with hypoxia (Sethi et al., 1999; Zhivotovsky, Joseph, & Orrenius, 1999).

5.2. The "reverse Warburg effect"

It was believed that, in a solid tumor, cancer cells are mainly glycolytic and secrete large amount of lactate, which is taken up by oxidative stromal cells and used for ATP production in the mitochondria. However, recently this concept about tumor metabolism has been challenged based on research in epithelial cancers (Nieman et al., 2011; Sotgia et al., 2011). The studies showed that in metastatic human breast cancer, stromal cells are glycolytic and produce lactate and ketone bodies which are transferred to the epithelial cancer cells, which are then utilized through an increased mitochondrial metabolism (Sotgia et al., 2011). These observations suggested a double-compartmented metabolic profile, where the tumor stroma with normal cells is catabolic while the cancer cells are anabolic (Martinez-Outschoorn, Sotgia, & Lisanti, 2012; Sotgia et al., 2011). Similar catabolic transformation was also observed in metastatic ovarian cancer cells, which present highly active lipid metabolism when cocultured with adipocytes (Nieman et al., 2011). In this case, adipocytes become highly catabolic, generating free fatty acids and act as an energy source for the cancer cells. Metastatic cancer cells showed increased levels of fatty acid binding protein 4 as compared to primary tumors and metabolize these fatty acids via mitochondrial β-oxidation. By contrast, such mitochondrial metabolism was not observed when ovarian cancer cells were cultured alone *in vitro* or at primary tumor sites *in vivo* (Nieman et al., 2011). These results highlight the importance of the tumor microenvironment and energy transfer in the "parasitic" cancer cell metabolism.

Moreover, there is evidence that this metabolic coupling mostly coincides with aggressive tumors in breast cancer lymph node metastases (Sotgia et al., 2012b). Double labeling experiments with glycolytic and oxidative (mitochondrial) markers directly showed that at least two different metabolic compartments coexist within primary tumors and their metastases

(Sotgia et al., 2012b). The tumor cell microenvironment contains supporting host cells, including fibroblast, adipocytes, smooth muscle cells, endothelia, and immune cells, which functionally promote tumor growth. Cancer cells extract energy from the surrounding cells by inducing catabolic processes, in forms of autophagy, mitophagy, and aerobic glycolysis. These processes provide the cancer cell mitochondria with high-energy fuels (lactate, ketone bodies, and glutamine) and promote proliferation (Martinez-Outschoorn et al., 2011).

5.3. Caveolin-1 and transformation of the tumor microenvironment

Caveolin-1 (Cav-1) is a scaffolding protein and the main component of the caveolae plasma membranes in most cell types. The expression of Cav-1 is found in most normal organs but decreased when tissue is isolated or grown in culture (Carver & Schnitzer, 2003). Due to its scaffolding function, Cav-1 is found to be involved in various signaling cascades that regulate cell cycle progression and proliferation. Studies have shown that loss of Cav-1 expression in tumor stroma is a robust biomarker for metastasis and aggressive phenotypes, which is also associated with oxidative stress, senescence, and inflammation in the tumor microenvironment (reviewed in Lisanti et al., 2011). Asterholm et al. showed that Cav-1 null mice display a gross defect in lipid metabolism caused by altered metabolic and mitochondrial function in adipose tissue (Asterholm, Mundy, Weng, Anderson, & Scherer, 2012). Although it is not clear if Cav-1 deficiency has a direct impact on mitochondrial function, this study has revealed elevated circulating levels of H_2O_2, increased buildup of lactate upon inhibition of gluconeogenesis, and high influx of glucose and branched chain amino acids in the Cav-1 null mice (Asterholm et al., 2012). These findings suggest that the overall metabolic environment of these mice potentially predisposes cells to malignant transformation or may facilitate tumor progression. In addition, high levels of oxidative stress in tumor stroma could also potentiate genome instability and DNA damage in adjacent cancer cells and promote a more malignant phenotype (Martinez-Outschoorn et al., 2010).

Besides interfering with mitochondrial function, loss of stromal Cav-1 also promotes fibrosis and extracellular matrix remodeling, both of which facilitate the process of tumor development and malignant transformation (Del Galdo et al., 2008; Sotgia et al., 2006). It has been suggested that the fundamental cause for diminished stromal Cav-1 is autophagy induced by oxidative stress. Meanwhile, loss of Cav-1 also drives the activation of transcription factors HIF-1α and NF-κB, forming a positive feedback mechanism for autophagy and mitophagy (Sotgia et al., 2012a). As a result, Cav-1 deficiency in tumor stroma induces metabolic reprogramming and drives a switch from oxidative phosphorylation to glycolysis, hence the "reverse Warburg effect."

As a summary, there is growing evidence suggesting the "reverse Warburg effect" in epithelial cancer: that tumor cells can reprogram their metabolism toward anabolic metabolism in the presence of catabolic supporting cells (microenvironment), leading to mitochondrial biogenesis and increased OXPHOS in tumor cells, driving distant metastasis (summarized in Fig. 4.2). This fact points to oxidative mitochondrial metabolism as a potential target for cancer treatment.

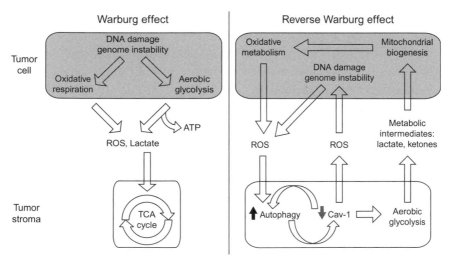

Figure 4.2 A comparison between Warburg effect and the reverse Warburg effect in tumor microenvironment. Warburg effect describes the phenomenon that cancer cells suppress oxidative respiration and utilize aerobic glycolysis to produce ATP for cellular function. Meanwhile, cancer cells secrete ROS and lactate to the microenvironment, which facilitate metastasis and could be taken up by stromal cells for energy production through the TCA cycle. On the contrary, the reverse Warburg effect occurs when tumor stroma loses Cav-1, due to oxidative stress-induced autophagy, which further exacerbates ROS production, and promotes aerobic glycolysis in stromal cells. As a result, tumor stroma provides high-energy metabolic intermediates to cancer cells, where mitochondrial biogenesis and oxidative metabolism take place.

6. DIAGNOSTIC AND THERAPEUTIC APPLICATIONS OF METABOLIC TRANSFORMATION

The metabolic transformation of cancer cells as well as tumor stroma provided multiple targets for early diagnosis and effective treatment in cancer patients. A prominent increase of glycolytic flux in the tumor is the basis of positron emission tomography (18-FDG PET), a medical imaging technique that relies on this phenomenon (Ter-Pogossian, Phelps, Hoffman, & Mullani, 1975). Meanwhile, glycolytic inhibitors, such as the glucose/mannose analog, 2-DG blocking glycolysis, and energy production, are extensively studied in preclinical and early stages of clinical trials (Dwarakanath et al., 2009). It is worth pointing out that in these studies, 2-DG is usually administrated in combination with other anticancer agents or treatments to abolish tumor progression at multiple levels.

6.1. Glycolytic inhibitors

The potential therapeutic application of targeting increased glycolytic flux has been under investigation for decades. Overexpression of glucose transporters GLUT1 and GLUT3 has been identified in many types of cancer, which warrant further studies of GLUT inhibitors in cancer treatment. Up to date, there have been a few agents with GLUT inhibitory effect entering different phases of clinical trials (Flaig et al., 2007; Stein et al., 2010). However, no specific GLUT1 or GLUT3 inhibitor has been discovered so far, aside from the tissue toxicity (of 2-DG in the brain) limits the development of treatment based on inhibiting glucose uptake (Tennant, Duran, & Gottlieb, 2010).

Hexokinase 2 (HK2) is the first enzyme in glycolysis. Overexpression of HK2 is widely associated with increased cell proliferation and resistance to apoptosis in several types of cancer. A group of specific inhibitors for HK2 have been identified, including 2-DG, lonidamine, and 3-bromopyruvate (3BP) (Ko, Pedersen, & Geschwind, 2001; Pelicano, Martin, Xu, & Huang, 2006). Early phase clinical trials of lonidamine have failed either due to lack of therapeutic benefits or severe complications in patients (Oudard et al., 2003) (Clinicaltrials.gov: NCT00237536, NCT00435448). 3BP shows promising anticancer activities in preclinical studies and is very likely to be adopted in clinical trials in the near future (Ganapathy-Kanniappan et al., 2010).

Dichloroacetate (DCA) is a PDK inhibitor and shifts aerobic glycolysis toward oxidative phosphorylation (Whitehouse & Randle, 1973). DCA is used in the clinics for treatment of metabolic disorders including lactic acidosis. Preclinical studies have shown that DCA could slow the growth of certain tumors *in vivo* (Bonnet et al., 2007). Currently, there are several phase I and II trials going on to evaluate the effect of DCA in brain cancer, refractory metastatic breast cancer, and non-small-cell lung carcinomas (Clinicaltrials.gov: NCT01111097, NCT01029925).

Similarly, the pan-agonist for nuclear receptor PPAR (peroxisome proliferator-activated receptor), bezafibrate, is a pharmaceutical agent used for decades to treat hyperlipidemia. It was reported earlier in *in vivo* studies that bezafibrate significantly enhanced the growth rate only of a specific type of tumor (MAC16) in cachexic mice. This tumor-stimulating effect of bezafibrate seemed to be related to its effect on activation of lipid catabolism (Mulligan & Tisdale, 1991). However, fibrates induce rodent-specific hepatomegaly, which could complicate liver metabolism in cachexia animals (Cattley, 2004). Recently, we showed that bezafibrate caused a metabolic shift in cultured cancer cells. It was found that bezafibrate-treated cancer cells had a decrease in glycolytic flux and lactate secretion. This shift could result in impairment of tumor development due to decreased cell growth and invasion (Wang & Moraes, 2011). Meanwhile, bezafibrate was reported to effectively decrease intestinal polyp formation in rodents (Niho et al., 2003). A follow-up study in patients with coronary artery disease taking bezafibrate daily indicates a potential preventive effect against the development of colon cancer (Tenenbaum et al., 2008).

In addition to the agents discussed earlier, more studies are trying to target other players in the glycolytic pathway including PKM2, LDH-5 (lactate dehydrogenase isoenzyme-5), and PFKFB3 (6-phosphofructo- 2-kinase/ fructose-2,6-biphosphatase 3) (reviewed in Porporato, Dhup, Dadhich, Copetti, & Sonveaux, 2011). Moreover, inhibitors for HIF-1α and c-Myc are also at initial stages of clinical trials. Although *in vivo* studies have provided large amount of evidence that HIF-1α and c-Myc inhibition could result in decreases in glycolysis and angiogenesis (Kim et al., 2007; Onnis, Rapisarda, & Melillo, 2009; Yeung et al., 2008), validation of these two molecules as "druggable" targets is still under investigation.

6.2. Targeting the tumor microenvironment

Tumor microenvironment plays an important role in tumor growth and metastasis and thus becomes a very promising target for anticancer treatment. Low pH or acidic environment could facilitate tumor growth, angiogenesis, and metastasis (Chiche, Brahimi-Horn, & Pouyssegur, 2010).

Dietary management and drug treatment that alleviating acidosis have been shown effective to reduce metastasis and inhibit tumor development (Robey et al., 2009; Supuran, 2008).

The "reverse Warburg effect" describes a tightly coupled metabolic regulation between cancer cells and stroma, but may also reveal the "Achilles heel" of a tumor. In this situation, bezafibrate and other reagents that are already adopted in clinics to treat metabolic disorders have become promising candidates for adjuvant cancer treatments. As mentioned earlier, the lipid-lowering PPAR ligand bezafibrate was shown to effectively decrease cancer cell invasion by stimulating oxidative metabolism and lowering lactate secretion, hence the extracellular acidosis (Wang & Moraes, 2011). It would be very interesting to look at if bezafibrate has a similar effect on tumor stroma and prevents cancer metastasis while simultaneously inhibiting cell proliferation. In addition, the oral antidiabetic drug metformin also exhibits a strong antiproliferative action in numerous cancer cell lines *in vitro* and *in vivo*. Mechanistically, metformin activates AMPK signaling by potently inhibiting mitochondrial Complex I and negatively modulating oxidative stress (Owen, Doran, & Halestrap, 2000). It has been shown that metformin interferes with the "reverse Warburg effect" through inhibition of stromal cell autophagy, blockage of cancer cell mitochondrial respiratory activity, and causing lactate accumulation in cultured cancer cells

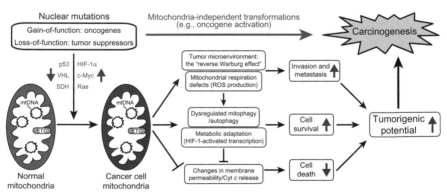

Figure 4.3 Mitochondrial modifications during carcinogenesis. Mutations in oncogenes and tumor suppressors cause not only cellular transformation but also changes in mitochondrial function, which involves metabolic adaptation, elevated mitophagy and resistance to apoptosis, mitochondrial defects-induced ROS production, and "reverse Warburg effect" in tumor stroma. These changes result in decreased cell death, increased cell survival, invasion, and metastasis, all of which contribute to the potential of tumorigenesis. Therefore, mitochondrial modifications play an important role and coordinate with other cellular transformations in the process of carcinogenesis. (For color version of this figure, the reader is referred to the online version of this chapter.)

(Buzzai et al., 2007). A combination treatment with 2-DG and metformin can induce p53-dependent apoptosis and has been shown to effectively deprive tumor bioenergetics in a broad range of epithelial type of cancers (Ben Sahra et al., 2010; Bonanni et al., 2012).

7. CONCLUDING REMARKS

Mitochondrial modifications play important roles in facilitating carcinogenesis and tumor progression at multiple stages. In this review, we went through the major metabolic modifications that contribute to the development or could serve as the treatment targets for cancer (Fig. 4.3). With a better understanding of the whole picture of mitochondrial transformation, we envision that tumor metabolism could become a major target in the treatment of cancer.

ACKNOWLEDGMENTS

Our work is supported in part by the National Institutes of Health Grants 1R01AG036871, 1R01NS079965, 5R01CA085700, and 5R01EY010804 and the Muscular Dystrophy Association. The authors declare no competing financial interests.

REFERENCES

Agani, F. H., Pichiule, P., Chavez, J. C., & LaManna, J. C. (2000). The role of mitochondria in the regulation of hypoxia-inducible factor 1 expression during hypoxia. *Journal of Biological Chemistry*, *275*, 35863–35867.

Ameln, A. K., Muschter, A., Mamlouk, S., Kalucka, J., Prade, I., Franke, K., et al. (2011). Inhibition of HIF prolyl hydroxylase-2 blocks tumor growth in mice through the antiproliferative activity of TGFbeta. *Cancer Research*, *71*, 3306–3316.

Andersen, S., Donnem, T., Stenvold, H., Al-Saad, S., Al-Shibli, K., Busund, L. T., et al. (2011). Overexpression of the HIF hydroxylases PHD1, PHD2, PHD3 and FIH are individually and collectively unfavorable prognosticators for NSCLC survival. *PLoS One*, *6*, e23847.

Asterholm, I. W., Mundy, D. I., Weng, J., Anderson, R. G., & Scherer, P. E. (2012). Altered mitochondrial function and metabolic inflexibility associated with loss of caveolin-1. *Cell Metabolism*, *15*, 171–185.

Bader, H. L., & Hsu, T. (2012). Systemic VHL gene functions and the VHL disease. *FEBS Letters*, *586*, 1562–1569.

Baysal, B. E., Ferrell, R. E., Willett-Brozick, J. E., Lawrence, E. C., Myssiorek, D., Bosch, A., et al. (2000). Mutations in SDHD, a mitochondrial complex II gene, in hereditary paraganglioma. *Science*, *287*, 848–851.

Bensaad, K., Tsuruta, A., Selak, M. A., Vidal, M. N., Nakano, K., Bartrons, R., et al. (2006). TIGAR, a p53-inducible regulator of glycolysis and apoptosis. *Cell*, *126*, 107–120.

Ben Sahra, I., Laurent, K., Giuliano, S., Larbret, F., Ponzio, G., Gounon, P., et al. (2010). Targeting cancer cell metabolism: the combination of metformin and 2-deoxyglucose induces p53-dependent apoptosis in prostate cancer cells. *Cancer Research*, *70*, 2465–2475.

Bertram, J. S. (2000). The molecular biology of cancer. *Molecular Aspects of Medicine*, *21*, 167–223.

Bonanni, B., Puntoni, M., Cazzaniga, M., Pruneri, G., Serrano, D., Guerrieri-Gonzaga, A., et al. (2012). Dual effect of metformin on breast cancer proliferation in a randomized presurgical trial. *Journal of Clinical Oncology, 30,* 2593–2600.

Bonnet, S., Archer, S. L., Allalunis-Turner, J., Haromy, A., Beaulieu, C., Thompson, R., et al. (2007). A mitochondria-K+ channel axis is suppressed in cancer and its normalization promotes apoptosis and inhibits cancer growth. *Cancer Cell, 11,* 37–51.

Buzzai, M., Jones, R. G., Amaravadi, R. K., Lum, J. J., DeBerardinis, R. J., Zhao, F., et al. (2007). Systemic treatment with the antidiabetic drug metformin selectively impairs p53-deficient tumor cell growth. *Cancer Research, 67,* 6745–6752.

Carver, L. A., & Schnitzer, J. E. (2003). Caveolae: mining little caves for new cancer targets. *Nature Reviews. Cancer, 3,* 571–581.

Cattley, R. C. (2004). Peroxisome proliferators and receptor-mediated hepatic carcinogenesis. *Toxicologic Pathology, 32*(Suppl. 2), 6–11.

Chaffer, C. L., & Weinberg, R. A. (2011). A perspective on cancer cell metastasis. *Science, 331,* 1559–1564.

Chandel, N. S., Maltepe, E., Goldwasser, E., Mathieu, C. E., Simon, M. C., & Schumacker, P. T. (1998). Mitochondrial reactive oxygen species trigger hypoxia-induced transcription. *Proceedings of the National Academy of Sciences of the United States of America, 95,* 11715–11720.

Chandel, N. S., McClintock, D. S., Feliciano, C. E., Wood, T. M., Melendez, J. A., Rodriguez, A. M., et al. (2000). Reactive oxygen species generated at mitochondrial complex III stabilize hypoxia-inducible factor-1α during hypoxia - A mechanism of 0-2 sensing. *Journal of Biological Chemistry, 275,* 25130–25138.

Chen, W. L., Kuo, K. T., Chou, T. Y., Chen, C. L., Wang, C. H., Wei, Y. H., et al. (2012). The role of cytochrome c oxidase subunit Va in non-small cell lung carcinoma cells: association with migration, invasion and prediction of distant metastasis. *BMC Cancer, 12,* 273.

Chiche, J., Brahimi-Horn, M. C., & Pouyssegur, J. (2010). Tumour hypoxia induces a metabolic shift causing acidosis: a common feature in cancer. *Journal of Cellular and Molecular Medicine, 14,* 771–794.

Christofk, H. R., Vander Heiden, M. G., Harris, M. H., Ramanathan, A., Gerszten, R. E., Wei, R., et al. (2008). The M2 splice isoform of pyruvate kinase is important for cancer metabolism and tumour growth. *Nature, 452,* 230–233.

Chua, Y. L., Dufour, E., Dassa, E. P., Rustin, P., Jacobs, H. T., Taylor, C. T., et al. (2010). Stabilization of Hypoxia-inducible Factor-1α Protein in Hypoxia Occurs Independently of Mitochondrial Reactive Oxygen Species Production. *Journal of Biological Chemistry, 285,* 31277–31284.

Cotterman, R., Jin, V. X., Krig, S. R., Lemen, J. M., Wey, A., Farnham, P. J., et al. (2008). N-Myc regulates a widespread euchromatic program in the human genome partially independent of its role as a classical transcription factor. *Cancer Research, 68,* 9654–9662.

Cuezva, J. M., Ortega, A. D., Willers, I., Sanchez-Cenizo, L., Aldea, M., & Sanchez-Arago, M. (2009). The tumor suppressor function of mitochondria: translation into the clinics. *Biochimica et Biophysica Acta, 1792,* 1145–1158.

Dang, C. V. (1999). c-myc target genes involved in cell growth, apoptosis, and metabolism. *Molecular and Cellular Biology, 19,* 1–11.

Dang, C. V., Li, F., & Lee, L. A. (2005). Could MYC induction of mitochondrial biogenesis be linked to ROS production and genomic instability? *Cell Cycle, 4,* 1465–1466.

DeBerardinis, R. J., Lum, J. J., Hatzivassiliou, G., & Thompson, C. B. (2008a). The biology of cancer: metabolic reprogramming fuels cell growth and proliferation. *Cell Metabolism, 7,* 11–20.

DeBerardinis, R. J., Lum, J. J., Hatzivassiliou, G., & Thompson, C. B. (2008b). The biology of cancer: metabolic reprogramming fuels cell growth and proliferation. *Cell Metabolism, 7,* 11–20.

DeBerardinis, R. J., Mancuso, A., Daikhin, E., Nissim, I., Yudkoff, M., Wehrli, S., et al. (2007). Beyond aerobic glycolysis: Transformed cells can engage in glutamine metabolism that exceeds the requirement for protein and nucleotide synthesis. *Proceedings of the National Academy of Sciences of the United States of America, 104*, 19345–19350.

Degenhardt, K., Mathew, R., Beaudoin, B., Bray, K., Anderson, D., Chen, G. H., et al. (2006). Autophagy promotes tumor cell survival and restricts necrosis, inflammation, and tumorigenesis. *Cancer Cell, 10*, 51–64.

DeHaan, C., Habibi-Nazhad, B., Yan, E., Salloum, N., Parliament, M., & Allalunis-Turner, J. (2004). Mutation in mitochondrial complex I ND6 subunit is associated with defective response to hypoxia in human glioma cells. *Molecular Cancer, 3*, 19.

Dekker, P. B. D., Hogendoorn, P. C. W., Kuipers-Dijkshoorn, N., Prins, F. A., van Duinen, S. G., Taschner, P. E. M., et al. (2003). SDHD mutations in head and neck paragangliomas result in destabilization of complex II in the mitochondrial respiratory chain with loss of enzymatic activity and abnormal mitochondrial morphology. *The Journal of Pathology, 201*, 480–486.

Del Galdo, F., Sotgia, F., de Almeida, C. J., Jasmin, J. F., Musick, M., Lisanti, M. P., et al. (2008). Decreased expression of caveolin 1 in patients with systemic sclerosis: crucial role in the pathogenesis of tissue fibrosis. *Arthritis and Rheumatism, 58*, 2854–2865.

Demaria, M., Giorgi, C., Lebiedzinskam, M., Esposito, G., D'Angeli, L., Bartoli, A., et al. (2010). A STAT3-mediated metabolic switch is involved in tumour transformation and STAT3 addiction. *Aging, 2*, 823–842.

Deng, X., Gao, F., Flagg, T., Anderson, J., & May, W. S. (2006). Bcl2's flexible loop domain regulates p53 binding and survival. *Molecular and Cellular Biology, 26*, 4421–4434.

Desagher, S., & Martinou, J. C. (2000). Mitochondria as the central control point of apoptosis. *Trends in Cell Biology, 10*, 369–377.

Downward, J. (2003). Targeting RAS signalling pathways in cancer therapy. *Nature Reviews. Cancer, 3*, 11–22.

Dwarakanath, B. S., Singh, D., Banerji, A. K., Sarin, R., Venkataramana, N. K., Jalali, R., et al. (2009). Clinical studies for improving radiotherapy with 2-deoxy-D-glucose: present status and future prospects. *Journal of Cancer Research and Therapeutics, 5*(Suppl. 1), S21–S26.

Ebert, B. L., Gleadle, J. M., O'Rourke, J. F., Bartlett, S. M., Poulton, J., & Ratcliffe, P. J. (1996). Isoenzyme-specific regulation of genes involved in energy metabolism by hypoxia: similarities with the regulation of erythropoietin. *Biochemical Journal, 313*(Pt 3), 809–814.

Enomoto, N., Koshikawa, N., Gassmann, M., Hayashi, J. I., & Takenaga, K. (2002). Hypoxic induction of hypoxia-inducible factor-1α and oxygen-regulated gene expression in mitochondrial DNA-depleted HeLa cells. *Biochemical and Biophysical Research Communications, 297*, 346–352.

Fantin, V. R., St-Pierre, J., & Leder, P. (2006). Attenuation of LDH-A expression uncovers a link between glycolysis, mitochondrial physiology, and tumor maintenance. *Cancer Cell, 9*, 425–434.

Flaig, T. W., Gustafson, D. L., Su, L. J., Zirrolli, J. A., Crighton, F., Harrison, G. S., et al. (2007). A phase I and pharmacokinetic study of silybin-phytosome in prostate cancer patients. *Investigational New Drugs, 25*, 139–146.

Formentini, L., Sanchez-Arago, M., Sanchez-Cenizo, L., & Cuezva, J. M. (2012). The mitochondrial ATPase inhibitory factor 1 triggers a ROS-mediated retrograde prosurvival and proliferative response. *Molecular Cell, 45*, 731–742.

Gaglio, D., Metallo, C. M., Gameiro, P. A., Hiller, K., Danna, L. S., Balestrieri, C., et al. (2011). Oncogenic K-Ras decouples glucose and glutamine metabolism to support cancer cell growth. *Molecular Systems Biology, 7*, 523.

Ganapathy-Kanniappan, S., Vali, M., Kunjithapatham, R., Buijs, M., Syed, L. H., Rao, P. P., et al. (2010). 3-bromopyruvate: a new targeted antiglycolytic agent and a promise for cancer therapy. *Current Pharmaceutical Biotechnology, 11*, 510–517.

Gao, P., Tchernyshyov, I., Chang, T. C., Lee, Y. S., Kita, K., Ochi, T., et al. (2009). c-Myc suppression of miR-23a/b enhances mitochondrial glutaminase expression and glutamine metabolism. *Nature, 458*, 762–765.

Gargini, R., Garcia-Escudero, V., & Izquierdo, M. (2011). Therapy mediated by mitophagy abrogates tumor progression. *Autophagy, 7*, 466–476.

Gasparre, G., Porcelli, A. M., Bonora, E., Pennisi, L. F., Toller, M., Iommarini, L., et al. (2007). Disruptive mitochondrial DNA mutations in complex I subunits are markers of oncocytic phenotype in thyroid tumors. *Proceedings of the National Academy of Sciences of the United States of America, 104*, 9001–9006.

Giaccia, A., Siim, B. G., & Johnson, R. S. (2003). HIF-1 as a target for drug development. *Nature Reviews. Drug Discovery, 2*, 803–811.

Gimenez-Roqueplo, A. P., Favier, J., Rustin, P., Mourad, J. J., Plouin, P. F., Corvol, P., et al. (2001). The R22X mutation of the SDHD gene in hereditary paraganglioma abolishes the enzymatic activity of complex II in the mitochondrial respiratory chain and activates the hypoxia pathway. *American Journal of Human Genetics, 69*, 1186–1197.

Gogvadze, V., Orrenius, S., & Zhivotovsky, B. (2008). Mitochondria in cancer cells: what is so special about them? *Trends in Cell Biology, 18*, 165–173.

Gottlieb, E., & Tomlinson, I. P. (2005). Mitochondrial tumour suppressors: a genetic and biochemical update. *Nature Reviews. Cancer, 5*, 857–866.

Graham, C. H., Forsdike, J., Fitzgerald, C. J., & Macdonald-Goodfellow, S. (1999). Hypoxia-mediated stimulation of carcinoma cell invasiveness via upregulation of urokinase receptor expression. *International Journal of Cancer, 80*, 617–623.

Guo, J. Y., Chen, H. Y., Mathew, R., Fan, J., Strohecker, A. M., Karsli-Uzunbas, G., et al. (2011). Activated Ras requires autophagy to maintain oxidative metabolism and tumorigenesis. *Genes & Development, 25*, 460–470.

Guppy, M., Greiner, E., & Brand, K. (1993). The role of the Crabtree effect and an endogenous fuel in the energy metabolism of resting and proliferating thymocytes. *European Journal of Biochemistry, 212*, 95–99.

Hanahan, D., & Weinberg, R. A. (2000). The hallmarks of cancer. *Cell, 100*, 57–70.

Hanahan, D., & Weinberg, R. A. (2011). Hallmarks of cancer: the next generation. *Cell, 144*, 646–674.

Heiden, M. G. V., Cantley, L. C., & Thompson, C. B. (2009). Understanding the Warburg Effect: The Metabolic Requirements of Cell Proliferation. *Science, 324*, 1029–1033.

Hempel, N., Ye, H., Abessi, B., Mian, B., & Melendez, J. A. (2009). Altered redox status accompanies progression to metastatic human bladder cancer. *Free Radical Biology & Medicine, 46*, 42–50.

Herrmann, P. C., Gillespie, J. W., Charboneau, L., Bichsel, V. E., Paweletz, C. P., Calvert, V. S., et al. (2003). Mitochondrial proteomc: altered cytochrome c oxidase subunit levels in prostate cancer. *Proteomics, 3*, 1801–1810.

Hockel, M., Schlenger, K., Hockel, S., & Vaupel, P. (1999). Hypoxic cervical cancers with low apoptotic index are highly aggressive. *Cancer Research, 59*, 4525–4528.

Hockel, M., & Vaupel, P. (2001). Tumor hypoxia: definitions and current clinical, biologic, and molecular aspects. *Journal of the National Cancer Institute, 93*, 266–276.

Hu, W., Zhang, C., Wu, R., Sun, Y., Levine, A., & Feng, Z. (2010). Glutaminase 2, a novel p53 target gene regulating energy metabolism and antioxidant function. *Proceedings of the National Academy of Sciences of the United States of America, 107*, 7455–7460.

Huang, L. E., Gu, J., Schau, M., & Bunn, H. F. (1998a). Regulation of hypoxia-inducible factor 1α is mediated by an O_2-dependent degradation domain via the ubiquitin-proteasome pathway. *Proceedings of the National Academy of Sciences of the United States of America, 95*, 7987–7992.

Huang, L. E., Gu, J., Schau, M., & Bunn, H. F. (1998b). Regulation of hypoxia-inducible factor 1α is mediated by an O_2-dependent degradation domain via the ubiquitin-proteasome pathway. *Proceedings of the National Academy of Sciences of the United States of America, 95*, 7987–7992.

Hung, W. Y., Huang, K. H., Wu, C. W., Chi, C. W., Kao, H. L., Li, A. F., et al. (2012). Mitochondrial dysfunction promotes cell migration via reactive oxygen species-enhanced β5-integrin expression in human gastric cancer SC-M1 cells. *Biochimica et Biophysica Acta, 1820*, 1102–1110.

Imanishi, H., Hattori, K., Wada, R., Ishikawa, K., Fukuda, S., Takenaga, K., et al. (2011). Mitochondrial DNA mutations regulate metastasis of human breast cancer cells. *PLoS One, 6*, e23401.

Ishikawa, K., Hashizume, O., Koshikawa, N., Fukuda, S., Nakada, K., Takenaga, K., et al. (2008a). Enhanced glycolysis induced by mtDNA mutations does not regulate metastasis. *FEBS Letters, 582*, 3525–3530.

Ishikawa, K., Takenaga, K., Akimoto, M., Koshikawa, N., Yamaguchi, A., Imanishi, H., et al. (2008b). ROS-generating mitochondrial DNA mutations can regulate tumor cell metastasis. *Science, 320*, 661–664.

Ivan, M., Kondo, K., Yang, H., Kim, W., Valiando, J., Ohh, M., et al. (2001). HIFα targeted for VHL-mediated destruction by proline hydroxylation: implications for O_2 sensing. *Science, 292*, 464–468.

Jaakkola, P., Molc, D. R., Tian, Y. M., Wilson, M. I., Gielbert, J., Gaskell, S. J., et al. (2001). Targeting of HIF-α to the von Hippel-Lindau ubiquitylation complex by O_2-regulated prolyl hydroxylation. *Science, 292*, 468–472.

Jang, A., & Hill, R. P. (1997). An examination of the effects of hypoxia, acidosis, and glucose starvation on the expression of metastasis-associated genes in murine tumor cells. *Clinical & Experimental Metastasis, 15*, 469–483.

Johnson, R. F., Witzel, I. I., & Perkins, N. D. (2011). p53-dependent regulation of mitochondrial energy production by the RelA subunit of NF-κB. *Cancer Research, 71*, 5588–5597.

Jokilehto, T., & Jaakkola, P. M. (2010). The role of HIF prolyl hydroxylases in tumour growth. *Journal of Cellular and Molecular Medicine, 14*, 758–770.

Jones, R. G., & Thompson, C. B. (2009). Tumor suppressors and cell metabolism: a recipe for cancer growth. *Genes & Development, 23*, 537–548.

Jung, H. J., Shim, J. S., Lee, J., Song, Y. M., Park, K. C., Choi, S. H., et al. (2010). Terpestacin Inhibits Tumor Angiogenesis by Targeting UQCRB of Mitochondrial Complex III and Suppressing Hypoxia-induced Reactive Oxygen Species Production and Cellular Oxygen Sensing. *Journal of Biological Chemistry, 285*, 11584–11595.

Kargiotis, O., Chetty, C., Gondi, C. S., Tsung, A. J., Dinh, D. H., Gujrati, M., et al. (2008). Adenovirus-mediated transfer of siRNA against MMP-2 mRNA results in impaired invasion and tumor-induced angiogenesis, induces apoptosis in vitro and inhibits tumor growth in vivo in glioblastoma. *Oncogene, 27*, 4830–4840.

KC, S., Carcamo, J. M., & Golde, D. W. (2006). Antioxidants prevent oxidative DNA damage and cellular transformation elicited by the over-expression of c-MYC. *Mutation Research, 593*, 64–79.

Kilbride, S. M., & Prehn, J. H. (2012). Central roles of apoptotic proteins in mitochondrial function. *Oncogene*, advance online publication. http://dx.doi.org/10.1038/onc.2012.348.

Kim, E. H., & Choi, K. S. (2008). A critical role of superoxide anion in selenite-induced mitophagic cell death. *Autophagy, 4*, 76–78.

Kim, J. H., Kim, H. Y., Lee, Y. K., Yoon, Y. S., Xu, W. G., Yoon, J. K., et al. (2011). Involvement of mitophagy in oncogenic K-Ras-induced transformation Overcoming a cellular energy deficit from glucose deficiency. *Autophagy, 7*, 1187–1198.

Kim, J. W., Gao, P., Liu, Y. C., Semenza, G. L., & Dang, C. V. (2007). Hypoxia-inducible factor 1 and dysregulated c-Myc cooperatively induce vascular endothelial growth factor and metabolic switches hexokinase 2 and pyruvate dehydrogenase kinase 1. *Molecular and Cellular Biology*, *27*, 7381–7393.

Kim, J. W., Tchernyshyov, I., Semenza, G. L., & Dang, C. V. (2006a). HIF-1-mediated expression of pyruvate dehydrogenase kinase: a metabolic switch required for cellular adaptation to hypoxia. *Cell Metabolism*, *3*, 177–185.

Kim, J. W., Tchernyshyov, I., Semenza, G. L., & Dang, C. V. (2006b). HIF-1-mediated expression of pyruvate dehydrogenase kinase: A metabolic switch required for cellular adaptation to hypoxia. *Cell Metabolism*, *3*, 177–185.

Ko, Y. H., Pedersen, P. L., & Geschwind, J. F. (2001). Glucose catabolism in the rabbit VX2 tumor model for liver cancer: characterization and targeting hexokinase. *Cancer Letters*, *173*, 83–91.

Koshikawa, N., Hayashi, J., Nakagawara, A., & Takenaga, K. (2009). Reactive oxygen species-generating mitochondrial DNA mutation up-regulates hypoxia-inducible factor-1α gene transcription via phosphatidylinositol 3-kinase-Akt/protein kinase C/histone deacetylase pathway. *Journal of Biological Chemistry*, *284*, 33185–33194.

Lee, H. M., Greeley, G. H., Jr., & Englander, E. W. (2008). Sustained hypoxia modulates mitochondrial DNA content in the neonatal rat brain. *Free Radical Biology & Medicine*, *44*, 807–814.

Lee, S. Y., Jeon, H. M., Ju, M. K., Kim, C. H., Yoon, G., Han, S. I., et al. (2012). Wnt/Snail signaling regulates cytochrome C oxidase and glucose metabolism. *Cancer Research*, *72*, 3607–3617.

Li, F., Wang, Y., Zeller, K. I., Potter, J. J., Wonsey, D. R., O'Donnell, K. A., et al. (2005). Myc stimulates nuclearly encoded mitochondrial genes and mitochondrial biogenesis. *Molecular and Cellular Biology*, *25*, 6225–6234.

Lisanti, M. P., Martinez-Outschoorn, U. E., Pavlides, S., Whitaker-Menezes, D., Pestell, R. G., Howell, A., et al. (2011). Accelerated aging in the tumor microenvironment: connecting aging, inflammation and cancer metabolism with personalized medicine. *Cell Cycle*, *10*, 2059–2063.

Lu, H., Forbes, R. A., & Verma, A. (2002). Hypoxia-inducible factor 1 activation by aerobic glycolysis implicates the Warburg effect in carcinogenesis. *Journal of Biological Chemistry*, *277*, 23111–23115.

Luo, W. B., Hu, H. X., Chang, R., Zhong, J., Knabel, M., O'Meally, R., et al. (2011). Pyruvate Kinase M2 Is a PHD3-Stimulated Coactivator for Hypoxia-Inducible Factor 1. *Cell*, *145*, 732–744.

Martinez-Outschoorn, U. E., Balliet, R. M., Rivadeneira, D. B., Chiavarina, B., Pavlides, S., Wang, C., et al. (2010). Oxidative stress in cancer associated fibroblasts drives tumor-stroma co-evolution: A new paradigm for understanding tumor metabolism, the field effect and genomic instability in cancer cells. *Cell Cycle*, *9*, 3256–3276.

Martinez-Outschoorn, U. E., Pestell, R. G., Howell, A., Tykocinski, M. L., Nagajyothi, F., Machado, F. S., et al. (2011). Energy transfer in "parasitic" cancer metabolism: mitochondria are the powerhouse and Achilles' heel of tumor cells. *Cell Cycle*, *10*, 4208–4216.

Martinez-Outschoorn, U. E., Sotgia, F., & Lisanti, M. P. (2012). Power surge: supporting cells "fuel" cancer cell mitochondria. *Cell Metabolism*, *15*, 4–5.

Matoba, S., Kang, J. G., Patino, W. D., Wragg, A., Boehm, M., Gavrilova, O., et al. (2006). p53 regulates mitochondrial respiration. *Science*, *312*, 1650–1653.

Maxwell, P. H., Wiesener, M. S., Chang, G. W., Clifford, S. C., Vaux, E. C., Cockman, M. E., et al. (1999). The tumour suppressor protein VHL targets hypoxia-inducible factors for oxygen-dependent proteolysis. *Nature*, *399*, 271–275.

Mayr, J. A., Meierhofer, D., Zimmermann, F., Feichtinger, R., Kogler, C., Ratschek, M., et al. (2008). Loss of complex I due to mitochondrial DNA mutations in renal oncocytoma. *Clinical Cancer Research*, *14*, 2270–2275.

Mazurek, S., Boschek, C. B., Hugo, F., & Eigenbrodt, E. (2005). Pyruvate kinase type M2 and its role in tumor growth and spreading. *Seminars in Cancer Biology*, *15*, 300–308.

Mizushima, N. (2007). Autophagy: process and function. *Genes & Development*, *21*, 2861–2873.

Moll, U. M., & Schramm, L. M. (1998). p53–an acrobat in tumorigenesis. *Critical Reviews in Oral Biology and Medicine*, *9*, 23–37.

Mulligan, H. D., & Tisdale, M. J. (1991). Effect of the lipid-lowering agent bezafibrate on tumour growth rate in vivo. *British Journal of Cancer*, *64*, 1035–1038.

Narita, M., Shimizu, S., Ito, T., Chittenden, T., Lutz, R. J., Matsuda, H., et al. (1998). *Proceedings of the National Academy of Sciences of the United States of America*, *95*, 14681–14686.

Nieman, K. M., Kenny, H. A., Penicka, C. V., Ladanyi, A., Buell-Gutbrod, R., Zillhardt, M. R., et al. (2011). Adipocytes promote ovarian cancer metastasis and provide energy for rapid tumor growth. *Nature Medicine*, *17*, 1498–1503.

Niho, N., Takahashi, M., Kitamura, T., Shoji, Y., Itoh, M., Noda, T., et al. (2003). Concomitant suppression of hyperlipidemia and intestinal polyp formation in Apc-deficient mice by peroxisome proliferator–activated receptor ligands. *Cancer Research*, *63*, 6090–6095.

Onnis, B., Rapisarda, A., & Melillo, G. (2009). Development of HIF-1 inhibitors for cancer therapy. *Journal of Cellular and Molecular Medicine*, *13*, 2780–2786.

Oudard, S., Carpentier, A., Banu, E., Fauchon, F., Celerier, D., Poupon, M. F., et al. (2003). Phase II study of lonidamine and diazepam in the treatment of recurrent glioblastoma multiforme. *Journal of Neuro-Oncology*, *63*, 81–86.

Owen, M. R., Doran, E., & Halestrap, A. P. (2000). Evidence that metformin exerts its anti-diabetic effects through inhibition of complex 1 of the mitochondrial respiratory chain. *Biochemical Journal*, *348*, 607–614, Pt. 3.

Owens, K. M., Kulawiec, M., Desouki, M. M., Vanniarajan, A., & Singh, K. K. (2011). Impaired OXPHOS Complex III in Breast Cancer. *PLoS One*, *6*, e23846.

Papandreou, I., Cairns, R. A., Fontana, L., Lim, A. L., & Denko, N. C. (2006a). HIF-1 mediates adaptation to hypoxia by actively downregulating mitochondrial oxygen consumption. *Cell Metabolism*, *3*, 187–197.

Papandreou, I., Cairns, R. A., Fontana, L., Lim, A. L., & Denko, N. C. (2006b). HIF-1 mediates adaptation to hypoxia by actively downregulating mitochondrial oxygen consumption. *Cell Metabolism*, *3*, 187–197.

Pavlides, S., Tsirigos, A., Migneco, G., Whitaker-Menezes, D., Chiavarina, B., Flomenberg, N., et al. (2010). The autophagic tumor stroma model of cancer Role of oxidative stress and ketone production in fueling tumor cell metabolism. *Cell Cycle*, *9*, 3485–3505.

Pelicano, H., Martin, D. S., Xu, R. H., & Huang, P. (2006). Glycolysis inhibition for anti-cancer treatment. *Oncogene*, *25*, 4633–4646.

Pollard, P. J., Briere, J. J., Alam, N. A., Barwell, J., Barclay, E., Wortham, N. C., et al. (2005). Accumulation of Krebs cycle intermediates and over-expression of HIF1α in tumours which result from germline FH and SDH mutations. *Human Molecular Genetics*, *14*, 2231–2239.

Porporato, P. E., Dhup, S., Dadhich, R. K., Copetti, T., & Sonveaux, P. (2011). Anticancer targets in the glycolytic metabolism of tumors: a comprehensive review. *Frontiers in Pharmacology*, *2*, 49.

Puigserver, P., Wu, Z., Park, C. W., Graves, R., Wright, M., & Spiegelman, B. M. (1998). A cold-inducible coactivator of nuclear receptors linked to adaptive thermogenesis. *Cell*, *92*, 829–839.

Robey, I. F., Baggett, B. K., Kirkpatrick, N. D., Roe, D. J., Dosescu, J., Sloane, B. F., et al. (2009). Bicarbonate increases tumor pH and inhibits spontaneous metastases. *Cancer Research, 69,* 2260–2268.

Sahin, E., Colla, S., Liesa, M., Moslehi, J., Muller, F. L., Guo, M., et al. (2011). Telomere dysfunction induces metabolic and mitochondrial compromise. *Nature, 470,* 359–365.

Sanchez–Cenizo, L., Formentini, L., Aldea, M., Ortega, A. D., Garcia-Huerta, P., Sanchez-Arago, M., et al. (2010). Up-regulation of the ATPase inhibitory factor 1 (IF1) of the mitochondrial H+-ATP synthase in human tumors mediates the metabolic shift of cancer cells to a Warburg phenotype. *Journal of Biological Chemistry, 285,* 25308–25313.

Selak, M. A., Armour, S. M., MacKenzie, E. D., Boulahbel, H., Watson, D. G., Mansfield, K. D., et al. (2005). Succinate links TCA cycle dysfunction to oncogenesis by inhibiting HIF-α prolyl hydroxylase. *Cancer Cell, 7,* 77–85.

Semenza, G. L., Jiang, B. H., Leung, S. W., Passantino, R., Concordet, J. P., Maire, P., et al. (1996). Hypoxia response elements in the aldolase A, enolase 1, and lactate dehydrogenase A gene promoters contain essential binding sites for hypoxia-inducible factor 1. *Journal of Biological Chemistry, 271,* 32529–32537.

Semenza, G. L., Roth, P. H., Fang, H. M., & Wang, G. L. (1994). Transcriptional regulation of genes encoding glycolytic enzymes by hypoxia-inducible factor 1. *Journal of Biological Chemistry, 269,* 23757–23763.

Sen, N., Satija, Y. K., & Das, S. (2011). PGC-1α, a Key Modulator of p53, Promotes Cell Survival upon Metabolic Stress. *Molecular Cell, 44,* 621–634.

Sethi, T., Rintoul, R. C., Moore, S. M., MacKinnon, A. C., Salter, D., Choo, C., et al. (1999). Extracellular matrix proteins protect small cell lung cancer cells against apoptosis: a mechanism for small cell lung cancer growth and drug resistance in vivo. *Nature Medicine, 5,* 662–668.

Shim, H., Dolde, C., Lewis, B. C., Wu, C. S., Dang, G., Jungmann, R. A., et al. (1997). c-Myc transactivation of LDH-A: implications for tumor metabolism and growth. *Proceedings of the National Academy of Sciences of the United States of America, 94,* 6658–6663.

Shimizu, S., Narita, M., & Tsujimoto, Y. (1999). Bcl-2 family proteins regulate the release of apoptogenic cytochrome c by the mitochondrial channel VDAC. *Nature, 399,* 483–487.

Simonnet, H., Demont, J., Pfeiffer, K., Guenaneche, L., Bouvier, R., Brandt, U., et al. (2003). Mitochondrial complex I is deficient in renal oncocytomas. *Carcinogenesis, 24,* 1461–1466.

Sotgia, F., Martinez-Outschoorn, U. E., Howell, A., Pestell, R. G., Pavlides, S., & Lisanti, M. P. (2012a). Caveolin-1 and cancer metabolism in the tumor microenvironment: markers, models, and mechanisms. *Annual Review of Pathology, 7,* 423–467.

Sotgia, F., Martinez-Outschoorn, U. E., Pavlides, S., Howell, A., Pestell, R. G., & Lisanti, M. P. (2011). Understanding the Warburg effect and the prognostic value of stromal caveolin-1 as a marker of a lethal tumor microenvironment. *Breast Cancer Research, 13,* 213.

Sotgia, F., Whitaker-Menezes, D., Martinez-Outschoorn, U. E., Flomenberg, N., Birbe, R. C., Witkiewicz, A. K., et al. (2012b). Mitochondrial metabolism in cancer metastasis: visualizing tumor cell mitochondria and the "reverse Warburg effect" in positive lymph node tissue. *Cell Cycle, 11,* 1445–1454.

Sotgia, F., Williams, T. M., Schubert, W., Medina, F., Minetti, C., Pestell, R. G., et al. (2006). Caveolin-1 deficiency (−/−) conveys premalignant alterations in mammary epithelia, with abnormal lumen formation, growth factor independence, and cell invasiveness. *American Journal of Pathology, 168,* 292–309.

Stein, M., Lin, H., Jeyamohan, C., Dvorzhinski, D., Gounder, M., Bray, K., et al. (2010). Targeting tumor metabolism with 2-deoxyglucose in patients with castrate-resistant prostate cancer and advanced malignancies. *Prostate, 70,* 1388–1394.

Sun, Y., Vashisht, A. A., Tchieu, J., Wohlschlegel, J. A., & Dreier, L. (2012). VDACs recruit Parkin to defective mitochondria to promote mitochondrial autophagy. *Journal of Biological Chemistry, 287,* 40652–40660.

Supuran, C. T. (2008). Development of small molecule carbonic anhydrase IX inhibitors. *BJU International, 101*(Suppl. 4), 39–40.

Taddei, M. L., Giannoni, E., Raugei, G., Scacco, S., Sardanelli, A. M., Papa, S., et al. (2012). Mitochondrial Oxidative Stress due to Complex I Dysfunction Promotes Fibroblast Activation and Melanoma Cell Invasiveness. *Journal of Signal Transduction, 2012*, 684592.

Telang, S., Lane, A. N., Nelson, K. K., Arumugam, S., & Chesney, J. (2007). The oncoprotein H-RasV12 increases mitochondrial metabolism. *Molecular Cancer, 6*, 77.

Tenenbaum, A., Boyko, V., Fisman, E. Z., Goldenberg, I., Adler, Y., Feinberg, M. S., et al. (2008). Does the lipid-lowering peroxisome proliferator-activated receptors ligand bezafibrate prevent colon cancer in patients with coronary artery disease? *Cardiovascular Diabetology, 7*, 18.

Tennant, D. A., Duran, R. V., & Gottlieb, E. (2010). Targeting metabolic transformation for cancer therapy. *Nature Reviews. Cancer, 10*, 267–277.

Ter-Pogossian, M. M., Phelps, M. E., Hoffman, E. J., & Mullani, N. A. (1975). A positron-emission transaxial tomograph for nuclear imaging (PETT). *Radiology, 114*, 89–98.

Tong, W. H., Sourbier, C., Kovtunovych, G., Jeong, S. Y., Vira, M., Ghosh, M., et al. (2011). The Glycolytic Shift in Fumarate-Hydratase-Deficient Kidney Cancer Lowers AMPK Levels, Increases Anabolic Propensities and Lowers Cellular Iron Levels. *Cancer Cell, 20*, 315–327.

Vafa, O., Wade, M., Kern, S., Beeche, M., Pandita, T. K., Hampton, G. M., et al. (2002). c-Myc can induce DNA damage, increase reactive oxygen species, and mitigate p53 function: a mechanism for oncogene-induced genetic instability. *Molecular Cell, 9*, 1031–1044.

van Waveren, C., Sun, Y. B., Cheung, H. S., & Moraes, C. T. (2006). Oxidative phosphorylation dysfunction modulates expression of extracellular matrix - remodeling genes and invasion. *Carcinogenesis, 27*, 409–418.

Vaux, D. L., & Korsmeyer, S. J. (1999). Cell death in development. *Cell, 96*, 245–254.

Villeneuve, C., Guilbeau-Frugier, C., Sicard, P., Lairez, O., Ordener, C., Duparc, T., et al. (2012). p53-PGC-1α Pathway Mediates Oxidative Mitochondrial Damage and Cardiomyocyte Necrosis Induced by Monoamine Oxidase-A Upregulation: Role in Chronic Left Ventricular Dysfunction in Mice. *Antioxidants & Redox Signaling, 18*, 5–18.

Walenta, S., Wetterling, M., Lehrke, M., Schwickert, G., Sundfor, K., Rofstad, E. K., et al. (2000). High lactate levels predict likelihood of metastases, tumor recurrence, and restricted patient survival in human cervical cancers. *Cancer Research, 60*, 916–921.

Wang, G. L., Jiang, B. H., Rue, E. A., & Semenza, G. L. (1995). Hypoxia-Inducible Factor-1 Is a Basic-Helix-Loop-Helix-Pas Heterodimer Regulated by Cellular 0-2 Tension. *Proceedings of the National Academy of Sciences of the United States of America, 92*, 5510–5514.

Wang, X., & Moraes, C. T. (2011). Increases in mitochondrial biogenesis impair carcinogenesis at multiple levels. *Molecular Oncology, 5*, 399–409.

Wang, G. L., & Semenza, G. L. (1993). General involvement of hypoxia-inducible factor 1 in transcriptional response to hypoxia. *Proceedings of the National Academy of Sciences of the United States of America, 90*, 4304–4308.

Warburg, O. (1956). On the origin of cancer cells. *Science, 123*, 309–314.

Warburg, O., Wind, F., & Negelein, E. (1927). The Metabolism of Tumors in the Body. *Journal of General Physiology, 8*, 519–530.

Whitehouse, S., & Randle, P. J. (1973). Activation of pyruvate dehydrogenase in perfused rat heart by dichloroacetate (Short Communication). *Biochemical Journal, 134*, 651–653.

Wolter, K. G., Hsu, Y. T., Smith, C. L., Nechushtan, A., Xi, X. G., & Youle, R. J. (1997). *The Journal of Cell Biology, 139*, 1281–1292.

Yen, K. E., Bittinger, M. A., Su, S. M., & Fantin, V. R. (2010). Cancer-associated IDH mutations: biomarker and therapeutic opportunities. *Oncogene, 29*, 6409–6417.

Yeung, S. J., Pan, J., & Lee, M. H. (2008). Roles of p53, MYC and HIF-1 in regulating glycolysis - the seventh hallmark of cancer. *Cellular and Molecular Life Sciences, 65*, 3981–3999.

Youle, R. J., & Narendra, D. P. (2011). Mechanisms of mitophagy. *Nature Reviews. Molecular Cell Biology*, *12*, 9–14.

Zhang, H., Gao, P., Fukuda, R., Kumar, G., Krishnamachary, B., Zeller, K. I., et al. (2007). HIF-1 inhibits mitochondrial biogenesis and cellular respiration in VHL-deficient renal cell carcinoma by repression of C-MYC activity. *Cancer Cell*, *11*, 407–420.

Zhivotovsky, B., Joseph, B., & Orrenius, S. (1999). Tumor radiosensitivity and apoptosis. *Experimental Cell Research*, *248*, 10–17.

CHAPTER FIVE

Human Polynucleotide Phosphorylase (*hPNPase^old-35*): Should I Eat You or Not—That Is the Question?

Upneet K. Sokhi*, **Swadesh K. Das***,†, **Santanu Dasgupta***,†,‡, **Luni Emdad***,†,‡, **Rita Shiang***, **Robert DeSalle**†,§,¶, **Devanand Sarkar***,†,‡, **Paul B. Fisher***,†,‡,1

*Department of Human and Molecular Genetics, Virginia Commonwealth University, School of Medicine, Richmond, Virginia, USA
†VCU Institute of Molecular Medicine, Virginia Commonwealth University, School of Medicine, Richmond, Virginia, USA
‡VCU Massey Cancer Center, Virginia Commonwealth University, School of Medicine, Richmond, Virginia, USA
§Sackler Institute for Comparative Genomics, American Museum of Natural History, New York, New York, USA
¶New York University, New York, New York, USA
1Corresponding author: e-mail address: pbfisher@vcu.edu

Contents

Advances in Cancer Research, Volume 119
ISSN 0065-230X
http://dx.doi.org/10.1016/B978-0-12-407190-2.00005-8

Abstract

RNA degradation plays a fundamental role in maintaining cellular homeostasis whether it occurs as a surveillance mechanism eliminating aberrant mRNAs or during RNA processing to generate mature transcripts. 3′-5′ exoribonucleases are essential mediators of RNA decay pathways, and one such evolutionarily conserved enzyme is polynucleotide phosphorylase (PNPase). The human homologue of this fascinating enzymatic protein (hPNPase^{old-35}) was cloned a decade ago in the context of terminal differentiation and senescence through a novel "overlapping pathway screening" approach. Since then, significant insights have been garnered about this exoribonuclease and its repertoire of expanding functions. The objective of this review is to provide an up-to-date perspective of the recent discoveries made relating to *hPNPase^{old-35}* and the impact they continue to have on our comprehension of its expanding and diverse array of functions.

1. RNA DEGRADATION PATHWAYS

Posttranscriptional control of gene expression occurs at multiple steps that include mRNA processing, its export from the nucleus to cytoplasm, mRNA localization, mRNA stability, translational regulation, and finally mRNA decay (Halbeisen, Galgano, Scherrer, & Gerber, 2008). A number of evolutionarily conserved *trans*-acting regulatory factors have been identified over the past years that are involved in each of these steps (Audic & Hartley, 2004; Glisovic, Bachorik, Yong, & Dreyfuss, 2008). Previously, RNA degradation was thought to be a random process, but studies over the past few years have proved quite the contrary. It is a well-controlled process involving the interplay of a number of proteins working together and contributing to the maintenance of cellular homeostasis (Andrade, Pobre, Silva, Domingues, & Arraiano, 2009). Two major pathways have been identified for eukaryotic mRNA turnover, deadenylation-dependent decay and deadenylation-independent decay (Newbury, 2006; Parker & Song, 2004; Shyu, Wilkinson, & van Hoof, 2008). Deadenylation-dependent decay is usually triggered by the presence of certain *cis*-elements in the mRNA sequence such as AU-rich elements in the 3′UTR and instability elements in the coding region to name a few (Garneau, Wilusz, & Wilusz, 2007; Hollams, Giles, Thomson, & Leedman, 2002). The first step in this process is the shortening of the poly(A) tail or deadenylation, which is brought about by certain deadenylases like Ccr4/Pop2/NOT complex,

Pan2/Pan3, or PARN [poly(A) ribonuclease]. Following this process, the mRNA can undergo 3'-5' decay by the exosome which is a complex of 3'-5' exoribonucleases or polynucleotide phosphorylase (PNPase, a phosphorolytic 3'-5' exoribonuclease), with the residual cap structure being hydrolyzed by the scavenger enzyme DcpS or 5'-3' exonucleolytic digestion by an exonuclease Xrn1p after decapping by the Dcp1/Dcp2 enzyme complex (Garneau et al., 2007; Meyer, Temme, & Wahle, 2004). Deadenylation–independent decay, mostly consisting of RNA surveillance mechanisms, may occur when there is a premature stop codon in the mRNA sequence (Nonsense-mediated decay or NMD) in which case 5'-3' digestion occurs after removal of the cap structure by the same enzymes discussed earlier (Doma & Parker, 2007; Shyu et al., 2008). In the event of the absence of a termination codon (Nonstop decay or NSD), the aberrant mRNA is recognized and degraded in a 3'-5' manner by the exosome. When there is a stall in translation elongation (No-go decay or NGD), cleavage occurs at the stall site by an endonuclease (IME1, PMR1, or RNase MRP) after which the fragments are digested by the 3' and 5' exoribonucleases (Doma & Parker, 2007). These processes are summarized in Fig. 5.1.

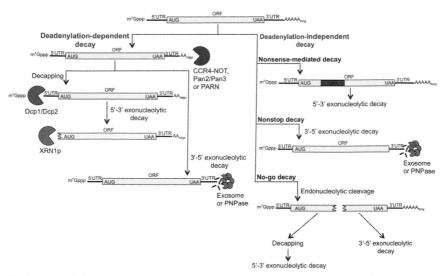

Figure 5.1 RNA decay mechanisms. A diagrammatic representation of various RNA degradation pathways. Refer to text for details. (For color version of this figure, the reader is referred to the online version of this chapter.)

2. EXORIBONUCLEASES

Ribonucleases (RNases) are the main players involved in RNA metabolism, performing important roles in RNA maturation, RNA end-turnover, and the degradation of aberrant RNAs or species no longer required by the cell (Arraiano et al., 2010). Based on their degradative properties, they are classified into exo- and endoribonucleases, with exoribonucleases being further classified into $5'$-$3'$ and $3'$-$5'$ exoribonucleases depending on the direction of degradation (Sarkar & Fisher, 2006c). Eukaryotes possess both these classes of enzymes, whereas prokaryotes like *Escherichia coli* lack the $5'$-$3'$ exoribonuclease activity. Extensive sequence homology studies have culminated in the various exoribonucleases in all kingdoms to be grouped into the following superfamilies (with representative examples from *E. coli* and some of their eukaryotic homologues): the RBN (RNase BN) family present only in eubacteria; the RNR (RNase II, RNase R, Rrp44, Dss I, Ssd I), DEDD (RNase D, RNase T, oligoribonuclease, Rrp6, Pan2, Rex1-4), and PDX (Rnase PH, PNPase, Rrp41-43, 45, 46, Mtr3) families present in bacteria, archaea, and eukaryotes; the RRP4 (Rrp4, 40, Csl4) family present in archaea and eukaryotes; and lastly, the 5PX (Xrn1, Rat1) family present only in eukaryotes (Ibrahim, Wilusz, & Wilusz, 2008; Zuo & Deutscher, 2001; Leszczyniecka, DeSalle, Kang, & Fisher, 2004). RNA degradation by most exoribonucleases occurs via hydrolytic cleavage of the $3'$ phosphodiester bond to generate nucleotide monophosphates, but certain members such as PNPase and RNase PH do so through a phosphorolytic mechanism that utilizes inorganic phosphate (Pi) to generate nucleotide diphosphates (Andrade et al., 2009).

3. PNPase

PNPase is an evolutionarily conserved phosphorolytic $3'$-$5'$ processive exoribonuclease that is present in all phyla extending from bacteria, plants, worms, flies, mice, and humans, although it is absent in fungi, trypanosomes, and Archaea (Sarkar & Fisher, 2006c). Studies in *E. coli* have shown that in the presence of high Pi this enzyme acts exclusively as an exoribonuclease, whereas in the event of low Pi, it acts as a polymerase resulting in the elongation of the poly(A) tail by adding nucleotides and generating Pi. PNPase was discovered by Severo Ochoa and Marianne Grunberg-Manago in 1955, and Ochoa received the Nobel Prize in 1959 for discovering its

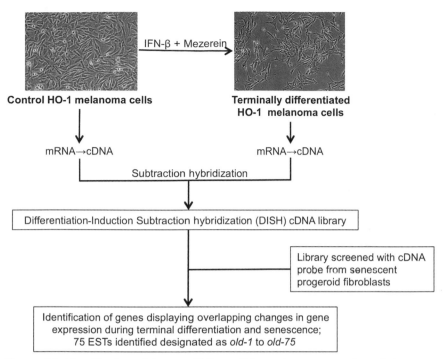

Figure 5.2 Identification of human PNPase. Schematic representation of the overlapping pathway screen that aided in the identification of transcripts upregulated in the processes of terminal differentiation and senescence. (For color version of this figure, the reader is referred to the online version of this chapter.)

polymerizing activity (Andrade et al., 2009; Grunberg–Manago, Ortiz, & Ochoa, 1955). The human homologue of PNPase (*hPNPase^old-35*) was first identified as a gene that was upregulated in terminally differentiated human melanoma cells and senescent progeroid fibroblasts in an "overlapping pathway screen" to identify differentially regulated genes in the context of terminal differentiation and senescence (Fig. 5.2; Leszczyniecka et al., 2002).

3.1. PNPase structure

PNPase from all diverse species shares a classical domain structure that contains a catalytic PDX domain characteristic of all PDX family members (Chen, Koehler, & Teitell, 2007; Sarkar & Fisher, 2006c). Almost all PNPases are composed of five conserved structural motifs; the amino-terminal domain is composed of two RNase PH domains, an α-helical domain that separates the RNase PH domains, and two carboxy-terminal RNA-binding domains KH (K-homology) and S1 (Chen et al., 2007;

Raijmakers, Egberts, van Venrooij, & Pruijn, 2002; Sarkar & Fisher, 2006c; Symmons, Jones, & Luisi, 2000) (Fig. 5.3). In plants, PNPase is translocated to the chloroplast stroma or mitochondrial matrix by an N-terminal chloroplast-transit peptide or a mitochondrial-targeting signal (MTS), respectively (Yehudai Resheff, Portnoy, Yogev, Adir, & Schuster, 2003). Human PNPase ($hPNPase^{old-35}$), a 783–amino acid-long ~86-kDa protein, also contains an N-terminal MTS that allows its primary location in the cell to be the mitochondrial where it assembles into a homotrimer or a dimer of two homotrimers (Piwowarski et al., 2003; Rainey et al., 2006). PNPase from the bacterium *Streptomyces antibioticus* was the first to be crystallized; it was observed that it forms a homotrimeric doughnut-shaped structure with a central channel formed by the six RNase PH domains containing the catalytic sites and through which a single-stranded RNA molecule can enter (Symmons et al., 2000; Symmons, Williams, Luisi, Jones, & Carpousis, 2002). Extensive deletion and mutation studies have been done in the past decade to understand the structural specificities of the phosphorolytic and RNA-binding properties of PNPase. The catalytic activity of the bacterial PNPase resides mainly in the second RNase PH domain, which also seems to possess RNA-binding ability as demonstrated by a G454D mutation, but both domains are required for proper enzymatic activity (Jarrige, Brchemier-Baey, Mathy, Duch, & Portier, 2002). In bacteria, deletion of the KH or S1 domain reduces the

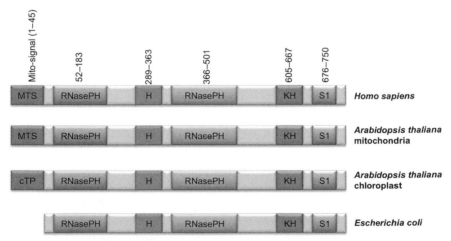

Figure 5.3 Domain organization of PNPases from different species. PNPases from humans, plants, and bacteria showing the presence of two conserved catalytic RNasePH domains and two RNA-binding domains KH and S1. (For color version of this figure, the reader is referred to the online version of this chapter.)

specific activity of the enzyme by ~19- or ~50-fold, respectively, whereas deleting both domains reduces the enzymatic activity to ~1% (Stickney, Hankins, Miao, & Mackie, 2005). X-ray crystallography studies using *E. coli* full-length and KH/S1-truncated PNPase further elucidated that the KH and S1 domains are important for proper trimer formation apart from RNA binding along with certain conserved residues in the first RNase PH domain that are important for proper RNA binding and processive degradation (Shi, Yang, Lin Chao, Chak, & Yuan, 2008). Recently, two missense mutations in humans were identified to cause mitochondrial disease phenotypes (Vedrenne et al., 2012; von Ameln et al., 2012). The p.Gln387Arg and p.Glu475Gly mutations are located in the second RNase PH domain and were found to disrupt trimerization, indicating that this domain and these particular residues are necessary for trimerization in human PNPase. In case of plants, chloroplast PNPase exists as a homomultimer (Baginsky et al., 2001) and both its RNase PH domains possess phosphorolytic and polymerizing activities (Yehudai Resheff et al., 2003). Mutation analysis of human PNPase showed that it possesses similar phosphorolytic properties to its homologues in different species with both the RNase PH domains possessing equal enzymatic activity and the presence of either one of them is sufficient for complete enzymatic activity as shown in our studies (Sarkar et al., 2005), though there are some differences in its RNA-binding affinities compared to other species of PNPases (Portnoy, Palnizky, Yehudai Resheff, Glaser, & Schuster, 2008). Also, human PNPase still retains enzymatic activity after deletion of both the KH and S1 domains in contrast to its bacterial homologue, which further strengthens the observation that the RNase PH domains may also play a role in RNA binding after all (Sarkar et al., 2005).

More recent advances in understanding the structural evolution of this interesting exoribonuclease come from the successful crystallization of an S1-domain truncated human PNPase (Lin, Wang, Yang, Hsiao, & Yuan, 2012). This study reveals a slight difference between the bacterial and human PNPases, in that the S1 domain was found to be dispensable for RNA binding and the 3 KH domains formed a pore on top of the homotrimer (formed by the six RNase PH domains) which, in turn, is solely responsible for binding RNA through its conserved GXXG motif (Lin et al., 2012). Although this is an exciting finding, it stands in contrast to previously published reports as mentioned earlier that the RNase PH domains also possess RNA-binding ability (Sarkar et al., 2005). Future studies crystallizing the full length and maybe a KH domain-truncated human PNPase may provide useful insights into the workings of this intriguing protein.

3.2. PNPase, degradosome, exosome

Apart from existing as an individual entity in the cytoplasm, PNPase is also found as part of a multiprotein complex associated with endonuclease RNase E, the DEAD-box helicase RhlB, and enolase, a glycolytic enzyme in bacteria (Carpousis, 2002; Carpousis, Van Houwe, Ehretsmann, & Krisch, 1994; Miczak, Kaberdin, Wei, & Lin Chao, 1996; Py, Causton, Mudd, & Higgins, 1994; Py, Higgins, Krisch, & Carpousis, 1996). This structure is called the degradosome and contains approximately 20% of the total PNPase present in the bacterial cell (Deutscher & Li, 2001). Yeast contains a similar multiprotein complex consisting of exoribonucleases called the exosome present in the nucleus and the cytoplasm, from which PNPase is absent (Min & Zassenhaus, 1991; Mitchell, Petfalski, Shevchenko, Mann, & Tollervey, 1997). The yeast exosome is made up of eight exoribonucleases (Rrp6p, Rrp41–46p, and Mtr3p) with some of them sharing homology with bacterial RNase PH and three RNA-binding subunits (Rrp4p, Rrp40p, and Csl4p) (Houseley, LaCava, & Tollervey, 2006). All of the yeast exosome subunits have human homologues, except Rrp43p and Mtr3p (Fritz, Bergman, Kilpatrick, Wilusz, & Wilusz, 2004; Raijmakers et al., 2002). There is high structural homology between the eukaryotic exosome and the homotrimeric PNPase, both looking like ring-shaped structures with a central catalytic core channel through which the RNA molecule enters for degradation (Symmons et al., 2002), although recent studies have shown that human PNPase possesses a KH pore versus a S1 pore in exosomes that facilitates RNA binding (Lin et al., 2012).

4. IDENTIFICATION AND REGULATION OF hPNPase^{old-35} EXPRESSION

Human PNPase is encoded by the gene *PNPT1* mapping to chromosome 2p15–2p16.1, consisting of 28 exons and spanning ~60 KB on the reverse strand (Leszczyniecka, Su, Kang, Sarkar, & Fisher, 2003). It was first identified as a gene that was upregulated in senescent progeroid fibroblasts and terminally differentiated HO-1 human melanoma cells in a quest to recognize molecular mediators of differentiation that would aid in improvement of cancer therapy (Leszczyniecka et al., 2002). Human melanoma is a well-established model for studying differentiation therapy of cancer, and terminal differentiation caused by combined treatment with recombinant human fibroblast interferon (IFN)-β and protein kinase C activator mezerein (MEZ) results in similar phenotypic characteristics as cellular senescence. These findings formed the basis for a screening strategy intended to identify overlapping gene expression changes associated with both terminal differentiation and senescence (Fig. 5.2). cDNA from terminally differentiated HO-1 cells was used to generate a differentiation-induction subtraction hybridization library, which was screened with a probe generated from the mRNA of senescent progeroid fibroblasts (Huang, Adelman, Jiang, Goldstein, & Fisher, 1999; Leszczyniecka et al., 2002). Of the 75 genes identified through this screening approach, designated $old-1$ to $old-75$, $old-35$ was one of the identified cDNAs that showed an elevated expression associated with both terminal differentiation and senescence (Leszczyniecka et al., 2002). $old-35$ exhibited substantial sequence homology to PNPase from other species and was thus identified as human polynucleotide phosphorylase or $hPNPase^{old-35}$. $hPNPase^{old-35}$ is expressed in all primary tissues (Leszczyniecka et al., 2002), and a $PNPT1$ knockout mouse is embryonic lethal indicating its importance in early mammalian development (Wang et al., 2010).

Bacterial PNPase has been shown to autocontrol its own expression at the posttranscriptional level by binding and degrading part of a stem-loop structure in its own mRNA 5′-leader sequence cleaved by RNase III, subsequently resulting in mRNA instability and reduced expression (Jarrige, Mathy, & Portier, 2001). On the other hand, not much is known about the transcriptional or posttranscriptional regulation of $hPNPase^{old-35}$, except that it is induced by type I interferons (IFN-α and -β) at the transcription level in both normal and cancer cells (Fig. 5.4) (Leszczyniecka et al., 2003). Since type I interferons are cytokines secreted by cells in response to viral infections, it could be speculated that $hPNPase^{old-35}$ may play a role in the cells' antiviral response through its RNA processing function. Induction of $hPNPase^{old-35}$ mRNA, which has a half-life of ~6 h (Leszczyniecka et al., 2003), was evident even with 1 U/ml of IFN-β and as early as 3 h in HO-1 cells (2000 U/ml IFN-β) proving that it is an early IFN response gene (Leszczyniecka et al., 2002). IFN-γ and TNF-α treatment resulted in minimal or no induction, whereas poly(I)/poly(C), which induces IFN-α and -β, also stimulated $hPNPase^{old-35}$ expression (Leszczyniecka et al., 2002). Promoter bashing experiments and sequence analysis resulted in the identification of an IFN-stimulated response element (ISRE) in the $hPNPase^{old-35}$ promoter, to which the ISGF3 complex shows increased binding upon IFN-β treatment (Leszczyniecka et al., 2003). Mutating this site abrogated promoter induction, demonstrating that IFN-β is a

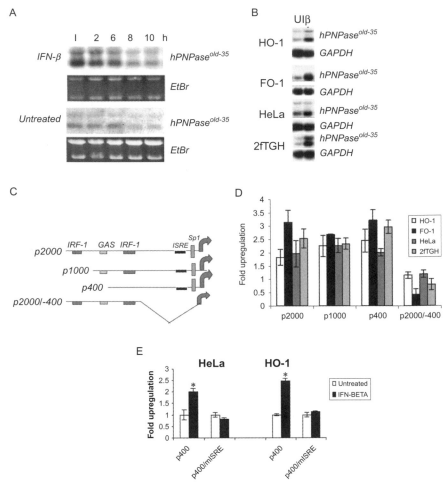

Figure 5.4 Type I interferon regulation of *hPNPase^old-35*. (A) HO-1 cells were treated or untreated with 2000 U/ml IFN-β overnight and transcription was blocked with 5 µg/ml actinomycin D. Half-life of *hPNPase^old-35* mRNA was found to be ~6 h as quantified by Northern blot analysis. (B) Northern blot showing induction of *hPNPase^old-35* mRNA by IFN-β (2000 U/ml, 7 h treatment) in HO-1, FO-1 (melanoma), HeLa (cervical), and 2fTGH (fibrosarcoma) cell lines. (C) Schematic representation of various deletions constructs of *hPNPase^old-35* promoter. (D) Luciferase assays in HO-1, FO-1, HeLa, and 2fTGH with *hPNPase^old-35* promoter constructs showing the importance of the ISRE and SP1 elements for *hPNPase^old-35* transcription. (E) Luciferase assay in HO-1 and HeLa cells with wt and mutant ISRE element showing loss of transcription induction by IFN-β in case of a mutant ISRE element (*, $P < 0.05$). *Figure adapted from Leszczyniecka et al. (2003).*

transcriptional activator of *hPNPase^old-35* whose activation depends on the JAK/STAT signal transduction pathways (Fig. 5.4) (Leszczyniecka et al., 2003). *In silico* analysis of the *hPNPase^old-35* promoter has highlighted several additional transcription factor binding sites, but at present, any connection of these regulatory elements to physiological conditions relating to *hPNPase^old-35* expression remains to be determined (Gewartowski et al., 2006). In any case, it would be relevant to determine if there are other regulatory factors controlling *hPNPase^old-35* expression levels and future endeavors toward achieving this goal are required to better understand *hPNPase^old-35* functions.

5. SUBCELLULAR LOCALIZATION OF hPNPase^old-35

Human PNPase possesses a mitochondrial-targeting signal (MTS) at its amino terminus. Immunofluorescence and cell fractionation studies demonstrated that endogenous and C-terminal myc-tagged hPNPase^old-35 localizes exclusively to the mitochondria (Piwowarski et al., 2003). It was expected that the location was obviously mitochondrial matrix, where an enzyme such as hPNPase^old-35 was predicted to perform mitochondrial RNA processing analogous to its bacterial and chloroplast counterparts. However, subsequent research involving subfractionation studies, protease protection assays, and carbonate extraction experiments proved that hPNPase^old-35 is actually a peripheral inner membrane (IM)-bound protein located in the mitochondrial inter membrane space (IMS) (Chen et al., 2006) and is imported to this location via a *i*-AAA protease Yme1-dependent mechanism (Rainey et al., 2006). hPNPase^old-35 first enters through the TOM complex at the outer mitochondrial membrane, and this is followed by its N-terminus extending through the TIM23 complex at the inner mitochondrial membrane into the matrix enabling its cleavage by the matrix-processing peptidase. Yme1 localized at the IM first facilitates release of the N-terminus into the IMS followed by pulling the C-terminus of hPNPase^old-35 across the TOM complex into the IMS (Rainey et al., 2006). While these studies are quite convincing, the topic of hPNPase^old-35 subcellular localization has been controversial, as our studies involving ectopic overexpression of hPNPase^old-35 have shown that it localizes to both mitochondria and cytoplasm. In our overexpression studies, N-terminal GFP-tagged hPNPase^old-35 was found to localize to the cytoplasm (Leszczyniecka et al., 2002), which is consistent with the findings of Chen et al. (2007). This observation might be due to blockage of the N-terminal MTS, although we did not see an effect of this on the growth inhibitory effects of hPNPase^{old-35}. C-terminal HA-tagged hPNPase^{old-35}, however, localizes in both the mitochondria and cytoplasm as documented in our studies. There may be shuttling mechanisms that facilitate the transport of hPNPase^{old-35} from the mitochondrial IMS to the cytoplasm, like the one suggested by Chen et al. where hPNPase^{old-35} migrated to the cytoplasm after mitochondrial outer membrane (OM) permeabilization (Chen et al., 2006), thus allowing it to degrade certain mRNA and miRNA species as reported previously in our studies (Das et al., 2010; Sarkar et al., 2003). Conversely, physiological conditions allowing OM permeabilization or RNA import pathways could also allow RNA substrates to enter the IMS and subsequently get processed by hPNPase^{old-35} (Wang et al., 2010; Wang, Shimada, Koehler, & Teitell, 2012). Recently, hPNPase^{old-35} has also been found in the nucleus of breast cancer cells exposed to ionizing radiation where it interacts with nuclear EGFR (Yu et al., 2012). Further studies are required to establish whether this is a global or cell line and condition-specific phenomena. Based on all this information, it would be safe to say hPNPase^{old-35} may have distinct roles based on its specific localization in distinct cellular compartments.

6. EVOLUTION OF PNPases

The function of PNPases is obviously very ancient and highly conserved over large evolutionary distances as deduced from the large degree of similarity of the PDX domains in these proteins and also as a result of the observation of the widespread phylogenetic distribution of the proteins in the tree of life. Several interesting evolutionary processes can be examined using the PNPases as a model. Specifically in the case of PNPases, we can address the questions of how the domains in the hPNPase^{old-35} protein evolved and when in evolutionary time did the *hPNPase^{old-35}* function evolve?

The repeated nature of the PDX domains in PNPases suggests a common ancestry for these domains. Indeed, there is significant sequence similarity of the PDX domains in PNPases and RNaseH to assume that these three domains have arisen by duplication events from a common ancestral PDX domain in the most recent common ancestor of life (MRCA). Figure 5.5 shows two possible scenarios that could have produced the structure of PDX domains in RNaseH and hPNPase^{old-35}. The left scenario (RNaseH "first") hypothesizes that RNaseH arose first and PNP1 and PNP2 arose as terminal duplications. The scenario on the right (PNP "first") hypothesizes that one of the PNP domains arose first and RNaseH and the second PNP domain arose as a result of the terminal duplication. The

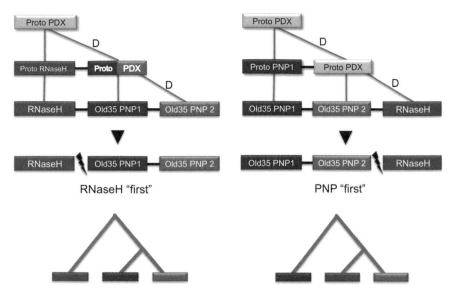

Figure 5.5 Scenarios for the evolution of PDX domain containing proteins. The scenario on the left shows the "RNaseH first" scenario and the one on the right shows the "PNPase first" scenario. Colors indicate the PDX domains discussed in the text (Red = RNaseH, Blue = PNPase1, Green = PNPase2). Gray indicates proto PDX domains. Slashes indicate the "unlinking" of the RNaseH gene from the PNPase gene. These duplication events occurred in the MRCA of the three major domains of life for both scenarios. The PNPase double PDX domain gene was then lost in the Archaea in both scenarios. The trees on the bottom indicate the expected phylogenetic relationships of the PDX domains in RNaseH and PNPase. (See Page 4 in Color Section at the back of the book.)

RNaseH domain was then "unlinked" to the PNP1–PNP2 linked domains in both scenarios. There are more complicated scenarios with several duplications and gene loss, but these two scenarios are equally parsimonious and more parsimonious than the more complicated scenarios.

This "who's on first" problem can be approached by a phylogenetic analysis of the three domains. The question then becomes which of these PDX domains are sister domains (i.e., share a more recent common ancestor than with the third domain). Figure 5.6 shows a cartoon of the results of phylogenetic analysis. This figure shows that the more basal PDX domain is the PNP1 domain and suggests that the right scenario in Fig. 5.5 best fits the data. This "PNPase first" scenario would suggest that a functional PNPase existed in the MRCA of all life on the planet.

Because RNaseH is found in all three of the major domains of life (Bacteria, Archaea, and Eukarya), and neither of the two PNPase PDX domains is found in Archaea, the "PNP first" scenario requires that we

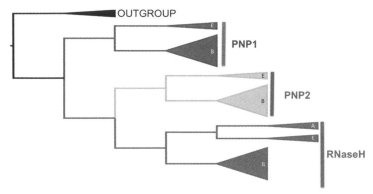

Figure 5.6 Phylogenetic tree showing the topology of relationships of PNP1, PNP2, and RNaseH. An important aspect of this is the choice of out-group for rooting the tree in the figure. Sequences from the RNase II and RNAP catalytic domains were used (Leszczyniecka et al., 2004) to root the overall PNPase tree. There is a great deal of divergence between PDX domains and the RNase II and RNAP domains, which could cause random connection of the out-group to the in-group tree; however, several mutational cost scenarios were used (Leszczyniecka et al., 2004) to analyze the amino acid sequence matrix and broadly congruent results were obtained regardless of cost scenario. (See Page 4 in Color Section at the back of the book.)

consider whether the terminally duplicated RNaseH domain in Bacteria and Eukarya was horizontally transferred to the archaeal lineage (most likely from the common ancestor of Eukarya), or whether the full complement of PNPase (with both PDX domains) and RNaseH was present in the MRCA of all life and, PNPase was simply lost (lineage extinction) in the archaeal lineage. Almeida, Leszczyniecka, Fisher, and DeSalle (2008) examined these possibilities using a "node height test." This test predicts that for HGT there should be no differences in the average substitution rates within and among the bacterial and eukaryal groups. On the other hand, the test predicts that for lineage extinction the substitution rates within Bacteria and Eukarya should be higher than the rates between the two major groups of life. As pointed out by Almeida et al. (2008), this test is only valid if there are homogeneous substitution rates across taxa. The results of the node height test indicate that the null hypothesis that HGT exists for PNPase can be rejected. Hence, the "PNPase first" scenario with subsequent lineage extinction of PNPase in Archaea is the supported scenario for how the domains evolved in the tree of life. $hPNPase^{old-35}$ function is indeed "old" as it more than likely existed in the MRCA of all life on this planet.

7. FUNCTIONS OF hPNPase^{old-35}

7.1. hPNPase^{old-35}, growth inhibition, senescence, and RNA degradation

As mentioned earlier, hPNPase^{old-35} was first identified as a gene that was upregulated in senescent progeroid fibroblasts and terminally differentiated HO-1 human melanoma cells (Leszczyniecka et al., 2002). Both senescence and terminal differentiation share certain common characteristics such as irreversible growth arrest notably occurring in the G1 phase of the cell cycle, with inhibition of both DNA synthesis and telomerase activity (Campisi, 1992; Fisher, Prignoli, Hermo, Weinstein, & Pestka, 1985; Hayflick, 1976). To elucidate the functional link between a 3′-5′ exoribonuclease such as hPNPase^{old-35} and phenomena such as senescence and terminal differentiation, replication-incompetent adenoviral-mediated hPNPase^{old-35}-overexpression (Ad.hPNPase^{old-35}) studies were performed (Sarkar et al., 2003). It was observed that slow and sustained overexpression of hPNPase^{old-35} results in inhibition of growth of multiple cell types and induces a senescent-like phenotype in HO-1 human melanoma cells and normal human melanocytes, ultimately resulting in apoptosis (Sarkar et al., 2003; Sarkar et al., 2005). The morphological features evident after about 5 days of adenoviral infection in HO-1 cells are very similar to terminally differentiated HO-1 cells treated with IFN-β and MEZ for the same period of time, which is accompanied by inhibition of telomerase activity (Sarkar et al., 2003). In separate studies, normal human melanocytes stained positive for SA-β-gal, a well-known biochemical marker for senescence, and resembled the morphology of cells entering a senescent stage by becoming large and flattened after Ad.hPNPase^{old-35} infection (Sarkar et al., 2003). This senescence-like phenotype could also be mediated by adenoviral overexpression of either of the two RNase PH domains of hPNPase^{old-35} alone, which was confirmed by deletion studies of the full-length protein (Sarkar et al., 2005). Cell-cycle analysis of different cell lines infected with Ad. hPNPase^{old-35} showed an initial G$_1$/S or G$_2$/M arrest with inhibition of DNA synthesis followed by induction of apoptosis (Sarkar et al., 2003; Van Maerken et al., 2009). This is accompanied by gene expression changes resembling those occurring during terminal differentiation and senescence-like upregulation of CDKI p27^{Kip1} levels and decreases in CDKI p21$^{CIP1/WAF-1/MDA-6}$ levels, both of which play significant roles in cell–cycle arrest (Sarkar et al., 2003). On the other hand, rapid overexpression of hPNPase^{old-35} promotes apoptosis without any cell-cycle changes implying that multiple intracellular targets and signaling pathways are involved in hPNPase^{old-35}-mediated inhibition of cell-cycle progression and apoptosis.

7.2. Molecular mechanisms of growth inhibition and senescence

Interestingly, some light was shed on one of the molecular mechanisms of *hPNPase*$^{old-35}$-induced growth inhibition and senescence when it was observed that Ad.*hPNPase*$^{old-35}$ overexpression could downregulate *c-myc* mRNA and protein expression (Sarkar et al., 2003). Interestingly, its expression is downregulated during both terminal differentiation of HO-1 melanoma cells and senescence. Myc is an important transcription factor that regulates numerous physiological processes such as cellular growth and proliferation, metabolism, and apoptosis (Dang, 1999). It is one of the key mediators that play a role in transition of cells from the G_1 to S phase of the cell cycle and is known to control p27^{Kip1} expression by multiple mechanisms. Rescue experiments involving ectopic overexpression of *c-myc* CDS (coding sequence) significantly protected HO-1 cells from Ad.*hPNPase*$^{old-35}$-mediated cell death, confirming that one mechanism by which Ad. *hPNPase*$^{old-35}$ causes growth inhibition is by downregulating *c-myc* expression (Fig. 5.7) (Sarkar et al., 2003). In addition, a recent study of breast cancer cells resistant to ionizing radiation had higher levels of *c-myc* mRNA. This was attributed to an inactive form of hPNPase which was phosphorylated at S776 in the nucleus by nEGFR-regulated DNAPK and resulting in its inability to degrade *c-myc* mRNA (Yu et al., 2012). Since *c-myc* overexpression provided only partial protection from Ad.*hPNPase*$^{old-35}$-mediated cell death, there might be other pathways involving genes that may be direct targets of hPNPase^{old-35} degradation, which is an interesting avenue of investigation we are currently pursuing.

Inhibition of growth by IFN-β is also characterized by downregulation of *c-myc* expression, which it is known to regulate posttranscriptionally (Dani et al., 1985; Knight, Anton, Fahey, Friedland, & Jonak, 1985). *hPNPase*$^{old-35}$ induction following IFN-β treatment has also been shown to downregulate *c-myc* mRNA and miR-221 levels, implicating its important role in IFN-β-mediated growth inhibition (Das et al., 2010; Sarkar, Park, & Fisher, 2006). The role of *hPNPase*$^{old-35}$ in IFN-β-mediated growth inhibition is further strengthened by the finding that transient or stable knockdown of *hPNPase*$^{old-35}$ in melanoma cells makes them resistant to IFN-β-mediated cell death (Sarkar et al., 2006). IFN-β has also been shown to reduce

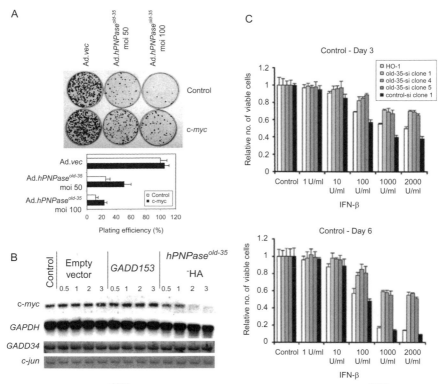

Figure 5.7 hPNPase*old-35*-mediated growth inhibition. (A) Ad.hPNPase*old-35* infection of HO-1 cells causes growth inhibition as represented by colony formation assays compared to cells infected with Ad.*vec*, which can be partially rescued by overexpressing c-myc by p290-myc. (B) Northern blot showing specific degradation of *c-myc* mRNA by hPNPase*old-35* in *in vitro* degradation assay. (C) HO-1 cells expressing hPNPase*old-35* shRNA are resistant to IFN-β-mediated growth inhibition as shown by MTT assay on day 3 and 6 posttreatment. *Figure adapted from Sarkar et al. (2003) and Sarkar et al. (2006).*

mitochondrial RNA levels (Shan, Vazquez, & Lewis, 1990), which can be another mechanism by which IFN-β-induced hPNPase[old-35] plays a role in growth inhibition via its RNA degrading activity (Fig. 5.7) (Piwowarski et al., 2003).

The mechanism by which hPNPase[old-35] mediates apoptosis has also been identified and is attributed to its ability to activate double-stranded RNA (dsRNA)-dependent protein kinase (PKR) and phosphorylation of eIF-2α. Activation of PKR results in induction of GADD153 and a decrease in the levels of the antiapoptotic protein Bcl-xL, ultimately culminating in apoptosis (Sarkar, Park, Barber, & Fisher, 2007).

7.3. RNA degradation by hPNPase^{old-35}

In vitro mRNA degradation assays further validated that full-length hPNPase^{old-35} or either of its two RNase PH domains can specifically degrade *c-myc* mRNA, whereas other mRNAs such as *c-jun, GAPDH,* or *GADD34* are not affected (Fig. 5.7) (Sarkar et al., 2003; Sarkar et al., 2005). The exact mechanism of how hPNPase^{old-35} specifically recognizes and degrades *c-myc* mRNA is still not clear, although our initial belief was that this may be mediated by a specific sequence in the 3'UTR of *c-myc* mRNA (as 3'UTRs contain various instability elements that act as important determinants of RNA turnover (López de Silanes, Quesada, & Esteller, 2007; Pesole et al., 2001), as *c-myc* RNA devoid its 3'UTR was resistant to hPNPase^{old-35} degradation (Sarkar et al., 2006; Sarkar et al., 2003). Also, no specific RNA-binding site has been described in PNPase from any other species either, which is further confounded by the fact that in *E. coli* PNPase degrades a family of CSP (Cold-shock proteins) that do not show any sequence similarity (Yamanaka & Inouye, 2001). Recent advances in studies involving substrate recognition by exoribonucleases have shown that RNA secondary structural elements may be primary determining factors in RNA binding and degradation rather than primary sequence. Future studies involving bioinformatics approaches and/or deletion/mutation analysis of *c-myc* 3'UTR mRNA sequence are required to better understand the importance of secondary structural elements in *c-myc* mRNA that allow its selective degradation by hPNPase^{old-35}.

7.4. miRNA regulation by *hPNPase*^{old-35}

miRNAs are small, single-stranded, noncoding RNAs, ~19–25 nucleotides in length that bind to a specific recognition sequence in the 3'UTR of their target mRNAs by complementary base pairing ultimately resulting in their cleavage and degradation or translational repression (Sun, Julie Li, Huang, Shyy, & Chien, 2010). Genes encoding miRNAs have been found to be located in intergenic regions, within introns of protein coding genes and also within introns and exons of nonprotein coding genes (Cai, Yu, Hu, & Yu, 2009). They are transcribed in the nucleus by RNA polymerase II or III as primary or pri-miRNA consisting of a 60–80 nucleotide long stem-loop structure. Precursor or pre-miRNA is generated by cleavage of pri-miRNA by an RNase complex DROSHA/DGCR8, which is then transported to the cytoplasm by Exportin-5. Further cleavage by an RNase Dicer results in the biogenesis of a double-stranded ~22 nucleotide miRNA which finally yields the mature miRNA after unwinding and stabilization by the Argonaute proteins (Das et al., 2011; Sun et al., 2010).

Regulation of gene expression by microRNAs was first discovered in *Caenorhabditis elegans* in 1993, and since then numerous miRNAs have been identified in plants and animal cells. The deregulation of miRNAs in various cancers and diseases has made them an active avenue of study that holds potential possibilities of therapeutic intervention (Sun et al., 2010). Stemming from this interest, recently our group identified another novel mechanism by which hPNPase^{old-35} causes growth inhibition by playing a significant role in posttranscriptional modification of miRNA biogenesis (Das et al., 2011; Das et al., 2010). In the quest to further understand hPNPase^{old-35} functions as an exoribonuclease, a miRNA Microarray analysis performed after Ad.*hPNPase^{old-35}* infection of melanoma cells resulted in the identification of specific miRNAs that were differentially regulated by *hPNPase^{old-35}*. Subsequent validation of downregulated miRNAs through immunoprecipitation, quantitative RT-PCR, Northern blotting analysis, and *in vitro* degradation assays showed that hPNPase^{old-35} could specifically bind and degrade certain mature miRNAs such as miR-221, miR-222, and miR-106b, while having no effect on others such as miR-184 and miR-let7a (Fig. 5.8) (Das et al., 2010). Also, this downregulation by hPNPase^{old-35} was limited to mature miRNAs only, primary and precursor miRNA were not affected, establishing a novel *hPNPase^{old-35}*-mediated posttranscriptional regulatory mechanism of miRNA biogenesis. Apart from *hPNPase^{old-35}* adenoviral overexpression studies, treatment of different melanoma cell lines with IFN-β (that induces *hPNPase^{old-35}*) also resulted in downregulation of miR-221 without having an effect on miR-184 or miR-let7a, which further strengthens the authenticity of this phenomenon in a more physiological setting. On the other hand, IFN-β treatment was unable to downregulate miR-221 in melanoma cells in which *hPNPase^{old-35}* was knocked down and, conversely, overexpression of miR-221 made HO-1 cells resistant to IFN-β-mediated growth arrest. Interestingly, miR-221 is a negative regulator of the *p27^{Kip1}* mRNA, and accordingly, this was identified as another pathway through which *hPNPase^{old-35}* suppresses growth by downregulating miR-221 resulting in a subsequent increase in p27^{Kip1} levels (Fig. 5.8) (Das et al., 2010). The specific binding of hPNPase^{old-35} to defined miRNAs can be sequence or secondary structure related, and this hypothesis needs to be further tested.

7.5. *hPNPase^{old-35}* and inflammation

A characteristic feature of aging is chronic inflammation, which is also a hallmark of progressive degenerative diseases such as Parkinson's disease, amyotrophic lateral sclerosis, atherosclerosis, Alzheimer's disease, and type

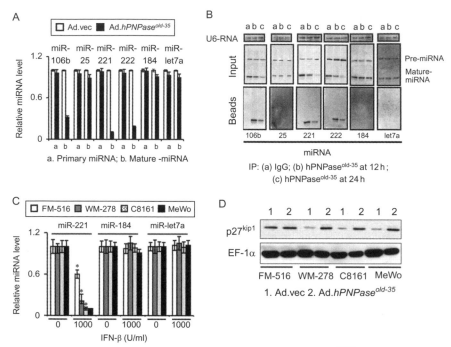

Figure 5.8 A novel mechanism of miRNA regulation by *hPNPase^old-35*. (A) qPCR analysis showing expression of various primary and mature miRNAs in HO-1 cells after Ad. *hPNPase^old-35* infection for 3 days. *hPNPase^old-35* can downregulate specific miRNAs posttranscriptionally. (B) Northern blot showing that hPNPase^old-35 binds specific miRNAs. Anti-PNPase^old-35 antibody was used to immunoprecipitate the miRNAs from Ad. *hPNPase^old-35*-infected cells after 12(b) and 24(c) h. (C) hPNPase^old-35 induced by IFN-β treatment in different melanoma cell lines is able to downregulate specific miRNAs after 24 h as shown by qPCR analysis (*, $P < 0.01$). (D) Downregulation of miR-221 by hPNPase^old-35 results in upregulation of its target p27^Kip1 at the protein level as shown by Western blot analysis in different melanoma cell lines. *Figure adapted from Das et al., 2010.*

2 diabetes (Finkel & Holbrook, 2000; Kiecolt Glaser et al., 2003). Aging is a phenomenon that is studied extensively, but at this point, not a lot is known about the relationship between cellular senescence and inflammation. Damage that is caused by oxidative stress plays an important role in aging–associated inflammation, which is evident in tissues collected from aged individuals or aged experimental animals in the form of oxidative damage in their DNA, protein, and lipids (Chen, 2000; Harman, 1956). One of the important generators of reactive oxygen species (ROS) in the cell is the mitochondrion, which causes cumulative oxidative damage over time (Finkel & Holbrook, 2000). ROS have been known to regulate activity of the transcription factor NF-κB, which can turn on the expression of

proinflammatory cytokines such as IL-6, IL-8, RANTES, and MMP-3 (Barnes & Karin, 1997; Schreck, Albermann, & Baeuerle, 1992). In our adenoviral overexpression studies, we found that hPNPase^old-35^ localization in the mitochondria plays an important role in inducing ROS that leads to degradation of IκB-α, followed by nuclear translocation of the p65 subunit of NF-κB and increased p50/p65 NF-κB DNA binding resulting in increased production of proinflammatory cytokines as mentioned earlier (Sarkar et al., 2004; Sarkar & Fisher, 2006a). The antioxidant N-acetyl-L-cysteine could inhibit all of these events linked to aging-related inflammation but does not protect cells from hPNPase^old-35^-induced growth inhibition (Sarkar et al., 2004; Sarkar & Fisher, 2006c). Since hPNPase^old-35^ expression is elevated during senescence and Ad. hPNPase^old-35^ infection increases ROS production in the mitochondria (one possibility being that ROS could be generated as a result of mitochondrial protein turnover of hPNPase^old-35^-substrates (Nagaike, Suzuki, Katoh, & Ueda, 2005; Nagaike, Suzuki, & Ueda, 2008; Wang, Shu, Lieser, Chen, & Lee, 2009), a hypothesis that needs to be further tested), we can conclude that mitochondria-localized hPNPase^old-35^ plays a role in regulating the inflammatory components of senescence (Sarkar et al., 2004; Sarkar & Fisher, 2006b). Also, given that hPNPase^old-35^ is an early type I IFN–responsive gene and activation of IFN-signaling pathway leading to the upregulation of IFN-regulated genes have been documented during the process of senescence (Leszczyniecka et al., 2002, 2003; Tahara et al., 1995), these findings indicate that IFN might play an important role in regulating inflammatory diseases of aging (Sarkar & Fisher, 2006b).

8. FUNCTIONS OF hPNPase^old-35^ IN MITOCHONDRIA

8.1. hPNPase^old-35^ and mitochondrial homeostasis

The subcellular localization of hPNPase^old-35^ in mitochondria has led to numerous studies aimed at trying to comprehend the role of hPNPase^old-35^, if any in maintaining mitochondrial homeostasis and this has been done by classical gene knockdown and overexpression studies. The level of hPNPase^old-35^ knockdown seems to play a critical role in deciding the observable changes in mitochondrial physiology and function (Chen et al., 2006; Wang et al., 2012). An ~75% reduction in hPNPase^old-35^ levels in HEK293T cells caused mitochondria to become filamentous and granular shaped, along with a three- to fourfold decrease in mitochondrial membrane potential ($\Delta\Psi$) and enzymatic activities of coupled respiratory complexes I and III, coupled complexes II and III, and individual complexes IV and V of the respiratory chain, accompanied by secondary changes such as lactate accumulation and reduction of steady-state ATP levels (Chen et al., 2006). Contradictory data are also available that show no change in mitochondrial morphology or function in HeLa cells after $hPNPase^{old-35}$ knockdown (Nagaike et al., 2005). Recent *in vivo* data, representing a more physiological setting, have confirmed a role for hPNPase^{old-35} in the maintenance of mito-chondrial homeostasis, wherein mitochondria of liver-specific *PNPT1* knockout mice show disordered, circular, and smooth cristae along with a 1.5- to 2-fold decrease in activity of respiratory chain complex IV and complexes II + III + IV (Wang et al., 2010).

Mitochondrial disorders are a diverse set of disorders and can comprise a spectrum of symptoms that include myopathy, encephalopathy, neuro-pathy, diabetes, and vision and hearing loss. Recently, mutations have been found in human PNPT1 in two disorders that have mitochondrial defect phenotypes. A homozygous p.Gln387Arg mutation was identified in a family with two affected individuals presenting with myopathy, encephalopathy, and neuropathy (Vedrenne et al., 2012). A second mutation, p.Glu475Gly, found in the same domain of PNPase causes autosomal recessive non-syndromic hearing loss in a consanguineous Moroccan family (von Ameln et al., 2012). The hearing loss is severe with early childhood onset, but there are no other phenotypic symptoms. These findings in humans support the importance of PNPase in mitochondrial function.

Overexpression of hPNPase^{old-35} is found to cause an increase in ROS accumulation over time (Sarkar et al., 2004), which could be due to increased respiratory chain activity caused by mitochondrial dysfunction and implicates a role of hPNPase^{old-35} in mitochondrial homeostasis maintenance. An exoribonuclease-independent role of hPNPase^{old-35} has also been highlighted in some studies, wherein overexpressed hPNPase^{old-35} plays a role in protecting cells from oxidized forms of RNA by limiting them from the translation mechanism (Hayakawa & Sekiguchi, 2006; Wu & Li, 2008).

8.2. hPNPase and RNA import into mitochondria

Different RNA import pathways have been identified in various organisms but not much is known about the components involved (Mukherjee, Basu, Home, Dhar, & Adhya, 2007; Tarassov, Entelis, & Martin, 1995). New advances continue to be made, with one of the recent discoveries being the involvement of PNPase in regulating translocation of RNA into the mitochondria. It has been recently reported that PNPase regulates the import of nucleus-encoded small RNAs including *RNase P* RNA, *MRP* RNA, and *5S* rRNA into the mitochondrial matrix by binding to specific stem-loop motifs in the imported RNAs (Wang et al., 2010; Wang et al., 2011).

Supporting this evidence is the proof that liver-specific *PNPT1* knockout mice mitochondria showed a significant reduction in the RNA component of the *RNase P* mtRNA complex (Wang et al., 2010). In addition, disruption of trimerization in the human disease-causing mutations described earlier leads to a decrease in RNA transport of *5S* rRNA, *MRP* RNA, and *RNAase P* RNA into the mitochondria using mitochondria isolated from skin fibroblast from patients and controls or in yeast cells, which do not have PNPase (Vedrenne et al., 2012; von Ameln et al., 2012). It should also be noted that PNPase mutations that inactivate RNA processing do not affect RNA import implying that these two functions are independent of each other. How PNPase makes the decision between RNA processing versus import remains an area that requires further experimentation (Wang et al., 2010; Wang et al., 2012).

8.3. Role of hPNPase in mtRNA processing

Polyadenylation plays an important role in mtRNA metabolism and depending on the organism can lead to quick decay or increased stability of the transcript (Gagliardi & Leaver, 1999; Temperley et al., 2003). In contrast to bacteria and some eukaryotes where polyadenylation is catalyzed by poly(A) polymerases (PAPs) or PNPase, human mtRNA is polyadenylated by mitochondria-specific PAP. In order to answer the question whether hPNPase acts as a polymerase or exoribonuclease in mitochondria, it was silenced by shRNA treatment, which resulted in an increase in the length of poly(A) tails of mtRNA without having any effect on steady-state levels of these mRNA species (Chen et al., 2006; Nagaike et al., 2005). Another group showed that stable silencing of hPNPase had varied effects on different mt-mRNAs and that polyadenylation of mtRNAs is performed by a polymerase distinct from both mtPAP and PNPase (Slomovic & Schuster, 2008). It is also suggested that hPNPase silencing causes ATP depletion, which can alter the length of mtRNA polyI(A) tails (Nagaike et al., 2005; Slomovic & Schuster, 2008). Other studies suggest a role for hPNPase in complex with hSUV3 (*human suppressor of Var1 3*) acting as a degradation complex on dsRNA in mammalian mitochondria, which hypothesizes a role for hPNPase in mitochondrial RNA degradation (Szczesny et al., 2010; Wang et al., 2009). This degradation activity is suppressed by the LRPPRC/SLIRP (leucine–rich pentatricopeptide repeat motif-containing protein/ stem-loop–interacting RNA-binding protein) complex, which also promotes polyadenylation of mRNAs in the mitochondria (Chujo et al., 2012). In the light of all this information, it is clear that a lot more needs to be resolved before a clear picture of the role of hPNPase in mtRNA processing can be depicted (Fig. 5.9).

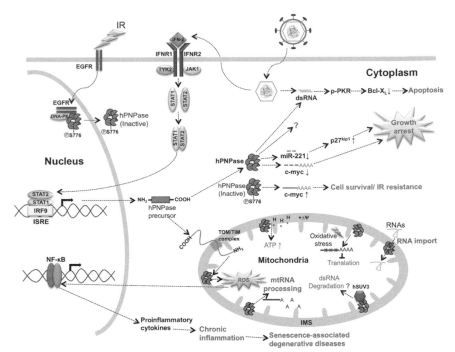

Figure 5.9 Functions of *hPNPase^old-35*. *hPNPase^old-35* is transcriptionally induced by the JAK/STAT pathway through IFN-β produced by the cell after viral infection, exposure to tumor cells, senescence-inducing signals or terminal-differentiation promoting signals. Next, hPNPase^old-35 is imported to the mitochondrial inter membrane space (IMS) by an YME-mediated mechanism or alternatively mobilized through an unknown mechanism to the cytoplasm where it causes growth arrest or apoptosis by targeting *c-myc* mRNA and miR-221 or activating PKR, respectively. Exposure to ionizing radiation inactivates hPNPase^old-35 by an EGFR-mediated mechanism, causing *c-myc* mRNA upregualtion and increasing radioresistance of cancer cells. In the mitochondria, hPNPase^old-35 maintains mitochondrial homeostasis, aids in RNA import, takes part in mtRNA processing/ degradation events, and increases ROS production that results in events leading to chronic inflammation. (For color version of this figure, the reader is referred to the online version of this chapter.)

9. CONCLUDING REMARKS

In summary, hPNPase has progressed from its initially perceived function only as a 3'-5' exoribonuclease to a molecule implicated in a multitude of diverse and important biological effects. hPNPase plays central roles in diverse physiological processes such as growth inhibition, senescence, mtRNA import, mitochondrial homeostasis, and RNA degradation, all while primarily being localized in the IMS. Further research is required to identify the mechanisms, if any, which allow hPNPase shuttling between the mitochondrial IMS and cytosol. How it decides to degrade one RNA species versus another is an additional question that needs attention. What provides substrate specificity, is it sequence, structure, or both? hPNPase also holds immense promise as a therapeutic agent due to its ability to degrade specific miRNA and mRNA species, and this property can be exploited in treating malignancies that are characterized by upregulation of harmful miRNA or mRNA molecules. At present still a lot remains unanswered regarding the biological consequences of hPNPase in normal versus cancer cells, animal models with either conditional overexpression or knockdown of hPNPase will be extremely useful. These models will prove helpful in understanding the various physiological processes hPNPase is a part of, like mtRNA import, and will also facilitate studies relating to viral infection, mitochondrial dysfunction, and aging, allowing the future development of suitable therapeutics.

ACKNOWLEDGMENTS

The present studies were supported in part by NIH Grant R01 CA097318 and the Samuel Waxman Cancer Research Foundation (SWCRF). D. S. and P. B. F. are SWCRF Investigators. D. S. is a Harrison Scholar in the VCU Massey Cancer Center (MCC) and a Blick Scholar in the VCU School of Medicine. P. B. F. holds the Thelma Newmeyer Corman Chair in Cancer Research. The excellent technical assistance of Leyla Peachy and Tiffany Simms is acknowledged.

REFERENCES

Almeida, F. C., Leszczyniecka, M., Fisher, P. B., & DeSalle, R. (2008). Examining ancient inter-domain horizontal gene transfer. *Evolutionary Bioinformatics Online, 4,* 109–119.

Andrade, J., Pobre, V., Silva, I., Domingues, S., & Arraiano, C. (2009). The role of 3'-5' exoribonucleases in RNA degradation. *Progress in Molecular Biology and Translational Science, 85,* 187–229.

Arraiano, C., Andrade, J., Domingues, S., Guinote, I., Malecki, M., Matos, R., et al. (2010). The critical role of RNA processing and degradation in the control of gene expression. *FEMS Microbiology Reviews, 34*(5), 883–923.

Audic, Y., & Hartley, R. (2004). Post-transcriptional regulation in cancer. *Biology of the Cell,* *96*(7), 479–498.

Baginsky, S., Shteiman Kotler, A., Liveanu, V., Yehudai Resheff, S., Bellaoui, M., Settlage, R. E., et al. (2001). Chloroplast PNPase exists as a homo-multimer enzyme complex that is distinct from the Escherichia coli degradosome. *RNA,* *7*(10), 1464–1475.

Barnes, P. J., & Karin, M. (1997). Nuclear factor-kappaB: A pivotal transcription factor in chronic inflammatory diseases. *The New England Journal of Medicine,* *336*(15), 1066–1071.

Cai, Y., Yu, X., Hu, S., & Yu, J. (2009). A brief review on the mechanisms of miRNA regulation. *Genomics, Proteomics & Bioinformatics,* *7*(4), 147–154.

Campisi, J. (1992). Gene expression in quiescent and senescent fibroblasts. *Annals of the New York Academy of Sciences,* *663*, 195–201.

Carpousis, A. J. (2002). The Escherichia coli RNA degradosome: Structure, function and relationship in other ribonucleolytic multienzyme complexes. *Biochemical Society Transactions,* *30*(2), 150–155.

Carpousis, A. J., Van Houwe, G., Ehretsmann, C., & Krisch, H. M. (1994). Copurification of E. coli RNAase E and PNPase: Evidence for a specific association between two enzymes important in RNA processing and degradation. *Cell,* *76*(5), 889–900.

Chen, Q. M. (2000). Replicative senescence and oxidant-induced premature senescence. Beyond the control of cell cycle checkpoints. *Annals of the New York Academy of Sciences,* *908*, 111–125.

Chen, H., Koehler, C., & Teitell, M. (2007). Human polynucleotide phosphorylase: Location matters. *Trends in Cell Biology,* *17*(12), 600–608.

Chen, H., Rainey, R., Balatoni, C., Dawson, D., Troke, J., Wasiak, S., et al. (2006). Mammalian polynucleotide phosphorylase is an intermembrane space RNase that maintains mitochondrial homeostasis. *Molecular and Cellular Biology,* *26*(22), 8475–8487.

Chujo, T., Ohira, T., Sakaguchi, Y., Goshima, N., Nomura, N., Nagao, A., et al. (2012). LRPPRC/SLIRP suppresses PNPase-mediated mRNA decay and promotes polyadenylation in human mitochondria. *Nucleic Acids Research,* *40*(16), 8033–8047.

Dang, C. V. (1999). c-myc target genes involved in cell growth, apoptosis, and metabolism. *Molecular and Cellular Biology,* *19*(1), 1–11.

Dani, C., Mechti, N., Piechaczyk, M., Lebleu, B., Jeanteur, P., & Blanchard, J. M. (1985). Increased rate of degradation of *c-myc* mRNA in interferon-treated Daudi cells. *Proceedings of the National Academy of Sciences of the United States of America,* *82*(15), 4896–4899.

Das, S. K., Bhutia, S. K., Sokhi, U. K., Dash, R., Azab, B., Sarkar, D., et al. (2011). Human polynucleotide phosphorylase (hPNPase(old-35)): An evolutionary conserved gene with an expanding repertoire of RNA degradation functions. *Oncogene,* *30*(15), 1733–1743.

Das, S. K., Sokhi, U. K., Bhutia, S., Azab, B., Su, Z. Z., Sarkar, D., et al. (2010). Human polynucleotide phosphorylase selectively and preferentially degrades microRNA-221 in human melanoma cells. *Proceedings of the National Academy of Sciences of the United States of America,* *107*(26), 11948–11953.

Deutscher, M. P., & Li, Z. (2001). Exoribonucleases and their multiple roles in RNA metabolism. *Progress in Nucleic Acid Research and Molecular Biology,* *66*, 67–105.

Doma, M., & Parker, R. (2007). RNA quality control in eukaryotes. *Cell,* *131*(4), 660–668.

Finkel, T., & Holbrook, N. J. (2000). Oxidants, oxidative stress and the biology of ageing. *Nature,* *408*(6809), 239–247.

Fisher, P. B., Prignoli, D. R., Hermo, H., Weinstein, I. B., & Pestka, S. (1985). Effects of combined treatment with interferon and mezerein on melanogenesis and growth in human melanoma cells. *Journal of Interferon Research,* *5*(1), 11–22.

Fritz, D., Bergman, N., Kilpatrick, W., Wilusz, C., & Wilusz, J. (2004). Messenger RNA decay in mammalian cells: The exonuclease perspective. *Cell Biochemistry and Biophysics,* *41*(2), 265–278.

Gagliardi, D., & Leaver, C. J. (1999). Polyadenylation accelerates the degradation of the mitochondrial mRNA associated with cytoplasmic male sterility in sunflower. *The EMBO Journal, 18*(13), 3757–3766.

Garneau, N., Wilusz, J., & Wilusz, C. (2007). The highways and byways of mRNA decay. *Nature Reviews. Molecular Cell Biology, 8*(2), 113–126.

Gewartowski, K., Tomecki, R., Muchowski, L., Dmochow Ska, A., Dzwonek, A., Malecki, M., et al. (2006). Up-regulation of human PNPase mRNA by beta-interferon has no effect on protein level in melanoma cell lines. *Acta Biochimica Polonica, 53*(1), 179–188.

Glisovic, T., Bachorik, J., Yong, J., & Dreyfuss, G. (2008). RNA-binding proteins and post-transcriptional gene regulation. *FEBS Letters, 582*(14), 1977–1986.

Grunberg-Manago, M., Ortiz, P. J., & Ochoa, S. (1955). Enzymatic synthesis of nucleic acid-like polynucleotides. *Science, 122*(3176), 907–910.

Halbeisen, R. E., Galgano, A., Scherrer, T., & Gerber, A. P. (2008). Post-transcriptional gene regulation: From genome-wide studies to principles. *Cellular and Molecular Life Sciences, 65*(5), 798–813.

Harman, D. (1956). Aging: A theory based on free radical and radiation chemistry. *Journal of Gerontology, 11*(3), 298–300.

Hayakawa, H., & Sekiguchi, M. (2006). Human polynucleotide phosphorylase protein in response to oxidative stress. *Biochemistry, 45*(21), 6749–6755.

Hayflick, L. (1976). The cell biology of human aging. *The New England Journal of Medicine, 295*(23), 1302–1308.

Hollams, E., Giles, K., Thomson, A., & Leedman, P. (2002). MRNA stability and the control of gene expression: Implications for human disease. *Neurochemical Research, 27*(10), 957–980.

Houseley, J., LaCava, J., & Tollervey, D. (2006). RNA-quality control by the exosome. *Nature Reviews. Molecular Cell Biology, 7*(7), 529–539.

Huang, F., Adelman, J., Jiang, H., Goldstein, N. I., & Fisher, P. B. (1999). Identification and temporal expression pattern of genes modulated during irreversible growth arrest and terminal differentiation in human melanoma cells. *Oncogene, 18*(23), 3546–3552.

Ibrahim, H., Wilusz, J., & Wilusz, C. (2008). RNA recognition by 3′-to-5′ exonucleases: The substrate perspective. *Biochimica Et Biophysica Acta, 1779*(4), 256–265.

Jarrige, A., Brchemier-Baey, D., Mathy, N., Duch, O., & Portier, C. (2002). Mutational analysis of polynucleotide phosphorylase from Escherichia coli. *Journal of Molecular Biology, 321*(3), 397–409.

Jarrige, A. C., Mathy, N., & Portier, C. (2001). PNPase autocontrols its expression by degrading a double-stranded structure in the pnp mRNA leader. *The EMBO Journal, 20*(23), 6845–6855.

Kiecolt Glaser, J., Preacher, K., MacCallum, R., Atkinson, C., Malarkey, W., & Glaser, R. (2003). Chronic stress and age-related increases in the proinflammatory cytokine IL-6. *Proceedings of the National Academy of Sciences of the United States of America, 100*(15), 9090–9095.

Knight, E., Anton, E. D., Fahey, D., Friedland, B. K., & Jonak, G. J. (1985). Interferon regulates *c-myc* gene expression in Daudi cells at the post-transcriptional level. *Proceedings of the National Academy of Sciences of the United States of America, 82*(4), 1151–1154.

Leszczyniecka, M., DeSalle, R., Kang, D. C. , & Fisher, P. B. (2004). The origin of polynucleotide phosphorylase domains. *Molecular Phylogenetics and Evolution, 31*(1), 123–130.

Leszczyniecka, M., Kang, D. C., Sarkar, D., Su, Z. Z., Holmes, M., Valerie, K., et al. (2002). Identification and cloning of human polynucleotide phosphorylase, hPNPase old-35, in the context of terminal differentiation and cellular senescence. *Proceedings of the National Academy of Sciences of the United States of America, 99*(26), 16636–16641.

Leszczyniecka, M., Su, Z. Z., Kang, D. C., Sarkar, D., & Fisher, P. B. (2003). Expression regulation and genomic organization of human polynucleotide phosphorylase, hPNPase(old-35), a type I interferon inducible early response gene. *Gene, 316*, 143–156.

Lin, C., Wang, Y., Yang, W., Hsiao, Y., & Yuan, H. (2012). Crystal structure of human polynucleotide phosphorylase: Insights into its domain function in RNA binding and degradation. *Nucleic Acids Research, 40*(9), 4146–4157.

López de Silanes, I., Quesada, M., & Esteller, M. (2007). Aberrant regulation of messenger RNA 3'-untranslated region in human cancer. *Cellular Oncology, 29*(1), 1–17.

Meyer, S., Temme, C., & Wahle, E. (2004). Messenger RNA turnover in eukaryotes: Pathways and enzymes. *Critical Reviews in Biochemistry and Molecular Biology, 39*(4), 197–216.

Miczak, A., Kaberdin, V. R., Wei, C. L., & Lin Chao, S. (1996). Proteins associated with RNase E in a multicomponent ribonucleolytic complex. *Proceedings of the National Academy of Sciences of the United States of America, 93*(9), 3865–3869.

Min, J. J., & Zassenhaus, H. P. (1991). Characterization of a novel NTP-dependent 3' exoribonuclease from yeast mitochondria. *SAAS Bulletin, Biochemistry and Biotechnology, 4*, 1–5.

Mitchell, P., Petfalski, E., Shevchenko, A., Mann, M., & Tollervey, D. (1997). The exosome: A conserved eukaryotic RNA processing complex containing multiple 3'–>5' exoribonucleases. *Cell, 91*(4), 457–466.

Mukherjee, S., Basu, S., Home, P., Dhar, G., & Adhya, S. (2007). Necessary and sufficient factors for the import of transfer RNA into the kinetoplast mitochondrion. *EMBO Reports, 8*(6), 589–595.

Nagaike, T., Suzuki, T., Katoh, T., & Ueda, T. (2005). Human mitochondrial mRNAs are stabilized with polyadenylation regulated by mitochondria-specific poly(A) polymerase and polynucleotide phosphorylase. *The Journal of Biological Chemistry, 280*(20), 19721–19727.

Nagaike, T., Suzuki, T., & Ueda, T. (2008). Polyadenylation in mammalian mitochondria: Insights from recent studies. *Biochimica Et Biophysica Acta, 1779*(4), 266–269.

Newbury, S. F. (2006). Control of mRNA stability in eukaryotes. *Biochemical Society Transactions, 34*(1), 30–34.

Parker, R., & Song, H. (2004). The enzymes and control of eukaryotic mRNA turnover. *Nature Structural & Molecular Biology, 11*(2), 121–127.

Pesole, G., Mignone, F., Gissi, C., Grillo, G., Licciulli, F., & Liuni, S. (2001). Structural and functional features of eukaryotic mRNA untranslated regions. *Gene, 276*(1–2), 73–81.

Piwowarski, J., Grzechnik, P., Dziembowski, A., Dmochowska, A., Minczuk, M., & Stepien, P. (2003). Human polynucleotide phosphorylase, hPNPase, is localized in mitochondria. *Journal of Molecular Biology, 329*(5), 853–857.

Portnoy, V., Palnizky, G., Yehudai Resheff, S., Glaser, F., & Schuster, G. (2008). Analysis of the human polynucleotide phosphorylase (PNPase) reveals differences in RNA binding and response to phosphate compared to its bacterial and chloroplast counterparts. *RNA, 14*(2), 297–309.

Py, B., Causton, H., Mudd, E. A., & Higgins, C. F. (1994). A protein complex mediating mRNA degradation in Escherichia coli. *Molecular Microbiology, 14*(4), 717–729.

Py, B., Higgins, C. F., Krisch, H. M., & Carpousis, A. J. (1996). A DEAD-box RNA helicase in the Escherichia coli RNA degradosome. *Nature, 381*(6578), 169–172.

Raijmakers, R., Egberts, W., van Venrooij, W., & Pruijn, G. J. M. (2002). Protein-protein interactions between human exosome components support the assembly of RNase PH-type subunits into a six-membered PNPase-like ring. *Journal of Molecular Biology, 323*(4), 653–663.

Rainey, R., Glavin, J., Chen, H., French, S., Teitell, M., & Koehler, C. (2006). A new function in translocation for the mitochondrial i-AAA protease Yme1: Import of polynucleotide phosphorylase into the intermembrane space. *Molecular and Cellular Biology*, *26*(22), 8488–8497.

Sarkar, D., & Fisher, P. B. (2006a). Human polynucleotide phosphorylase (hPNPase old-35): An RNA degradation enzyme with pleiotrophic biological effects. *Cell Cycle*, *5*(10), 1080–1084.

Sarkar, D., & Fisher, P. B. (2006b). Molecular mechanisms of aging-associated inflammation. *Cancer Letters*, *236*(1), 13–23.

Sarkar, D., & Fisher, P. B. (2006c). Polynucleotide phosphorylase: An evolutionary conserved gene with an expanding repertoire of functions. *Pharmacology & Therapeutics*, *112*(1), 243–263.

Sarkar, D., Lebedeva, I. V., Emdad, L., Kang, D. C., Baldwin, A. S., & Fisher, P. B. (2004). Human polynucleotide phosphorylase (hPNPaseold-35): A potential link between aging and inflammation. *Cancer Research*, *64*(20), 7473–7478.

Sarkar, D., Leszczyniecka, M., Kang, D. C., Lebedeva, I. V., Valerie, K., Dhar, S., et al. (2003). Down-regulation of myc as a potential target for growth arrest induced by human polynucleotide phosphorylase (hPNPaseold-35) in human melanoma cells. *The Journal of Biological Chemistry*, *278*(27), 24542–24551.

Sarkar, D., Park, E. S., Barber, G., & Fisher, P. B. (2007). Activation of double-stranded RNA dependent protein kinase, a new pathway by which human polynucleotide phosphorylase (hPNPase(old-35)) induces apoptosis. *Cancer Research*, *67*(17), 7948–7953.

Sarkar, D., Park, E. S., Emdad, L., Randolph, A., Valerie, K., & Fisher, P. B. (2005). Defining the domains of human polynucleotide phosphorylase (hPNPaseOLD-35) mediating cellular senescence. *Molecular and Cellular Biology*, *25*(16), 7333–7343.

Sarkar, D., Park, E. S., & Fisher, P. B. (2006). Defining the mechanism by which IFN-beta downregulates c-myc expression in human melanoma cells: Pivotal role for human polynucleotide phosphorylase (hPNPaseold-35). *Cell Death and Differentiation*, *13*(9), 1541–1553.

Schreck, R., Albermann, K., & Baeuerle, P. A. (1992). Nuclear factor kappa B: An oxidative stress-responsive transcription factor of eukaryotic cells (a review). *Free Radical Research Communications*, *17*(4), 221–237.

Shan, B., Vazquez, E., & Lewis, J. A. (1990). Interferon selectively inhibits the expression of mitochondrial genes: A novel pathway for interferon-mediated responses. *The EMBO Journal*, *9*(13), 4307–4314.

Shi, Z., Yang, W., Lin Chao, S., Chak, K., & Yuan, H. (2008). Crystal structure of Escherichia coli PNPase: Central channel residues are involved in processive RNA degradation. *RNA*, *14*(11), 2361–2371.

Shyu, A., Wilkinson, M., & van Hoof, A. (2008). Messenger RNA regulation: To translate or to degrade. *The EMBO Journal*, *27*(3), 471–481.

Slomovic, S., & Schuster, G. (2008). Stable PNPase RNAi silencing: Its effect on the processing and adenylation of human mitochondrial RNA. *RNA*, *14*(2), 310–323.

Stickney, L., Hankins, J., Miao, X., & Mackie, G. (2005). Function of the conserved S1 and KH domains in polynucleotide phosphorylase. *Journal of Bacteriology*, *187*(21), 7214–7221.

Sun, W., Julie Li, Y., Huang, H., Shyy, J. Y., & Chien, S. (2010). microRNA: A master regulator of cellular processes for bioengineering systems. *Annual Review of Biomedical Engineering*, *12*, 1–27.

Symmons, M. F., Jones, G. H., & Luisi, B. F. (2000). A duplicated fold is the structural basis for polynucleotide phosphorylase catalytic activity, processivity, and regulation. *Structure*, *8*(11), 1215–1226.

Symmons, M., Williams, M., Luisi, B., Jones, G., & Carpousis, A. (2002). Running rings around RNA: A superfamily of phosphate-dependent RNases. *Trends in Biochemical Sciences*, *27*(1), 11–18.

Szczesny, R., Borowski, L., Brzezniak, L., Dmochowska, A., Gewartowski, K., Bartnik, E., et al. (2010). Human mitochondrial RNA turnover caught in flagranti: Involvement of hSuv3p helicase in RNA surveillance. *Nucleic Acids Research, 38*(1), 279–298.

Tahara, H., Kamada, K., Sato, E., Tsuyama, N., Kim, J. K., Hara, E., et al. (1995). Increase in expression levels of interferon-inducible genes in senescent human diploid fibroblasts and in SV40-transformed human fibroblasts with extended lifespan. *Oncogene, 11*(6), 1125–1132.

Tarassov, I., Entelis, N., & Martin, R. P. (1995). An intact protein translocating machinery is required for mitochondrial import of a yeast cytoplasmic tRNA. *Journal of Molecular Biology, 245*(4), 315–323.

Temperley, R., Seneca, S., Tonska, K., Bartnik, E., Bindoff, L., Lightowlers, R., et al. (2003). Investigation of a pathogenic mtDNA microdeletion reveals a translation-dependent deadenylation decay pathway in human mitochondria. *Human Molecular Genetics, 12*(18), 2341–2348.

Van Maerken, T., Sarkar, D., Speleman, F., Dent, P., Weiss, W., & Fisher, P. B. (2009). Adenovirus-mediated hPNPase(old-35) gene transfer as a therapeutic strategy for neuroblastoma. *Journal of Cellular Physiology, 219*(3), 707–715.

Vedrenne, V., Gowher, A., De Lonlay, P., Nitschke, P., Serre, V., Boddaert, N., et al. (2012). Mutation in PNPT1, which encodes a polyribonucleotide nucleotidyl-transferase, impairs RNA import into mitochondria and causes respiratory-chain deficiency. *American Journal of Human Genetics, 91*(5), 912–918.

von Ameln, S., Wang, G., Boulouiz, R., Rutherford, M. A., Smith, G. M., Li, Y., et al. (2012). Mutation in PNPT1, encoding mitochondrial-RNA-import protein PNPase, causes hereditary hearing loss. *American Journal of Human Genetics, 91*(5), 919–927.

Wang, G., Chen, H., Oktay, Y., Zhang, J., Allen, E., Smith, G., et al. (2010). PNPASE regulates RNA import into mitochondria. *Cell, 142*(3), 456–467.

Wang, G., Shimada, E., Koehler, C., & Teitell, M. (2012). PNPASE and RNA trafficking into mitochondria. *Biochimica Et Biophysica Acta, 1819*(9–10), 998–1007.

Wang, D., Shu, Z., Lieser, S., Chen, P., & Lee, W. (2009). Human mitochondrial SUV3 and polynucleotide phosphorylase form a 330-kDa heteropentamer to cooperatively degrade double-stranded RNA with a 3′-to-5′ directionality. *The Journal of Biological Chemistry, 284*(31), 20812–20821.

Wu, J., & Li, Z. (2008). Human polynucleotide phosphorylase reduces oxidative RNA damage and protects HeLa cell against oxidative stress. *Biochemical and Biophysical Research Communications, 372*(2), 288–292.

Yamanaka, K., & Inouye, M. (2001). Selective mRNA degradation by polynucleotide phosphorylase in cold shock adaptation in Escherichia coli. *Journal of Bacteriology, 183*(9), 2808–2816.

Yehudai Resheff, S., Portnoy, V., Yogev, S., Adir, N., & Schuster, G. (2003). Domain analysis of the chloroplast polynucleotide phosphorylase reveals discrete functions in RNA degradation, polyadenylation, and sequence homology with exosome proteins. *The Plant Cell, 15*(9), 2003–2019.

Yu, Y., Chou, R., Wu, C., Wang, Y., Chang, W., Tseng, Y., et al. (2012). Nuclear EGFR suppresses ribonuclease activity of polynucleotide phosphorylase through DNAPK-mediated phosphorylation at serine 776. *The Journal of Biological Chemistry, 287*(37), 31015–31026.

Zuo, Y., & Deutscher, M. P. (2001). Exoribonuclease superfamilies: Structural analysis and phylogenetic distribution. *Nucleic Acids Research, 29*(5), 1017–1026.

CHAPTER SIX

FOXM1 (Forkhead box M1) in Tumorigenesis: Overexpression in Human Cancer, Implication in Tumorigenesis, Oncogenic Functions, Tumor-Suppressive Properties, and Target of Anticancer Therapy

Inken Wierstra[1]

Wißmannstr. 17, D-30173 Hannover, Germany
[1]Corresponding author: e-mail address: iwiwiwi@web.de

Contents

The present chapter is Part II of a two-part review on the transcription factor FOXM1. Part I of this FOXM1 review is published in Volume 118 of Advances in Cancer Research: Inken Wierstra. The Transcription Factor FOXM1 (Forkhead box M1): Proliferation-Specific Expression, Transcription Factor Function, Target Genes, Mouse Models, and Normal Biological Roles. Advances in Cancer Research, 2013, volume 118, pages 97–398.

Advances in Cancer Research, Volume 119
ISSN 0065-230X
http://dx.doi.org/10.1016/B978-0-12-407190-2.00016-2
191

Abstract

FOXM1 (Forkhead box M1) is a typical proliferation-associated transcription factor and is also intimately involved in tumorigenesis. FOXM1 stimulates cell proliferation and cell cycle progression by promoting the entry into S-phase and M-phase. Additionally, FOXM1 is required for proper execution of mitosis. In accordance with its role in stimulation of cell proliferation, FOXM1 exhibits a proliferation-specific expression pattern and its expression is regulated by proliferation and anti-proliferation signals as well as by proto-oncoproteins and tumor suppressors. Since these factors are often mutated, overexpressed, or lost in human cancer, the normal control of the *foxm1* expression by them provides the basis for deregulated FOXM1 expression in tumors. Accordingly, FOXM1 is overexpressed in many types of human cancer. FOXM1 is intimately involved in tumorigenesis, because it contributes to oncogenic transformation and participates in tumor initiation, growth, and progression, including positive effects on angiogenesis, migration, invasion, epithelial–mesenchymal transition, metastasis, recruitment of tumor-associated macrophages, tumor-associated lung inflammation, self-renewal capacity of cancer cells, prevention of premature cellular senescence, and chemotherapeutic drug resistance. However, in the context of urethane-induced lung tumorigenesis, FOXM1 has an unexpected tumor suppressor role in endothelial cells because it limits pulmonary inflammation and canonical Wnt signaling in epithelial lung cells, thereby restricting carcinogenesis. Accordingly, FOXM1 plays a role in homologous recombination repair of DNA double-strand breaks and maintenance of genomic stability, that is, prevention of polyploidy and aneuploidy. The implication of FOXM1 in tumorigenesis makes it an attractive target for anticancer therapy, and several antitumor drugs have been reported to decrease FOXM1 expression.

1. FOXM1 OVEREXPRESSION IN TUMOR CELLS

1.1. The overexpression of FOXM1 in human tumors

1.1.1 FOXM1 is overexpressed in numerous human cancers

FOXM1 (Forkhead box M1) is overexpressed in a large variety of human tumors, namely, in hepatocellular carcinoma (HCC), HBV (hepatitis B virus)-associated HCC (HBV-HCC), intrahepatic cholangiocarcinoma (ICC), pancreatic ductal adenocarcinoma (PDA), gastric cancer, colon adenocarcinoma, colorectal adenoma, colorectal cancer (CRC), esophageal squamous cell carcinoma (ESCC), esophageal adenocarcinoma, anaplastic thyroid carcinoma (ATC), lung adenocarcinoma (non-small cell lung cancer (NSCLC)), lung squamous cell carcinoma (NSCLC), malignant pleural mesothelioma (MM), anaplastic astrocytoma and glioblastoma, glioblastoma multiforme (GBM), classic and desmoplastic medulloblastoma, meningioma, head and neck squamous cell carcinoma (HNSCC), basal cell carcinoma (BCC), breast carcinoma, mammary ductal carcinoma *in situ* (DCIS), mammary invasive ductal carcinoma (IDC), triple negative breast cancer (TNBC), ovarian cancer, serous ovarian carcinoma, cervical cancer, prostate adenocarcinoma, testicular cancer, uterine cancer, bladder carcinoma, renal cell carcinoma, clear cell renal cell carcinoma (ccRCC), laryngeal squamous cell carcinoma (LSCC), oropharyngeal squamous cell carcinoma (OPSCC), nasopharyngeal carcinoma (NPC), diffuse large B-cell lymphoma (DLBCL), and acute myeloid leukemia (AML) (Table 6.1) (Banz et al., 2009; Bektas et al., 2008; Bellelli et al., 2012; Calvisi et al., 2009; Chan et al., 2008; Chen et al., 2012a; Chu et al., 2012; Craig et al., 2013; Dibb et al., 2012; Elgaaen et al., 2012; Francis et al., 2009; Garber et al., 2001; Gemenetzidis et al., 2009; Gialmanidis et al., 2009; Guerra et al., 2013; He et al., 2012; Hodgson et al., 2009; Hui et al., 2012; Iacobuzio-Donahue et al., 2003; Janus et al., 2011; Jiang et al., 2011; Kalin et al., 2006; Kalinina et al., 2003; Kim et al., 2006a; Kretschmer et al., 2011; Laurendeau et al., 2010; Li et al., 2009, 2013; Lin et al., 2010a,b; Liu et al., 2006; Llaurado et al., 2012; Lok et al., 2011; Nakamura et al., 2004, 2009, 2010a,b; Newick et al., 2012; Obama et al., 2005; Okabe et al., 2001; Park et al., 2012; Pellegrino et al., 2010; Perou et al., 2000; Pignot et al., 2012; Pilarsky et al., 2004;

Table 6.1 Overexpression of FOXM1 in human tumors

Tumor with FOXM1 overexpression	References
Hepatocellular carcinoma (HCC)	Okabe et al. (2001), Kalinina et al. (2003), Calvisi et al. (2009), Lin et al. (2010b), Pellegrino et al. (2010), Wu et al. (2010a), Sun et al. (2011b), Qu et al. (2013), and references in Pilarsky et al. (2004)
HBV-associated HCC (HBV-HCC)	Xia et al. (2012b)
Intrahepatic cholangiocarcinoma (ICC)	Obama et al. (2005) and Yokomine et al. (2009)
Pancreatic adenocarcinoma	Iacobuzio-Donahue et al. (2003), Nakamura et al. (2004), Rodriguez et al. (2005), and references in Pilarsky et al. (2004)
Pancreatic ductal adenocarcinoma (PDA)	Xia et al. (2012a)
Gastric cancer	Li et al. (2009), Zeng et al. (2009), Park et al. (2012), and references in Pilarsky et al. (2004)
Colon adenocarcinoma	Yoshida et al. (2007), Park et al. (2012), and references in Pilarsky et al. (2004)
Colorectal adenoma and colorectal carcinoma (CRC)	Uddin et al. (2011)
Colorectal cancer (CRC)	Chu et al. (2012) and Li et al. (2013)
Esophageal squamous cell carcinoma (ESCC)	Hui et al. (2012)
Esophageal adenocarcinoma	Dibb et al. (2012) and references in Gemenetzidis et al. (2009)
Laryngeal squamous cell carcinoma (LSCC)	Jiang et al. (2011)
Oropharyngeal squamous cell carcinoma (OPSCC)	Janus et al. (2011)
Nasopharyngeal carcinoma (NPC)	Chen et al. (2012a,b)
Anaplastic thyroid carcinoma (ATC)	Salvatore et al. (2007) and Bellelli et al. (2012)

Table 6.1 Overexpression of FOXM1 in human tumors—cont'd

Tumor with FOXM1 overexpression	References
Lung adenocarcinoma (NSCLC)	Garber et al. (2001), Kim et al. (2006a,b,c), Gialmanidis et al. (2009), and Park et al. (2012)
Lung squamous cell carcinoma (NSCLC)	Kim et al. (2006a,b,c), Gialmanidis et al. (2009), and references in Pilarsky et al. (2004)
Lung cancer (NSCLC)	Takahashi et al. (2006)
Malignant pleural mesothelioma (MM)	Romagnoli et al. (2009), Newick et al. (2012), and references in Gemenetzidis et al. (2009)
Anaplastic astrocytoma (WHO grade III) and glioblastoma (WHO grade IV)	van den Boom et al. (2003)
Glioblastoma multiforme (GBM, WHO grade IV)	Rickman et al. (2001), Liu et al. (2006), Hodgson et al. (2009), and Lin et al. (2010a)
Classic and desmoplastic medulloblastoma	Schüller et al. (2006)
Meningioma	Laurendeau et al. (2010)
Head and neck squamous cell carcinoma (HNSCC)	Gemenetzidis et al. (2009) and Waseem et al. (2010)
Basal cell carcinoma (BCC)	Teh et al. (2002)
Breast carcinoma	Perou et al. (2000), Sorlie et al. (2001), Wonsey and Follettie (2005), Bektas et al. (2008), Francis et al. (2009), Park et al. (2012), and references in Pilarsky et al. (2004)
Mammary ductal carcinoma *in situ* (DCIS) and mammary invasive ductal carcinoma (IDC)	Kretschmer et al. (2011)
Triple negative breast cancer (TNBC)	Craig et al. (2013)
Ovarian cancer	Banz et al. (2009), Lok et al. (2011), Llaurado et al. (2012), Park et al. (2012), and references in Pilarsky et al. (2004)

Continued

Table 6.1 Overexpression of FOXM1 in human tumors—cont'd

Tumor with FOXM1 overexpression	References
Serous ovarian carcinoma	The Cancer Genome Atlas Research Network (2011) and Elgaaen et al. (2012)
Cervical cancer	Pilarsky et al. (2004), Rosty et al. (2005), Santin et al. (2005), Chan et al. (2008), and He et al. (2012)
Prostate adenocarcinoma	Kalin et al. (2006) and references in Pilarsky et al. (2004)
Testicular cancer	Pilarsky et al. (2004)
Uterine cancer	Pilarsky et al. (2004)
Bladder carcinoma	Pignot et al. (2012) and references in Pilarsky et al. (2004)
Renal cell carcinoma	References in Pilarsky et al. (2004)
Clear cell renal cell carcinoma (ccRCC)	Xue et al. (2012)
Diffuse large B-cell lymphoma (DLBCL)	Uddin et al. (2012)
Acute myeloid leukemia (AML)	Nakamura et al. (2009, 2010a,b) and Zhang et al. (2012d)

HBV, hepatitis B virus; NSCLC, non-small cell lung cancer; WHO, world health organization.

Qu et al., 2013; Rickman et al., 2001; Rodriguez et al., 2005; Romagnoli et al., 2009; Rosty et al., 2005; Salvatore et al., 2007; Santin et al., 2005; Schüller et al., 2006; Sorlie et al., 2001; Sun et al., 2011b; Takahashi et al., 2006; Teh et al., 2002; The Cancer Genome Atlas Research Network, 2011; Uddin et al., 2011, 2012; van den Boom et al., 2003; Waseem et al., 2010; Wonsey and Follettie, 2005; Wu et al., 2010a; Xia et al., 2012a,b; Xue et al., 2012; Yokomine et al., 2009; Yoshida et al., 2007; Zeng et al., 2009; Zhang et al., 2012d).

In particular, the FOXM1 expression level increases with tumor grade or stage (Carr et al., 2012; Chan et al., 2008, 2012; Chen et al., 2013b; Chu et al., 2012; He et al., 2012; Janus et al., 2011; Jiang et al., 2011; Kalin et al., 2006; Laurendeau et al., 2010; Li et al., 2012a, 2013; Lin et al., 2010b; Liu et al., 2006; Llaurado et al., 2012; Lok et al., 2011; Nishidate et al., 2004; Qu et al., 2013; Rickman et al., 2001; Sotiriou et al., 2006; The Cancer Genome Atlas Research Network, 2011; van den Boom et al., 2003; Wonsey and Follettie, 2005; Wu et al., 2013; Xia et al., 2012a; Xu et al., 2012c; Xue et al., 2012; Yang et al., 2009).

Moreover, the FOXM1 expression level is inversely correlated with patient survival times so that a high FOXM1 expression correlates with poor prognosis (Calvisi et al., 2009; Chen et al., 2009; Chu et al., 2012; Dibb et al., 2012; Ertel et al., 2010; Fournier et al., 2006; Freije et al., 2004; Garber et al., 2001; He et al., 2012; Jiang et al., 2011; Li et al., 2009, 2013; Liu et al., 2006; Martin et al., 2008; Okada et al., 2013; Park et al., 2012; Pellegrino et al., 2010; Priller et al., 2011; Qu et al., 2013; Sun et al., 2011a,b; Takahashi et al., 2006; Wang et al., 2013; Wu et al., 2013; Xia et al., 2012a,b,c; Xue et al., 2012; Yang et al., 2009; Yau et al., 2011; Yu et al., 2011; Zhang et al., 2012e).

Furthermore, a high FOXM1 expression correlates with incidence of metastases (Chu et al., 2012; Garber et al., 2001; Jiang et al., 2011; Li et al., 2009, 2013; Mito et al., 2009; Xia et al., 2012a; Xu et al., 2012c; Xue et al., 2012; Yang et al., 2009).

1.1.2 FOXM1 overexpression and clinicopathological characteristics

Metastatic prostate cancers showed a higher FOXM1 expression than the primary organ-confined prostate tumors (Chandran et al., 2007; LaTulippe et al., 2002).

In colon cancer, the FOXM1 expression level was significantly higher in lymph node metastases than in primary colon tumors (Li et al., 2013).

A high FOXM1 expression correlated with decreased MFS (metastasis-free survival) in human malignant fibrous histiocytoma (MFH, undifferentiated pleomorphic sarcoma) and with the development of lung metastases in the $LSL\text{-}Kras^{G12D}$; $p53^{Flox/Flox}$ mouse model of soft tissue sarcoma (Mito et al., 2009).

A strong FOXM1 expression was correlated with an increased incidence of lymph node metastasis in primary gastric tumors (Li et al., 2009), PDA (Xia et al., 2012a), CRC (Chu et al., 2012), NSCLC (Xu et al., 2012c) and LSCC (Jiang et al., 2011).

In ccRCC (Xue et al., 2012) and CRC (Li et al., 2013), a high FOXM expression level correlated with lymph node metastasis as well as with distant metastasis.

In lung adenocarcinoma and pulmonary SCC (squamous cell carcinoma), high FOXM1 expression was found in an invasive subgroup characterized by poor prognosis, high incidence of metastases, and poor tumor differentiation (Garber et al., 2001; Yang et al., 2009).

HCC patients with FOXM1-positive tumors showed shorter survival times and more aggressive clinicopathological tumor phenotypes (larger tumor size, multiple tumor numbers, bilobar tumor distribution, poor tumor differentiation, advanced tumor stage, and macrovascular invasion) than patients with FOXM1-negative tumors (Sun et al., 2011a).

HBV-HCC patients with FOXM1-positive tumors had shorter overall survival times and higher recurrence rates than patients with FOXM1-negative tumors, and the FOXM1-positive group showed multiple malignant clinicopathological characteristics (larger maximal tumor size, loss of tumor encapsulation, microvascular invasion, malignant differentiation, advanced TNM (tumor-node-metastasis) stage) (Xia et al., 2012b). The same malignant clinicopathological characteristics were also observed among patients with FOXM1-positive HCC (Xia et al., 2012b).

In PTC (papillary thyroid carcinoma), FOXM1 expression was more frequently detected in the aggressive tall-cell variant compared with the classical and follicular variants (Ahmed et al., 2012).

In pulmonary NE (neuroendocrine) tumors, FOXM1 expression was more frequently detected in the aggressive high-grade NE tumors (LCNEC (high-grade large cell neuroendocrine carcinoma) and SCLC (small cell lung cancer)) than in the carcinoid tumors (TC (low-grade typical carcinoid) and AC (intermediate-grade atypical carcinoid)) and the frequency of FOXM1 expression increased with the aggressiveness of the four subtypes, that is, from TC (0% FOXM1-positive, least aggressive) to AC to LCNEC to SCLC (94.4% FOXM1-positive, most aggressive) (Ha et al., 2012).

For human AML, the *foxm1* mRNA level was approximately 21-fold higher in *de novo* AML patients than in healthy controls and the *foxm1* mRNA expression was dramatically reduced after complete remission achievement in induction chemotherapy, namely, by 92% compared to AML and thus to only approximately double the amount in healthy individuals (Zhang et al., 2012d).

In GBM, the *foxm1* mRNA and protein expression was significantly higher in recurrent tumors than in the primary tumors (Zhang et al., 2012e).

In NSCLC, a high FOXM1 expression level was correlated with recurrence after tumor resection, that is, with shorter disease-free survival (Xu et al., 2012c).

In general, a high FOXM1 expression level correlates with tumor progression and aggressiveness.

1.1.3 Amplification of the foxm1 locus in human tumors

The *foxm1* locus at 12p13.33 is amplified in the three most common NHL (Non-Hodgkin's lymphoma) entities, namely, in DLBCL, FL (follicular lymphoma), and B-CLL (B-cell chronic lymphocytic leukemia) (Green et al., 2011). As expected, the *foxm1* transcript level is higher in NHL samples with the 12p13.33 amplification than in those without this amplification (Green et al., 2011).

The *foxm1* gene is amplified in primary breast cancer, as evidenced by somatically acquired copy number aberrations (Curtis et al., 2012). As expected, this *foxm1* amplification is associated with FOXM1 over-expression in primary breast tumors, which suggests that FOXM1 may be a putative breast cancer driver (Curtis et al., 2012).

1.1.4 FOXM1 is found in several cancer signatures

FOXM1 is part of the breast tumor proliferation cluster (Perou et al., 2000; Sorlie et al., 2001), which includes genes whose elevated expression correlates with increased proliferation rates of tumors (Chung et al., 2002). Like many other members of this proliferation signature (Whitfield et al., 2006), FOXM1 is a cell cycle-regulated gene (Bar-Joseph et al., 2008; Whitfield et al., 2002, 2006).

FOXM1 belongs to a group of 100 transcription factors, which are enriched in high-grade breast cancers, that is, in poorly differentiated mammary tumors (Ben-Porath et al., 2010).

FOXM1 is part of the SV40 (simian virus 40) T/t-antigen signature of epithelial tumors, which was identified by comparison of gene expression profiles from mouse models of T/t-antigen-induced breast, prostate, and lung cancers (Deeb et al., 2007). The SV40 T/t-antigen gene signature is highly predictive of survival in breast cancer patients such that expression of this signature correlates with poor prognosis (Deeb et al., 2007).

FOXM1 is part of the CCPC (cervical cancer proliferation cluster), which was defined in HPV16/HPV18-positive invasive cervical carcinoma (Rosty et al., 2005). The expression level of the CCPC genes was positively correlated with the *E7* (*early ORF* (*open reading frame*) 7) mRNA level and to a lower extent with the HPV DNA load (Rosty et al., 2005). The average expression level of all CCPC genes was higher in tumors with an unfavorable outcome (i.e., early relapse) than in tumors with a favorable course so that it may be indicative of poor disease prognosis (Rosty et al., 2005).

FOXM1 is part of the so-called CIN25 signature, that is, the 25 top-scoring genes, when a signature of CIN (chromosomal instability) is inferred from gene expression profiles building on the observation of significant correspondence between copy number alterations and gene expression changes in the affected regions (Carter et al., 2006). Net overexpression of the CIN25 signature is predictive of poor clinical outcome in six different types of cancer (breast cancer, lung cancer, medulloblastoma, glioma, mesothelioma, and lymphoma), and the expression level of the CIN25 signature is higher in metastatic specimens than in primary tumors (Carter et al., 2006).

FOXM1 belongs to the meta-signature of undifferentiated cancer, that is, a transcriptional profile of genes overexpressed in undifferentiated cancers compared to well-differentiated cancers of the same origin, which is common to multiple types of cancer (Rhodes et al., 2004). This signature may be important for the ability of cancer cells to avoid differentiation or to dedifferentiate so that it may play a role in cancer progression (Rhodes et al., 2004).

FOXM1 belongs to the RB (retinoblastoma protein, RB1 (retinoblastoma 1), p105)-loss signature that consists of 159 genes, which were identified in common among at least two of the following three model systems, namely, (1) upregulated by RB deletion in fibroblastic models, (2) repressed by the activation of RB, or (3) upregulated with acute RB deletion in murine liver (Ertel et al., 2010). In human breast cancer, the expression of the RB-loss signature is elevated in tumor samples relative to normal controls and a high RB-loss signature expression is associated with poor disease outcome in ER (estrogen receptor)-positive breast cancer (Ertel et al., 2010).

FOXM1 is part of the prognostic gene expression signature CINSARC (complexity index in sarcomas), which is composed of 67 genes related to mitosis and chromosome integrity and which predicts negative metastatic outcome in non-translocation-related sarcomas such that sarcoma patients with a high expression level of the CINSARC signature show shorter MFS (metastasis-free survival) times (Chibon et al., 2010). Also in three other cancers, the CINSARC signature predicts poor clinical outcome so that a higher expression of the CINSARC signature correlates with a lower overall survival rate for lymphoma patients and with lower MFS rates for patients with breast carcinomas and GISTs (gastrointestinal stromal tumors) (Chibon et al., 2010).

1.2. The normal control of the FOXM1 expression by proto-oncoproteins and tumor suppressors as well as by other proliferation and anti-proliferation signals implies the deregulation of the FOXM1 expression in tumor cells

1.2.1 General considerations

FOXM1 displays a strictly proliferation-specific expression pattern (Costa et al., 2001, 2003; Kalin et al., 2011; Koo et al., 2011; Laoukili et al.,

2007; Murakami et al., 2010; Raychaudhuri and Park, 2011; Wierstra and Alves, 2007c). Accordingly, its expression is upregulated by proliferation signals, but downregulated by anti-proliferation signals (Koo et al., 2011; Laoukili et al., 2007; Raychaudhuri and Park, 2011; Wang et al., 2010a; Wierstra and Alves, 2007c). In particular, the expression of FOXM1 is increased by proto-oncoproteins (c-Myc (MYC), c-Myb (MYB), AKT, H-Ras, N-Ras, EGFR (EGF (epidermal growth factor) receptor, HER1 (human EGF receptor 1), ErbB1), HER2/Neu (ErbB2), Notch1, c-Rel, STAT3 (signal transducer and activator of transcription 3), cyclin D1/ Cdk4 (cyclin-dependent kinase 4), Wnt3a (wingless type 3a), c-Met (HGFR (HGF (hepatocyte growth factor) receptor)), ETV5 (ETS variant gene 5), YAP (Yes-associated protein), GLI-1 (glioma transcription factor-1)), whereas the FOXM1 expression is decreased by tumor suppressors (RB, PTEN (phosphatase and tensin homolog deleted on chromosome ten), Cdh1 (Cdc20 (cell division cycle 20) homolog 1, FZR1 (Fizzy-related protein 1)), GATA3 (GATA-binding protein 3), KLF4 (Krüppel-like factor 4), p53 (TP53, tumor protein p53)), potential tumor suppressors (Mad1 (Max (Myc-associated factor X) dimerizer 1, MXD1 (Max dimerization protein 1)), FOXO3a (Forkhead box O3a)), and tumor suppressor miRNAs (miR-31, miR-34, miR-200b) (Fig. 6.1). This normal control of the FOXM1 expression by proto-oncoproteins and tumor suppressors implies the deregulation of the FOXM1 expression in tumor cells (Wierstra and Alves, 2007c) because proto-oncogenes are often activated and tumor suppressor genes are often inactivated in cancer cells (Balmain et al., 2003; Benvenuti et al., 2005; Bishop, 1995; Bocchetta and Carbone, 2004; Boulikas, 1995; Futreal et al., 2004; Hanahan and Weinberg, 2000, 2011; Harris and McCormick, 2010; Hesketh, 1997; Lee and Muller, 2010; Massagué, 2004; Polsky and Cordon-Cardo, 2003; Ponder, 2001; Sherr, 2004; Vogelstein and Kinzler, 2004; Weinberg, 2006; Yeang et al., 2008). Also proliferation signals and anti-proliferation signals are frequently activated or inactivated in cancer, respectively (Balmain et al., 2003; Benvenuti et al., 2005; Bishop, 1995; Bocchetta and Carbone, 2004; Boulikas, 1995; Futreal et al., 2004; Hanahan and Weinberg, 2000, 2011; Harris and McCormick, 2010; Hesketh, 1997; Lee and Muller, 2010; Massagué, 2004; Polsky and Cordon-Cardo, 2003; Ponder, 2001; Sherr, 2004; Vogelstein and Kinzler, 2004; Weinberg, 2006; Yeang et al., 2008), so that the normal control of the FOXM1 expression by other proliferation signals (E2F-1, ERK2 (extracellular signal-regulated kinase 2), MEK1/2 (MAPK (mitogen-activated protein kinase)/ERK kinase 1/2), cyclin D3/Cdk6, B-Myb (MYBL2 (MYB-like 2)), HSF1 (heat shock factor 1),

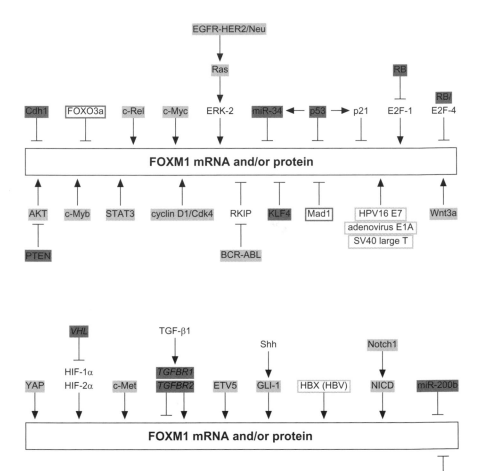

Figure 6.1 Proto-oncoproteins and tumor suppressors, which control the FOXM1 (Forkhead box M1) expression. The figure depicts the control of the FOXM1 mRNA or/and protein expression by cellular proto-oncoproteins (light gray), tumor suppressors (dark gray), candidate tumor suppressors (dark gray border), and viral oncoproteins (light gray border). HIF-1α and HIF-2α increase the *foxm1* mRNA expression, but a regulation of the FOXM1 expression by VHL (von Hippel–Lindau) (italics) has not been shown so far. TGF-β1 was reported to decrease the *foxm1* mRNA level and to increase the FOXM1 protein level, but a regulation of the FOXM1 expression by TGFBR1 (TGF-β type I receptor) or TGFBR2 (TGF-β type II receptor) (italics) has not been shown so far. Proteolytic cleavage of Notch1 generates the NICD (Notch intracellular domain).

PLK1 (Polo-like kinase 1), Shh (Sonic hedgehog)) and other anti-proliferation signals (Mxi1 (Max interactor 1), SPDEF (SAM-pointed domain-containing ETS transcription factor, PDEF (prostate-derived ETS transcription factor)), p21 (CDKN1A (Cdk inhibitor 1A), CIP1 (Cdk-interacting protein 1), WAF1 (wild-type p53-activated fragment)), DKK1 (Dickkopf 1), C/EBPα (CCAAT/enhancer-binding protein α)) (Fig. 6.2) may provide an additional basis for deregulated expression of FOXM1 in tumors (Wierstra and Alves, 2007c). Thus, the antagonistic regulation of the FOXM1 expression by proto-oncoproteins and tumor suppressors (Fig. 6.1) as well as by other proliferation and anti-proliferation signals (Fig. 6.2) in normal cells should predetermine the overexpression of FOXM1 in tumor cells (Wierstra and Alves, 2007c). In fact, FOXM1 is overexpressed in a large variety of human cancers (Table 6.1).

The overexpression of FOXM1 in human tumors may result either from mutations in the *foxm1* locus itself or from alterations in those signal trans-duction chains, which control the FOXM1 expression at the level of tran-scription, mRNA stability, translation, or protein stability. Since the signaling pathways controlling the normal FOXM1 expression include numerous proto-oncoproteins and tumor suppressors (Fig. 6.1), their fre-quent activation or inactivation in human cancer, respectively, offers many opportunities for the development of FOXM1 overexpression in tumors. Also tumor-derived alterations of the upstream regulators and downstream effectors (Fig. 6.1) of these proto-oncoproteins and tumor suppressors might lead to FOXM1 overexpression in tumors.

1.2.2 Possible deregulation of the foxm1 expression by transcription factors that are either proto-oncoproteins or tumor suppressors

Among the proto-oncoproteins, which increase the FOXM1 expression (Fig. 6.1), are the transcription factors c-Myc (Blanco-Bose et al., 2008; Delpuech et al., 2007; Huynh et al., 2010; Keller et al., 2007), c-Myb (Lefebvre et al., 2010), ETV5 (Llaurado et al., 2012), STAT3 (Mencalha et al., 2012), and c-Rel (Gieling et al., 2010) as well as the transcriptional coactivator YAP (interacts with DNA-binding transcription factors (e.g., TEAD (TEA domain family member))) (Mizuno et al., 2012), which all occupy the *foxm1* promoter *in vivo* (Table 6.2). The proto-oncoproteins, which elevate the FOXM1 expression, include also the transcription factor GLI-1 (Calvisi et al., 2009; Teh et al., 2002) and the transcriptional coactivator Notch1 (NICD (Notch intracellular domain) interacts with the DNA-binding transcription factor CSL (CBF1 (C-promoter-binding

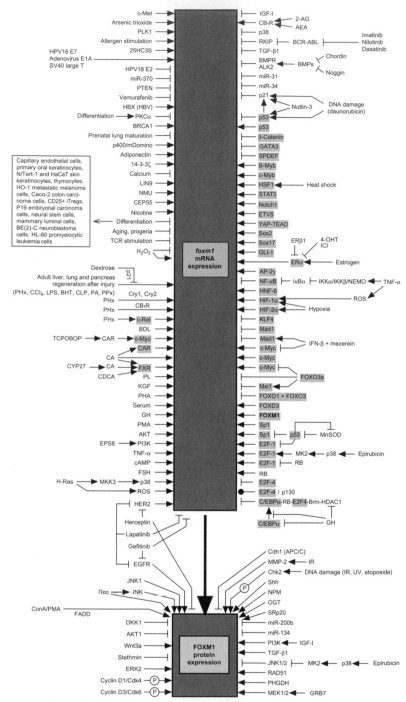

Figure 6.2 Control of the FOXM1 expression. Shown are the factors which regulate the expression of *foxm1* at the mRNA or/and protein level. Transcription factors are marked in gray and their effects on the *foxm1* expression are summarized in Table 6.2. The top left box lists those differentiation systems in which the *foxm1* expression was downregulated

(Continued)

Figure 6.2—Cont'd during differentiation. Phosphorylation events are depicted as an encircled P (i.e., a white circle with a P). CAR (constitutive androstane receptor) and FXR (farnesoid X receptor) are nuclear receptors. Cholic acid (CA) is a bile acid. Nuclear receptor (CAR, FXR)-dependent bile acid (CA) signaling is required for normal liver regeneration after PHx (partial hepatectomy). The xenobiotic and nongenotoxic carcinogen TCPOBOP (1,4-bis[2-(3,5-dichloropyridyloxy)]benzene), a halogenated hydrocarbon, is a synthetic ligand for CAR. **References**: Korver et al. (1997a) (serum), Korver et al. (1997b) (serum); Korver et al. (1997c) (PHA, PMA, differentiation of thymocytes); Ye et al. (1997) (KGF, H_2O_2, PHx, differentiation of Caco-2 colon carcinoma cells toward intestinal enterocytes); Ye et al. (1999) (PHx); Chaudhary et al. (2000) (FSH, cAMP); Ly et al. (2000) (aging, progeria); Kalinichenko et al. (2001) (BHT); Leung et al. (2001) (serum); Matsushima-Nishiu et al. (2001) (PTEN); Wang et al. (2001a) (CCl_4); Wang et al. (2001b) (PHx); Teh et al. (2002) (GLI-1); Wang et al. (2002a) (PHx); Wang et al. (2002b) (PHx); Fernandez et al. (2003) (c-Myc); Kalinichenko et al. (2003) (BHT); Krupczak-Hollis et al. (2003) (PHx, GH); Matsuo et al. (2003) (PHx, Cry1, Cry2); Ahn et al. (2004a) (differentiation of neural stem cells); Cicatiello et al. (2004) (estrogen); Huang et al. (2004) (TCR stimulation); Ledda-Columbano et al. (2004) (TCPOBOP); Thierry et al. (2004) (HPV18 E2); Bae et al. (2005) (BRCA1); Laoukili et al. (2005) (serum, FOXO3a); Takahashi et al. (2005) (FGF-2 (fibroblast growth factor-2)-induced differentiation of capillary endothelial cells), Untergasser et al. (2005) (TGF-β1); Bhonde et al. (2006) (p53); Huang et al. (2006) (CA, CAR, FXR); Karadedou (2006) (estrogen, ICI); Spurgers et al. (2006) (p53); Takahashi et al. (2006) (NMU); Tan et al. (2006) (PHx, HNF-6); Wierstra and Alves (2006d) (TPA (12-O-tetradecanoylphorbol-13-acetate)-induced differentiation of human promyelocytic HL-60 leukemia cells toward macrophages); Zhao et al. (2006) (TNF-α, H_2O_2, LPS); Chang et al. (2007) (miR-34); Delpuech et al. (2007) (c-Myc, Mxi1, FOXO3a); He et al. (2007) (miR-34); Keller et al. (2007) (c-Myc); Markey et al. (2007) (RB); Masumoto et al. (2007) (GH); Osborn et al. (2007) (ConA/PMA, FADD); Schüller et al. (2007) (Shh); Wang et al. (2007b) (E2F-1, p130, C/EBPα, E2F-4, RB, Brm, GH); Woodfield et al. (2007) (AP-2γ); Ackermann Misfeldt et al. (2008) (PPx); Blanco-Bose et al. (2008) (TCPOBOP, c-Myc); Halasi and Gartel (2009) (FOXM1); Hallstrom et al. (2008) (E2F-1); Laoukili et al. (2008) (Cdh1, APC/C); Park et al. (2008c) (Cdh1, APC/C); Ueno et al. (2008) (BMPs, BMPR, ALK2, Noggin, Chordin); Wang et al. (2008b) (serum, JNK1); Wang et al. (2008c) (HDAC1, C/EBPα, E2F-4, RB, Brm, GH); Knight et al. (2009) (LIN9); Zou et al. (2008) (FOXO3a); Adam et al. (2009) (p38); Barsotti and Prives (2009) (p53, p21, Nutlin-3, daunorubicin, E2F-1, pocket proteins); Calvisi et al. (2009) (ERK2, GLI-1); Chen et al. (2009) (CEP55); Chetty et al. (2009) (MMP-2, IR); Chua et al. (2009) (H_2O_2); Francis et al. (2009) (HER2, lapatinib); Gemenetzidis et al. (2009) (nicotine); Huynh et al. (2009) (differentiation of HO-1 human metastatic melanoma cells induced by IFN-β (interferon-β)+mezerein); Lange et al. (2009) (Sox17); McGovern et al. (2009) (gefitinib, EGFR, FOXO3a); Pandit et al. (2009) (p53); Park et al. (2009) (serum, H_2O_2, H-Ras, ROS, JNK, AKT1); Penzo et al. (2009) (TNF-α, IKKα, IKKβ, NEMO, IκBα, NF-κB); Weymann et al. (2009) (PHx, dextrose, p21); Xia et al. (2009) (HIF-1α, hypoxia); Behren et al. (2010) (H-Ras, MKK3, p38); Berger et al. (2010) (adiponectin); Caldwell et al. (2010) (OGT); Carr et al. (2010) (Herceptin); Chen et al. (2010) (PHx, CCl_4, FXR, CDCA, C/EBPα); Christoffersen et al. (2010) (miR-34); Fujii et al. (2010) (p400/mDomino); Gemenetzidis et al. (2010) (differentiation of primary oral keratinocytes); Gieling et al. (2010) (PHx, c-Rel); Huynh et al. (2010)

(IFN-β + mezerein, c-Myc, Mad1, PKCα, differentiation of HO-1 human metastatic melanoma cells induced by IFN-β + mezerein); Jia et al. (2010) (SRp20); Jin et al. (2010) (C/EBPα); Lefebvre et al. (2010) (c-Myb); Lorvellec et al. (2010) (B-Myb); Meng et al. (2010) (FXR, CCl$_4$); Millour et al. (2010) (ERα, estrogen, 4-OHT, ICI); Pellegrino et al. (2010) (Ha-Ras); Petrovich et al. (2010) (Sp1); Ren et al. (2010) (CCl$_4$); Sanchez-Calderon et al. (2010) (IGF-I); Takemura et al. (2010) (RKIP, BCR-ABL, imatinib, nilotinib, Dasatinib); Tan et al. (2010a,b) (p53, IR, etoposide); Wang et al. (2010c) (EPS8, PI3K, FOXM1, nicotine); Wang et al. (2010d) (AKT, PTEN, Notch1); Wu et al. (2010a) (p53); Xie et al. (2010) (FOXM1, RA (retinoic acid)-induced neural differentiation of mouse P19 embryonal carcinoma cells); Zhang et al. (2010a) (PL); Anders et al. (2011) (cyclin D1/Cdk4, cyclin D3/Cdk6, Cdh1); Bao et al. (2011) (miR-200b); Bergamaschi et al. (2011) (14-3-3τ); Bhat et al. (2011) (NPM); Bräuning et al. (2011) (TCPOBOP, β-catenin); Horimoto et al. (2011) (ERα, ERβ1); Kim et al. (2011) (H$_2$O$_2$); Liu et al. (2011) (PA); Lok et al. (2011) (p53); Meng et al. (2011) (PHx, FXR, CA, CYP27); Millour et al. (2011) (E2F-1, RB, p53, p21); Mitra et al. (2011) (stathmin); Mukhopadhyay et al. (2011) (PHx, CB$_1$R, AEA, 2-AG); Plank et al. (2011) (FOXD3); Prots et al. (2011) (IL-4 (interleukin-4)-induced differentiation of CD25 + iTregs (induced regulatory T-cells)); Rovillain et al. (2011) (E2F, p53, p21 HPV16 E7, Adenovirus E1A, SV40 large T antigen); Takizawa et al. (2011) (CAR, TCPOBOP); Tompkins et al. (2011) (Sox2); Wang et al. (2011b) (RA-induced differentiation of BE(2)-C neuroblastoma cells toward neuronal lineage); Wang et al. (2011c) (calcium); Yang et al. (2011) (BDL); Zhang et al. (2011b) (Wnt3a, DKK1); Aburto et al. (2012) (IGF-I, PI3K); Balli et al. (2012) (BHT (FOXM1 in macrophages)); Bellelli et al. (2012) (E2F-4, p21, p53); Bonet et al. (2012) (vemurafenib); Bose et al. (2012) (PKC, differentiation of HaCaT keratinocytes (spontaneously immortalized human skin keratinocytes) induced by suspension culture in methylcellulose (anchorage deprivation), differentiation of N/Tert-1 keratinocytes (hTERT- and p16 loss-immortalized normal human skin keratinocytes) induced by methylcellulose suspension culture, PMA (phorbol-12-myristate-13-acetate)-induced differentiation of N/Tert-1 keratinocytes, differentiation of N/Tert-1 keratinocytes induced by high cell density); Carr et al. (2012) (GATA3, RB, differentiation of mammary luminal cells); Chan et al. (2012) (GRB7, MEK1/2); Chen et al. (2013a) (LIN9); Dai et al. (2013) (heat shock, HSF1); Dean et al. (2012) (RB); Demirci et al. (2012) (c-Met); de Olano et al. (2012) (E2F-1, H$_2$O$_2$, p38, MK2, JNK1/2, epirubicin); Dibb et al. (2012) (PLK1); Down et al. (2012) (FOXM1, B-Myb, p130, serum); Halasi and Gartel (2012) (IR); He et al. (2013) (FOXO3a); Ho et al. (2012) (N-Ras, AKT); Huang and Zhao (2012) (CLP); Karadedou et al. (2012) (FOXO3a, lapatinib); Lehmann et al. (2012) (PHx); Liu et al. (2013a) (PHGDH); Llaurado et al. (2012) (ETV5); Li et al. (2012a) (KLF4); Li et al. (2013) (TGF-β1, miR-134); Lin et al. (2013) (miR-31); Lynch et al. (2012) (OGT); Mencalha et al. (2012) (STAT3, imatinib); Mizuno et al. (2012) (YAP, TEAD); Newick et al. (2012) (serum); Raghavan et al. (2012) (HIF-1α, HIF-2α, hypoxia); Ren et al. (2013) (SPDEF, allergen stimulation); Sadavisam et al. (2012) (B-Myb); Sengupta et al. (2013) (FOXO1 + FOXO3); Stoyanova et al. (2012) (p21); Xia et al. (2012b) (HBX (HBV)); Xia et al. (2012c) (TNF-α, HIF-1α, ROS); Xu et al. (2012a) (gefitinib); Xu et al. (2012b) (prenatal lung maturation); Zhang et al. (2012a) (FXR, CDCA, LPS); Zhang et al. (2012b) (FXR, PHx, CCl$_4$); Zhang et al. (2012c) (25HC3S). Zhang et al. (2012d) (miR-370); Zhang et al. (2012e) (RAD51); Chen et al. (2013b) (MnSOD, p53, E2F-1, Sp1); Kurinna et al. (2013) (p53, PHx); Liu et al. (2013b) (arsenic trioxide); Qu et al. (2013) (p53). **Abbreviations**: P, phosphorylation;

(Continued)

Figure 6.2—Cont'd ALK2, activin receptor-like kinase 2; BMP7 (bone morphogenetic protein 7) receptor; AEA, arachidonoyl ethanolamide, anandamide; 2-AG, 2-arachidonoylglycerol; AP-2γ, activator protein-2γ; APC/C, anaphase-promoting complex/cyclosome; BDL, bile duct ligation in liver; BHT, BHT (butylated hydroxytoluene) lung injury; BCR-ABL, breakpoint cluster region-Abelson; BMP, bone morphogenetic protein; BMPR, BMP receptor; BRCA1, breast cancer-associated gene 1; Brm, Brahma, SMARCA4 (SWI/SNF-related, matrix-associated, actin-dependent regulator of chromatin, subfamily A, member 4); CA, cholic acid; cAMP, cyclic adenosine monophosphate; CAR, constitutive androstane receptor; CB_1R, cannabinoid type 1 receptor; CCl_4, CCl_4 (carbon tetrachloride) liver injury; CDCA, chenodeoxycholic acid; Cdh1, Cdc20 homolog 1; Cdk, cyclin-dependent kinase; C/EBPα, CCAAT/enhancer-binding protein α; CEP55, centrosomal protein 55-kDa; Chk2, checkpoint kinase 2; CLP, CLP (cecal ligation and puncture)-induced lung vascular injury; ConA, concanavalin A; Cry, Cryptochrome; CYP27, sterol 27-hydroxylase; Dasatinib, Sprycel, BMS354825; DKK1, Dickkopf 1; EGFR, EGF (epidermal growth factor) receptor, ErbB1; EPS8, EGFR pathway substrate 8; ER, estrogen receptor; ERK2, extracellular signal-regulated kinase 2; ETV5, ETS variant 5; ERM; FOX, Forkhead box; FADD, Fas-associated death domain; FSH, follicle stimulating hormone; FXR, farnesoid X receptor; gefitinib, Iressa, ZD1839; GH, growth hormone; GLI-1, glioma transcription factor-1; GRB7, growth factor receptor-bound protein 7; HBV, hepatitis B virus; HBX, HBV X protein; 25HC3S, 5-cholesten-3b, 25-diol-3-sulfate; HDAC1, histone deacetylase 1; HER2, human EGF receptor 2, ErbB2, Neu; Herceptin, Trastuzumab; HIF, hypoxia-inducible factor; HNF-6, hepatocyte nuclear factor-6; HPV, human papillomavirus; HSF1, heat shock factor 1; ICI, ICI182780, fulvestrant, Faslodex; IGF-I, insulin-like growth factor-I; IκB, inhibitor of NF-κB; IKK, IκB kinase; imatinib, Gleevec, STI571, imatinib mesylate; IR, ionizing radiation, γ-irradiation; JNK, c-Jun N-terminal kinase; KGF, keratinocyte growth factor; KLF4, Krüppel-like factor 4; lapatinib, Tykerb, GW572016; LIN9, cell lineage-abnormal 9; LPS, LPS (lipopolysaccharide)-induced acute vascular lung injury; Mad1, Max dimerizer 1, MXD1 (Max dimerization protein 1); c-Met, HGFR (HGF (hepatocyte growth factor) receptor); MK2, MAPKAPK2 (MAPK (mitogen-activated protein kinase)-activated protein kinase 2); MKK3, MAPK kinase kinase 3; MMP-2, matrix metalloproteinase-2; MnSOD, manganese superoxide dismutase; Mxi1, Max interactor 1; c-Myb, MYB; B-Myb, MYBL2 (MYB-like 2); c-Myc, MYC; NEMO, NF-κB essential modifier, IKKγ; NF-κB, nuclear factor-κB; nilotinib, Tasigna, AMN107; NMU, neuromedin U; NPM, nucleophosmin, B23; OGT, O-GlcNAc (O-linked β-N-acetylglucosamine) transferase; 4-OHT, 4-hydroxytamoxifen; p21, CDKN1A (Cdk inhibitor 1A); p130, RBL2 (retinoblastoma-like 2); p53, TP53, tumor protein p53; PA, *Pseudomonas aeruginosa*-induced lung alveolar injury; PHA, phytohemagglutinin; PHx, partial hepatectomy; PKCα, protein kinase Cα; PHGDH, phosphoglycerate dehydrogenase; PI3K, phosphatidylinositol-3 kinase; PL, placental lactogen; PLK1, Polo-like kinase 1; PMA, phorbol-12-myristate-13-acetate; PPx, partial pancreatectomy; PTEN, phosphatase and tensin homolog on chromosome 10; RB, retinoblastoma protein, RB1 (retinoblastoma 1), p105; c-Rel, REL; RKIP, Raf kinase inhibitor protein; ROS, reactive oxygen species; Shh, Sonic hedgehog; Sox, SRY box, SRY-related HMG (high mobility group)-box; Sp1, specificity protein 1; SPDEF, SAM-pointed domain-containing ETS transcription factor, PDEF, prostate-derived ETS transcription factor; SRp20, SFRS3 (splicing factor, arginine/serine-rich 3); STAT3, signal transducer and activator of transcription 3; SV40, simian virus 40; TCPOBOP, 1,4-bis[2-(3,5-dichloropyridyloxy)]benzene; TEAD, TEAD1 (TEA domain family member 1); TCR, T-cell receptor; TGF-β1, transforming growth factor-β1; TNF-α, tumor necrosis factor-α; vemurafenib, PLX4032, RG7204, RO5185426; Wnt, wingless type; YAP, Yes-associated protein.

Table 6.2 Transcription factors, which regulate the *foxm1* expression

Transcription factor	Binding to *foxm1* promoter		Regulation of endogenous *foxm1* expression		Regulation of *foxm1* promoter		References
	Method[a]	Comment[b]	Expression[c]	Manipulation of transcription factor[d]	Manipulation of transcription factor-binding site[e]	Manipulation of transcription factor[d]	
AP-2γ			T	↑ siRNA			Woodfield et al. (2007)
CAR			T	↑ KO			Huang et al. (2006)
β-Catenin[f]			T, P	↓ KO			Bräuning et al. (2011)
C/EBPα	C		P	↓ P	wt, P	↓ OE, P	Wang et al. (2007b, 2008c), Chen et al. (2010), and Jin et al. (2010)
CREB[g]	C				wt, del, P	↑ B	Xia et al. (2012b)
E2F-1	C		T, P	↑ OE, dn, siRNA	wt, P	↑ OE, P	Wang et al. (2007b), Hallstrom et al. (2008), Barsotti and Prives (2009), Millour et al. (2011), de Olano et al. (2012), and Chen et al. (2013b)
E2F-4	E, C	S, P	T	↓ OE			Wang et al. (2007b, 2008c) and Bellelli et al. (2012)

E2F			T	dn	P	↓		Wang et al. (2007b), Millour et al. (2011), and Rovillain et al. (2011)
ERα	E, C, D	S, C, P	T, P	siRNA, L, A	wt, del, P	↑	OE, L, A	Millour et al. (2010) and Horimoto et al. (2011)
ERβ1[h]	C		T, P	OE	wt, P	↓	OE	Horimoto et al. (2011)
ETV5	C		T, P	siRNA		↑		Llaurado et al. (2012)
FOXD3			T	KO		↑		Plank et al. (2011)
FOXM1	C		T, P	OE	wt	↑	OE	Halasi and Gartel (2009), Wang et al. (2010c), Xie et al. (2010), and Down et al. (2012)
FOXO3a			T, P	OE, ca, siRNA	wt	↓	ca	Laoukili et al. (2005), Delpuech et al. (2007), Zou et al. (2008), McGovern et al. (2009), and Karadedou et al. (2012)
FOXO1 + FOXO3[i]			T	DKO		↓		Sengupta et al. (2013)
FXR/RXRα[j]	E, C	IV, S, C	T	ca, KO, L	P, H	↑	L	Huang et al. (2006), Chen et al. (2010), Meng et al. (2010, 2011), and Zhang et al. (2012a,b)

Continued

Table 6.2 Transcription factors, which regulate the *foxm1* expression—cont'd

Transcription factor	Binding to *foxm1* promoter		Regulation of endogenous *foxm1* expression		Regulation of *foxm1* promoter		References
	Method[a]	Comment[b]	Expression[c]	Manipulation of transcription factor[d]	Manipulation of transcription factor-binding site[e]	Manipulation of transcription factor[d]	
GATA3			T	↓ siRNA			Carr et al. (2012)
GLI-1			T, P	↑ OE, siRNA			Teh et al. (2002) and Calvisi et al. (2009)
HIF-1α	C		T, P	↑ siRNA	wt, del, P	↑ siRNA	Xia et al. (2009, 2012c) and Raghavan et al. (2012)
HIF-2α			T	↑ siRNA			Raghavan et al. (2012)
HNF-6			T	↑ OE			Tan et al. (2006)
HSF1	C		T, P	↑ ca, siRNA, KO	wt, del, P	↑ HS	Dai et al. (2013)
KLF4	C		P	↓ OE, KO	wt	↓ OE, siRNA	Li et al. (2012a)
Mad1	C		T	↓ OE			Huynh et al. (2010)
Mxi1	C		T	↓ siRNA			Delpuech et al. (2007)
B-Myb	C		T	↑ siRNA, KO			Lorvellec et al. (2010), Down et al. (2012), and Sadavisam et al. (2012)
c-Myb	C		T, P	↑ siRNA			Lefebvre et al. (2010)

Factor						References
c-Myc	C, D	T, P	↑ siRNA, tg, KO			Fernandez et al. (2003), Delpuech et al. (2007), Keller et al. (2007), Blanco–Bose et al. (2008), and Huynh et al. (2010)
Notch1	C	T, P	↑ OE, siRNA			Wang et al. (2010d)
p53[k]	C	T, P	↓ OE, P, siRNA, wt, KO, CC, SM, GSE		↓ OE, P	Bhonde et al. (2006), Spurgers et al. (2006), Barsotti and Prives (2009), Pandit et al. (2009), Lok et al. (2011), Millour et al. (2011), Rovillain et al. (2011), Chen et al. (2013b), Kurinna et al. (2013), and Qu et al. (2013)
p53		T, P	↑ siRNA, KO			Tan et al. (2010a) and Kurinna et al. (2013)
c–Rel	C	T, P	↑ KO			Gieling et al. (2010)
Sox2		T	↑ tg	I?	↑ OE	Tompkins et al. (2011)
Sox17		T	↑ OE			Lange et al. (2009)
Sp1[l]	C	P	↑ siRNA	wt	↑ OE	Petrovich et al. (2010) and Chen et al. (2013b)

Continued

Table 6.2 Transcription factors, which regulate the *foxm1* expression—cont'd

Transcription factor	Binding to *foxm1* promoter		Regulation of endogenous *foxm1* expression		Regulation of *foxm1* promoter		References
	Method[a]	Comment[b]	Expression[c]	Manipulation of transcription factor[d]	Manipulation of transcription factor-binding site[e]	Manipulation of transcription factor[d]	
SPDEF			T	↓ KO			Ren et al. (2013)
STAT3	E, C	S, C	T	↑ IS3	wt	↑ IS3	Mencalha et al. (2012)
TEAD			T		wt	↑ OE, del	Mizuno et al. (2012)
YAP	C		T	↑ siRNA	wt	↑ OE, ca, P	Mizuno et al. (2012)

AP-2γ, activator protein-2γ; CAR, constitutive androstane receptor; C/EBPα, CCAAT/enhancer-binding protein α; CREB, cAMP (cyclic adenosine monophosphate) response element-binding protein; ER, estrogen receptor; ETV5, ETS variant 5, ERM, Ets-related molecule; FOX, Forkhead box; FXR, farnesoid X receptor; GATA3, GATA-binding protein 3; GLI-1, glioma transcription factor-1; HIF, hypoxia-inducible factor; HNF-6, hepatocyte nuclear factor-6; HSF1, heat shock factor 1; KLF4, Krüppel-like factor 4; Mad1, Max dimerizer 1, MXD1, Max dimerization protein 1; Mxi1, Max interactor 1; c-Myb, MYB; B-Myb, MYBL2, MYB-like 2; c-Myc, MYC; p53, TP53, tumor protein p53; c-Rel, REL; RXRα, retinoid X receptor α; Sox, SRY (sex-determing region Y) box, SRY-related HMG (high mobility group)-box; Sp1, specificity protein 1; SPDEF, SAM-pointed domain-containing ETS transcription factor, PDEF, prostate-derived ETS transcription factor; STAT3, signal transducer and activator of transcription 3; TEAD, TEAD1, TEA domain family member 1; YAP, Yes-associated protein.

[a]**Method**: E = EMSA = electrophoretic mobility shift assay; C = ChIP = chromatin immunoprecipitation assay; D = DNAP = DNA precipitation assay = DAPA = DNA affinity pull-down assay.

[b]**Comment**: IV = *in vitro* = purified or *in vitro* transcribed/translated transcription factor; S = supershift experiments; C = competition experiments; P = binding site was point-mutated.

[c]**Expression**: T = transcript = mRNA = endogenous mRNA level affected; P = protein = endogenous protein level affected.

[d]**Manipulation of transcription factor**: OE = overexpression of wild-type; dn = dominant-negative form; ca = constitutively active form; del = analyzed with deletion mutant of the transcription factor; P = analyzed with point mutant of the transcription factor; siRNA = siRNA-mediated or shRNA-mediated knockdown; tg = transgenic cells/mice; KO = knockout cells/mice; CC = p53-deficient colon carcinoma cell lines; SM = small-molecule activator of p53 (Nutlin-3); L = activation by ligand; A = ERα antagonist (4-OHT (4-hydroxytamoxifen), ICI (ICI182780, fulvestrant, Faslodex)); GSE = genetic suppressor element (100–300 nt (nucleotide) long rat *p53* cDNA segment); B = HBX = HBV (hepatitis B virus) X protein; DKO = cardiomyocyte-specific conditional *foxo1/foxo3* double knockout; HS = heat shock; IS3 = small-molecule inhibitor of STAT3 (LLL-3).

"**Manipulation of transcription factor-binding site**: wt = "wild-type" *foxm1* promoter; del = analyzed with deletion mutant of *foxm1* promoter; P = transcription factor-binding site was point-mutated; H = transcription factor-binding site in context of heterologous core promoter; I? = 0.78 kb (kilobase) *foxm1* reporter construct containing a regulatory sequence from the first intron.

[f]The negative effect of β-catenin on the *foxm1* mRNA and protein expression was observed in female mice in the presence of TCPOBOP (1,4-bis[2-(3,5-dichloropyridyloxy)]benzene), but β-catenin had no negative effect on the *foxm1* mRNA level in the absence of TCPOBOP (Bräuning et al., 2011).

[g]The activation of the *foxm1* promoter by HBX (HBV (hepatitis B virus) X protein) was considerably reduced if a putative CREB-binding site in the *foxm1* promoter was point-mutated, but an effect of CREB on the *foxm1* promoter or the *foxm1* expression was not shown (Xia et al., 2012b).

[h]ER-β1, which displaces ERα from the *foxm1* promoter, represses the *foxm1* promoter and downregulates the *foxm1* mRNA and protein expression only in ERα-positive, but not in ERα-negative breast carcinoma cell lines (Horimoto et al., 2011).

[i]The *foxm1* mRNA level was increased in the neonatal hearts of mice with a cardiomyocyte-specific conditional *foxo1*/*foxo3* double knockout compared to the neonatal hearts of control mice, indicating that FOXO1 + FOXO3 decrease the *foxm1* mRNA expression (Sengupta et al., 2013).

[j]The FXR/RXRα binding site is located in intron 3 of the *foxm1* gene.

[k]The point-mutated versions of the *foxm1* promoter had point mutations in the two putative E2F binding sites (but not in any putative p53 binding site).

[l]Overexpression of MnSOD (manganese superoxide dismutase) increased the FOXM1 protein level in H460 lung cancer cells and shRNA against Sp1 abolished this MnSOD-induced increase in the FOXM1 protein level, which indicates an upregulation of the FOXM1 protein expression by Sp1. However, the effect of the Sp1 knockdown on the FOXM1 protein expression in the absence of exogenous MnSOD was not analyzed.

factor 1, RBP-Jκ (recombination signal sequence-binding protein Jκ)), Su(H) (Suppressor of Hairless), Lag-1 (lymphocyte activation gene-1, CCL4L1 (chemokine, CC motif, ligand 4-like 1)))) (Wang et al., 2010d), which both upregulate the *foxm1* mRNA level (Table 6.2).

Another positive regulator of the FOXM1 expression is the transcription factor E2F-1 (Barsotti and Prives, 2009; Chen et al., 2013b; de Olano et al., 2012; Hallstrom et al., 2008; Millour et al., 2011), which *in vivo* occupies the *foxm1* promoter (Table 6.2) and which is inhibited by the tumor suppressor RB (Fig. 6.1) (Beijersbergen and Bernards, 1996; Burkhart and Sage, 2008; Chan et al., 2001; Chinnam and Goodrich, 2011; Classon and Harlow, 2002; Cobrinik, 1996, 2005; DeGregori, 2004; Du and Pogoriler, 2006; Dyson, 1998; Fiorentino et al., 2013; Frolov and Dyson, 2004; Giacinti and Giordano, 2006; Gordon, and Du, 2011; Grana et al., 1998; Halaban, 2005; Henley and Dick, 2012; Herwig and Strauss, 1997; Johnson and Schneider-Broussard, 1998; Kaelin, 1999; Khidr and Chen, 2006; Knudsen and Knudsen, 2006; Macaluso et al., 2006; Müller and Helin, 2000; Munro et al., 2012; Nevins, 1998; Poznic, 2009; Sellers and Kaelin, 1996; Seville et al., 2005; Singh et al., 2010; Sun et al., 2007; Talluri and Dick, 2012; Zheng and Lee, 2001; Zhu, 2005). Additional positive regulators of the FOXM1 expression are the transcription factors HIF-1α (hypoxia-inducible factor-1α) (Raghavan et al., 2012; Xia et al., 2009) and HIF-2α (Raghavan et al., 2012), which upregulate the *foxm1* mRNA level (Table 6.2) and which are antagonized by the tumor suppressor VHL (von Hippel–Lindau) (Fig. 6.1) (Brahimi-Horn and Pouyssegur, 2005, 2006, 2009; Kaelin, 2007, 2008; Kaelin and Radcliffe, 2008; Keith et al., 2012; Majumdar et al., 2010; Semenza, 2010, 2011, 2012).

Among the tumor suppressors, which decrease the FOXM1 expression (Fig. 6.1), are the transcription factors p53 (Barsotti and Prives, 2009; Bellelli et al., 2012; Bhonde et al., 2006; Chen et al., 2013b; Lok et al., 2011; Millour et al., 2011; Pandit et al., 2009; Qu et al., 2013; Rovillain et al., 2011; Spurgers et al., 2006), and KLF-4 (Li et al., 2012a), which both occupy the *foxm1* promoter *in vivo* (Table 6.2). The tumor suppressors, which diminish the FOXM1 expression, include also the transcription factor GATA3 (Carr et al., 2012), which downregulates the *foxm1* mRNA level (Table 6.2).

The candidate tumor suppressors, which reduce the FOXM1 expression, include the transcription factors Mad1 (Huynh et al., 2010) and FOXO3a (Delpuech et al., 2007; He et al., 2013; Karadedou et al., 2012;

Laoukili et al., 2005; McGovern et al., 2009; Zou et al., 2008), which *in vivo* occupy or repress the *foxm1* promoter, respectively (Table 6.2).

Hence, the expression of *foxm1* seems to be regulated by the Myc/Max (Myc-associated factor X)/Mad network, the Notch pathway (Notch1, NICD), the Hippo pathway (YAP), the JAK (Janus kinase)/STAT pathway (STAT3), and the NF-κB (nuclear factor-κB) pathway (c-Rel), which all have been implicated in tumorigenesis (Aggarwal et al., 2009a,b; Albihn et al., 2010; Alitalo et al., 1987; Aster et al., 2008; Basseres and Baldwin, 2006; Chaturvedi et al., 2011; Ecker et al., 2009; Guo et al., 2011; Haura et al., 2005; Inoue et al., 2007; Jatiani et al., 2011; Junttila and Westermarck, 2008; Karin, 2006a,b; Kim et al., 2006b; Koch and Radtke, 2007; Marcu et al., 1992; Meyer and Penn, 2008; Naugler and Karin, 2008; Nesbit et al., 1999; Oster et al., 2002; Pan, 2010; Perkins, 2012; Ponzielli et al., 2005; Popescu and Zimonjic, 2002; Ranganathan et al., 2011; Spencer and Groudine, 1991; Stanger, 2012; Staudt, 2010; Talora et al., 2008; Turkson, 2004; Vita and Henriksson, 2006; Yu and Jove, 2004; Yu et al., 2009; Zeng and Hong, 2008; Zhao et al., 2010).

1.2.3 Possible deregulation of the foxm1 expression by other proto-oncoproteins and tumor suppressors

The *foxm1* mRNA expression is increased by the proto-oncoproteins H-Ras (Behren et al., 2010; Park et al., 2009), N-Ras (Ho et al., 2012), HER2/Neu (Francis et al., 2009), AKT (Ho et al., 2012; Wang et al., 2010d), and c-Met (Demirci et al., 2012), but decreased by the tumor suppressor RB (Dean et al., 2012; Markey et al., 2007), the tumor suppressor miRNAs of the miR-34 family (Chang et al., 2007; Christoffersen et al., 2010; He et al., 2007), and the tumor suppressor miRNA miR-31 (Fig. 6.1) (Lin et al., 2013).

The FOXM1 protein level is upregulated by the proto-oncoproteins cyclin D1/Cdk4 (Anders et al., 2011), Wnt3a (Zhang et al., 2011b), and EGFR (McGovern et al., 2009), but downregulated by the tumor suppressors Cdh1 (Anders et al., 2011; Laoukili et al., 2008; Park et al., 2008c) and PTEN (Wang et al., 2010d) as well as by the tumor suppressor miRNA miR-200b (Bao *et al.*, 2011).

Moreover, the expression of FOXM1 is subject to control by additional proteins with known roles in tumorigenesis, so that the FOXM1 expression is upregulated by those factors which promote oncogenesis, but down-regulated by those factors which impair carcinogenesis:

On the one hand, the cell cycle kinase cyclin D3/Cdk6 (Anders et al., 2011), the serine/threonine kinase ERK2 (Calvisi et al., 2009), and the secreted ligand Shh (Schüller et al., 2007) elevate the FOXM1 protein level (Fig. 6.1). Likewise, the serine/threonine kinase PLK1 (Dibb et al., 2012) as well as the transcription factors B-Myb (Down et al., 2012; Lorvellec et al., 2010; Sadavisam et al., 2012), HSF1 (Dai et al., 2013), and ERα (Horimoto et al., 2011; Millour et al., 2010) enhance the *foxm1* mRNA expression (Fig. 6.2).

On the other hand, the *foxm1* mRNA expression is reduced by the CKI (Cdk inhibitor) p21 (Barsotti and Prives, 2009; Bellelli et al., 2012; Rovillain et al., 2011) as well as by the transcription factors E2F-4 (Bellelli et al., 2012), Mxi1 (Delpuech et al., 2007), SPDEF (Ren et al., 2013), and C/EBPα (Wang et al., 2007b). Similarly, the FOXM1 protein level is diminished by the secreted Wnt inhibitor DKK1 (Fig. 6.1) (Zhang et al., 2011b).

Thus, the FOXM1 expression is controlled by some RTKs (receptor tyrosine kinases) (c-Met, EGFR HER2/Neu), which are often activated in human cancers (Fig. 6.1) (Blume-Jensen and Hunter, 2001; Hunter, 1998, 1999, 2009; Kolibaba and Druker, 1997; Sawyers, 2003; Scheijen and Griffin, 2002; Zhang et al., 2009). Furthermore, the FOXM1 expression seems to be controlled by the Ras/MAPK pathway (Ras, ERK2) and the PI3K (phosphatidylinositol-3 kinase)/AKT pathway (AKT, PTEN), which both are frequently activated in tumors (Bunney and Katan, 2010; Calvo et al., 2010; Chalhoub and Baker, 2009; Chung and Kondo, 2011; Cully et al., 2006; De Luca et al., 2012; Dhillon et al., 2007; Downward, 2003; Engelman, 2009; Fresno Vara et al., 2004; Hay, 2005; Ji et al., 2011; Jiang and Liu, 2008; Liu et al., 2009a; Lopez-Bergami, 2011; Markman et al., 2010; Maurer et al., 2011; McCormick, 2011; McCubrey et al., 2007, 2011; Mitsiades et al., 2004; Platanias, 2003; Pratilas and Solit, 2010; Pylayeva-Gupta et al., 2011; Roberts and Der, 2007; Sansal and Sellers, 2004; Schubbert et al., 2007; Sebolt-Leupold and Herrera, 2004; Steelman et al., 2004, 2011a,b; Vakiani and Solit, 2011; Yap et al., 2011; Yuan and Cantley, 2008). Also the Wnt pathway (Wnt3a, DKK1) and the Hedgehog pathway (Shh, GLI-1), which are often activated in cancer cells (Barakat et al., 2010; Barker and Clevers, 2006; Clevers, 2006; de Lau et al., 2007; Fodde and Brabletz, 2007; Harris et al., 2011; Javelaud et al., 2012; Karim et al., 2004; Klaus and Birchmeier, 2008; Lai et al., 2009; Ng and Curran, 2011; Nusse and Varmus, 2012; Onishi and Katano, 2011; Polakis, 2007, 2012; Reya and Clevers, 2005; Scales and de Sauvage, 2009; Taipale and Beachy, 2001; Teglund and Toftgard, 2010; Yang et al., 2010), appear to regulate the FOXM1 expression.

1.2.4 Possible deregulation of the foxm1 expression by alterations in the RB and p53 tumor suppressor pathways

The two major tumor suppressor pathways are the RB pathway and the p53 pathway (Hanahan and Weinberg, 2000, 2011; Sherr, 2004; Sherr and McCormick, 2002; Stein and Pardee, 2004; Vogelstein and Kinzler, 2004; Weinberg, 2006). The RB and p53 tumor suppressor pathways are inactivated in virtually every human cancer cell (Fanciulli, 2006; Hall and Peters, 1996; Sherr and McCormick, 2002; Stein and Pardee, 2004; Weinberg, 2006), which might contribute to FOXM1 overexpression in many human tumors (Table 6.1) because the expression of *foxm1* is repressed by the two key tumor suppressors RB and p53 (Fig. 6.1):

First, *foxm1* is a direct p53 target gene (Kurinna et al., 2013). p53 represses the *foxm1* promoter (Chen et al., 2013b; Millour et al., 2011) and reduces the *foxm1* mRNA and protein expression (Barsotti and Prives, 2009; Bellelli et al., 2012; Bhonde et al., 2006; Chen et al., 2013b; Lok et al., 2011; Millour et al., 2011; Pandit et al., 2009; Qu et al., 2013; Rovillain et al., 2011; Spurgers et al., 2006). In accordance with *foxm1* being a direct target gene of p53 (Kurinna et al., 2013), p53 can down-regulate the *foxm1* mRNA level independently of p21 (Barsotti and Prives, 2009), which is encoded by the direct p53 target gene *p21* (El-Deiry et al., 1993).

Second, RB decreases the *foxm1* mRNA and protein expression because loss of RB increased the *foxm1* mRNA level (Markey et al., 2007) and because RB knockdown increased the FOXM1 protein level (Dean et al., 2012). In fact, *foxm1* is a direct target gene of E2F-1 (Chen et al., 2013b; Millour et al., 2011; Wang et al., 2007b) and E2F-4 (Wang et al., 2007b, 2008c) and also RB occupies the *foxm1* promoter *in vivo* (Millour et al., 2011; Wang et al., 2007b, 2008c). In accordance with the repression of the *foxm1* mRNA (Markey et al., 2007) and protein (Dean et al., 2012) expression by RB, E2F-1 transactivates the *foxm1* promoter (de Olano et al., 2012; Millour et al., 2011) and upregulates the *foxm1* mRNA and protein expression (Barsotti and Prives, 2009; Chen et al., 2013b; de Olano et al., 2012; Hallstrom et al., 2008) whereas E2F-4 downregulates the *foxm1* mRNA level (Bellelli et al., 2012).

This negative regulation of the *foxm1* expression by the two key tumor suppressors RB and p53 implies the control of the *foxm1* expression by the upstream regulators of RB and p53 in the RB pathway (p16 ⊣ cyclin D1/

Cdk4 ⊣RB) or the p53 pathway (ARF ⊣MDM2 ⊣p53), respectively, that is, by the tumor suppressors p16 (CDKN2A (Cdk inhibitor 2A), INK4A (inhibitor of Cdk4)) and ARF (alternative reading frame, CDKN2A (Cdk inhibitor 2A), p14, p19) as well as by the proto-oncoproteins cyclin D1, Cdk4, and MDM2 (mouse double minute 2).

1.3. Upregulation of the FOXM1 expression by the oncoproteins of tumor viruses

1.3.1 Viral oncoproteins stimulate the expression of foxm1

Some viral oncoproteins have been reported to enhance the *foxm1* mRNA expression (Fig. 6.1), namely, HBX (HBV X protein) of HBV (Xia et al., 2012b), E7 of HPV16, E1A (early region 1A)-12S of adenovirus, and the large T antigen of SV40 (Rovillain et al., 2011), so that they may cause FOXM1 overexpression in those tumor cells which are infected with these DNA viruses.

Cervix cancer often results from HPV16 infection (Baseman and Koutsky, 2005; Burk et al., 2009; Doorbar, 2006; Fehrmann and Laimins, 2003; Feller et al., 2009; Hebner and Laimins, 2006; Jayshree et al., 2009; Kisseljov et al., 2008; Lehoux et al., 2009; Lizano et al., 2009; Longworth and Laimins, 2004; McLaughlin-Drubin and Münger, 2009b, 2012; Moody and Laimins, 2010; Münger, 2002; Münger et al, 2004; Narisawa-Saito and Kiyono, 2007; Roden and Wu, 2006; Schiffman et al., 2007; Stanley et al., 2007; Stubenrauch and Laimins, 1999; Subramanya and Grivas, 2008; Woodman et al., 2007; zur Hausen, 1996, 2002, 2009). In accordance with the stimulation of the *foxm1* mRNA expression by HPV16 E7 (Fig. 6.1) (Rovillain et al., 2011), FOXM1 is over-expressed in cervical cancer, in particular in late-stage tumors (Table 6.1) (Chan et al., 2008; Pilarsky et al., 2004; Rosty et al., 2005; Santin et al., 2005), and the FOXM1 expression level correlated with the E7 expression level in HPV16-infected invasive cervical carcinoma (Rosty et al., 2005).

HCC can be caused by HBV infection, which is named HBV-HCC (Anzola, 2004; Bharadwaj et al., 2013; Chisari, 2000; De Mitri et al., 2010; Feitelson, 1999; Fung et al., 2009; Lee and Lee, 2007; Lupberger and Hildt, 2007; Neuveut et al., 2010; Park et al., 2007; Robinson, 1994). In accordance with the HBX-mediated upregulation of the *foxm1* mRNA and protein levels (Fig. 6.1), FOXM1 overexpression was observed in HBV-HCC (Table 6.1) (Xia et al., 2012b).

1.3.2 Control of the foxm1 promoter by the viral oncoprotein HBX

The oncoprotein HBX of the human cancer virus HBV (*Hepadnaviridae* family) activates the *foxm1* promoter and increases the *foxm1* mRNA and protein expression (Fig. 6.1) (Xia et al., 2012b). This activation of the *foxm1* promoter by HBX depends on a putative CREB (cAMP (cyclic adenosine monophosphate) response element-binding protein) binding site in the *foxm1* promoter and HBX induces the *in vivo* association of CREB with the *foxm1* promoter, which is not *in vivo* occupied by CREB in the absence of HBX (Xia et al., 2012b). HBX can activate ERK1/2 and CREB (Andrisani and Barnabas, 1999; Benhenda et al., 2009; Bouchard and Schneider, 2004; Diao et al., 2001; Ma et al., 2011; Martin-Vilchez et al., 2011; Matsuda and Ichida, 2009; Murakami, 2001; Ng and Lee, 2011; Wei et al., 2010; Zhang et al., 2006b), the latter of which is phosphorylated and activated by the ERK1/2 substrates RSK1/2 (90-kDa ribosomal S6 kinase 1/2, p90RSK) and MSK1/2 (mitogen- and stress-activated kinase 1/2) (Andrisani, 1999; Johannessen and Moens, 2007; Johannessen et al., 2004; Lonze and Ginty, 2002; Mayr and Montminy, 2001; Montminy, 1997; Shaywitz and Greenberg, 1999). Accordingly, HBX induced the activating phosphorylation of CREB and ERK1/2 and both events were reduced by the MEK1/2 inhibitor U0126 (Xia et al., 2012b). Since U0126 prevented the HBX-induced *in vivo* association of CREB with the *foxm1* promoter and diminished the HBX-induced FOXM1 protein expression, HBX might activate the *foxm1* promoter via the pathway HBX → ERK1/2 → RSK1/2 or MSK1/2 → CREB → *foxm1* promoter(Xia et al., 2012b).

1.3.3 Control of the foxm1 expression by the oncoproteins of small DNA tumor viruses

HPV16 (*Papillomaviridae* family), adenovirus (*Adenoviridae* family), and SV40 (*Polyomaviridae* family) are small DNA tumor viruses (Butel, 2000; Howley and Livingston, 2009; Javier and Butel, 2008; Klein, 2002; Levine, 1994; McLaughlin-Drubin and Münger, 2008; Talbot and Crawford, 2004). Since their related oncoproteins E7, E1A, and large T each inactivate the key tumor suppressor RB (Ahuja et al., 2005; Ali and DeCaprio, 2001; Ben-Israel and Kleinberger, 2002; Blanchette and Branton, 2009; Caracciolo et al., 2006; Chakrabarti and Krishna, 2003; Cheng et al., 2009; Chinnadurai, 2011; Cress and Nevins, 1996; D'Abramo and Archambault, 2011; DeCaprio, 2009; Duensing and Münger, 2004; Duensing et al., 2009; Endter and Dobner, 2004; Fehrmann and Laimins, 2003; Felsani et al., 2006; Ferrari et al., 2009; Frisch and Mymryk, 2002; Gallimore and Turnell, 2001; Ganguly and Parihar, 2009; Ghittoni et al., 2010; Hamid et al., 2009; Helt and Galloway, 2003; Jansen-Dürr, 1996; Jones and Münger, 1996; Klingelhutz and Roman, 2012; Korzeniewski et al., 2011; Lavia et al., 2003; Lee and Cho, 2002; Lehoux et al., 2009; Liu and Marmorstein, 2006; Lizano et al., 2009; Longworth and Laimins, 2004; McCance, 2005; McLaughlin-Drubin and Münger, 2009a,b; Moens et al., 2007; Moody and Laimins, 2010; Münger, 2003; Münger and Howley, 2002; Münger et al., 2001, 2004; Narisawa-Saito and Kiyono, 2007; O'Shea, 2005a,b; Paggi et al., 2003; Pelka et al., 2008; Pim and Banks, 2010; Pipas, 2009; Poznic, 2009; Randow and Lehner, 2009; Saenz-Robles and Pipas, 2009; Saenz-Robles et al., 2001; Sang et al., 2002; Scheffner and Whitaker, 2003; Sullivan and Pipas, 2002; Tan et al., 2012; Vousden, 1995; Weinberg, 1997; White and Khalili, 2004, 2006; Wise-Draper and Wells, 2008; Yugawa and Kiyono, 2009; Zwerschke and Jansen-Dürr, 2000) and since RB, which *in vivo* occupies the *foxm1* promoter (Millour et al., 2011; Wang et al., 2007b, 2008c), represses the mRNA and protein expression of the direct E2F target gene *foxm1* (Fig. 6.2) (Dean et al., 2012; Markey et al., 2007), one may speculate that E7, E1A, and large T effectuate FOXM1 overexpression through inactivation of RB. In addition, SV40 large T inactivates the key tumor suppressor p53 (Ahuja et al., 2005; Ali and DeCaprio, 2001; Cheng et al., 2009; Levine, 2009; Liu and Marmorstein, 2006; Moens et al., 2007; O'Shea and Fried, 2005; Pipas, 2009; Pipas and Levine, 2001; Saenz-Robles and Pipas, 2009; Saenz-Robles et al., 2001; Sullivan and Pipas, 2002; White and Khalili, 2004), which *in vivo* occupies (Kurinna et al., 2013) and represses (Chen et al., 2013b; Millour et al., 2011) the *foxm1* promoter and which down-regulates the *foxm1* mRNA and protein levels (Barsotti and Prives, 2009; Bellelli et al., 2012; Bhonde et al., 2006; Chen et al., 2013b; Lok et al., 2011; Millour et al., 2011; Pandit et al., 2009; Qu et al., 2013; Rovillain et al., 2011; Spurgers et al., 2006), so that SV40 large T might also achieve FOXM1 overexpression through inactivation of p53.

In contrast, the *foxm1* mRNA level was decreased by the viral replication protein E2 of HPV18 (Fig. 6.2) (Thierry et al., 2004), another high-risk HPV that can lead to cervical cancer (zur Hausen, 2002).

1.4. The oncogenic translocation product BCR-ABL may elevate the *foxm1* expression

The chimeric fusion protein BCR-ABL (breakpoint cluster region-Abelson) is the product of the t(9;22)(q34;q11) reciprocal chromosome translocation, which involves the *BCR* gene on chromosome 22 and the proto-oncogene *c-abl* on chromosome 9. This t(9;22)(q34;q11) translocation, the so-called Philadelphia chromosome, is found in the majority of patients with CML (chronic myeloid leukemia) and in a significant fraction of patients with AML (Arlinghaus and Sun, 2004; Burke and Carroll, 2010; Chopra et al., 1999; Cilloni and Saglio, 2012; Goldman and Melo, 2008; Hantschel and Superti-Furga, 2004; Hazlehurst et al., 2009; Ren, 2005; Rumpold and Webersinke, 2011; Salesse and Verfaillie, 2002; Sattler and Griffin, 2001, 2003; Van Etten, 2004; Wong and Witte, 2004).

The chimeric oncoprotein BCR-ABL seems to increase the *foxm1* mRNA expression (Fig. 6.1) because the *foxm1* mRNA level was decreased by siRNA against BCR-ABL in CFU-GM (colony forming unit-granulocyte, macrophage), CFU-GEMM (colony forming unit-granulocyte, erythroid, macrophage, megakaryocyte), and BFU-E (burst forming unit-erythroid) hematopoietic progenitor cells derived from patients with CML, which is characterized by the t(9;22)(q34;q11) translocation that generates BCR-ABL (Takemura et al., 2010). Accordingly, also the BCR-ABL kinase inhibitors Imatinib (STI571, Gleevec), Nilotinib (AMN107, Tasigna), and Dasatinib (BMS354825, Sprycel) diminished the *foxm1* mRNA expression in these cells (Fig. 6.2) (Takemura et al., 2010).

BCR-ABL downregulates the mRNA level of the metastasis suppressor RKIP (Raf kinase inhibitor protein) (Fig. 6.1) (Takemura et al., 2010), which inhibits Ras/MAPK signaling (by binding to Raf-1 and inhibiting the activation of MEK by Raf-1) and NF-κB signaling (by binding to NIK (NF-κB-inducing kinase, MAP3K14 (MAPK kinase kinase 14)) and TAK1 (TGF-β (transforming growth factor-β)-activated kinase 1, MAP3K7 (MAPK kinase kinase 7)) and inhibiting the activation of IKK (IκB (inhibitor of NF-κB) kinase) by NIK and TAK1), but stimulates GPCR (G-protein-coupled receptor) signaling (by binding to and inhibiting GRK2 (GPCR kinase 2)) (Granovsky and Rosner, 2008; Hagan et al., 2006; Keller, 2004; Keller et al., 2004a,b, 2005; Klysik et al., 2008; Odabaei et al., 2004; Trakul and Rosner, 2005; Zeng et al., 2008). RKIP reduces the *foxm1* mRNA and protein expression (Fig. 6.1) (Takemura et al., 2010). Since siRNA-mediated knockdown of RKIP alleviated the negative effect of the BCR-ABL inhibitor Imatinib on the *foxm1* mRNA level in hematopoietic progenitor cells (CFU-GM, CFU-GEMM, BFU-E) from CML patients, the chimeric oncoprotein BCR-ABL might enhance the *foxm1* expression at least partially through RKIP via the pathway BCR-ABL ⊣RKIP ⊣*foxm1* (Fig. 6.2) (Takemura et al., 2010).

1.5. Positive FOXM1 autoregulation

FOXM1 enhances its own expression in a positive feedback loop (Fig. 6.2), because the *foxm1* promoter is *in vivo* occupied (Down et al., 2012) and activated (Wang et al., 2010c) by FOXM1 itself (Table 6.2). Accordingly, exogenous FOXM1 increased the endogenous *foxm1* mRNA and protein expression (Halasi and Gartel, 2009; Xie et al., 2010). This positive FOXM1 autoregulation will amplify the FOXM1 overexpression in tumor cells, because it will raise the FOXM1 level further once the *foxm1* expression is deregulated.

1.6. Unexpected findings

Surprisingly, the tumor suppressor BRCA1 (breast cancer-associated gene 1) enhances the *foxm1* mRNA expression (Bae et al., 2005) and both the tumor suppressor Chk2 (checkpoint kinase 2) (Tan et al., 2007) and the anti-proliferation signal RAD51 (Zhang et al., 2012e) upregulate the FOXM1 protein level (Fig. 6.2), although the opposite is expected in each case.

First, siRNA-mediated knockdown of BRCA1 decreased the *foxm1* mRNA level in human DU-145 prostate cancer cells, indicating that BRCA1 increases the *foxm1* mRNA expression (Bae et al., 2005).

Second, Chk2 phosphorylates FOXM1 at S_{375} (aa (amino acid) numbering refers to the splice variant FOXM1c) in response to IR (ionizing radiation, γ-irradiation) and etoposide (Tan et al., 2007). IR (Chetty et al., 2009; Tan et al., 2007), etoposide, and UV (ultraviolet radiation) (Tan et al., 2007), that is, three DNA-damaging treatments, increase the FOXM1 protein level through Chk2, probably by stabilizing the FOXM1 protein (Tan et al., 2007).

Third, the FOXM1 protein level was decreased by shRNA against RAD51, but increased by RAD51 overexpression in recurrent GBM1 cells, indicating that RAD51 augments the FOXM1 protein expression (Zhang et al., 2012e).

The tumor suppressors BRCA1 and Chk2 as well as the anti-proliferation signal RAD51 play pivotal roles in the HR (homologous recombination) repair of DSBs (DNA double-strand breaks) or in the DSB-induced DNA damage response (Abraham, 2001; Ahn et al., 2004b; Antoni et al., 2007; Bartek and Lukas, 2003; Bartek and Lukas, 2007; Bartek et al., 2001; Bartek et al., 2004; Baumann and West, 1998; Branzei and Foiani, 2008; Caestecker and Van de Walle, 2013; Chen and Poon, 2008; D'Andrea, 2003; Daboussi et al., 2002; Deckbar et al., 2011; Finn et al., 2012; Forget and Kowalczykowski, 2010; Harper and Elledge, 2007; Harrison and Haber, 2006; Henning and Stürzbecher, 2003; Huen et al., 2010;

Ishikawa et al., 2006; Jackson, 2009; Jackson and Bartek, 2009; Kastan, 2008; Kastan and Bartek, 2004; Kawabata et al., 2005; Kennedy and D'Andrea, 2005; Li and Greenberg, 2012; Li and Zou, 2005; Liang et al., 2009; Lukas et al., 2004; Masson and West, 2001; McGowan, 2002; McKinnon and Caldecott., 2007; Morris, 2010; Mullan et al., 2006; Murray et al., 2007; Nakanishi et al., 2006; Nyberg et al., 2002; Ohta et al., 2011; Perona et al., 2008; Reinhardt and Yaffe, 2009; Richardson, 2005; Rouse and Jackson, 2002; Roy et al., 2011; Samuel et al., 2002; Sancar et al., 2004; Shinohara and Ogawa, 1999; Silver and Livingston, 2012; Smith et al., 2010b; Stracker et al., 2009; Su, 2006; Sung et al., 2003; Thacker, 2005; Tutt and Ashworth, 2002; Venkitaraman, 2004; Vispe and Defais, 1997; Walworth, 2000; Wang, 2007; Wang et al., 2009a,b; Wu et al., 2010b; Yang and Xia, 2010; Yang and Zou, 2006; Yarden and Papa, 2006; Yoshida and Miki, 2004; Zhou and Elledge, 2000).

FOXM1 promotes HR repair of DSBs (Monteiro et al., 2012; Park et al., 2012) so that FOXM1-deficient cells exhibit an increase in the number of DNA breaks (Chetty et al., 2009; Kwok et al., 2010; Monteiro et al., 2012; Tan et al., 2007) whereas overexpression of FOXM1 and a N-terminally truncated FOXM1 mutant (ΔN-FOXM1) decreased the numbers of epirubicin-induced and cisplatin-induced DNA breaks in MCF-7 breast cancer cells, respectively (Kwok et al., 2010; Monteiro et al., 2012). This role of FOXM1 in HR (Monteiro et al., 2012; Park et al., 2012) correlates with the upregulation of the *foxm1* mRNA expression by BRCA1 (Bae et al., 2005), with the upregulation of the FOXM1 protein expression by RAD51 (Zhang et al., 2012e), and with the putative stabilization of the FOXM1 protein through its Chk2-mediated phosphorylation in response to DNA damage (Tan et al., 2007). Thus, the stimulation of HR by FOXM1 (Monteiro et al., 2012; Park et al., 2012) could reconcile the apparent contradiction that the tumor suppressors Chk2 (Tan et al., 2007) and BRCA1 (Bae et al., 2005) and the anti-proliferation signal RAD51 (Zhang et al., 2012e) augment the FOXM1 expression.

2. FOXM1 IN TUMORIGENESIS

2.1. FOXM1 is involved in tumorigenesis

FOXM1 is a typical proliferation-associated transcription factor (Kalin et al., 2011; Koo et al., 2011; Laoukili et al., 2007; Raychaudhuri and Park, 2011; Wierstra and Alves, 2007c) and it stimulates cell proliferation and cell cycle progression (Fig. 6.3), which represent (one of) its main function(s) (Costa,

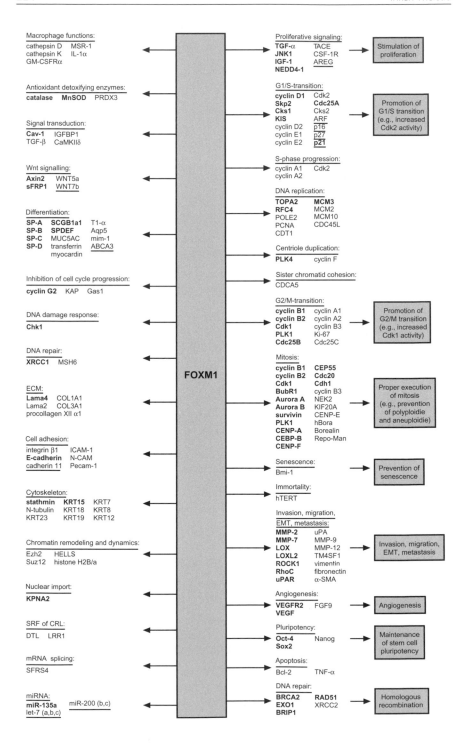

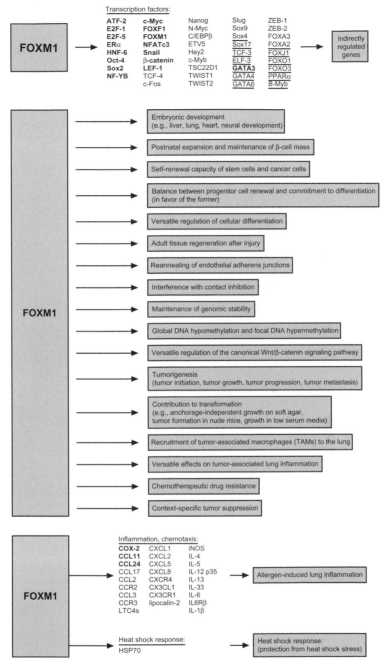

Figure 6.3 FOXM1 target genes and biological functions of FOXM1. Listed are direct FOXM1 target genes (bold) and possibly indirect FOXM1 target genes (normal). It is indicated whether FOXM1 upregulates (not underlined) or downregulates (underlined) the expression of a target gene. Contradictory findings have been reported for the

(Continued)

Figure 6.3—Cont'd regulation of the *E-cadherin* (*epithelial cadherin, CDH1* (*cadherin 1, type*)) expression by FOXM1: On the one hand, FOXM1 transactivated the *E-cadherin* promoter and increased the E-cadherin protein expression (see figure). On the other hand, FOXM1 decreased the *E-cadherin* mRNA and protein expression (not shown). In addition, the E-cadherin protein level was unaffected by FOXM1. Demonstrated biological functions of FOXM1 are shown in gray boxes. The figure does not show the following findings: (1) FOXM1 is important for the pregnancy-induced expansion of maternal β-cell mass. (2) In accordance with the antioxidant detoxifying enzymes *catalase, MnSOD* (*manganese superoxide dismutase*) and *PRDX3* (*peroxiredoxin 3*) being FOXM1 target genes, FOXM1 downregulates the intracellular ROS (reactive oxygen species) level. (3) FOXM1 prevents premature oxidative stress-induced senescence by decreasing the ROS level and the level of active p38. (4) FOXM1 supports formation of the premetastatic niche in an *ARF* (*alternative reading frame, CDKN2A* (*Cdk* (*cyclin-dependent kinase*) *inhibitor 2A*), *p14, p19*)-null background. (5) FOXM1 plays a role in radiation-induced and bleomycin-induced pulmonary fibrosis. The FOXM1 target genes are grouped according to their functions. Many FOXM1 target genes have additional functions, which are not shown in the figure. **Abbreviations**: ABCA3, ATP-binding cassette, subfamily A, member 3; ARF, alternative reading frame, CDKN2A, Cdk (cyclin-dependent kinase) inhibitor 2A, p14, p19; ATF-2, activating transcription factor-2; Aqp5, aquaporin 5; AREG, amphiregulin; Aurora B, Aurora B kinase; Axin2, axis inhibitor 2, conductin; Bcl-2, B-cell lymphoma-2; Bmi-1, B lymphoma Mo-MLV (Moloney-murine leukemia virus) insertion region-1, PCGF4, Polycomb group RING (really interesting new gene) finger protein 4; hBora, human Bora; Borealin, Dasra B, CDCA8, cell division cycle-associated protein 8; BRCA2, breast cancer-associated gene 2, FANCD1, Fanconi anemia complementation group D1; BRIP1, BRCA1-interacting protein 1, Bach1, BRCA1-associated C-terminal helicase 1; BubR1, Bub1-related kinase; CaMKIIδ, calcium/calmodulin-dependent protein kinase IIδ; Cav-1, caveolin-1; CCL, chemokine, CC motif, ligand; CCR, chemokine, CC motif, receptor; Cdc, cell division cycle; Cdc20, p55CDC, FZY, Fizzy; CDCA5, cell division cycle-associated protein 5, sororin; CDC45L, CDC45-like, CDC45, cell division cycle 45; Cdh1, Cdc20 homolog 1, FZR1, Fizzy-related protein 1; Cdk, cyclin-dependent kinase; CDT1, chromatin licensing and DNA replication factor 1; C/EBPβ, CCAAT/enhancer-binding protein β; CENP, centromere protein; CEP55, centrosomal protein 55-kDa; Chk, checkpoint kinase; Cks, CDC2-associated protein, Cdk subunit; COL1A1, collagen, type I, α-1; COL3A1, collagen, type III, α-1; COX-2, cyclooxygenase-2; CRL, Cullin-RING E3 ubiquitin ligase; CSF-1R, colony-stimulating factor-1 receptor; CXCL, chemokine, CXC motif, ligand; CXCR, chemokine, CXC motif, receptor; CX3CL, chemokine, CX3C motif, ligand; CX3CR, chemokine, CX3C motif, receptor; cyclin F, CCNF, FBXO1, FBX1, F-box only protein 1; DTL, denticleless, CDT2, DCAF2, Ddb1 (damage-specific DNA-binding protein 1)- and Cul4 (Cullin 4)-associated factor 2; E-cadherin, epithelial cadherin, CDH1, cadherin 1, type 1; ECM, extracellular matrix; ELF-3, E74-like factor-3; EMT, epithelial–mesenchymal transition; ERα, estrogen receptor α; ETV5, Ets variant gene 5, Erm, Ets-related molecule; EXO1, exonuclease 1; Ezh2, enhancer of Zeste homolog 2, KMT6, K-methyltransferase 6; FGF9, fibroblast growth factor 9; c-Fos, FOS; FOX, Forkhead box; Gas1, growth arrest-specific 1; GATA, GATA-binding protein; GM-CSFRα, granulocyte-macrophage colony-stimulating factor receptor α; CSF2RA, colony-stimulating factor 2 receptor α; HELLS, helicase, lymphoid-specific, LSH, lymphoid-specific helicase, SMARCA6, SWI/SNF-related, matrix-associated, actin-dependent regulator of chromatin, subfamily A, member 6; Hey2, Hairy/enhancer of split-related with YRPW motif 2; HNF-6, hepatocyte nuclear factor-6; HSP70, heat shock protein 70; ICAM-1, intercellular cell adhesion molecule-1; IGF-1, insulin-like growth factor-1; IGFBP1, IGF-binding protein 1; IL, interleukin; IL6Rβ, IL-6 receptor β; IL-12 p35, IL-12, 35-kDa subunit, IL12A, interleukin 12A; iNOS, inducible nitric oxide synthase, NOS2A, nitric oxide synthase 2A, NOS2; JNK1, c-Jun (JUN) N-terminal kinase 1; KAP, Cdk-associated protein phosphatase, CDKN3, Cdk inhibitor 3; Ki-67, Mki67, proliferation-related Ki-67 antigen; KIF20A, kinesin family member 20A; KIS, kinase interacting stathmin; KPNA2, karyopherin α-2; KRT, cytokeratin; Lama, laminin α; LEF-1, lymphoid enhancer factor-1; LOX, lysyl oxidase; LOXL2, LOX oxidase-like 2; LRR1, leucinerich repeat protein 1, PPIL5, peptidyl-prolyl isomerase-like 5; LTC4s, leukotriene synthase; MCM, minichromosome maintenance; mim-1, myb-induced myeloid protein-1; MMP, matrix metalloproteinase; MnSOD, manganese superoxide dismutase; MSH6, mutS homolog 6; MSR-1, macrophage scavenger receptor-1; MUC5AC, Mucin 5, subtypes A and C, tracheobronchial; B-Myb, MYBL2, MYB-like 2; c-Myb, MYB; c-Myc, MYC; N-Myc, MYCN, Myc oncogene neuroblastoma-derived; N-CAM, neural cell adhesion molecule; NEDD4-1, neural precursor cell-expressed, developmentally down-regulated 4-1; NEK2, NIMA (never in mitosis A)-related kinase 2; NFATc3, nuclear factor of activated T-cell c3; NF-YB, nuclear factor-Y B; Oct-4, octamer-binding transcription factor 4, POU5F1, POU domain, class 5, transcription factor 1; p16, CDKN2A, Cdk inhibitor 2A, INK4A, inhibitor of Cdk4 A; p21, CDKN1A, Cdk inhibitor 1A, CIP1, Cdk-interacting protein 1, WAF1, wild-type p53-activated fragment 1; p27, CDKN1B, Cdk inhibitor 1B, KIP1, kinase inhibitor protein 1; PCNA, proliferating cell nuclear antigen; Pecam-1, platelet endothelial cell adhesion molecule 1; PLK, Polo-like kinase; POLE2, DNA polymerase ε 2; PPARα, peroxisome proliferator-activated receptor α; PRDX3, peroxiredoxin 3; Repo-Man, recruits PP1 (protein phosphatase 1) onto mitotic chromatin at anaphase protein, CDCA2, cell division cycle-associated protein 2; RFC4, replication factor C, subunit 4; RhoC, Ras homolog gene family, member C; ROCK1, Rho-associated coiled-coil-containing protein kinase 1; SCGB1a1, secretoglobin, family 1A, member 1, CCSP, Clara cell secretory protein, CC10, Clara cell-specific 10-KD protein; sFRP1, secreted Frizzled-related protein 1; SFRS4, splicing factor, arginine/serine-rich 4, SRp75, splicing factor, arginine/serine-rich 75-KD; Skp2, S-phase kinase-associated protein 2; Slug, SNAI2, Snail2; Snail, SNAI1, Snail1; α-SMA, α-smooth muscle actin; Sox, SRY (sex-determining region Y)-box, SRY-related HMG (high mobility group)-box; SP, surfactant protein; SPDEF, SAM-pointed domain-containing ETS transcription factor, PDEF, prostate-derived ETS transcription factor; SRF, substrate recognition factor; survivin, BIRC5, baculoviral IAP repeat-containing 5; Suz12, suppressor of Zeste 12; T1-α, PDPN, podoplanin; TACE, TNF-α converting enzyme, ADAM-17, a disintegrin and metalloprotease domain-17; TCF, T-cell factor; hTERT, human telomerase reverse transcriptase; TGF, transforming growth factor; TM4SF1, transmembrane 4 superfamily member 1, TAL6, TAAL6, tumor-associated antigen L6; TNF-α, tumor necrosis factor-α; TOP2A, TOPO-2α, DNA topoisomerase 2α; TSC22D1, TSC22 (TGF-β-stimulated clone 22) domain family, member 1, TGF-β-stimulated clone 22 domain 1; TWIST2, Dermo1, Dermis-expressed protein 1; uPA, urokinase-type plasminogen activator; uPAR, uPA receptor; VEGF, vascular endothelial growth factor; VEGFR2, VEGF receptor type II; WNT, wingless type; XRCC, X-ray cross-completing group; ZEB-1, zinc finger E-box-binding homeobox-1, δEF1, δ-crystallin enhancer-binding factor 1; ZEB-2, zinc finger E-box-binding homeobox-2, SIP1, Smad-interacting protein 1.

2005; Costa et al., 2003, 2005b; Kalin et al., 2011; Koo et al., 2011; Laoukili et al., 2007; Mackey et al., 2003; Minamino and Komuro, 2006; Raychaudhuri and Park, 2011; Wierstra and Alves, 2007c).

As expected, FOXM1 is also implicated in tumorigenesis (Fig. 6.3) (Costa et al., 2005a; Gong and Huang, 2012; Kalin et al., 2011; Koo et al., 2011; Laoukili et al., 2007; Myatt and Lam, 2007; Raychaudhuri and Park, 2011; Wierstra and Alves, 2007c). Actually, FOXM1 contributes to oncogenic transformation and participates in both tumor initiation and tumor progression (Fig. 6.3) (Costa et al., 2005a; Gong and Huang, 2012; Kalin et al., 2011; Koo et al., 2011; Laoukili et al., 2007; Myatt and Lam, 2007; Raychaudhuri and Park, 2011; Wierstra and Alves, 2007c). FOXM1 promotes anchorage-independent growth on soft agar, tumor formation in nude mice, and growth in low-serum media, demonstrating that FOXM1 contributes to oncogenic transformation and enhances tumorigenicity (Fig. 6.3) (Ahmad et al., 2010; Bhat et al., 2011; Chan et al., 2008; Chen et al., 2011a, 2012a, 2013b; Dai et al., 2010; He et al., 2012; Kalin et al., 2006; Kalinichenko et al., 2004; Kim et al., 2006a; Li et al., 2009, 2013; Liu et al., 2006, 2013b; Park et al., 2008a, 2009, 2011, 2012; Radhakrishnan et al., 2006; Uddin et al., 2011; Wang et al., 2008b, 2011b; Wu et al., 2010a; Xie et al., 2010; Yoshida et al., 2007; Zhang et al., 2008, 2011b, 2012d,e). Moreover, FOXM1 stimulates migration and invasion (Ahmad et al., 2010; Balli et al., 2012; Bao et al., 2011; Behren et al., 2010; Bellelli et al., 2012; Chen et al., 2009, 2013b; Chu et al., 2012; Dai et al., 2007; He et al., 2012; Li et al., 2011, 2013; Lok et al., 2011; Lynch et al., 2012; Mizuno et al., 2012; Park et al., 2008a, 2011; Uddin et al., 2011, 2012; Wang et al., 2007a, 2008b, 2010c, 2013; Wu et al., 2010a; Xia et al., 2012b; Xue et al., 2012) as well as EMT (epithelial–mesenchymal transition) (Balli et al., 2013; Bao et al., 2011; Li et al., 2012c; Park et al., 2011) and metastasis (Fig. 6.3) (Li et al., 2009, 2013; Park et al., 2011; Xia et al., 2012b). For example, FOXM1 supports formation of the premetastatic niche in an *ARF*-null background (Park et al., 2011). Also angiogenesis is promoted by FOXM1 (Fig. 6.3) (Kim et al., 2005; Li et al., 2009; Lynch et al., 2012; Wang et al., 2007a; Xue et al., 2012; Zhang et al., 2008). In addition, FOXM1 plays a role in recruitment of TAMs (tumor-associated macrophages) to the lung (Fig. 6.3) and in tumor-associated pulmonary inflammation (Balli et al., 2012). Furthermore, the self-renewal capacity of cancer cells is supported by FOXM1 (Fig. 6.3) (Bao et al., 2011; Wang et al., 2011b; Zhang et al., 2011b). In addition, FOXM1 counteracts premature senescence (Fig. 6.3) (Anders et al., 2011; Li et al., 2008a; Park et al., 2009; Qu et al., 2013; Rovillain et al., 2011; Wang et al., 2005; Zeng et al., 2009;

Zhang et al., 2006a), which represents an important tumor suppression mechanism (Adams, 2009; Braig and Schmitt, 2006; Caino et al., 2009; Campisi, 2000, 2001, 2003, 2005a,b; Campisi, 2011; Campisi and d'Adda di Fagagna, 2007; d'Adda di Fagagna, 2008; Cichowski and Hahn, 2008; Collado and Serrano, 2005, 2006, 2010; Collado et al., 2007; Coppe et al., 2010; Courtois-Cox et al., 2008; Di Micco et al., 2007; Evan and d'Adda di Fagana, 2009; Evan et al., 2005; Ewald et al., 2010; Freund et al., 2010; Fridman and Tainsky, 2008; Hanahan and Weinberg, 2011; Hemann and Narita, 2007; Kuilman et al., 2010; Lanigan et al., 2011; Lleonart et al., 2009; Lowe et al., 2004; Mathon and Lloyd, 2001; McDuff and Turner, 2011; Mooi and Peeper, 2006; Nardella et al., 2011; Ogrunc and d'Adda di Fagana, 2011; Ohtani et al., 2009; Pazolli and Stewart, 2008; Prieur and Peeper, 2008; Rangarajan and Weinberg, 2003; Roninson, 2003; Sage, 2005; Schmitt, 2003, 2007; Sharpless and DePinho, 2004; Shay and Roninson, 2004; Stewart and Weinberg, 2002, 2006). Since FOXM1 overexpression can confer chemotherapeutic drug resistance (Fig. 6.3), downregulation of FOXM1 may resensitize tumor cells to anticancer drugs (Koo et al., 2011; Raychaudhuri and Park, 2011; Teh et al., 2012; Wilson et al., 2012; Zhao and Lam, 2012). Hence, FOXM1 is not only a proliferation-associated transcription factor but it is also intimately involved in oncogenesis (Fig. 6.3) (Kalin et al., 2011; Koo et al., 2011; Laoukili et al., 2007; Raychaudhuri and Park, 2011; Wierstra and Alves, 2007c).

In fact, studies with *foxm1* knockout or transgenic mice revealed that FOXM1 contributes to tumor initiation, tumor growth, and tumor progression in different types of tumors (i.e., liver, colon, lung, and prostate) (Fig. 6.3), at least in part due to its ability to promote cell proliferation (Gusarova et al., 2007; Kalin et al., 2006; Kalinichenko et al., 2004; Kim et al., 2006a,b,c; Wang et al., 2008a, 2009a; Yoshida et al., 2007). In particular, FOXM1 is essential for the development (initiation) of DEN/PB (diethylnitrosamine/phenobarbital)–induced hepatic adenomas and HCC so that mice with a hepatocyte-specific conditional *foxm1* knockout are highly resistant to developing these carcinogen-induced liver tumors (Costa et al., 2005a; Kalinichenko et al., 2004).

Its implication in tumorigenesis makes FOXM1 an attractive target for anticancer therapy (Adami and Ye, 2007; Ahmad et al., 2012; Costa et al., 2005a; Gartel, 2008, 2010, 2012; Halasi and Gartel, 2013; Katoh et al., 2013; Koo et al., 2011; Li et al., 2012b; Myatt and Lam, 2007; Raychaudhuri and Park, 2011; Teh, 2012a,b; Wang et al., 2010a; Wierstra and Alves, 2007c; Wilson et al., 2011; Zhao and Lam, 2012). Accordingly, several trials to interfere with the expression or activity of FOXM1 have been published (see below).

2.2. Oncogenic transformation

FOXM1 contributes to oncogenic transformation and enhances tumorigenicity because it promotes anchorage-independent growth on soft agar (Ahmad et al., 2010; Bhat et al., 2011; Chan et al., 2008; Chen et al., 2012a, 2013b; Dai et al., 2010; Kalin et al., 2006; Kalinichenko et al., 2004; Kim et al., 2006a; Liu et al., 2006, 2013b; Park et al., 2008a, 2009, 2011; Radhakrishnan et al., 2006; Uddin et al., 2011; Wang et al., 2008b, 2011b; Wu et al., 2010a; Yoshida et al., 2007; Zhang et al., 2012d,e), growth in low-serum media (Park et al., 2008a), and tumor formation in nude mice (Fig. 6.3) (Bhat et al., 2011; Chen et al., 2011a, 2012a; Dai et al., 2010; He et al., 2012; Li et al., 2009, 2013; Liu et al., 2006, 2013b; Park et al., 2011, 2012; Uddin et al., 2011; Wang et al., 2011b; Xie et al., 2010; Zhang et al., 2008, 2011b, 2012e).

The anchorage-independent growth on soft agar was increased by FOXM1 overexpression (Ahmad et al., 2010; Chan et al., 2008; Dai et al., 2010; Kalinichenko et al., 2004; Liu et al., 2006; Park et al., 2008a, 2009, 2011; Radhakrishnan et al., 2006; Uddin et al., 2011; Wang et al., 2011b), but decreased by siRNA/shRNA-mediated knockdown of FOXM1 (Ahmad et al., 2010; Bhat et al., 2011; Chan et al., 2008; Chen et al., 2012a, 2013b; Kalin et al., 2006; Kim et al., 2006a; Liu et al., 2006, 2013b; Wang et al., 2008b, 2011b; Wu et al., 2010a; Yoshida et al., 2007; Zhang et al., 2012d,e) and by Cre-mediated conditional deletion of floxed *foxm1* alleles (Park et al., 2009).

Also the formation of tumors in athymic nude mice was enhanced by FOXM1 overexpression (Dai et al., 2010; He et al., 2012; Li et al., 2009, 2013; Liu et al., 2006; Park et al., 2011; Uddin et al., 2011), but reduced by siRNA/shRNA-mediated knockdown of FOXM1 (Bhat et al., 2011; Chen et al., 2011a, 2012a; He et al., 2012; Li et al., 2009, 2013; Liu et al., 2006, 2013b; Park et al., 2012; Wang et al., 2011b; Xie et al., 2010; Zhang et al., 2008, 2011a,b). In this instance, FOXM1 increased the incidence of xenograft tumors (Dai et al., 2010; Liu et al., 2006; Park et al., 2011; Wang et al., 2011b; Zhang et al., 2008, 2011a,b) as well as the size or mass of the xenograft tumors (Bhat et al., 2011; Chen et al., 2011a, 2012a; Dai et al., 2010; He et al., 2012; Li et al., 2009, 2013; Liu et al., 2006, 2013b; Park et al., 2011, 2012; Uddin et al., 2011; Xie et al., 2010), which indicates a positive effect of FOXM1 on both tumor initiation and tumor growth.

NIH3T3 fibroblasts with stable overexpression of a ca (constitutively active) N-terminally truncated FoxM1B mutant grew efficiently in low-serum media (2% serum), whereas NIH3T3 fibroblasts with stable overexpression of wild-type FoxM1B did not (Park et al., 2008a). Thus, FoxM1B can stimulate growth in low-serum media, but only if the auto-inhibitory FOXM1 N-terminus is deleted (Park et al., 2008a), that is, if FOXM1 is released from cellular constraints, which keep it in check under normal conditions. The ability of ca FoxM1B-overexpressing cells to grow in low-serum media shows that FoxM1B overexpression can reduce the requirement for growth factors, provided that its auto-inhibitory N-terminus is removed (Park et al., 2008a) or otherwise disabled.

In summary, FOXM1 promotes anchorage-independent growth on soft agar, tumor formation in nude mice, and growth in low-serum media, indicating that it contributes to oncogenic transformation and enhances tumorigenicity (Fig. 6.3) (Ahmad et al., 2010; Bhat et al., 2011; Chan et al., 2008; Chen et al., 2011a, 2012a, 2013b; Dai et al., 2010; He et al., 2012; Kalin et al., 2006; Kalinichenko et al., 2004; Kim et al., 2006a; Li et al., 2009, 2013; Liu et al., 2006, 2013b; Park et al., 2008a, 2009, 2011, 2012; Radhakrishnan et al., 2006; Uddin et al., 2011; Wang et al., 2008b, 2011b; Wu et al., 2010a; Xie et al., 2010; Yoshida et al., 2007; Zhang et al., 2008, 2011b, 2012d, 2012e).

Remarkably, FOXM1c disposes of only a weak transformation potential (see below) (Lüscher-Firzlaff et al., 1999). Accordingly, several studies found that overexpression of FOXM1 alone does not result in tumors. Hence, FOXM1 is no oncogene as often stated in the literature, but it rather facilitates or reinforces the tumorigenesis induced by oncoproteins, loss of tumor suppressors, and other insults.

The transforming oncoprotein E7 of the small DNA tumor virus HPV16 binds directly to FOXM1c (Lüscher-Firzlaff et al., 1999) and enhances the transcriptional activity of FOXM1c (Lüscher-Firzlaff et al., 1999; Wierstra and Alves, 2006d, 2008).

HPV16 E7 can cooperate with an activated form of the cellular oncoprotein Ras in the transformation of primary REFs (rat embryo fibroblasts) (Lüscher-Firzlaff et al., 1999; Nishikawa et al., 1991; Yamashita et al., 1993, 2001). FOXM1c stimulated this E7/Ha-Ras-dependent co-transformation only 1.5- to 2-fold (Lüscher-Firzlaff et al., 1999). In contrast to the co-expression of HPV16 E7 and activated Ha-Ras (Lüscher-Firzlaff et al., 1999; Nishikawa et al., 1991; Yamashita et al., 1993, 2001), neither the combination of FOXM1c with activated Ha-Ras nor the combination of FOXM1c with HPV16 E7 was sufficient to achieve a (significant) transformation of primay REFs (Lüscher-Firzlaff et al., 1999).

Thus, FOXM1c is no oncoprotein because it failed to transform primary REFs in cooperation with activated Ha-Ras (Lüscher-Firzlaff et al., 1999). Instead, FOXM1c possesses at best a low transformation potential, which is only uncovered in the presence of two already cooperating, transforming oncoproteins (i.e., activated Ha-Ras and HPV16 E7) (Lüscher-Firzlaff et al., 1999).

2.3. Mouse model of tumorigenesis

2.3.1 FOXM1 contributes to tumor initiation as well as to tumor progression

FOXM1 contributes to both tumor initiation and tumor progression (Fig. 6.3) (Gusarova et al., 2007; Kalin et al., 2006; Kalinichenko et al., 2004; Kim et al., 2006a,b,c; Wang et al., 2008a, 2009a; Yoshida et al., 2007).

The *in vivo* role of FOXM1 in tumor initiation and tumor progression was analyzed with *FoxM1B*-transgenic mice (Kalin et al., 2006; Wang et al., 2008a; Yoshida et al., 2007) and *foxm1* knockout mice (Gusarova et al., 2007; Kalinichenko et al., 2004; Kim et al., 2006a,b,c; Wang et al., 2009a,b; Yoshida et al., 2007). These studies demonstrated an involvement of FOXM1 in the initiation, growth, and progression of tumors in various organs (Fig. 6.3), namely, liver tumors (Gusarova et al., 2007; Kalinichenko et al., 2004), colon tumors (Yoshida et al., 2007), lung tumors (Kim et al., 2006a,b,c; Wang et al., 2008a, 2009a), and prostate tumors (Kalin et al., 2006). The contribution of FOXM1 to tumor initiation, tumor growth, and tumor progression is at least partially based on its ability to stimulate cell proliferation (Gusarova et al., 2007; Kalin et al., 2006; Kalinichenko et al., 2004; Kim et al., 2006a,b,c; Wang et al., 2008a, 2009a; Yoshida et al., 2007).

Especially intriguing are the essential requirement of FOXM1 for the development (initiation) of DEN/PB-induced HCC and hepatic adenomas and the resulting complete resistance of hepatocyte-specific conditional *foxm1* knockout mice to developing these carcinogen-induced liver tumors (Costa et al., 2005a; Kalinichenko et al., 2004).

FOXM1 is also required for the growth and progression of liver, lung, and colon tumors (Gusarova et al., 2007; Kim et al., 2006a,b,c; Wang et al., 2009a,b; Yoshida et al., 2007). In addition, FoxM1B overexpression accelerates the growth and progression of prostate, lung, and colon tumors (Kalin et al., 2006; Wang et al., 2008a; Yoshida et al., 2007).

2.3.2 foxm1 *kockout mice are highly resistant to liver tumor induction by DEN/PB*

The DEN/PB liver tumor induction protocol causes the development of HCC and hepatic adenomas (Lee, 2000; Pitot et al., 1996; Sargent et al., 1996; Tamano et al., 1994). In the study of Kalinichenko et al. (2004), mice were given a single dose of the tumor initiator DEN at postnatal day 14 and then they were continously admistered the tumor promoter PB from 4 weeks after birth onwards. DEN/PB treatment for 6 weeks induced abundant nuclear FOXM1 protein in hepatocytes, whereas no FOXM1 protein was detected in hepatocyte nuclei of untreated livers (Kalinichenko et al., 2004). The high levels of nuclear FOXM1 protein persisted in hyperproliferative DEN/PB-induced hepatic adenomas and HCC (Kalinichenko et al., 2004). Thus, nuclear FOXM1 protein expression in hepatocytes is induced prior to liver tumor formation and continues during liver tumor progression (Kalinichenko et al., 2004).

In contrast to the development of HCC and hepatic adenomas in DEN/PB-treated control animals, DEN/PB exposure completely failed to induce HCC or hepatic adenomas in conditional *foxm1* knockout mice with a hepatocyte-specific *foxm1* deletion by 6 weeks after birth (Kalinichenko et al., 2004; Postic and Magnuson, 2000). This total resistance of *foxm1* knockout mice to developing carcinogen-induced liver tumors demonstrates that FOXM1 is essentially required for the development (initiation) of DEN/PB-induced HCC and hepatic adenomas (Kalinichenko et al., 2004).

The failure of hepatocyte-specific conditional *foxm1* knockout mice to develop liver tumors after DEN/PB treatment was not due to an increase in hepatocyte apoptosis (Kalinichenko et al., 2004). Instead, the *foxm1*-deficient hepatocytes in non-tumor regions of *foxm1* knockout mice proliferated only marginally following DEN/PB exposure whereas both liver tumor cells and hepatocytes in non-tumor regions of control animals underwent extensive proliferation (Kalinichenko et al., 2004). This suggests that *foxm1* deficiency prevents hepatocyte proliferation, which is required for tumor formation, explaining why *foxm1* knockout mice are highly resistant to developing DEN/PB-induced HCC and hepatic adenomas (Kalinichenko et al., 2004).

In summary, mouse livers, in which *foxm1* was knocked out upon tumor induction by DEN/PB, developed neither hepatic adenomas nor HCC, demonstrating that FOXM1 is essential for liver tumor initiation (Fig. 6.3) (Costa et al., 2005a; Kalinichenko et al., 2004). The total resistance of *foxm1* knockout mice to liver tumor induction by DEN/PB results at least in part from the failure of *foxm1*-deficient hepatocytes to proliferate in response to DEN/PB treatment, because this failure prevents the extensive proliferation required for tumor formation (Costa et al., 2005a; Kalinichenko et al., 2004).

2.3.3 FOXM1 is involved in liver, lung, colon, and prostate tumor growth and progression

Foxm1 knockout mice revealed that FOXM1 is required for tumor growth and progression of lung adenomas (Kim et al., 2006a,b,c; Wang et al., 2009a,b), colorectal tumors (Yoshida et al., 2007), HCC, and hepatic adenomas (Gusarova et al., 2007). Conversely, *FoxM1B*-transgenic mice revealed that FoxM1B overexpression accelerates tumor growth and progression of lung adenomas (Wang et al., 2008a), invasive colon adenocarcinomas (Yoshida et al., 2007), and prostate carcinomas (Kalin et al., 2006). In detail, the following mouse models have been studied:

The *foxm1* knockout before colon tumor induction by AOM/DSS (azoxymethane/dextran sodium sulfate) reduced the number and size of colorectal tumors as well as the incidence of invasive adenocarcinomas, which shows that FOXM1 is required for colon tumor growth and progression (Yoshida et al., 2007).

The *foxm1* knockout in preexisting DEN/PB-induced liver tumors decreased the number and size of hepatic adenomas and HCC, indicating that FOXM1 is required for liver tumor growth and progression (Gusarova et al., 2007).

Three findings demonstrate that FOXM1 is required for lung tumor growth and progression: First, the *foxm1* knockout before lung tumor induction by urethane diminished the number and size of lung adenomas (Kim et al., 2006a,b,c; Wang et al., 2009a,b). Second, the *foxm1* knockout before lung tumor induction by MCA/BHT (3-methylcholanthrene/butylated hydroxytoluene) reduced the number of lung adenomas (Wang et al., 2009a,b). Third, the *foxm1* knockout after chemically induced (urethane or MCA/BHT) lung tumor initiation decreased the number of lung adenomas (Wang et al., 2009a,b).

In murine AOM/DSS-induced colon tumorigenesis, the ubiquitous expression of a human *foxm1b* transgene in all cell types increased the number and size of colorectal tumors as well as the incidence of invasive colon adenocarcinomas, demonstrating that FoxM1B overexpression accelerates colon tumor growth and progression (Yoshida et al., 2007).

In murine MCA/BHT-induced lung tumorigenesis, the same ubiqui-tous expression of a human *foxm1b* transgene increased the number and size of lung adenomas, which indicates that FoxM1B overexpression accelerates lung tumor growth and progression (Wang et al., 2008a).

In TRAMP and LADY transgenic mouse models of prostate cancer, the same expression of a human *foxm1b* transgene in all cell types increased the incidence of prostate carcinomas and the prostate weight, showing that FoxM1B overexpression accelerates prostate tumor growth and progression (Kalin et al., 2006). In TRAMP (transgenic adenocarcinoma of the mouse prostate) transgenic mice, the rat *probasin* promoter drives prostate epithelial expression of the SV40 large T and small t antigens, two oncoproteins of the transforming small DNA tumor virus SV40 (Abate-Shen and Shen, 2002; Abdulkadir and Kim, 2005; Ahmad et al., 2008; Hensley and Kyprianou, 2012; Huss et al., 2001; Kasper, 2005; Navone et al., 1999; Valkenburg and Williams, 2011; Wang, 2011). The TRAMP model recapitulates mul-tiple stages of human prostate cancer, because it results in the development of prostate epithelial cell hyperplasia and PIN (prostatic intraepithelial neopla-sia), which progresses to histologic invasive prostatic carcinomas (Berman-Booty et al., 2012; Gingrich et al., 1996; Greenberg et al., 1995). In LADY transgenic mice, only the SV40 large T antigen is expressed under the con-trol of the rat *probasin* promoter (Abate-Shen and Shen, 2002; Abdulkadir and Kim, 2005; Ahmad et al., 2008; Hensley and Kyprianou, 2012; Huss et al., 2001; Kasper, 2005; Valkenburg and Williams, 2011; Wang, 2011). LADY mice develop multifocal, low-grade PIN that progresses to high-grade PIN and early invasive prostate carcinomas with progressive neuroen-docrine differentiation (Kasper et al., 1998; Masumori et al., 2001).

In summary, FOXM1 contributes to the growth and progression of liver, prostate, lung, and colon tumors (Fig. 6.3) (Gusarova et al., 2007; Kalin et al., 2006; Kim et al., 2006a,b,c; Wang et al., 2008a, 2009a; Yoshida et al., 2007).

In particular, the *foxm1*$^{-/-}$ cells in lung tumors (Kim et al., 2006a,b,c; Wang et al., 2009a,b), colon adenomas, colon adenocarcinomas (Yoshida et al., 2007), hepatic adenomas, and HCC (Gusarova et al., 2007; Kalinichenko et al., 2004) display a reduced proliferation so that *foxm1* defi-ciency probably prevents their proliferative expansion, which is required for tumor growth and progression. This view is supported by the increased pro-liferation of tumor cells in prostate carcinomas (Kalin et al., 2006), lung ade-nomas (Wang et al., 2008a), colon adenomas, and colon adenocarcinomas (Yoshida et al., 2007) of *FoxM1B*-transgenic mice, which suggests that

FoxM1B overexpression enhances their proliferative expansion resulting in the observed acceleration of tumor growth and progression.

In accordance with the involvement of FOXM1 in lung tumor growth and progression (Kim et al., 2006a,b,c; Wang et al., 2008a, 2009a), conditional *FoxM1B/K-Ras*-double-transgenic mice showed that the expression of a ca N-terminally truncated FoxM1B mutant (FoxM1B-ΔN) in the respiratory epithelium accelerates the growth of lung tumors, which were induced by oncogenic (activated) K-Ras (Wang et al., 2010e).

In a *K-Ras*-transgenic mouse model of lung cancer (Fisher et al., 2001; Johnson et al., 2001), the lung epithelial-specific expression of FoxM1B-ΔN significantly increased the size of K-Ras-induced lung tumors compared to those in mice expressing only activated K-Ras, which indicates that FoxM1B-ΔN accelerates lung tumor growth induced by oncogenic K-Ras (Wang et al., 2010e). However, the expression of FOXM1B-ΔN alone did not cause lung tumors, but resulted in the formation of epithelial hyperplasia (Wang et al., 2010e). In contrast, the expression of oncogenic K-Ras alone was sufficient to induce the formation of lung adenomas (Fisher et al., 2001; Johnson et al., 2001; Wang et al., 2010e). Hence, in the respiratory epithelium, FOXM1B-ΔN cooperates with activated K-Ras to accelerate lung tumor growth (Wang et al., 2010e).

2.3.4 Summary
In summary, several *foxm1* knockout or transgenic mouse models revealed that FOXM1 contributes to tumor initiation, growth, and progression in different types of tumors (liver, colon, lung, and prostate) (Fig. 6.3), at least in part due to its ability to promote cell proliferation (Gusarova et al., 2007; Kalin et al., 2006; Kalinichenko et al., 2004; Kim et al., 2006a,b,c; Wang et al., 2008a, 2009a; Yoshida et al., 2007). Intriguingly, FOXM1 is essential for the development (initiation) of DEN/PB-induced hepatic adenomas and HCC, because mice with a hepatocyte-specific conditional *foxm1* knockout are highly resistant to developing these carcinogen-induced liver tumors (Costa et al., 2005a; Kalinichenko et al., 2004).

2.4. Angiogenesis
FOXM1 stimulates angiogenesis (Fig. 6.3) (Kim et al., 2005; Li et al., 2009; Lynch et al., 2012; Wang et al., 2007a; Xue et al., 2012; Zhang et al., 2008).

First, FOXM1 promotes angiogenesis during embryonic development (Kim et al., 2005):

Pecam-1-stained lung sections of *foxm1* knockout (*foxm1*$^{-/-}$) and wild-type mouse embryos at 15.5 dpc (days post coitum) and 17.5 dpc revealed that the *foxm1*$^{-/-}$ lungs exhibited a significantly reduced number of large pulmonary blood vessels (arteries and veins) compared to wild-type lungs (Kim et al., 2005). This defect in formation of peripheral pulmonary capillaries from lung mesenchyme shows that FOXM1 is essential for proper development of the pulmonary microvasculature during embryonic lung morphogenesis (Kim et al., 2005).

Second, FOXM1 stimulates angiogenesis in the context of tumorigenesis (Li et al., 2009; Lynch et al., 2012; Wang et al., 2007a; Xue et al., 2012; Zhang et al., 2008):

Conditioned media collected from human SWI1783 anaplastic astrocytoma cells stably overexpressing FoxM1B displayed a higher angiogenic potential in endothelial cell tube formation assays with HUVECs (human umbilical vein endothelial cells) than media from control SWI1783 cells, demonstrating that FoxM1B overexpression promotes angiogenesis (Zhang et al., 2008). Conversely, siRNA-mediated silencing of FOXM1 reduced the angiogenic ability of culture supernatants harvested from human U-87MG and HFU-251 glioblastoma cells (Zhang et al., 2008) or human BxPC-3 and HPAC pancreatic cancer cells (Wang et al., 2007a), or human Caki-1 and 786-O RCC cells (Xue et al., 2012) in HUVEC tube formation assays, which verifies the stimulation of angiogenesis by FOXM1. Also the angiogenic ability of human NCI-N87 gastric cancer cells was increased by FoxM1B overexpression, but decreased by siRNA-mediated knockdown of FOXM1, as measured by tube formation of HUVECs cultured with conditioned medium (Li et al., 2009).

Culture supernatants collected from human PC3-ML prostate cancer cells with stable expression of shRNA against OGT (O-GlcNAc (O-linked β-N-acetylglucosamine) transferase) exhibited a diminished angiogenic potential in HUVEC tube formation assays compared to control cells, but overexpression of a N-terminally truncated FOXM1c mutant was able to fully overcome this inhibition of angiogenesis mediated by OGT silencing, which again indicates that FOXM1 promotes angiogenesis (Lynch et al., 2012).

The tumor microvessel density of gastric tumors produced by NCI-N87 cells after their injection into the stomach wall was increased by FOXM1 overexpression, but decreased by FOXM1 knockdown with siRNA, pointing to a positive effect of FOXM1 on tumor vascularization and thus on tumor angiogenesis (Li et al., 2009). Accordingly, siRNA-mediated depletion of FOXM1 also resulted in a lower tumor microvessel density of brain tumors produced by intracranially injected U-87MG cells (Zhang et al., 2008).

In summary, FOXM1 promotes angiogenesis (Fig. 6.3), namely, during embryonic development (Kim et al., 2005) and in the context of tumorigenesis (Li et al., 2009; Lynch et al., 2012; Wang et al., 2007a; Xue et al., 2012; Zhang et al., 2008).

2.5. Migration and invasion

FOXM1 promotes migration and invasion (Fig. 6.3) (Ahmad et al., 2010; Balli et al., 2012; Bao et al., 2011; Behren et al., 2010; Bellelli et al., 2012; Chen et al., 2009, 2013b; Chu et al., 2012; Dai et al., 2007; He et al., 2012; Li et al., 2011, 2013; Lok et al., 2011; Lynch et al., 2012; Mizuno et al., 2012; Park et al., 2008a, 2011; Uddin et al., 2011, 2012; Wang et al., 2007a, 2008b, 2010c, 2013; Wu et al., 2010a; Xia et al., 2012b; Xue et al., 2012). The stimulation of migration and invasion by FOXM1 has been reported in a large variety of tumor cells:

Overexpression of FOXM1 increased the migration and invasion abilities of SKBR3 and SUM102 breast cancer cells (Ahmad et al., 2010), HCC94 cervical cancer cells (He et al., 2012), A2780cp and OVCA433 ovarian cancer cells (Lok et al., 2011), HCT-15 colon adenocarcinoma cells (Uddin et al., 2011), and SW480 colon cancer cells (Li et al., 2013). Conversely, siRNA/shRNA-mediated depletion of FOXM1 decreased the migration and invasion abilities of primary ATC cells (Bellelli et al., 2012), 786-O and Caki-1 RCC cells (Xue et al., 2012), MDA-MB-231 and SUM149 breast cancer cells (Ahmad et al., 2010), SiHa cervical cancer cells (He et al., 2012), SUDHL4 DLBCL cells (Uddin et al., 2012), SW480 and LoVo CRC cells (Chu et al., 2012), SW620 colon cancer cells (Li et al., 2013), HCCLM3, HepG2, and Huh-7 HCC cells (Xia et al., 2012b), BxPC-3 and HPAC pancreatic cancer cells (Wang et al., 2007a), U2OS osteosarcoma cells (Wang et al., 2008b), A549 lung adenocarcinoma cells, and cisplatin-resistant A549/DDP cells (Wang et al., 2013).

FOXM1 overexpression also enhanced the motility of AsPC-1 pancreatic cancer cells (Bao et al., 2011) and HN4 HNSCC cells (Wang et al., 2010c). Conversely, FOXM1 knockdown by siRNA/shRNA reduced the migratory capacity of NCI-H290 MM cells (Mizuno et al., 2012). Likewise, deletion of *foxm1* from macrophages decreased their migration ability (Balli et al., 2012).

Overexpression of FOXM1 also augmented the invasiveness of androgen-independent LNCaP-AI prostate cancer cells (Li et al., 2011) and of $ARF^{-/-}$ HCC cells isolated from mice bearing DEN/PB-induced HCC (Park et al., 2011). Conversely, silencing of FOXM1 with siRNA/ shRNA diminished the invasive capacity of HF-U251 and U-87MG glioma cells (Dai et al., 2007), MHCC-97H HCC cells (Wu et al., 2010a), Ha-Ras-transformed NIH3T3 fibroblasts, and NIH3T3 fibroblasts stably expressing a MKK3 mutant (Behren et al., 2010).

In addition, stable overexpression of a ca N-terminally truncated FoxM1B mutant in NIH3T3 fibroblasts increased their invasion ability (Park et al., 2008a).

Furthermore, stable expression of shRNA against OGT reduced the invasiveness of PC3-ML prostate cancer cells, but overexpression of a ca N-terminally truncated FOXM1c mutant near-completely restored their invasive capacity (Lynch et al., 2012). Conversely, the invasion ability of unmanipulated PC3-ML cells was decreased by shRNA against FOXM1 (Lynch et al., 2012).

Moreover, overexpression of MnSOD (manganese superoxide dismutase) augmented the migration and invasion abilities of H358 and CL1-0 lung cancer cells, and shRNA-mediated knockdown of FOXM1 abolished this MnSOD-induced increase in their motiliy and invasiveness (Chen et al., 2013b).

The migration and invasion abilities of FaDU oral cancer cells were increased by FOXM1 overexpression, but decreased by FOXM1 knockdown with siRNA (Chen et al., 2009). Overexpression of FOXM1 also augmented the migratory and invasive capacities of human SMMC7721 HCC cells (Xia et al., 2012b) as well as the invasiveness of human SW1783 and Hs683 glioma cells (Dai et al., 2007). These positive effects of FOXM1 on migration and invasion depend on the direct FOXM1 target genes *MMP-2* (*matrix metalloproteinase-2*) (Chen et al., 2009; Dai et al., 2007), *RhoC* (*Ras homolog gene family, member C*), and *ROCK1* (*Rho-associated coiled-coil-containing protein kinase 1*) (Fig. 6.3; Table 6.3) (Xia et al., 2012b). In addition, the positive effect of FOXM1 on invasion depends on the direct FOXM1 target gene *MMP-7* (Fig. 6.3; Table 6.3) (Xia et al., 2012b).

First, siRNA against MMP-2 abolished the FOXM1-induced increase in the migration and invasion of FaDU cells (Chen et al., 2009). Accordingly, the FOXM1-induced increase in the invasion of SW1783 and Hs683 cells was abrogated by the MMP-2 inhibitor OA-Hy (Dai et al., 2007).

Table 6.3 FOXM1 target genes

| Gene | FOXM1 binding to promoter | | Regulation of endogenous gene expression by FOXM1 | | Regulation of promoter by FOXM1 | | References |
	Method[a]	Comment[b]	Expression[c]	Manipulation of FOXM1[d]	Manipulation of FOXM1-binding site[e]	Manipulation of FOXM1[d]	
Cyclin D1	C		T, P	↑ OE, P, siRNA, tg, KO	wt	↑ OE	Ye et al. (1999), Wang et al. (2001a,b, 2007a, 2010c), Liu et al. (2006, 2013), Tan et al. (2006), Petrovic et al. (2008), Chan et al. (2008), Ustiyan et al. (2009), Xia et al. (2009), Dai et al. (2010), Millour et al. (2010), Ren et al. (2010), Wu et al. (2010a,b), Xie et al. (2010), Bao et al. (2011), Raghavan et al. (2012), Zhang et al. (2011b), Xue et al. (2012), and Qu et al. (2013)
Cyclin D1			T	↓ siRNA, KO			Balli et al. (2011) and Wang et al. (2012)
Cyclin D2			T	↑ siRNA			Zhao et al. (2006) and Wang et al. (2010c)
Cyclin E			T, P	↑ OE, siRNA, tg			Wang et al. (2001a), Kalinichenko et al. (2003), Dai et al. (2010), and Davis et al. (2010)
Cyclin E1			T	↑ OE			Davis et al. (2010)
Cyclin E2			T	↑ OE, tg, KO	wt	↑ OE	Wang et al. (2008a), Davis et al. (2010), and Anders et al. (2011)
Cyclin A			P	↑ OE, ca, siRNA, KO			Laoukili et al. (2008), Davis et al. (2010), Xue et al. (2010), and Faust et al. (2012)
Cyclin A1			T	↑ OE			Davis et al. (2010)

Cyclin A2		T, P	↑	OE, ca, siRNA, tg, KO	Wang et al. (2001a,b, 2002a, 2008b, 2010c), Kalinichenko et al. (2003), Krupczak-Hollis et al. (2004), Kim et al. (2006a,b,c), Kalin et al. (2006), Zhao et al. (2006), Yoshida et al. (2007), Alvarez-Fernandez et al. (2010), Davis et al. (2010), and Huang and Zhao (2012)
Cyclin B1	C	T, P	↑	OE, dn, ca, wt, del, siRNA, tg, KO	Ye et al. (1999), Leung et al. (2001), Wang et al. (2001a,b, 2002a, 2005, 2007a, 2009, 2010c, 2012), Kalinichenko et al. (2003), Krupczak-Hollis et al. (2004), Kim et al. (2006a,b,c), Laoukili et al. (2005, 2008), Ma et al. (2005, 2010), Kalin et al. (2006), Zhao et al. (2006), Schüller et al. (2007), Yoshida et al. (2007), Chan et al. (2008), Fu et al. (2008), Li et al. (2008a), Park et al. (2008a), Ueno et al. (2008), Ustiyan et al. (2009), Xia et al. (2009), Gemenetzidis et al. (2009), Alvarez-Fernandez et al. (2010), Davis et al. (2010), Nakamura et al. (2010a), Wu et al. (2010a), Xue et al. (2010, 2012), Bolte et al. (2011), Hedge et al. (2011), Liu et al. (2011), Chen et al. (2011a, 2012a, 2013a), Bellelli et al. (2012), Down et al. (2012), Huang and Zhao (2012), Mencalha et al. (2012), Sadavisam et al. (2012), Xu et al. (2012a), and Qu et al. (2013)

Continued

Table 6.3 FOXM1 target genes—cont'd

Gene	FOXM1 binding to promoter Method[a]	Comment[b]	Regulation of endogenous gene expression by FOXM1 Expression[c]	Manipulation of FOXM1[d]	Regulation of promoter by FOXM1 Manipulation of FOXM1-binding site[e]	Manipulation of FOXM1[d]	References
Cyclin B1			P	↓ siRNA			Priller et al. (2011)
Cyclin B2	C		T, P	↑ OE, tg, KO			Wang et al. (2001b, 2002a), Laoukili et al. (2005), Davis et al. (2010), Lefebvre et al. (2010), and Chen et al. (2013a)
Cyclin B3			T	↑ OE, siRNA			Ueno et al. (2008)
Cyclin F			T	↑ OE, siRNA, tg			Wang et al. (2001a,b), Kalinichenko et al. (2003), Zhao et al. (2006), Chen et al. (2013a), and Huang and Zhao (2012)
Cdk1	C		T, P	↑ OE, siRNA, tg, KO			Ye et al. (1999), Wang et al. (2002a), Kalinichenko et al. (2003), Krupczak-Hollis et al. (2004), Zhao et al. (2006), Calvisi et al. (2009), Davis et al. (2010), Xie et al. (2010), Xue et al. (2010), Chen et al. (2013a), and Dai et al. (2013)
Cdk2			T, P	↑ OE, siRNA, KO			Wang et al. (2007a), Ahmad et al. (2010), Davis et al. (2010), and Xue et al. (2010, 2012)
Cdc25A[f]	C		T, P	↑ OE, siRNA, tg, KO	wt, del, P	↑ OE	Wang et al. (2001a, 2002b, 2005, 2007a), Davis et al. (2010), Sullivan et al. (2012), and Liu et al. (2013b)

Cdc25B	C	T, P	↑ OE, dn, siRNA, tg, KO, CI	wt	↑ OE, del	Wang et al. (2001a,b, 2002a,b, 2005), Kalinichenko et al. (2004), Krupczak-Hollis et al. (2004), Madureira et al. (2006), Radhakrishnan et al. (2006), Tan et al. (2006, 2007, 2010), Park et al. (2008a), Ramakrishna et al. (2007), Schüller et al. (2007), Chan et al. (2008), Kalin et al. (2008), Li et al. (2008a), Ueno et al. (2008), Ustiyan et al. (2009), Chen et al. (2010), Davis et al. (2010), Nakamura et al. (2010a), Xie et al. (2010), Zhou et al. (2010a,b), Bergamaschi et al. (2011), Bolte et al. (2011), Hedge et al. (2011), Balli et al. (2012), Dai et al. (2013), Mencalha et al. (2012), and Wang and Gartel (2012)
Cdc25C		T, P	↑ OE, siRNA, tg, KO			Madureira et al. (2006), Zhao et al. (2006), Wang et al. (2008a), Davis et al. (2010), Liu et al. (2011b), and Huang and Zhao (2012)
Cdc20	C	T, P	↑ OE, siRNA, tg			Wang et al. (2001b, 2002a, 2010c), Davis et al. (2010), Chen et al. (2013a), and Dai et al. (2013)
PLK1	C	T, P	↑ OE, dn, ca, P, siRNA, KO	wt	↑ OE, del, KO	Krupczak-Hollis et al. (2004), Kim et al. (2005), Laoukili et al. (2005), Wang et al. (2005, 2012), Madureira et al. (2006), Gusarova et al. (2007), Fu et al. (2008), Chen et al. (2009, 2012a, 2013a), Ustiyan et al. (2009), Alvarez-Fernandez et al. (2010), Davis et al. (2010), Pellegrino et al. (2010), Bolte et al. (2011, 2012), Anders et al. (2011), Bellelli et al. (2012), Dibb et al. (2012), Down et al. (2012), Faust et al. (2012), Monteiro et al. (2012), and Sadavisam et al. (2012)

Continued

Table 6.3 FOXM1 target genes—cont'd

Gene	FOXM1 binding to promoter		Regulation of endogenous gene expression by FOXM1		Regulation of promoter by FOXM1		References
	Method[a]	Comment[b]	Expression[c]	Manipulation of FOXM1[d]	Manipulation of FOXM1-binding site[e]	Manipulation of FOXM1[d]	
BUBR1	C		↑ T, P	siRNA	del	↑ siRNA	Lefebvre et al. (2010) and Wan et al. (2012)
Aurora A	C		↑ T	siRNA			Lefebvre et al. (2010), Raghavan et al. (2012), Down et al. (2012), Mencalha et al. (2012), and Sadavisam et al. (2012)
Aurora B	C		↑ T, P	OE, dn, ca, del, P, siRNA, KO	wt	↑ OE, P	Krupczak-Hollis et al. (2004), Kim et al. (2005), Wang et al. (2005), Gusarova et al. (2007), Park et al. (2008a), Fu et al. (2008), Chen et al. (2009), Davis et al. (2010), Nakamura et al. (2010a), Zhou et al. (2010a), Bergamaschi et al. (2011), Bellelli et al. (2012), Bonet et al. (2012), Down et al. (2012), Wang and Gartel (2012), and Xu et al. (2012a)
Survivin[g]	C		↑ T, P	OE, P, siRNA, KO, CI	wt, P	↑ siRNA	Wang et al. (2005, 2007a), Radhakrishnan et al. (2006), Gusarova et al. (2007), Yoshida et al. (2007), Chen et al. (2009, 2012a), Dai et al. (2010), Davis et al. (2010), Nakamura et al. (2010a), Bergamaschi et al. (2011), Down et al. (2012), and Xu et al. (2012a)
hBora			↑ T	siRNA			Alvarez-Fernandez et al. (2010)
CENP-A	C		↑ T, P	OE, P, siRNA, KO			Wang et al. (2005), Wonsey and Follettie (2005), Chen et al. (2009, 2013a), Davis et al. (2010), and Zhou et al. (2010a)

CENP-B	C	T	↑	OE, siRNA, CI	Wang et al. (2005), Radhakrishnan et al. (2006), Zhou et al. (2010a,b), and Chen et al. (2012a)
CENP-E		T	↑	OE	Davis et al. (2010)
CENP-F	C	T, P	↑	OE, siRNA, KO wt ↑ OE, KO	Laoukili et al. (2005, 2008), Davis et al. (2010), Anders et al. (2011), and Chen et al. (2013a)
NEK2		T	↑	OE, siRNA, KO	Laoukili et al. (2005), Wonsey and Follettie (2005), and Davis et al. (2010)
KIF20A		T	↑	OE, siRNA	Wonsey and Follettie (2005) and Davis et al. (2010)
Stathmin	C	T	↑	OE, tg	Carr et al. (2010) and Park et al. (2011)
KAP		T	↑	siRNA	Wonsey and Follettie (2005)
ARF		T	↓	OE, siRNA, KO	Wang et al. (2005) and Li et al. (2008a)
p21		T, P	↓	OE, siRNA, tg, KO, CI	Wang et al. (2001a, 2002b, 2005, 2007a), Kalinichenko et al. (2003), Kalin et al. (2006), Kim et al. (2006a,b,c), Radhakrishnan et al. (2006), Ramakrishna et al. (2007), Tan et al. (2007), Chan et al. (2008), Li et al. (2008a), Ahmad et al. (2010), Xia et al. (2009), Nakamura et al. (2010a), Bolte et al. (2011), Sengupta et al. (2013), Xue et al. (2012), and Qu et al. (2013)

Continued

Table 6.3 FOXM1 target genes—cont'd

Gene	FOXM1 binding to promoter		Regulation of endogenous gene expression by FOXM1		Regulation of promoter by FOXM1		References
	Method[a]	Comment[b]	Expression[c]	Manipulation of FOXM1[d]	Manipulation of FOXM1-binding site[e]	Manipulation of FOXM1[d]	
p21	C		T, P	↑ OE, tg	wt, del	↑ OE	Tan et al. (2010a) and Wang et al. (2010e)
p27			T, P	↓ OE, siRNA, KO	wt	↓ siRNA	Wang et al. (2002a, 2005, 2007a), Kalinichenko et al. (2004), Kalin et al. (2006), Kim et al. (2006a,b,c), Liu et al. (2006), Zhao et al. (2006), Chan et al. (2008), Petrovic et al. (2008), Ahmad et al. (2010), Zeng et al. (2009), Nakamura et al. (2010a), Wu et al. (2010a), Chen et al. (2011a), Wang and Gartel (2012), Sengupta et al. (2013), Xue et al. (2012), Zhang et al. (2012d), and Qu et al. (2013)
p53			P	↓ OE, siRNA			Li et al. (2008a)
p53			P	↑ siRNA			Chetty et al. (2009)
Bmi-1[h]			T, P	↑ OE, siRNA, KO	wt, del, P	↑ OE	Li et al. (2008a) and Wang et al. (2011b)
Skp2	C		T, P	↑ OE, dn, del, siRNA, KO	wt	↑ OE	Wang et al. (2005), Liu et al. (2006), Park et al. (2008a, 2009), Calvisi et al. (2009), Zeng et al. (2009), Davis et al. (2010), Nakamura et al. (2010a), Xie et al. (2010), Zhou et al. (2010a,b), Anders et al. (2011), Bellelli et al. (2012), Chen et al. (2012a), Mencalha et al. (2012), Xu et al. (2012a), and Zhang et al. (2012d)

Cks1	C		T, P	↑ OE, siRNA, KO		Wang et al. (2005), Calvisi et al. (2009), and Davis et al. (2010)
Cks2			T	↑ OE		Davis et al. (2010)
KIS[i,j]	C		T, P	↑ OE, siRNA, KO	P, art	Petrovic et al. (2008) and Nakamura et al. (2010a)
hTERT			T	↑ siRNA		Zeng et al. (2009) and Zhang et al. (2012d)
MCM3	C		T	↑ siRNA		Lefebvre et al. (2010)
Gas1			T, P	↑ OE, KO		Laoukili et al. (2005)
BRCA2	C		T, P	↑ OE, ca, siRNA, KO	wt	Tan et al. (2007, 2010a), Kwok et al. (2010), and Zhang et al. (2012e)
XRCC1	C		T, P	↑ OE, ca, siRNA, KO	wt	Tan et al. (2007, 2010a), Chetty et al. (2009), and Kwok et al. (2010)
BTG2			P	↓ OE, siRNA		Park et al. (2009)
Chk1	E, C	S, C	T, P	↑ siRNA, tg	wt, del	Chetty et al. (2009) and Tan et al. (2010a)
Chk2			P	↑ siRNA		Chetty et al. (2009) and Zhang et al. (2012e)
Histone H2B/a				↑ OE, del	wt, H	Wierstra and Alves (2006d)
Hsp70				↑ OE, del	wt, del, H	Wierstra and Alves (2006d)

Continued

Table 6.3 FOXM1 target genes—cont'd

Gene	FOXM1 binding to promoter		Regulation of endogenous gene expression by FOXM1		Regulation of promoter by FOXM1		References
	Method[a]	Comment[b]	Expression[c]	Manipulation of FOXM1[d]	Manipulation of FOXM1-binding site[e]	Manipulation of FOXM1[d]	
c-Myc[k]	E, C	IV, S, C	T, P	↑ OE, P, siRNA, tg, KO	wt, del, P, H	↑ OE, dn, del	Wierstra and Alves (2006d, 2007a,b, 2008), Li et al. (2008a), Ustiyan et al. (2009), Zeng et al. (2009), Ren et al. (2010), Wang et al. (2010e), Green et al. (2011), Hedge et al. (2011), and Zhang et al. (2011b, 2012d)
c-Myc			T	↓ siRNA, KO			Balli et al. (2011) and Wang et al. (2012)
c-Myc[l]	C	P					Zhang et al. (2011b)
N-Myc			T	↑ KO			Bolte et al. (2011)
c-Fos					wt, H	↑ OE, del	Wierstra and Alves (2006d)
ERα	C, D, F	IV, C, P	T, P	↑ OE, siRNA	wt, del, P	↑ OE	Madureira et al. (2006), Millour et al. (2010), Littler et al. (2010), and Hedge et al. (2011)
ERα			T	↓ OE, KO			Carr et al. (2012)
E2F-1	C		T	↑ siRNA			Wang et al. (2007a), Ahmad et al. (2010), and Lefebvre et al. (2010)
E2F-5	C		T	↑ siRNA			Lefebvre et al. (2010)
NF-YB	C		T	↑ siRNA			Lefebvre et al. (2010)
TCF-4			T, P	↑ OE, tg, KO			Yoshida et al. (2007) and Zhang et al. (2011b)

TCF-4			T	↓ siRNA, KO			Wang et al. (2012)
ERM			T	↑ OE, KO			Laoukili et al. (2005)
NFATc3	C		T	↑ siRNA, KO			Ramakrishna et al. (2007) and Bolte et al. (2011)
C/EBPβ			T	↑ tg			Ye et al. (1999)
HNF-6	C				wt, del	↑ OE	Tan et al. (2006)
FOXF1	C		T	↑ ca, siRNA, KO	wt	↑ OE	Kim et al. (2005) and Balli et al. (2011, 2013)
FOXM1	C		T	↑ OE	wt	↑ OE	Halasi and Gartel (2008), Wang et al. (2010c), Xie et al. (2010), and Down et al. (2012)
PPARα			T	↓ siRNA, KO			Wang et al. (2009a,b)
ATF-2	C		T, P	↑ siRNA			Wang et al. (2008b)
JNK1	C		T, P	↑ siRNA, KO	wt	↑ OE	Wang et al. (2008b, 2012), Ustiyan et al. (2009), and Balli et al. (2013)
TOP2A	C		T, P	↑ siRNA, KO	wt	↑ OE	Wang et al. (2009a,b)
MMP-2	E, C	S, C	T, P	↑ OE, siRNA	wt, P	↑ OE, siRNA	Dai et al. (2007), Wang et al. (2007a, 2008b), Ahmad et al. (2010), Chen et al. (2009, 2011a, 2013b), Wu et al. (2010a), Ahmad et al. (2011), Uddin et al. (2011, 2012), He et al. (2012), Lynch et al. (2012), and Xue et al. (2012)

Continued

Table 6.3 FOXM1 target genes—cont'd

Gene	FOXM1 binding to promoter		Regulation of endogenous gene expression by FOXM1		Regulation of promoter by FOXM1		References
	Method[a]	Comment[b]	Expression[c]	Manipulation of FOXM1[d]	Manipulation of FOXM1-binding site[e]	Manipulation of FOXM1[d]	
MMP-9			T, P	↑ OE, siRNA			Wang et al. (2007a, 2008b), Ahmad et al. (2010), Ahmad et al. (2011), Lok et al. (2011), Uddin et al. (2011, 2012), He et al. (2012), and Xue et al. (2012)
MMP-9			T	↓ KO			Bolte et al. (2012)
MMP-12			T	↑ tg, KO			Wang et al. (2008a) and Balli et al. (2012)
uPA			T, P	↑ siRNA			Ahmad et al. (2010), Wu et al. (2010a), and Bellelli et al. (2012)
uPAR	C		T, P	↑ OE, siRNA	wt	↑ OE, siRNA	Wang et al. (2007a), Ahmad et al. (2010), Lok et al. (2011), and Li et al. (2013)
TGF-α	C				wt	↑ OE	Tan et al. (2006)
DUSP1			P	↓ OE, siRNA			Calvisi et al. (2009)
PTEN			P	↓ OE, siRNA			Dai et al. (2010)
NEDD4-1	C		T, P	↑ OE, siRNA			Dai et al. (2010) and Kwak et al. (2012)
VEGFR2	C		T	↑ siRNA, KO	wt	↑ OE	Kim et al. (2005) and Balli et al. (2011)
VEGF	E, C, D	S, C, P	T, P	↑ OE, siRNA	wt, del, P	↑ OE, del, P, siRNA	Zhang et al. (2008), Calvisi et al. (2009), Li et al. (2009), Gemenetzidis et al. (2009), Bao et al. (2011), Karadedou et al. (2011), Chen et al. (2011a), Lynch et al. (2012), and Xue et al. (2012)

Gene			Type	Expression			Reference
EPO			P	↑ OE, siRNA			Calvisi et al. (2009)
TM4SF1			T	↑ OE, KO			Laoukili et al. (2005)
Pecam-1			T	↑ KO			Kim et al. (2005)
Lama2			T	↑ KO			Kim et al. (2005)
Lama4	E	S, C	T	↑ KO	wt	↑ OE	Kim et al. (2005)
Procollagen type XII α1			T	↑ KO			Kim et al. (2005)
Integrin β1			T	↑ KO			Kim et al. (2005)
E-cadherin[m]	E, C	IV, C	P	↑ OE	wt	↑ OE, del	Ye et al. (1997), Zhou et al. (2010a,b), Bao et al. (2011), and Wierstra (2011a)
E-cadherin			T, P	↓ OE, ca, siRNA, tg			Park et al. (2011), Li et al. (2012c), and Balli et al. (2013)
N-CAM			T	↑ siRNA			Ueno et al. (2008)
N-tubulin			T	↑ siRNA			Ueno et al. (2008)
Tubulin β III			P	↓ siRNA			Wang et al. (2011b)
Catalase[n]	C		T, P	↑ OE, siRNA			Park et al. (2009)
MnSOD	C		T, P	↑ OE, siRNA			Park et al. (2009)
PRDX3			T, P	↑ siRNA			Park et al. (2009)

Continued

Table 6.3 FOXM1 target genes—cont'd

Gene	FOXM1 binding to promoter		Regulation of endogenous gene expression by FOXM1		Regulation of promoter by FOXM1			References
	Method[a]	Comment[b]	Expression[c]	Manipulation of FOXM1[d]	Manipulation of FOXM1-binding site[e]	Manipulation of FOXM1[d]		
Cox-2	C		↑	siRNA, tg, KO	wt	↑ OE		Wang et al. (2008a), Balli et al. (2012), and Ren et al. (2013)
Lipocalin-2			↑	tg				Wang et al. (2008a)
MSR-1			↑	tg				Wang et al. (2008a)
Cathepsin D			↑	tg				Wang et al. (2008a)
Cathepsin K			↑	tg				Wang et al. (2008a)
CCL3			↑	tg, KO				Wang et al. (2008a) and Balli et al. (2012)
CXCL1			↑	tg				Wang et al. (2008a)
CXCL5°			↑	ca, siRNA, tg, KO	wt	↑ OE		Wang et al. (2008a, 2010c) and Balli et al. (2013)
CXCL8			↑	siRNA				Wang et al. (2010c)
SP-A	C		↑	KO	wt	↑ OE		Kalin et al. (2008)
SP-B	C		↑	KO	wt	↑ OE		Kalin et al. (2008)
SP-B			↓	KO				Ustiyan et al. (2012)
SP-C	C		↑	KO				Kalin et al. (2008)

SP-C		T			↓ KO	Ustiyan et al. (2012)
SP-D	C	T			↑ KO	Kalin et al. (2008)
Aqp5		T			↑ KO	Kalin et al. (2008) and Liu et al. (2011)
T1-α		T			↑ KO	Kalin et al. (2008)
CEP55	C	T			↑ OE	Gemenetzidis et al. (2009)
HELLS[p]		T			↑ OE	Gemenetzidis et al. (2009)
TACE		T			↑ KO	Kim et al. (2005)
Mim-1				wt	↑ OE, del	Wierstra and Alves (2007b)
Transferrin				wt	↑ dn, as	Chaudhary et al. (2000)
Oct-4	C	T, P	OE, siRNA	wt, del, P	↑ OE, siRNA	Xie et al. (2010) and Wang et al. (2011b)
Sox2[q]	C	T, P	OE, P, siRNA, KO	wt, P, H	↑ OE	Xie et al. (2010), Wang et al. (2011b), Zhang et al. (2011b), and Ustiyan et al. (2012)
Sox2		T	siRNA, tg		↓	Wang et al. (2010e, 2012)
Nanog		T, P	OE, siRNA		↑	Xie et al. (2010) and Wang et al. (2011b)
GATA4		T	siRNA		↓	Xie et al. (2010)
SSEA-1		P	OE, siRNA		↑	Xie et al. (2010) and Zhang et al. (2011b)
β-Catenin	C	T, P	siRNA, KO		↑	Mirza et al. (2010) and Liu et al. (2011)
PCNA		T, P	OE, siRNA		↑	Davis et al. (2010) and Raghavan et al. (2012)

Continued

Table 6.3 FOXM1 target genes—cont'd

Gene	FOXM1 binding to promoter Method[a]	Comment[b]	Regulation of endogenous gene expression by FOXM1 Expression[c]	Manipulation of FOXM1[d]	Regulation of promoter by FOXM1 Manipulation of FOXM1-binding site[e]	Manipulation of FOXM1[d]	References
Borealin	T		↑	OE, siRNA			Davis et al. (2010) and Bergamaschi et al. (2011)
Ki-67	T		↑	OE			Davis et al. (2010)
Repo-Man	T		↑	OE			Davis et al. (2010)
Sox9	T		↑	tg			Wang et al. (2010e)
Sox4	T		↓	tg, KO			Wang et al. (2010e, 2012)
Sox17	T		↓	siRNA, tg			Wang et al. (2010e, 2012)
FOXJ1	T		↓	tg			Wang et al. (2010e)
CCR2	T		↑	KO	wt, del	↑ OE	Ren et al. (2010, 2013)
Vimentin	T, P		↑	OE, ca, siRNA, tg			Gemenetzidis et al. (2010), Park et al. (2011), Bao et al. (2011), Li et al (2012c), and Balli et al. (2013)
Vimentin	T		↓	KO			Bolte et al. (2012)
KRT7	T		↑	OE			Gemenetzidis et al. (2010)
KRT8	T		↑	OE			Gemenetzidis et al. (2010)
KRT12	T		↑	OE			Gemenetzidis et al. (2010)

KRT15	C	T	↑	OE	Gemenetzidis et al. (2010) and Bose et al. (2012)
KRT18		T	↑	OE	Gemenetzidis et al. (2010)
KRT18		T	↓	OE, KO	Carr et al. (2012)
KRT19		T	↑	OE	Gemenetzidis et al. (2010)
KRT23		T	↑	OE	Gemenetzidis et al. (2010)
α-SMA		T, P	↑	OE, ca, siRNA, tg, KO	Park et al. (2011) and Balli et al. (2013)
α-SMA		T	↓	KO	Bolte et al. (2012)
Snail	C	T, P	↑	OE, ca, siRNA, tg, KO wt, P ↑ OE	Park et al. (2011), He et al. (2012), and Balli et al. (2013)
Slug		T, P	↑	OE, del	Bao et al. (2011) and Balli et al. (2013)
LOX	C	T	↑	OE, tg	Park et al. (2011)
LOXL2	C	T	↑	OE, tg	Park et al. (2011)
sFRP1	C	T	↑	KO	Balli et al. (2011)
RelA (p65)		P	↑	OE	Bao et al. (2011)
Hes-1		P	↑	OE	Bao et al. (2011)
ZEB-1		T, P	↑	OE, ca, siRNA, KO	Bao et al. (2011) and Balli et al. (2013)

Continued

Table 6.3 FOXM1 target genes—cont'd

Gene	FOXM1 binding to promoter		Regulation of endogenous gene expression by FOXM1		Regulation of promoter by FOXM1		References
	Method[a]	Comment[b]	Expression[c]	Manipulation of FOXM1[d]	Manipulation of FOXM1-binding site[e]	Manipulation of FOXM1[d]	
ZEB-2	T, P			↑ OE, ca, siRNA, KO			Bao et al. (2011) and Balli et al. (2013)
CD44	P			↑ OE			Bao et al. (2011)
EpCAM	P			↑ OE			Bao et al. (2011)
let-7a	T			↓ OE			Bao et al. (2011)
let-7b	T			↓ OE			Bao et al. (2011)
let-7c	T			↓ OE			Bao et al. (2011)
miR-200b	T			↓ OE			Bao et al. (2011)
miR-200c	T			↓ OE			Bao et al. (2011)
Ezh2	T			↑ siRNA			Wang et al. (2011b)
Suz12	T			↑ siRNA			Wang et al. (2011b)
miR-135a	T	C		↑ OE, siRNA			Liu et al. (2011)
NF-M	P			↓ siRNA			Wang et al. (2011b)
Nestin	P			↑ OE, P, siRNA, KO			Wang et al. (2011b) and Zhang et al. (2011b)
CaMKIIδ	T			↑ KO			Bolte et al. (2011)

Hey2		T	↑ KO			Bolte et al. (2011, 2012)
Myocardin		T	↑ KO			Bolte et al. (2011)
IL-1β		T	↓ KO			Bolte et al. (2011)
IL-1β		T	↑ ca, KO			Balli et al. (2013)
Axin2	C	T, P	↑ OE, P, siRNA, KO	wt	↑ OE	Zhang et al. (2011b) and Wang et al. (2012)
LEF-1[r]	C	P	↑ OE, P, siRNA, KO	wt, P	↑ OE	Zhang et al. (2011b)
CD133		P	↑ OE, P, siRNA			Zhang et al. (2011b)
Musashi-1		P	↑ OE, P, siRNA			Zhang et al. (2011b)
GFAP		P	↑ OE, P, siRNA			Zhang et al. (2011b)
DTL		T	↑ KO	wt	↑ OE	Anders et al. (2011)
MSH6		T	↑ KO	wt	↑ OE	Anders et al. (2011)
LRR1		T	↑ KO	wt	↑ OE	Anders et al. (2011)
CDCA5				wt	↑ OE	Anders et al. (2011)
XRCC2		T	↑ KO	wt	↑ OE	Anders et al. (2011)
IGFBP1				wt	↑ OE	Anders et al. (2011)
Cyclin G2	C			wt	↑ OE	Anders et al. (2011) and Chen et al. (2013a)
TSC22D1				wt	↑ OE	Anders et al. (2011)

Continued

Table 6.3 FOXM1 target genes—cont'd

	FOXM1 binding to promoter		Regulation of endogenous gene expression by FOXM1		Regulation of promoter by FOXM1		
Gene	Method[a]	Comment[b]	Expression[c]	Manipulation of FOXM1[d]	Manipulation of FOXM1-binding site[e]	Manipulation of FOXM1[d]	References
CDT1			T	↑ KO			Anders et al. (2011)
SFRS4			T	↑ KO			Anders et al. (2011)
MCM2			T	↑ KO			Anders et al. (2011)
MCM10			T	↑ KO			Anders et al. (2011)
c-Myb			T	↑ KO			Anders et al. (2011)
CXCL2			T	↑ OE, KO			Balli et al. (2012) and Huang and Zhao (2012)
iNOS			T	↑ KO			Balli et al. (2012)
IL-6			T	↑ OE, KO			Balli et al. (2012) and Huang and Zhao (2012)
IL6Rβ			T	↑ KO			Balli et al. (2012)
CCL11		C	T, P	↑ KO			Balli et al. (2012) and Ren et al. (2013)
CX3CL1			T	↑ KO			Balli et al. (2012) and Ren et al. (2013)
CX3CR1			T	↑ KO			Balli et al. (2012) and Ren et al. (2013)
CSF-1R			T	↑ KO			Balli et al. (2012)
CXCR4			T	↑ KO			Balli et al. (2012)

Cav-1	C	T, P	↑ OE, siRNA	wt	↑ OE, siRNA	Huang et al. (2012)
RACGAP1	C					Sadavisam et al. (2012)
B-Myb		T, P	↓ siRNA			Sadavisam et al. (2012)
BCL-2		T, P	↑ siRNA			Halasi and Gartel (2012)
p16^s		T, P	↓ OE			Teh et al. (2012)
POLE2		T	↑ siRNA			Park et al. (2012)
CDC45L		T	↑ siRNA			Park et al. (2012)
RFC4	C	T	↑ siRNA	wt	↑ OE, siRNA	Park et al. (2012)
PLK4	C	T	↑ siRNA	wt	↑ OE, siRNA	Park et al. (2012)
EXO1	C	T	↑ siRNA	wt	↑ OE, siRNA	Park et al. (2012)
MMP-7	C	T, P	↑ OE, siRNA	wt, P	↑ OE	Xia et al. (2012b)
RhoC	C	T, P	↑ OE, siRNA	wt, P	↑ OE	Xia et al. (2012b)
ROCK1	C	T, P	↑ OE, siRNA	wt, P	↑ OE	Xia et al. (2012b)
p107		P	↑ KO			Xue et al. (2010)
Menin		P	↓ KO			Zhang et al. (2010a)
Wnt5a		T	↑ KO			Wang et al. (2012)
FGF9		T	↑ KO			Wang et al. (2012)
GATA6		T	↓ KO			Wang et al. (2012)

Continued

Table 6.3 FOXM1 target genes—cont'd

Gene	FOXM1 binding to promoter		Regulation of endogenous gene expression by FOXM1		Regulation of promoter by FOXM1		References
	Method[a]	Comment[b]	Expression[c]	Manipulation of FOXM1[d]	Manipulation of FOXM1-binding site[e]	Manipulation of FOXM1[d]	
Wnt7b			T	↓ KO			Wang et al. (2012)
TCF-3			T	↑ KO			Wang et al. (2012)
AREG			T	↓ OE, KO			Carr et al. (2012)
Cadherin 11			T	↓ OE, KO			Carr et al. (2012)
GATA3	C		T, P	↓ OE, KO			Carr et al. (2012)
GATA3[f]			T	↑ OE			Carr et al. (2012)
α–Casein			P	↑ KO			Carr et al. (2012)
β–Casein			P	↑ KO			Carr et al. (2012)
SCGB1A1	C		T	↑ KO	wt	↑ OE	Ustiyan et al. (2012)
ABCA3			T	↓ KO			Ustiyan et al. (2012)
IL-13			T	↓ KO			Ustiyan et al. (2012)
IL-13			T, P	↑ KO			Ren et al. (2013)
FOXA3			T	↓ KO			Ustiyan et al. (2012)
FOXA3			T	↑ KO			Ren et al. (2013)
FOXA2			T	↓ KO			Ren et al. (2013)

ELF-3		T	↓ KO			Ustiyan et al. (2012)
RAD51	C	T, P	↑ siRNA	wt, del	↑ siRNA	Zhang et al. (2012e)
Cdh1	C	T	↑ siRNA			Chen et al. (2013a)
KPNA2	C	T	↑ siRNA			Chen et al. (2013a)
BUB3	C					Chen et al. (2013a)
ETV4	C					Chen et al. (2013a)
PRC1	C					Chen et al. (2013a)
PTMS	C					Chen et al. (2013a)
UBE2C	C					Chen et al. (2013a)
UBE2S	C					Chen et al. (2013a)
TACO1	C					Chen et al. (2013a)
BRIP1	C	T, P	↑ OE, del, siRNA, KO	wt, del, P	↑ del	Monteiro et al. (2012)
IGF-1	C	T	↑ KO	wt, P	↑ OE	Sengupta et al. (2013)
FOXO1	C	T	↓ KO			Sengupta et al. (2013)
FOXO3	C	T	↓ KO			Sengupta et al. (2013)
Fibronectin		T	↓ KO			Bolte et al. (2012)

Continued

Table 6.3 FOXM1 target genes—cont'd

Gene	FOXM1 binding to promoter		Regulation of endogenous gene expression by FOXM1		Regulation of promoter by FOXM1		References
	Method[a]	Comment[b]	Expression[c]	Manipulation of FOXM1[d]	Manipulation of FOXM1-binding site[e]	Manipulation of FOXM1[d]	
Fibronectin			T, P	↑ ca, siRNA			Balli et al. (2013)
IL-12 p35			T	↑ KO			Ren et al. (2013)
IL-12 p40			P	↑ KO			Ren et al. (2013)
IL-4			T, P	↑ KO			Ren et al. (2013)
IL-5			T, P	↑ KO			Ren et al. (2013)
IL-1α			T	↑ KO			Ren et al. (2013)
IL-33			T	↑ KO			Ren et al. (2013)
LTC4s			T	↑ KO			Ren et al. (2013)
MUC5AC			T	↑ KO			Ren et al. (2013)
CCL2			T, P	↑ ca, KO	wt	↑ OE	Ren et al. (2013) and Balli et al. (2013)
CCL24		C	T	↑ KO			Ren et al. (2013)
CCR3			T	↑ KO			Ren et al. (2013)
CCL17			T	↑ KO			Ren et al. (2013)

Target						Reference
GM-CSFRα		T	↑ KO			Ren et al. (2013)
SPDEF	C	T	↑ KO	wt	↑ OE	Ren et al. (2013)
ICAM-1		T	↑ OE			Huang and Zhao (2012)
TNF-α		T	↑ OE			Huang and Zhao (2012)
Integrin β3	C					Malin et al. (2007)
CDC6	P		↑ siRNA			Liu et al. (2013b)
COL1A1		T	↑ ca, KO			Balli et al. (2013)
COL3A1		T	↑ ca, KO			Balli et al. (2013)
TWIST1		T	↑ KO			Balli et al. (2013)
TWIST2		T	↑ ca, siRNA, KO			Balli et al. (2013)
TGF-β		T	↑ KO			Balli et al. (2013)
PTTG1	C					Lefebvre et al. (2010)
FANCI	C					Lefebvre et al. (2010)
XBP-1	C					Hedge et al. (2011)
GREB1	C					Hedge et al. (2011)

Continued

Table 6.3 FOXM1 target genes—cont'd

Gene	FOXM1 binding to promoter		Regulation of endogenous gene expression by FOXM1		Regulation of promoter by FOXM1		References
	Method[a]	Comment[b]	Expression[c]	Manipulation of FOXM1[d]	Manipulation of FOXM1-binding site[e]	Manipulation of FOXM1[d]	
Cdx-2	E	IV, C					Ye et al. (1997)
Fabpi	E	IV, C					Ye et al. (1997)
MHC II			P	↑ KO			Ren et al. (2013)
CD86			P	↑ KO			Ren et al. (2013)

ABCA3, ATP-binding cassette, subfamily A, member 3; ARF, alternative reading frame, CDKN2A, Cdk (cyclin-dependent kinase) inhibitor 2A, p14, p19; ATF-2, activating transcription factor-2; Aqp5, aquaporin 5; AREG, amphiregulin; Aurora B, Aurora B kinase; Axin2, axis inhibitor 2, conductin; Bcl-2, B-cell lymphoma-2; Bmi-1, B lymphoma Mo-MLV (Moloney-murine leukemia virus) insertion region-1, PCGF4, Polycomb group RING (really interesting new gene) finger protein 4; hBora, human Bora; Borealin; Dasra B, CDCA8, cell division cycle-associated protein 8; BRCA2, breast cancer-associated gene 2, FANCD1, Fanconi anemia complementation group D1; BRIP1, BRCA1-interacting protein 1, Bach1, BRCA1-associated C-terminal helicase 1; BTG2, B-cell translocation gene 2, TIS21, TPA (12-O-tetradecanoylphorbol-13-acetate)-inducible sequence 21, APRO1, antiproliferative 1; BUB3, budding uninhibited by benzimidazoles 3; BubR1, Bub1-related kinase; CaMKIIδ, calcium/calmodulin-dependent protein kinase IIδ; Cav-1, caveolin-1; CCL, chemokine, CC motif, ligand; CCR, chemokine, CC motif receptor; CD133, CD133 antigen, PROM1, prominin 1; Cdc, cell division cycle; Cdc20, p55CDC; FZY, Fizzy; CDCA5, cell division cycle-associated protein 5, sororin; CDC45L, CDC45-like, cell division cycle 45; Cdh1, Cdc20 homolog 1, FZR1, Fizzy-related protein 1; Cdk, cyclin-dependent kinase; CDT1, chromatin licensing and DNA replication factor 1; Cdx-2, caudal type homeobox-2; C/EBPβ, CCAAT/enhancer-binding protein β; CENP, centromere protein; CEP55, centrosomal protein 55-kDa; Chk, checkpoint kinase; Cks, CDC2-associated protein, Cdk subunit; COL1A1, collagen, type I, α-1; COL3A1, collagen, type III, α-1; COX-2, cyclooxygenase-2; CSF-1R, colony-stimulating factor-1 receptor; CXCL, chemokine, CXC motif, ligand; CXCR, chemokine, CXC motif receptor; CX3CL, chemokine, CX3C motif, ligand; CX3CR, chemokine, CX3C motif, receptor; cyclin F, CCNF, FBXO1, FBX1, F-box only protein 1; DTL, denticleless, CDT2, DCAF2, Ddb1 (damage-specific DNA-binding protein 1)- and Cul4 (Cullin 4)-associated factor 2; DUSP1, dual-specificity phosphatase 1; MKP-1, MAPK (mitogen-activated protein kinase) phosphatase-1; E-cadherin, epithelial cadherin, CDH1, cadherin 1, type 1; ELF-3, E74-like factor-3; EpCAM, epithelial cell adhesion molecule; EPO, erythropoietin; ERα, estrogen receptor α; ETV5, Ets variant gene 5, Erm, Ets-related molecule; EXO1, exonuclease 1; Ezh2, enhancer of Zeste homolog 2, KMT6, K-methyltransferase 6; Fabpi, fatty acid-binding protein intestinal; FANCI, Fanconi anemia complementation group I; FGF9, fibroblast growth factor 9; c-Fos, FOS, Forkhead box; Gas1, growth arrest-specific 1; GATA, GATA-binding protein; GFAP, glial fibrillary acidic protein; GM-CSFRα, granulocyte-macrophage colony-stimulating factor receptor α, CSF2RA, colony-stimulating factor 2 receptor α; GREB1, growth regulation by estrogen in breast cancer 1; HELLS, helicase, lymphoid-specific, LSH, lymphoid-specific helicase, SMARCA6, SWI/SNF-related, matrix-associated, actin-dependent regulator of chromatin, subfamily A, member 6; Hes-1, Hairy/enhancer of split-1; Hey2, Hairy/enhancer of split-related with YRPW motif 2; HNF-6, hepatocyte nuclear factor-6; HSP70, heat shock protein 70; ICAM-1, intercellular cell adhesion molecule-1; IGF-1, insulin-like growth factor-1; IGFBP1, IGF-binding protein 1; IL, interleukin; IL6Rβ, IL-6 receptor β; IL-12 p35, IL-12, 35-kDa subunit, IL12A, interleukin 12A; IL-12 p40, IL-12, 40-kDa subunit, IL12B; iNOS, inducible nitric oxide synthase, NOS2A, nitric oxide synthase, NOS2; JNK1, c-Jun (JUN) N-terminal kinase 1; KAP, Cdk-associated protein phosphatase, CDKN3, Cdk inhibitor 3; Ki-67, Mki67, proliferation-related Ki-67 antigen; KIF20A, kinesin family member 20A; KIS, kinase interacting stathmin; KPNA2, karyopherin α-2; KRT, cytokeratin; Lama, laminin α; LEF-1, lymphoid enhancer factor-1; LOX, lysyl oxidase; LOXL2, LOX-like 2; LRR1, leucine-rich repeat protein 1, PPIL5, peptidyl-prolyl isomerase-like 5; LTC4s, leukotriene synthase; MCM, minichromosome maintenance; Menin, MEN1, multiple endocrine neoplasia 1; MHC II, major histocompatibility complex class II; mim–1, myb-induced myeloid protein–1; MMP, matrix metalloproteinase; MnSOD, manganese superoxide dismutase; MSH6, mutS homolog 6; MSR–1, macrophage scavenger receptor–1; MUC5AC, Mucin 5, subtypes A and C, tracheobronchial; B-Myb, MYBL2, MYB-like 2; c-Myb, MYB; c-Myc, MYC; N-Myc, MYCN, Myc oncogene neuroblastoma-derived; N-CAM, neural cell adhesion molecule; NEDD4-1, neural precursor cell-expressed, developmentally down-regulated 4-1; NEK2, NIMA (never in mitosis A)-related kinase 2; NFATc3, nuclear factor of activated T-cell c3; NF-M, neurofilament protein, medium polypeptide; NF-YB, nuclear factor-Y B; Oct-4, octamer-binding transcription factor 4, POU5F1, POU domain, class 5, transcription factor 1; p16, CIDKN2A, Cdk inhibitor 2A, INK4A, inhibitor of Cdk4 A; p21, CDKN1A, Cdk inhibitor 1A, CIP1, Cdk-interacting protein 1, WAF1, wild-type p53-activated fragment 1; p27, CDKN1B, Cdk inhibitor 1B, KIP1, kinase inhibitor protein 1; p53, TP53, tumor protein p53; p107, RBL1, retinoblastoma-like 1; PCNA, proliferating cell nuclear antigen; Pecam–1, platelet endothelial cell adhesion molecule 1; PLK, Polo-like kinase; POLE2, DNA polymerase ε 2; PPARα, peroxisome proliferator-activated receptor α; PRC1, protein regulating cytokinesis 1; PRIDX3, peroxiredoxin 3; PTEN, phosphatase and tensin homolog deleted on chromosome ten; PTMS, parathymosin; PTTG1, pituitary tumor-transforming gene 1, securin; RACGAP1, Rac GTPase-activating protein 1; RelA, RELA, p65; Repo-Man, recruits PP1 (protein phosphatase 1) onto mitotic chromatin at anaphase protein, CDCA2, cell division cycle-associated protein 2; RFC4, replication factor C, subunit 4; RhoC, Ras homolog gene family, member C; ROCK1, Rho-associated coiled-coil-containing protein kinase 1; SCGB1a1, secretoglobin, family 1A, member 1, CCSP, Clara cell secretoy protein, CC10, Clara cell-specific 10-KD protein; sFRP1, secreted Frizzled-related protein 1; SFRS4, splicing factor, arginine/serine-rich 4, SRp75, splicing factor, arginine/serine-rich 75-KD; Skp2, S-phase kinase-associated protein 2; Slug, SNAI2; Snail, SNAI1; Snail1; α-SMA, α-smooth muscle actin; Sox, SRY (sex-determing region Y)-box, SRY-related HMG (high mobility group)-box; SP, surfactant protein; SPDEF, SAM-pointed domain-containing ETS transcription factor, PDEF, prostate-derived ETS transcription factor; SSEA–1, stage-specific embryonic antigen–1; survivin, BIRC5, baculoviral IAP repeat-containing 5; Suz12, suppressor of Zeste 12; T1-α, PDPN, podoplanin; TACE, TNF-α converting enzyme, ADAM-17, a disintegrin and metalloprotease domain-17; TACO1, translational activator of mitochondrially encoded cytochrome c oxidase subunit 1; TCF, T-cell factor; hTERT, human telomerase reverse transcriptase; TGF, transforming growth factor; TM4SF1, transmembrane 4 superfamily member 1, TAL6, TAAL6, tumor-associated antigen L6; TNF-α, tumor necrosis factor-α; TOP2A, TOPO-2α, DNA topoisomerase 2α; TSC22D1, TSC22 (TGF-β-stimulated clone 22) domain family, member 1, TGF-β-stimulated clone 22 domain 1; TWIST2, Dermo1, Dermis-expressed protein 1; UBE2C, ubiquitin-conjugating enzyme E2C; uPA, urokinase-type plasminogen activator; uPAR, uPA receptor; VEGF, vascular endothelial growth factor; VEGFR2, VEGF receptor type II; WNT, wingless type; XBP-1, X box-binding protein-1; XRCC, X-ray cross-completing group; ZEB-1, zinc finger E-box-binding homeobox-1; δEF1, δ-crystallin enhancer-binding factor 1; ZEB-2, zinc finger E-box-binding homeobox-2, SIP1, Smad-interacting protein 1.

[a]**Method**. E = EMSA = electrophoretic mobility shift assay; C = ChIP = chromatin immunoprecipitation assay; D = DNAP = DNA precipitation assay = DAPA = DNA affinity pull-down assay; F = fluorescence anisotropy assay.

[b]**Comment**. IV = *in vitro* = purified FOXM1; S = supershift experiments; C = competition experiments; P = FOXM1-binding site was point-mutated.

[c]**Expression**: T = transcript = mRNA = endogenous mRNA level affected; P = protein = endogenous protein level affected.

[d]**Manipulation of FOXM1**: OE = overexpression of wild-type FOXM1; ca = constitutively active form of FOXM1; del = analyzed with deletion mutant of FOXM1; dn = dominant-negative form of FOXM1; siRNA = siRNA-mediated or shRNA-mediated knockdown of FOXM1; as = addition of antisense oligonucleotide to FOXM1; tg = *foxm1* transgenic cells/mice; KO = *foxm1* knockout cells/mice; CI = chemical inhibitor of FOXM1 (siomycin A).

[e]**Manipulation of FOXM1-binding site**: wt = "wild-type" promoter; del = analyzed with deletion mutant of promoter; P = FOXM1-binding site was point-mutated; H = FOXM1-binding site in context of heterologous core promoter; art = analyzed with an artificial construct.

[f]The point-mutated versions of the *Cdc25A* promoter had point mutations either in three putative FOXM1 binding sites or in two known E2F binding sites.

[g]The point-mutated version of the *survivin* promoter had point mutations in the four B-Myb (MYBL2 (MYB-like 2)) binding sites (but not in the two FOXM1 binding sites).

[h]*Bmi-1* is probably an indirect FOXM1 target gene because the transactivation of the *Bmi-1* promoter by FOXM1c depends on the presence of an E-box in the *Bmi-1* promoter (Li et al., 2008a), which is bound by c-Myc (Guney et al., 2006; Guo et al., 2007; Hydbring et al., 2010), and because shRNA-mediated knockdown of c-Myc attenuated the

Continued

stimulation of Bmi-1 protein expression by FOXM1c (Li et al., 2008a). Since *Bmi-1* is a direct c-Myc target gene (Guney et al., 2006; Guo et al., 2007; Hydbring et al., 2010) and since *c-myc* is a direct FOXM1c target gene (Hedge et al., 2011; Wierstra and Alves, 2006d; Zhang et al., 2011b), FOXM1 may indirectly activate the *Bmi-1* transcription through c-Myc (Li et al., 2008a).

[i]The 1.1 kb (kilobases) proximal human *KIS* promoter is not FOXM1-responsive, but FOXM1 activated an artificial construct, in which the 23 kb putative FOXM1 binding site from intron 6 of the *KIS* genes was inserted upstream of the 1.1 kb proximal *KIS* promoter (Petrovic et al., 2007).

[j]Putative FOXM1 binding site is positioned in intron 6.

[k]In EMSAs (electrophoretic mobility shift assays), the purified DBD (DNA-binding domain) of FOXM1c bound directly to the TATA-box of the *c-myc* P1 promoter and to the TATA-box of the *c-myc* P2 promoter (Wierstra and Alves, 2006d).

[l]The point-mutated transfected *c-myc* promoter had point mutations in TBE1 and TBE2, two known TCF-4/LEF-1 (T-cell factor-4/lymphoid enhancer factor-1) binding sites (but not in the *c-myc* P1 or P2 TATA-boxes).

[m]In the presence of genistein, FOXM1 had a positive effect on the E-cadherin protein expression because the E-cadherin protein level was higher in cells stably overexpressing FOXM1 than in parental cells (Bao et al., 2011). However, in the absence of genistein, the E-cadherin protein expression was unaffected by FOXM1 because the E-cadherin protein level was similar in FOXM1-overexpressing and parental cells (Bao et al., 2011).

[n]Putative FOXM1 binding site is positioned in intron 1.

[o]*Cxcl5* seems to be an indirect FOXM1 target gene because FOXM1 transactivates the *cxcl5* promoter (Balli et al., 2013; Wang et al., 2010c) and increases the *cxcl5* mRNA expression (Balli et al., 2013; Wang et al., 2008a, 2010c), but an *in vivo* occupancy of the *cxcl5* promoter by FOXM1 was not detected (Wang et al., 2010c).

[p]*HELLS* is probably an indirect FOXM1 target gene because FOXM1 increased the *HELLS* mRNA level, but did not occupy the *HELLS* promoter *in vivo* (Gemenetzidis et al., 2009).

[q]FOXM1 activated the 5.7 kb mouse *sox2* promoter (Ustiyan et al., 2012). Additionally, FOXM1 activated an artificial reporter construct, in which a far upstream region (−15178/−14836) of the *sox2* promoter containing the putative FOXM1 binding site (−15008/−14991) was inserted into the vector pGL3 (Wang et al., 2011b).

[r]The point-mutated version of the *LEF-1* promoter had point mutations in the three TBEs/WREs (TCF (T-cell factor)-binding elements/Wnt (wingless type)-responsive elements) (but not in any putative FOXM1 binding site).

[s]*p16* is probably an indirect FOXM1 target gene because the *p16* promoter was not *in vivo* occupied by FOXM1 (Gemenetzidis et al., 2009), which decreased the *p16* mRNA and protein expression (Teh et al., 2012).

[t]FOXM1 overexpression increased the *GATA3* mRNA level only in RB-deficient cells, namely, in the RB (retinoblastoma protein, RB1 (retinoblastoma 1), p105) mutant breast cancer cell line MDA-MB-468 and in MCF-7 breast cancer cells depleted of RB by shRNA (Carr et al., 2012).

Second, siRNA against either ROCK1 or RhoC inhibited the FOXM1-enhanced migration and invasion of SMMC7721 cells (Xia et al., 2012b). Accordingly, the FOXM1-enhanced migration and invasion of SMMC7721 cells was attenuated by the ROCK inhibitor Y2632 (Xia et al., 2012b).

Third, siRNA against MMP-7 inhibited the FOXM1-enhanced invasion of SMMC7721 cells (Xia et al., 2012b).

Thus, FOXM1 promotes migration and invasion by activating its direct target genes *MMP-2* (Chen et al., 2009; Dai et al., 2007), *ROCK1*, and *RhoC* (Fig. 6.3; Table 6.3) (Xia et al., 2012b). Additionally, FOXM1 stimulates invasion through activation of its direct target gene *MMP-7* (Fig. 6.3; Table 6.3) (Xia et al., 2012b).

In summary, there is convincing evidence that FOXM1 promotes migration and invasion (Fig. 6.3) (Ahmad et al., 2010; Balli et al., 2012; Bao et al., 2011; Behren et al., 2010; Bellelli et al., 2012; Chen et al., 2009, 2013b; Chu et al., 2012; Dai et al., 2007; He et al., 2012; Li et al., 2011, 2013; Lok et al., 2011; Lynch et al., 2012; Mizuno et al., 2012; Park et al., 2008a, 2011; Uddin et al., 2011, 2012; Wang et al., 2007a, 2008b, 2010c, 2013; Wu et al., 2010a; Xia et al., 2012b; Xue et al., 2012).

2.6. Epithelial–mesenchymal transition (EMT)

2.6.1 FOXM1 stimulates EMT

FOXM1 favors EMT (Fig. 6.3) (Balli et al., 2013; Bao et al., 2011; Li et al., 2012c; Park et al., 2011). Four studies reported a positive effect of FOXM1 on EMT:

2.6.1.1 ARF-deficient HCC cells (and MDCK cells)

Park et al. (2011) reported that FoxM1B induces an EMT-like phenotype in MDCK (Maden–Darby Canine kidney) cells, which were infected with retrovirus expressing FoxM1B, and in HCC cell lines isolated from *FoxM1B*-transgenic $ARF^{-/-}$ mice with DEN/PB-induced HCC. Both the FoxM1B-overexpressing MDCK cells and the *FoxM1B*-transgenic $ARF^{-/-}$ HCC cells exhibited a spindle-shape morphology indicative of EMT, which was not observed in control MCDK cells or in HCC cell lines isolated from $ARF^{-/-}$ mice with DEN/PB-induced HCC (Park et al., 2011). The *FoxM1B*-transgenic $ARF^{-/-}$ HCC cells also displayed some of those gene expression changes, which characterize EMT, namely, a downregulation of the epithelial marker protein E-cadherin (epithelial cadherin, CDH1 (cadherin 1, type 1)) and an upregulation of the mesenchymal marker proteins vimentin and α-SMA (α-smooth muscle actin) compared to $ARF^{-/-}$ HCC cells (Park et al., 2011). Additionally, the protein level of the EMT-promoting trancription factor Snail (SNAI1, Snail) was higher in *FoxM1B*-transgenic $ARF^{-/-}$ HCC cells than in $ARF^{-/-}$ HCC cells (Park et al., 2011). The same gene expression changes (except for an upregulation of the *Snail* mRNA level instead of the Snail protein level) were also detected in $ARF^{-/-}$ HCC cells, which were infected with adenovirus expressing FoxM1B, in comparison to $ARF^{-/-}$ HCC cells, confirming that they are induced by FoxM1B overexpression (Park et al., 2011). However, none of the EMT-characteristic gene expression changes was shown in MDCK cells.

Thus, FoxM1B induces an EMT-like phenotype in an *ARF*-null background, that is, in *FoxM1B*-transgenic $ARF^{-/-}$ HCC cell lines isolated from mice bearing DEN/PB-induced HCC and in corresponding $ARF^{-/-}$ HCC cell lines with adenovirus-mediated overexpression of FoxM1B (Park et al., 2011). Moreover, FoxM1B induces an EMT-like morphology in FoxM1B-overexpressing MDCK cells, which express only a negligible amount of ARF protein (Park et al., 2011) so that they resemble ARF-deficient cells.

2.6.1.2 AsPC-1 pancreatic cancer cells

Bao et al. (2011) reported that FOXM1 led to acquisition of an EMT phenotype in human AsPC-1 pancreatic cancer cells stably overexpressing FOXM1. Stable overexpression of FOXM1 in AsPC-1 cells resulted in a spindle-shaped mesenchymal morphology, whereas the parental AsPC-1 cells displayed the normal epithelial morphology (Bao et al., 2011). In addition, the FOXM1-overexpressing AsPC-1 cells exhibited EMT-typical gene expression changes compared to parental AsPC-1 cells, namely, an upregulation of the mesenchymal marker protein vimentin as well as an increased protein expression of the EMT-promoting transcription factors Slug (SNAI2, Snail2), ZEB-1 (zinc finger E-box-binding homeobox-1, δEF1 (δ-crystallin enhancer-binding factor 1)), and ZEB-2 (SIP1 (Smad-interacting protein 1)) as well as a decreased expression of the EMT-inhibiting miRNAs miR-200b and miR-200c (Bao et al., 2011). However, the expression of the epithelial marker protein E-cadherin was similar in FOXM1-overexpressing and parental AsPC-1 cells so that the FOXM1-overexpressing AsPC-1 cells did not show a downregulation of E-cadherin (Bao et al., 2011), which is mandatory for EMT (Agiostratidou et al., 2008; Baum et al., 2008; Berx et al., 2007; Bonnomet et al., 2010; Brabletz et al., 2005a,b; Christiansen and Rajasekran, 2006; Gavert and Ben-Ze'ev, 2010; Gavert and Ben-Ze'ev,

2008; Gotzmann et al., 2004; Guarino et al., 2007; Heuberger and Birchmeier, 2010; Huber et al., 2005; Hugo et al., 2007; Iwatsuki et al., 2010; Jechinger et al., 2003; Kalluri and Weinberg, 2009; Kang and Massagué, 2004; Klymkowsky and Savagner, 2009; Larue and Bellacosa, 2005; Le Bras et al., 2012; Lee et al., 2006; Levayer and Lecuit, 2008; Lopez-Novoa and Nieto, 2009; Martinez Arias, 2001; Mathias and Simpson, 2009; Micalizzi et al., 2010; Mimeault and Batra, 2007; Moreno-Bueno et al., 2008; Moustakas and Heldin, 2007; Nelson and Nusse, 2004; Nieto, 2011; Ouyang et al., 2010; Peinado et al., 2004, 2007; Polyak and Weinberg, 2009; Roussos et al., 2010; Savagner, 2001; Savagner, 2010; Schmalhofer et al., 2009; Singh and Settleman, 2010; Tarin, 2005; Thiery, 2002, 2003; Thiery and Sleeman, 2006; Thiery et al., 2009; Thompson and Newgreen, 2005; Tse and Kalluri, 2007; Tsuji et al., 2009; Turley et al., 2008; Voulgari and Pintzas, 2009; Yang and Weinberg, 2008; Yilmaz and Christofori, 2009, 2010; Zavadil and Böttinger, 2005). This unaltered E-cadherin expression strongly argues against EMT so that it appears questionable whether the AsPC-1 cells stably overexpressing FOXM1 had really acquired an EMT phenotype.

This doubt is underscored by the finding that, in the presence of the soybean isoflavone genistein (a natural cancer-chemo-preventive agent), the E-cadherin protein level was higher in FOXM1-overexpressing AsPC-1 cells than in parental AsPC-1 cells (Bao et al., 2011), which indicates an upregulation of the E-cadherin protein expression by FOXM1.

2.6.1.3 TGF-β1-induced EMT in A549 NSCLC cells

Li et al. (2012c) and Balli et al. (2013) reported that FOXM1 is involved in the TGF-β1-induced EMT in human A549 NSCLC cells.

TGF-β1 induces EMT in A549 cells so that TGF-β1 changes their morphology from epithelial-like appearance to mesenchymal-like spindle-cell shape (Kasai et al., 2005; Li et al., 2012c). The TGF-β1-induced EMT includes the downregulation of the E-cadherin protein level (Balli et al., 2013; Kasai et al., 2005; Li et al., 2012c) as well as the upregulation of the protein levels of vimentin (Kasai et al., 2005; Li et al., 2012c), fibronectin (Balli et al., 2013; Kasai et al., 2005), α-SMA, Snail, ZEB-1, and ZEB-2 (Balli et al., 2013).

FOXM1 is involved in the TGF-β1-induced EMT in A549 cells because siRNA against FOXM1 seems to change their morphology from TGF-β1-induced mesenchymal-like spindle-cell shape back to epithelial-like appearance (Li et al., 2012c) and because siRNA against FOXM1 reverses the TGF-β1-induced downregulation of the E-cadherin protein level (Balli et al., 2013; Li et al., 2012c) as well as the TGF-β1-induced upregulation of the protein levels of vimentin (Li et al., 2012c), fibronectin, α-SMA, Snail, ZEB-1, and ZEB-2 (Balli et al., 2013). In accordance with this involvement of FOXM1 in TGF-β1-induced EMT (Balli et al., 2013; Li et al., 2012c), TGF-β1 increases the FOXM1 protein expression in A549 cells in the course of TGF-β1-induced EMT (Fig. 6.2) (Li et al., 2012c).

miR-134 inhibits the TGF-β1-induced EMT in A549 cells because miR-134 changes their morphology from TGF-β1-induced mesenchymal-like spindle-cell shape back to epithelial-like appearance and because miR-134 reverses the TGF-β1-induced downregulation of the E-cadherin protein level and upregulation of the vimentin protein level (Li et al., 2012c). In accordance with this inhibition of TGF-β1-induced EMT by miR-134, TGF-β1 decreases the miR-134 expression in A549 cells in the course of TGF-β1-induced EMT (Li et al., 2012c).

miR-134 reduces the FOXM1 protein expression by targeting the 3′-UTR (untranslated region) of the *foxm1* mRNA (Li et al., 2012c). In the presence of miR-134, overexpression of a miR-134-resistant FOXM1 protein without wild-type 3′-UTR rescues the TGF-β1-induced EMT in A549 cells, because it restores the TGF-β1-induced downregulation of the E-cadherin protein level and upregulation of the vimentin protein level (Li et al., 2012c). This finding demonstrates again that FOXM1 is involved in the TGF-β1-induced EMT, and it indicates that miR-134 inhibits the TGF-β1-induced EMT at least partially through downregulation of FOXM1 (Li et al., 2012c).

Thus, FOXM1 is involved in the TGF-β1-induced EMT in A549 cells, where TGF-β1 seems to induce EMT via the pathway TGF-β1 ⊣miR-134 ⊣FOXM1 → EMT (Li et al., 2012c).

TGF-β failed to decrease the E-cadherin protein expression in A549 cells if FOXM1 was depleted by siRNA (Balli et al., 2013; Li et al., 2012c), but overexpression of Snail restored the TGF-β-induced downregulation of the E-cadherin protein level in these FOXM1 knockdown cells (Balli et al., 2013). Since *Snail* is a direct FOXM1 target gene (Fig. 6.3; Table 6.3) and since TGF-β increased the *in vivo* occupancy of the Snail promoter by FOXM1 in A549 cells (Balli et al., 2013), TGF-β seems to downregulate the E-cadherin protein expression via the pathway TGF-β-→ FOXM1 → Snail ⊣E-cadherin in A549 cells (Balli et al., 2013).

FOXM1 is involved in the TGF-β1-induced EMT in A549 cells (Balli et al., 2013; Li et al., 2012c). However, despite high levels of FOXM1 in

A549 cells, they did not undergo EMT in the absence of TGF-β, which demonstrates that FOXM1 alone is unable to induce EMT (Balli et al., 2013). Accordingly, FOXM1 alone is unable to downregulate the E-cadherin protein expression and to upregulate the α-SMA, fibronectin, Snail, ZEB-1, and ZEB-2 protein expression because their protein levels are unaffected by siRNA-mediated silencing of FOXM1 in the absence of TGF-β in A549 cells (Balli et al., 2013).

2.6.1.4 Mouse models of radiation-induced and bleomycin-induced pulmonary fibrosis

Balli et al. (2013) reported that a ca N-terminally truncated FoxM1B mutant (FoxM1B-ΔN) induces EMT during radiation-induced and bleomycin-induced pulmonary fibrosis.

In these two mouse models, pulmonary fibrosis is caused by radiation-induced or bleomycin-induced lung injury after thoracic irradiation (IR) or intratracheal administration of the glycopeptide antibiotic bleomycin, respectively (Chen and Stubbe, 2005; Moeller et al., 2008; Moore and Hogaboam, 2008; Mouratis and Aidini, 2011).

After irradiation, the lungs of transgenic *epiFoxm1-ΔN* mice, which conditionally overexpress FoxM1B-ΔN in the respiratory epithelium, displayed in comparison with the lungs of control mice a decrease in the mRNA expression of the epithelial marker E-cadherin, but increases in the mRNA expression of the mesenchymal markers vimentin and fibronectin and of the EMT-transcription factors Snail, Slug, TWIST2 (Dermo1 (Dermis-expressed protein 1)), ZEB-1, and ZEB-2 (Balli et al., 2013). Conversely, in irradiated *epiFoxm1 KO* knockout mice with a conditional deletion of *foxm1* from the pulmonary epithelium, the mRNA levels of the EMT-transcription factors Snail, Slug, TWIST1, TWIST2, ZEB-1, and ZEB-2 in lung were decreased compared to control mice (Balli et al., 2013). Thus, following IR, EMT-characteristic gene expression changes were enhanced by FoxM1B-ΔN overexpression, but reduced by ablation of *foxm1*, suggesting that ca FoxM1B promotes EMT whereas *foxm1* deficiency impairs EMT during radiation-induced pulmonary fibrosis (Balli et al., 2013).

When primary type II lung epithelial cells were isolated followed by flow cytometry-based cell sorting for GFP (green fluorescent protein), the FoxM1B-ΔN-expressing cell population (GFP-positive) of bleomycin-treated *epiFoxm1-ΔN* lungs contained increased *vimentin* and *Snail* mRNA levels, but a decreased *E-cadherin* mRNA level, compared to control cells (GFP-negative) of control lungs (Balli et al., 2013). Hence, in response to bleomycin treatment, FoxM1B-ΔN overexpression reinforced EMT-characteristic gene expression changes in the FoxM1B-ΔN-expressing population of alveolar type II epithelial cells, which suggests that ca FoxM1B promotes EMT during bleomycin-induced pulmonary fibrosis (Balli et al., 2013).

Pulmonary fibrosis is characterized by excess deposition of ECM (extracellular matrix) components (e.g., collagen) so that normal lung parenchyma is replaced with permanent scar tissue (Chapman, 2011; King et al., 2011; Kisseleva and Brenner, 2008a,b; Thannickal et al., 2004; Wynn, 2007, 2008, 2011). Pulmonary fibrosis includes the activation of myofibroblasts, which produce collagen and other ECM proteins (Desmouliere et al., 2005; Hinz et al., 2007; Kis et al., 2011; Phan, 2002, 2012; Scotton and Chambers, 2007; Tomasek et al., 2002; Willis et al., 2006). Myofibroblasts are mesenchymal cells.

EMT may contribute to pulmonary fibrosis because lung epithelial cells may undergo EMT thereby generating myofibroblasts (Chapman, 2011; Corvol et al., 2009; Coward et al., 2010; Gharaee-Khermani et al., 2009; Kalluri and Neilson, 2003; Willis and Borok, 2007; Willis et al., 2006). However, the contribution of EMT to pulmonary fibrosis is controversial and myofiboblasts can also originate from three other sources (Günther et al., 2012; Hardie et al., 2010; Karge and Borok, 2012; King et al., 2011; Kisseleva and Brenner, 2008a,b; Scotton and Chambers, 2007; Wynn, 2007, 2008, 2011). First, resident stromal fibroblasts in the lung, that is, local mesenchymal cells, can proliferate and transdifferentiate into myofibroblasts (Desmouliere et al., 2005; Hinz et al., 2007; Phan, 2012; Wynn, 2004; Wynn and Barron, 2010). Second, circulating fibrocytes, that is, bone marrow-derived mesenchymal progenitors, can be recruited to the lung where they can further proliferate and differentiate into myofibroblasts (Andersson-Sjöland et al., 2011; Bellini and Mattoli, 2007; Herzog and Bucala, 2010; Mattoli et al., 2009; Quan et al., 2006; Strieter et al., 2009a,b). Third, endothelial cells can undergo EndoMT (endothelial–mesenchymal transition) to generate myofibroblasts (Kis et al., 2011; Piera-Velazquez et al., 2011).

Since myofibroblast activation is part of the pathology of pulmonary fibrosis, radiation-induced and bleomycin-induced pulmonary fibrosis entail the abundant expression of mesenchymal marker proteins (Desmouliere et al., 2005; Hinz et al., 2007; Kis et al., 2011; Phan, 2002, 2012; Scotton and Chambers, 2007; Tomasek et al., 2002; Willis et al., 2006). Therefore, it is difficult to distinguish whether the increased mRNA levels of mesenchymal proteins (vimentin, fibronectin) in irradiated or bleomycin-treated *epiFoxm1-ΔN* lungs compared to control lungs result from EMT or from proliferation of myofibroblasts, which were derived from resident stromal fibroblasts, from circulating fibrocytes, or through EndoMT. Lineage tracing experiments and colocalization studies performed by Balli et al. (2013) indicated that FoxM1B-ΔN-expressing epithelial cells underwent EMT after irradiation or bleomycin exposure. However, the percentage of fibroblasts positive for both β-galactosidase (indicative of FoxM1B-ΔN expression) and α-SMA or vimentin was low, namely only, 14–18.5% in *epiFoxm1-ΔN* lungs compared to 3–10% in control lungs, indicating a minor contribution of EMT to the generation of myofibroblasts in irradiated or bleomycin-treated *epiFoxm1-ΔN* lungs (Balli et al., 2013). Nonetheless, the percentage of cells undergoing EMT was increased in *epiFoxm1-ΔN* mice so that expression of a ca FoxM1B mutant in alveolar type II epithelial cells increased the contribution of EMT to radiation-induced and bleomycin-induced pulmonary fibrosis (Balli et al., 2013).

In summary, overexpression of ca FoxM1B in respiratory epithelial cells promotes EMT during radiation-induced and bleomycin-induced pulmonary fibrosis (Balli et al., 2013).

Balli et al. (2013) did not analyze the effects of FoxM1B-ΔN overexpression and *foxm1* knockout on the expression of the epithelial marker E-cadherin, mesenchymal markers (vimentin, fibronectin), and EMT-transcription factors (Snail, Slug, TWIST1, TWSIT2, ZEB-1, ZEB-2) without ionizing radiation or bleyomycin treatment, so that it remains unknown whether FoxM1B-ΔN alone can bring about EMT-characteristic gene expression changes and whether FoxM1B-ΔN alone is capable of effectuating EMT. Moreover, Balli et al. (2013) did not analyze whether also wild-type FoxM1B stimulates EMT during radiation-induced and bleomycin-induced pulmonary fibrosis so that the role of endogenous FOXM1 in the pathogenesis of pulmonary fibrosis is unclear.

2.6.1.5 Summary
In summary, FOXM1 favors EMT (Fig. 6.3) (Balli et al., 2013; Bao et al., 2011; Li et al., 2012c; Park et al., 2011).

Yet, the available data for the role of FOXM1 in EMT may be interpreted with some caution:

The four above-mentioned studies analyzed the positive effect of FOXM1 on EMT in cancer cell lines, namely, in DEN/PB-induced HCC cells (Park et al., 2011), in pancreatic cancer cells (Bao et al., 2011), and in NSCLC cells

(Balli et al., 2013; Li et al., 2012c), where not only FOXM1 contributes to EMT, but also all those transformation events which gave rise to these cancer cell lines. Although Park et al. (2011) used MDCK cells, they did unfortunately not show the EMT-characteristic gene expression changes in this normal polarized epithelial cell line.

Park et al. (2011) performed their experiments with DEN/PB-induced $ARF^{-/-}$ HCC cells in an ARF-null background, and also MDCK cells actually represent ARF-deficient cells (see above). Li et al. (2012c) and Balli et al. (2013) carried out their experiments with NSCLC cells in the presence of the EMT inducer TGF-β1. Bao et al. (2011) did not find the mandatory E-cadherin downregulation in their experiments with pancreatic cancer cells.

Balli et al. (2013) investigated the contribution of FOXM1 to EMT during radiation-induced and bleomycin-induced pulmonary fibrosis, where EMT is triggered by IR-induced and bleomycin-induced lung injury, but not by FOXM1. The EMT in these two mouse models is caused by the cytokine TGF-β, which is released in reponse to lung injury (Chapman, 2011; Coward et al., 2010; Fernandez and Eickelberg, 2012; Flanders, 2004; Gharaee-Khermani et al., 2009; Järvinen and Laiho, 2012; Karge and Borok, 2012; Kisseleva and Brenner, 2008a; Willis and Borok, 2007; Willis et al., 2006; Wynn, 2008, 2011), so that Balli et al. (2013) studied the role of FOXM1 in TGF-β-induced EMT. Furthermore, Balli et al. (2013) analyzed a ca FoxM1B mutant, which may exert effects on EMT that are not accomplished by wild-type FOXM1.

Consequently, FOXM1 favors EMT (Fig. 6.3) (Balli et al., 2013; Bao et al., 2011; Li et al., 2012c; Park et al., 2011), but the exact role of FOXM1 in EMT warrants further investigation.

In particular, FOXM1 can either upregulate or downregulate the expression of the epithelial marker E-cadherin and of the mesenchymal markers α-SMA, vimentin, and fibronectin depending on the cellular context (Table 6.3). This opposite regulation of key players in EMT by FOXM1 points to a highly context-specific role of FOXM1 in EMT.

2.6.2 Antagonistic regulation of the E-cadherin expression by FOXM1

As described above, FOXM1 downregulates the E-cadherin expression in the course of EMT (Balli et al., 2013; Li et al., 2012c; Park et al., 2011):

FoxM1B-transgenic $ARF^{-/-}$ HCC cell lines isolated from mice with DEN/PB-induced HCC expressed less E-cadherin protein than the corresponding $ARF^{-/-}$ HCC cell lines, indicating that FoxM1B represses the E-cadherin protein expression (Park et al., 2011). Accordingly, adenovirus-mediated overexpression of FoxM1B in $ARF^{-/-}$ HCC cells decreased the E-cadherin protein level compared to $ARF^{-/-}$ HCC cells (Park et al., 2011).

Thus, FoxM1β downregulates the E-cadherin protein expression in an *ARF*-null background (Park et al., 2011).

TGF-β1 downregulated the E-cadherin protein level in A549 cells, but siRNA against FOXM1 abolished this TGF-β1-induced decrease in the E-cadherin protein level, which indicates that FOXM1 represses the E-cadherin protein expression (Balli et al., 2013; Li et al., 2012c).

miR-134 abrogated the TGF-β1-induced decrease in the E-cadherin protein level in A549 cells so that E-cadherin protein was abundantly expressed in the presence of both TGF-β1 and miR-134 (Li et al., 2012c). In such A549 cells cotreated with TGF-β1 and miR-134, over-expression of a miR-134-resistant FOXM1 protein without wild-type 3′-UTR reduced the E-cadherin protein level (Li et al., 2012c).

Hence, FOXM1 downregulates the E-cadherin protein expression in the context of TGF-β1-induced EMT (Balli et al., 2013; Li et al., 2012c).

However, without TGF-β treatment, the E-cadherin protein level was unaffected by siRNA-mediated knockdown of FOXM1 in A549 cells, demonstrating that FOXM1 alone is unable to decrease the E-cadherin protein expression (Balli et al., 2013).

In the context of radiation-induced pulmonary fibrosis, the *E-cadherin* mRNA level in lungs was decreased in irradiated *epiFoxm1-ΔN* mice, which conditionally overexpress a ca FoxM1B mutant (FoxM1B-ΔN) in the respiratory epithelium, compared to control mice, which indicates a repression of the *E-cadherin* mRNA expression by FoxM1B-ΔN (Balli et al., 2013).

In the context of bleomycin-induced pulmonary fibrosis, the *E-cadherin* mRNA level was decreased in the FoxM1B-ΔN-expressing cell population (GFP-positive) of bleomycin-treated *epiFoxm1-ΔN* lungs (which was obtained through isolation of primary type II lung epithelial cells followed by flow cytometry-based cell sorting for GFP) compared to control cells (GFP-negative) of control lungs, again indicating a repression of the *E-cadherin* mRNA expression by FoxM1B-ΔN (Balli et al., 2013).

In clear contrast to this repression of the E-cadherin expression by FOXM1 (Balli et al., 2013; Li et al., 2012c; Park et al., 2011), FOXM1c transactivated the murine *E-cadherin* promoter in RK-13 rabbit kidney cells

(Wierstra, 2011a), which originate from a 5-week-old rabbit (ATCC (American Type Culture Collection) CCL-37).

E-cadherin is a direct FOXM1c target gene (Fig. 6.3; Table 6.3) because the purified DBD (DNA-binding domain) of FOXM1c bound to a FOXM1 consensus binding site in the murine E-cadherin promoter in EMSAs (electrophoretic mobility shift assays) in vitro (Wierstra, 2011a). This FOXM1 binding site is perfectly conserved in the human E-cadherin gene (Wierstra, 2011a). Accordingly, the E-cadherin promoter was in vivo occupied by FOXM1 in human HEK293T embryonic kidney cells (Zhou et al., 2010b) and human FOXM1 than in parental AsPC-1 cells (Bao et al., 2011).

In the absence of genistein, FOXM1 had no effect on the E-cadherin promoter by FOXM1c (Wierstra, 2011a), FOXM1 upregulated the E-cadherin protein expression in AsPC-1 cells in the presence of genistein because the E-cadherin protein level was higher in AsPC-1 cells stably overexpressing FOXM1 than in parental AsPC-1 cells (Bao et al., 2011).

In the absence of genistein, FOXM1 had no effect on the E-cadherin protein expression in AsPC-1 cells because similar E-cadherin protein levels were found in FOXM1-overexpressing and parental AsPC-1 cells (Bao et al., 2011).

In summary, on the one hand, FOXM1c transactivates the E-cadherin promoter (Wierstra, 2011a) and FOXM1 increases the E-cadherin protein expression in genistein-treated AsPC-1 cells (Table 6.3) (Bao et al., 2011). On the other hand, the E-cadherin protein expression is decreased by FoxM1B in an ARF-null background (Park et al., 2011) and by FOXM1 in the context of TGF-β1-induced EMT (Table 6.3) (Balli et al., 2013; Li et al., 2012c). Additionally, a ca FoxM1B mutant decreases the E-cadherin mRNA expression during radiation-induced and bleomycin-induced pulmonary fibrosis (Balli et al., 2013).

These contradictory findings raise the question as to how FOXM1 can decrease the E-cadherin level (Balli et al., 2013; Li et al., 2012c; Park et al., 2011) and effectuate EMT (Balli et al., 2013; Bao et al., 2011; Li et al., 2012c; Park et al., 2011) although it transactivates the E-cadherin promoter (Wierstra, 2011a).

The answer to this question may be that FOXM1 increases the expression of several proteins (e.g., c-Myc, c-Fos (FOS), β-catenin/TCF-4 (T-cell factor-4), β-catenin/LEF-1 (lymphoid enhancer factor-1), Snail, c-Myb, LOXL2 (LOX (lysyl oxidase)-like 2), Cav-1 (caveolin-1), Suz12 (suppressor of Zeste 12), MMP-7, Bmi-1 (B lymphoma Mo-MLV (Moloney-murine leukemia virus) insertion region-1, PCGF4 (Polycomb group RING (really interesting new gene) finger protein 4)), Slug, ZEB-1, ZEB-2, TWIST1, TWIST2, TGF-β) (Fig. 6.3; Table 6.3), which are known to promote EMT and to decrease the E-cadherin expression (Agiostratidou et al., 2007; Baranwal and Alahari, 2009; Barrallo-Gimeno and Nieto, 2005; Beavon, 2000; Berx and van Roy, 2009; Berx et al., 2007; Bonnomet et al., 2010; Brabletz et al., 2005a,b; Cannito et al., 2010; Cano et al., 2010; Chen et al., 2012b; Christiansen and Rajasekran, 2006; Conacci-Sorrell et al., 2002; Davidson et al., 2012; Fuxe et al., 2010; Gavert and Ben-Ze'ev, 2007, 2008; Guarino et al., 2007; Hajra and Fearon, 2002; Heuberger and Birchmeier, 2010; Howard et al., 2008; Huber et al., 2005; Hugo et al., 2007, 2011; Iwatsuki et al., 2010; Jing et al., 2011; Kalluri and Weinberg, 2009; Kang and Massagué, 2004; Klymkowsky and Savagner, 2009; Kowalczyk and Nanes, 2012; Le Bras et al., 2012; Lee et al., 2006; Lee et al., 2012; Martinez Arias, 2001; Micalizzi et al., 2010; Mimeault and Batra, 2007; Moreno-Bueno et al., 2008; Moustakas and Heldin, 2007; Nelson and Nusse, 2004; Nieto, 2002; Nieto, 2011; Ouyang et al., 2010; Paredes et al., 2012; Peinado et al., 2004, 2005, 2007; Polyak and Weinberg, 2009; Said and Williams, 2011; Salon et al., 2005; Sato et al., 2012; Schmalhofer et al., 2009; Singh and Settleman, 2010; Stemmler, 2008; Strathdee, 2002; Takebe et al., 2011; Thiery, 2002, 2003; Thiery and Sleeman, 2006; Thiery et al., 2009; Tse and Kalluri, 2007; van Roy and Berx, 2008; Voulgari and Pintzas, 2009; Voutsadakis, 2012a,b; Wang and Zhou, 2011; Wendt et al., 2012; Wu and Zhou, 2010; Wu et al., 2012; Xiao and He, 2010; Yang and Weinberg, 2008; Yilmaz and Christofori, 2009, 2010; Zavadil and Böttinger, 2005), so that FOXM1 may achieve EMT (Fig. 6.4) and may indirectly downregulate the E-cadherin expression (Fig. 6.5) by activating the corresponding genes.

Additionally, FOXM1 decreases the expression of miRNAs (miR-200b, miR-200c) (Fig. 6.3; Table 6.3), which are known to inhibit EMT and to increase the E-cadherin expression (Brabletz and Brabletz, 2010; Cano and Nieto, 2008; de Krijger et al., 2011; Dykxhoorn, 2010; Feng et al., 2012; Gregory et al., 2008; Hill et al., 2013; Howe et al., 2012; Inui et al., 2010; Korpal and Kang, 2008; Ma and Weinberg, 2008a,b; Mongroo and Rustgi, 2010; Nicoloso et al., 2009; Peter, 2009; Petrocca and Lieberman, 2009; Reshmi et al., 2011; Spizzo et al., 2009; Sreekumar et al., 2011; Valastyan and Weinberg, 2011; Valastyn and Weinberg, 2009; Ventura and Jacks, 2009; Wang and Wang, 2012;

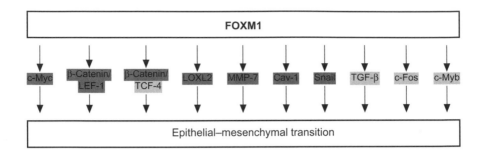

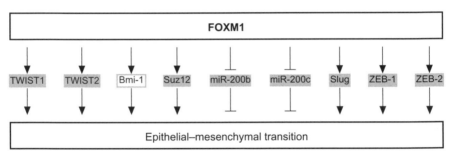

Figure 6.4 FOXM1-regulated genes with an impact on epithelial–mesenchymal transition (EMT). Direct (dark gray), possibly indirect (light gray), and probably indirect (light gray border) FOXM1 target genes are indicated. JNK1, the protein product of a direct FOXM1 target gene (Fig. 6.3; Table 6.3), is implicated in the induction of EMT by TGF-β (not shown).

Wright et al., 2010; Zhang and Ma, 2012; Zhang et al., 2010b), so that FOXM1 may give effect to EMT (Fig. 6.4) and may indirectly down-regulate the E-cadherin expression (Fig. 6.5) by repressing the transcription of these miRNAs.

Consequently, via its target genes FOXM1 can indirectly reduce the E–cadherin level (Fig. 6.5) (Balli et al., 2013; Li et al., 2012c; Park et al., 2011) and stimulate EMT (Fig. 6.4) (Balli et al., 2013; Bao et al., 2011; Li et al., 2012c; Park et al., 2011) in spite of the transactivation of the *E-cadherin* promoter by FOXM1c (Wierstra, 2011a).

This assumption is corroborated by the given fact that the proto-oncoprotein c–Myc shows exactly the same contradictory behavior as FOXM1c:

On the one hand, c–Myc transactivates the *E-cadherin* promoter (Batsche and Cremisi, 1999; Batsche et al, 1998; Liu et al., 2009c), which is *in vivo* occupied by c–Myc (Liu et al., 2009c).

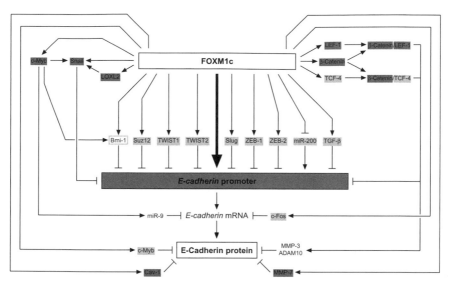

Figure 6.5 FOXM1-regulated genes, which control the expression of *E-cadherin*. Direct (dark gray), possibly indirect (light gray), and probably indirect (light gray border) FOXM1 target genes are indicated. FOXM1c itself binds to and transactivates the *E-cadherin* promoter. Please note that the figure does not show the complex interplay between the regulators of the *E-cadherin* expression. JNK1, the protein product of a direct FOXM1 target gene (Fig. 6.3; Table 6.3), is implicated in the repression of the *E-cadherin* expression by TGF-β (not shown). ADAM10, a disintegrin and metalloproteinase domain 10, KUZ, Kuzbanian; MMP-3, matrix metalloproteinase-3.

On the other hand, c–Myc decreases the E–cadherin expression post-transcriptionally (Cowling and Cole, 2007) by activating the transcription of miR-9, which targets the *E-cadherin* mRNA to suppress E–cadherin expression (Fig. 6.5) (Ma et al., 2010). In accordance with this down-regulation of the E–cadherin protein level by c–Myc (Amatangelo et al., 2012; Cho et al., 2010; Cowling and Cole, 2007; Liu et al., 2009b; Smith et al., 2009; Zhu et al., 2012), c–Myc can induce EMT (Fig. 6.4) (Amatangelo et al., 2012; Cho et al., 2010; Cowling and Cole, 2007; Liu et al., 2009b; Smith et al., 2009; Trimboli et al., 2008; Zhu et al., 2012).

In contrast to the general downregulation of the E–cadherin protein level by c–Myc (Amatangelo et al., 2012; Cho et al., 2010; Cowling and Cole, 2007; Liu et al., 2009b; Smith et al., 2009; Zhu et al., 2012), one study reported an upregulation of the E–cadherin expression by c–Myc (Celia-Terrassa et al., 2012), which is in accordance with the transactivation of the *E-cadherin* promoter by c–Myc (Batsche and Cremisi, 1999; Batsche et al, 1998; Liu et al., 2009c): In PC–3/Mc prostate cancer cells, c–Myc knockdown downregulated the *E-cadherin* mRNA and protein levels, indicating that c-Myc upregulates the *E-cadherin* mRNA and protein expression (Celia-Terrassa et al., 2012).

Yet, these PC-3/Mc cells represent an exceptional context because E-cadherin had a positive effect on their self-renewal and their metastasis to lung since both the spheroid-forming potential of PC-3Mc cells (i.e., tumor-initiating capability) and the lung colonization after intravein injection of PC-3/Mc cells into NOD-SCID (non-obese diabetic-severe combined immunodeficiency) mice were increased by E-cadherin overexpression, but decreased by E-cadherin knockdown (Celia-Terrassa et al., 2012).

In summary, despite the transactivation of the *E-cadherin* promoter by FOXM1c (Wierstra, 2011a), FOXM1 may indirectly decrease the E-cadherin level (Fig. 6.5) and may effectuate EMT (Fig. 6.4) via its other target genes (Fig. 6.3; Table 6.3).

Since FOXM1c binds to and transactivates the *E-cadherin* promoter (Wierstra, 2011a), it should upregulate the expression of its direct target gene *E-cadherin* in biological contexts where this direct positive effect of FOXM1c on the *E-cadherin* transcription is not canceled by indirect negative effects of FOXM1 on the E-cadherin expression, which are mediated by the products of other FOXM1 target genes (Fig. 6.5) (e.g., c-Myc, c-Fos, β-catenin/TCF-4, β-catenin/LEF-1, Snail, LOXL2, Cav-1, Suz12, MMP-7, Bmi-1, miR-200b, miR-200c, Slug, ZEB-1, ZEB-2). This suggests a context-dependent outcome of FOXM1 action, resulting in EMT (Balli et al., 2013; Bao et al., 2011; Li et al., 2012c; Park et al., 2011) and E-cadherin downregulation (Balli et al., 2013; Li et al., 2012c; Park et al., 2011), if the effects of FOXM1 on other FOXM1 target genes (Figs. 6.4 and 6.5) exceed the FOXM1c-mediated transactivation of the *E-cadherin* promoter (Wierstra, 2011a), but resulting in E-cadherin upregulation (Bao et al., 2011), if the direct transactivation of the *E-cadherin* promoter by FOXM1c (Wierstra, 2011a) prevails.

In fact, there is *in vivo* evidence for a biological role of FOXM1 in the activation of the E-cadherin expression because FOXM1 is required for the preservation of epithelial junctions in bronchioles of the developing mouse lung, which was revealed by a Clara cell-specific *foxm1* knockout (Ustiyan et al., 2012). The conditional deletion of *foxm1* from airway Clara cells in $CCSP\text{-}Foxm1^{-/-}$ mice from E16.5 onwards resulted in disrupted epithelial junctions in the $CCSP\text{-}Foxm1^{-/-}$ bronchioles at P30 (Ustiyan et al., 2012). While E–cadherin and β–catenin were present in epithelial adherens junctions of control bronchioles, epithelial adherens junctions were disrupted in *foxm1*-deficient bronchiolar epithelium, causing abnormal localization of E–cadherin and β–catenin in basal membranes (Ustiyan et al., 2012). Although Ustiyan et al. (2012) did not compare the total E–cadherin expression in the lungs of $CCSP\text{-}Foxm1^{-/-}$ and control mice, their findings argue for a positive effect of FOXM1 on the E–cadherin level and thus for the transactivation of the *E-cadherin* promoter by FOXM1c *in vivo* (Wierstra, 2011a).

This requirement of FOXM1 for the preservation of epithelial adherens junctions in bronchioles of the developing mouse lung (Ustiyan et al., 2012) is in line with the role of FOXM1 in endothelial repair during lung regeneration after acute vascular lung injury, where FOXM1 mediates the reannealing of endothelial adherens junctions (made of VE (vascular endothelial)-cadherin) (Fig. 6.3) in order to restore the restrictive endothelial barrier (Mirza et al., 2010). Additionally, FOXM1 plays a similar role for recovery of vascular integrity during endothelial repair upon lung regeneration following CLP (cecal ligation and puncture)-induced lung vascular injury, where FOXM1 is necessary and sufficient for the restoration of endothelial barrier integrity (Huang and Zhao, 2012).

In conclusion, FOXM1c transactivates the *E-cadherin* promoter (Wierstra, 2011a) and FOXM1 increases the E-cadherin protein expression in genistein-treated AsPC-1 cells (Bao et al., 2011) whereas the E-cadherin protein expression is decreased by FoxM1B in an *ARF*-null background (Park et al., 2011) and by FOXM1 in the context of TGF-β1-induced EMT (Balli et al., 2013; Li et al., 2012c) and whereas a ca FoxM1B mutant decreases the *E-cadherin* mRNA expression during radiation-induced and bleomycin-induced pulmonary fibrosis (Table 6.3) (Balli et al., 2013).

The upregulation of E-cadherin by FOXM1 (Fig. 6.3; Table 6.3) (Bao et al., 2011; Wierstra, 2011a) could explain the requirement of FOXM1 for preservation of epithelial adherens junctions in bronchioles of the developing mouse lung (Ustiyan et al., 2012). The opposite FOXM1-mediated downregulation of E-cadherin (Table 6.3) (Balli et al., 2013; Li et al., 2012c; Park et al., 2011) plays a role for the observed stimulation of EMT by FOXM1 (Fig. 6.3) (Balli et al., 2013; Bao et al., 2011; Li et al., 2012c; Park et al., 2011).

2.6.3 Why does the proliferation-associated, tumorigenesis-promoting transcription factor FOXM1c transactivate the promoter of the tumor suppressor gene E-cadherin?

E-cadherin (Stemmler, 2008; van Roy and Berx, 2008) is a Ca^{2+}-dependent single-pass transmembrane glycoprotein that mediates cell–cell adhesion in adherens junctions of epithelial cells (Baum and Georgiou, 2011; Borghi and Nelson, 2009; Cavey and Lecuit, 2009; Gates and Peifer, 2005; Green et al., 2010; Harder and Margolis, 2008; Harris and Tepass, 2010; Hartsock and Nelson, 2008; Koch et al., 2004; Leckband and Prakasam, 2006; Leckband and Sivasankar, 2012; Meng and Takeichi, 2009; Nelson, 2008; Niessen and Gottardi, 2008; Oda and Takeichi, 2011; Perez-Moreno et al., 2003; Pokutta and Weis, 2007; Rudini and Dejana, 2008; Shapiro and Weis, 2009; Tepass, 2002; Watanabe et al., 2009; Weis and Nelson, 2006; Yamada et al., 2005). It is critical for intercellular adhesion, structural integrity, cell identity, and tissue polarity of epithelia (Borghi and Nelson, 2009; Foty and Steinberg, 2004; Giehl and Menke, 2008; Gooding et al., 2004; Goodwin and Yap, 2004; Gumbiner, 2005; Halbleib and Nelson, 2006; Harris and Tepass, 2010; Hartsock and Nelson, 2008; Jeanes et al., 2008; Leckband and Prakasam, 2006; Lien et al., 2006; Meng and Takeichi, 2009; Nelson, 2008; Niessen and Gottardi, 2008; Patel et al., 2003; Perez-Moreno et al., 2003; Pokutta and Weis, 2007; Saito et al., 2012; Shapiro and Weis, 2009; Stemmler, 2008; Stepniak et al., 2009; Tian et al., 2011; Troyanovsky, 2005; van Roy and Berx, 2008; Wheelock and Johnson, 2003). Moreover, *E-cadherin* is a tumor suppressor gene so that its expression is frequently lost or reduced in human tumors, which correlates with poor prognosis (Agiostratidou et al., 2008; Baranwal and Alahari, 2009; Beavon, 2000; Berx and van Roy, 2009; Birchmeier, 1995; Buda and Pignatelli, 2011; Cavallaro and Christofori, 2004; Cavallaro et al., 2002; Christofori and Semb, 1999; Conacci-Sorrell et al., 2002; Giehl and Menke, 2008; Hajra and Fearon, 2002; Howard et al., 2008; Jankowski et al., 1997; Jeanes et al., 2008; Le Bras et al., 2012; Margineanu et al., 2008; Masterson and O'Dea, 2007; Moreno-Bueno et al., 2008; Paredes et al., 2012; Peinado et al., 2004, 2007; Salon et al., 2005; Schmalhofer et al., 2009; Sobrinho-Simeos and Oliveira, 2002; Stemmler, 2008; Strathdee, 2002; van Roy and Berx, 2008; Wheelock et al., 2007; Wood and Leong, 2003).

In clear contrast, the typical proliferation-associated transcription factor FOXM1 is intimately involved in tumorigenesis because it contributes to oncogenic transformation and participates in tumor initiation, tumor growth, and tumor progression (Fig. 6.3) (Costa et al., 2005a; Gong and Huang, 2012; Kalin et al., 2011; Koo et al., 2011; Laoukili et al., 2007; Myatt and Lam, 2007; Raychaudhuri and Park, 2011; Wierstra and Alves, 2007c). Accordingly, FOXM1 is overexpressed in many human cancers (Table 6.1), which correlates with poor prognosis (Calvisi et al., 2009; Chen et al., 2009; Chu et al., 2012; Dibb et al., 2012; Ertel et al., 2010; Fournier et al., 2006; Freije et al., 2004; Garber et al., 2001; He et al., 2012; Jiang et al., 2011; Li et al., 2009; Li et al., 2013; Liu et al., 2006; Martin et al., 2008; Okada et al., 2013; Park et al., 2012; Pellegrino et al., 2010; Priller et al., 2011; Qu et al., 2013; Sun et al., 2011a,b; Takahashi et al., 2006; Wang et al., 2013; Wu et al., 2013; Xia et al., 2012a,b,c; Xue et al., 2012; Yang et al., 2009; Yau et al., 2011; Yu et al., 2011; Zhang et al., 2012e).

Because of these opposite biological roles of FOXM1 and E-cadherin, FOXM1c is not expected to activate the *E-cadherin* transcription. Therefore, the striking contradiction that the proliferation-associated, tumorigenesis-promoting transcription factor FOXM1c transactivates the promoter of the tumor suppressor gene *E-cadherin* (Wierstra, 2011a) asks for those conditions under which cells could benefit from an upregulation of the E-cadherin expression by FOXM1c.

The following explanations might reconcile the apparent contradiction that the transcription of the tumor suppressor gene *E-cadherin* is activated by the tumorigenesis-promoting transcription factor FOXM1c (Wierstra, 2011a):

Since FOXM1 is essential for embryonic development (see Part I of this two-part review, that is, see Wierstra, 2013b) (Kalin et al., 2011; Kim et al., 2005; Korver et al., 1998; Krupczak-Hollis et al., 2004; Ramakrishna et al., 2007), the FOXM1c-induced *E-cadherin* expression (Wierstra, 2011a) may play a role during embryogenesis.

E-cadherin is essential for intercellular adhesion and tissue polarity of epithelia (Borghi and Nelson, 2009; Foty and Steinberg, 2004; Giehl and Menke, 2008; Gooding et al., 2004; Goodwin and Yap, 2004; Gumbiner, 2005; Halbleib and Nelson, 2006; Harris and Tepass, 2010; Hartsock and Nelson, 2008; Jeanes et al., 2008; Leckband and Prakasam, 2006; Lien et al., 2006; Meng and Takeichi, 2009; Nelson, 2008; Niessen and Gottardi, 2008; Patel et al., 2003; Perez-Moreno et al., 2003; Pokutta and Weis, 2007; Saito et al., 2012; Shapiro and Weis, 2009; Stemmler, 2008; Stepniak et al., 2009; Tian et al., 2011; Troyanovsky, 2005; van Roy and Berx, 2008; Wheelock and Johnson,

2003) so that the *de novo* synthesis of E-cadherin is required during each normal division of epithelial cells in order to attain the normal E-cadherin level in both daughter cells. Consequently, the FOXM1c-induced *E-cadherin* expression (Wierstra, 2011a) may be important in the course of normal epithelial proliferation to ensure the maintainance of the normal E-cadherin level, for example, during permanent tissue renewal (e.g., skin, lining of gut and lung) or during accidental tissue repair after injury.

In contrast to the general downregulation or loss of the E-cadherin expression in cancer cells (Agiostratidou et al., 2008; Baranwal and Alahari, 2009; Beavon, 2000; Berx and van Roy, 2009; Birchmeier, 1995; Buda and Pignatelli, 2011; Cavallaro and Christofori, 2004; Cavallaro et al., 2002; Christofori and Semb, 1999; Conacci-Sorrell et al., 2002; Giehl and Menke, 2008; Hajra and Fearon, 2002; Howard et al., 2008; Jankowski et al., 1997; Jeanes et al., 2008; Le Bras et al., 2012; Margineanu et al., 2008; Masterson and O'Dea, 2007; Moreno-Bueno et al., 2008; Paredes et al., 2012; Peinado et al., 2004, 2007; Salon et al., 2005; Schmalhofer et al., 2009; Sobrinho-Simeos and Oliveira, 2002; Stemmler, 2008; Strathdee, 2002; van Roy and Berx, 2008; Wheelock et al., 2007; Wood and Leong, 2003), there are a few exceptional tumor types, in which the expression of E-cadherin is not reduced or even elevated. For example, the E-cadherin expression is maintained in oral SCC (Ziober et al., 2001), in inflammatory breast cancer (Charafe-Jauffret et al., 2004, 2007; Colpaert et al., 2003; Garcia et al., 2007; Kleer et al., 2001; Nguyen et al., 2006; Tomlinson et al., 2001), and in biphasic synovial sarcoma (Kanemitsu et al., 2007; Laskin and Miettinen, 2002; Saito et al., 2000; Sato et al., 1999). Notably, an increase in the E-cadherin expression is frequently found in ovarian epithelial cancer (Gallo et al., 2010; Reddy et al., 2005; Sundfeldt, 2003).

Since FOXM1 is overexpressed in ovarian cancer (Table 6.1) (Banz et al., 2009; Elgaaen et al., 2012; Llaurado et al., 2012; Lok et al., 2011; Park et al., 2012; Pilarsky et al., 2004; The Cancer Genome Atlas Research Network, 2011), the FOXM1c-induced *E-cadherin* expression (Wierstra, 2011a) may account for the increased E-cadherin level in ovarian epithelial cancer. The FOXM1c-induced *E-cadherin* expression (Wierstra, 2011a) may also sustain the moderate E-cadherin level in oral SCC because FOXM1 is overexpressed in oral SCC, too (Table 6.1) (Gemenetzidis et al., 2009; Waseem et al., 2010). Whether FOXM1 is overexpressed in inflammatory breast cancer or biphasic synovial sarcoma is unknown.

The E-cadherin expression is enhanced in ovarian epithelial cancer (Gallo et al., 2010; Reddy et al., 2005; Sundfeldt, 2003), and this high

E-cadherin level favors ovarian tumor development and is beneficial for the growth and survival of ovarian cancer cells (e.g., Auersperg et al., 1999; Ong et al., 2000; Reddy et al., 2005). Similarly, the E-cadherin expression is maintained in oral SCC (Ziober et al., 2001), where E-cadherin supports the growth and survival of oral SCC cells (e.g., Kantak and Kramer, 1998; Shen and Kramer, 2004). Likewise, E-cadherin seems to be involved in the pathology of inflammatory breast cancer (Alpaugh and Barsky, 2002; Alpaugh et al., 2002; Dong et al., 2007; Tomlinson et al., 2001), in accordance with its persistent expression (Charafe-Jauffret et al., 2004, 2007; Colpaert et al., 2003; Garcia et al., 2007; Kleer et al., 2001; Nguyen et al., 2006; Tomlinson et al., 2001).

E-cadherin is reexpressed in metastases (e.g., Brabletz et al., 2001; Bukholm et al., 2000; De Marzo et al., 1999; Imai et al., 2004; Kowalski et al., 2003; Oltean et al., 2006; Rubin et al., 2001; van der Wurff et al., 1994) in the course of MET (mesenchymal–epithelial transition). Since the reexpression of E-cadherin is an integral part of MET (Barker and Clevers, 2001; Bednarz-Knoll et al., 2012; Biddle and Mackenzie, 2012; Bonnomet et al., 2010; Brabletz, 2012; Brabletz et al., 2005a; Buda and Pignatelli, 2011; Cannito et al., 2010; Chaffer et al., 2007; Christiansen and Rajasekran, 2006; Floor et al., 2011; Gao et al., 2012; Gavert and Ben-Ze'ev, 2007; Giannoni et al., 2012; Gunasinghe et al., 2012; Hayashida et al., 2011; Hugo et al., 2007; Iwatsuki et al., 2010; Jing et al., 2011; Lee et al., 2006; May et al., 2011; Nieto, 2011; Polyak and Weinberg, 2009; Said and Williams, 2011; Takebe et al., 2011; Thiery, 2002; Tse and Kalluri, 2007; van der Horst et al., 2012; van der Pluijm, 2011; Wells et al., 2008; Yang and Weinberg, 2008; Yao et al., 2011), the FOXM1c-induced *E-cadherin* expression (Wierstra, 2011a) may play a role during MET at the metastatic site.

E-cadherin is a potent and famous tumor suppressor (Beavon, 2000; Berx and van Roy, 2009; Birchmeier, 1995; Cavallaro et al., 2002; Christofori and Semb, 1999; Conacci-Sorrell et al., 2002; Hajra and Fearon, 2002; Howard et al., 2008; Jankowski et al., 1997; Jeanes et al., 2008; Margineanu et al., 2008; Paredes et al., 2012; Salon et al., 2005; Stemmler, 2008; Strathdee, 2002; van Roy and Berx, 2008; Wood and Leong, 2003). In addition, E-cadherin can also exert pro-tumorigenic effects in some special contexts so that it is more and more realized to play a supportive role for tumorigenesis, too (Rodriguez et al., 2012). In particular, proteolytic cleavage of E-cadherin results in the shedding of the E-cadherin ectodomain and generates extracellular E-cadherin fragments, called sE-cad (soluble E-cadherin), which can cause cells to become motile and invasive (David and Rajasekaran, 2012;

Grabowska and Day, 2012). This dualism of E-cadherin implies that the FOXM1c-induced *E-cadherin* expression (Wierstra, 2011a) can be beneficial for tumor cells under some circumstances.

2.7. Metastasis

FOXM1 promotes metastasis (Fig. 6.3) (Li et al., 2009, 2013; Park et al., 2011; Xia et al., 2012b). Four studies reported positive effects of FOXM1 on metastasis:

FoxM1B-overexpressing human NCI-N87 gastric cancer cells injected into the stomach wall of nude mice produced larger gastric tumors than control cells, indicating that FoxM1B overexpression enhances gastric tumor growth (Li et al., 2009).

The primary gastric tumors produced by FoxM1B-overexpressing NCI-N87 cells formed liver metastases whereas those formed by control cells did not metastasize to liver (Li et al., 2009). Thus, overexpression of FoxM1B promotes the metastasis of gastric cancer cells to the liver (Li et al., 2009).

Subcutaneously injected human SW480 and SW620 colon cancer cells formed tumors in nude mice (Li et al., 2013). The tumor weight was increased by overexpression of FoxM1B in SW480 cells, but decreased by siRNA-mediated silencing of FOXM1 in SW620 cells, showing that FOXM1 enhances their tumorigenicity (Li et al., 2013).

SW480 and SW620 cells, which were orthotopically implanted into the cecum of nude mice, metastasized to liver (Li et al., 2013). Both the number of liver metastases and the incidence of liver metastasis were raised by FoxM1B overexpression in SW480 cells, but diminished by FOXM1 knockdown in SW620 cells (Li et al., 2013). Hence, FOXM1 promotes liver metastasis of colon cancer cells (Li et al., 2013).

After orthotopic transplantation into the livers of nude mice, human HCCLM3 HCC cells with high metastatic potential metastasized to the lung and within the liver in 100% of cases whereas human SMMC7721 HCC cells with low metastatic potential produced distant lung metastases and intrahepatic metastases only in 17% of cases (Xia et al., 2012b). The incidence of both distant lung metastasis and intrahepatic metastasis was augmented by overexpression of FOXM1 in SMMC7721 cells (to 67%), but reduced by shRNA-mediated depletion of FOXM1 in HCCLM3 cells (to 33% or 50%) (Xia et al., 2012b). Thus, FOXM1 promotes the metastasis of HCC cells to the lung and their intrahepatic metastasis (Xia et al., 2012b).

The development of HCC can be caused by the DEN/PB liver tumor induction protocol with DEN as tumor initiator and PB as tumor promoter (Lee, 2000; Pitot et al., 1996; Sargent et al., 1996; Tamano et al., 1994). DEN/PB treatment generated liver tumors with 71% penetrance in wild-type mice, with 85% penetrance in $ARF^{-/-}$ mice, and with 100% penetrance in both *FoxM1B*-transgenic mice and *FoxM1B*-transgenic $ARF^{-/-}$ mice so that FoxM1B overexpression reinforced the development of primary DEN/PB-induced liver tumors independently of the status of the tumor suppressor gene *ARF* (Park et al., 2011).

In comparison with $ARF^{-/-}$ mice, the *FoxM1B*-transgenic $ARF^{-/-}$ mice exhibited markedly increased mortality and liver tumor burden, indicating that FoxM1B overexpression exacerbates the aggressiveness of HCC in an *ARF*-null background (Park et al., 2011).

Accordingly, HCC cells isolated from *FoxM1B*-transgenic $ARF^{-/-}$ mice produced more and larger tumors after their subcutaneous injection into nude mice than HCC cells derived from $ARF^{-/-}$ mice, which shows that FoxM1B enhances the tumorigenicity of HCC cells in an *ARF*-null background (Park et al., 2011).

The DEN/PB-induced HCC formed more lung metastases in *FoxM1B*-transgenic $ARF^{-/-}$ mice than in $ARF^{-/-}$ mice (Park et al., 2011). In contrast, the DEN/PB-induced HCC did not at all metastasize to lung in *FoxM1B*-transgenic mice and wild-type mice (Park et al., 2011). Therefore, FoxM1B overexpression can enhance lung metastasis of HCC, but only in an *ARF*-null background, which clearly points to the predominant importance of *ARF* deficiency in this mouse model (Park et al., 2011).

Intravenous injection of HCC cells isolated from *FoxM1B*-transgenic $ARF^{-/-}$ mice into nude mice via the tail vein resulted in considerably more lung nodules than injection of HCC cells derived from $ARF^{-/-}$ mice (Park et al., 2011). Hence, FoxM1B promotes the metastasis of ARF-deficient HCC cells to the lung (Park et al., 2011).

Accordingly, following injection into nude mice via the tail vein, $ARF^{-/-}$ HCC cells, which were infected with retrovirus expressing FoxM1B, formed more lung nodules than $ARF^{-/-}$ HCC cells, confirming that FoxM1B promotes lung metastasis of ARF-deficient HCC cells (Park et al., 2011). This positive effect of FoxM1B on the metastatic potential of ARF-deficient HCC cells depends on the direct FOXM1 target genes *LOX*, *LOXL2*, and *stathmin* (Fig. 6.3; Table 6.3) because shRNA against either LOX or LOXL2 or stathmin reduced the number of lung nodules formed by the FoxM1B-overexpressing $ARF^{-/-}$ HCC cells (Park et al., 2011). Thus, FoxM1B promotes the metastasis of ARF-deficient HCC cells to

, and *stathmin* (Fig. 6.3; Table 6.3) (Park et al., 2011).

FOXM1 supports the formation of the premetastatic niche in an *ARF*-null background (Park et al., 2011).

Premetastatic niche formation includes the recuitment of bone marrow-derived cells to the prospective metastatic site (Carlini et al., 2011; Comen, 2012; Duda and Jain, 2010; Kaplan et al., 2006; Rucci et al., 2011; Zoccoli et al., 2012). Lung tumor sections revealed infiltration of Cd11b$^+$ (ITGAM (integrin α-M)) myeloid cells and c-Kit$^+$ myeloid progenitor cells into metastatic lung tumors of *FoxM1B*-transgenic *ARF*$^{-/-}$ mice, but not in lungs of *ARF*$^{-/-}$ mice (Park et al., 2011). A substantial number of Cd11b$^+$ cells was also recruited to the tumor-free lungs of *FoxM1B*-transgenic *ARF*$^{-/-}$ mice (Park et al., 2011). These findings indicate that FoxM1B overexpression facilitates the formation of the premetastatic niche in an *ARF*-null background (Park et al., 2011).

Accordingly, subcutaneous injection of *ARF*$^{-/-}$ HCC cells, which were infected with retrovirus expressing FoxM1B, increased the number of infiltrated Cd11b$^+$ cells in the lungs of nude mice compared to injection of *ARF*$^{-/-}$ HCC cells, verifying that FoxM1B supports premetastatic niche formation in an *ARF*-null background (Park et al., 2011). This positive effect of FoxM1B on premetastatic niche formation depends on the direct FOXM1 target gene *LOX* (Fig. 6.3; Table 6.3) because the number of Cd11b$^+$ cells recruited to the lungs of nude mice after subcutaneous injection of FoxM1B-overexpressing *ARF*$^{-/-}$ HCC cells was diminished by shRNA against LOX and by the LOX inhibitor BAPN (beta-aminopropionitrile) (Park et al., 2011). Hence, FoxM1B supports the formation of the premetastatic niche in an *ARF*-null background at least in part through activation of its direct target gene *LOX* (Fig. 6.3; Table 6.3) (Park et al., 2011).

In summary, FOXM1 promotes metastasis (Fig. 6.3) (Li et al., 2009, 2013; Park et al., 2011; Xia et al., 2012b). The ability of FOXM1 to facilitate premetastatic niche formation in an *ARF*-null background suggests that FOXM1 promotes metastasis not only by enhancing the motility and invasiveness of tumor cells in the primary tumor, but also by making distant target organs more amenable to arriving disseminated cancer cells (Park et al., 2011).

2.8. Recruitment of tumor-associated macrophages (TAMs)

FOXM1 plays a role in the recruitment of TAMs (Fig. 6.3) (Balli et al., 2012):

Deletion of *foxm1* from macrophages and granulocytes in *macFoxm1* −/− mice reduced the number and size of lung tumors induced by MCA/BHT, that is, *foxm1* deficiency in myeloid cells impaired lung tumorigenesis after MCA/BHT treatment (Balli et al., 2012).

FOXM1 expression in macrophages is required for BHT-induced macrophage recruitment to the lung and for BHT-induced pulmonary inflammation (Balli et al., 2012). In fact, *macFoxm1*−/− mice are resistant to BHT-induced lung inflammation (Balli et al., 2012). However, the macrophage-specific ablation of *foxm1* did not affect the BHT-induced systemic inflammation (Balli et al., 2012). After lung tumor induction by MCA/BHT, *macFoxm1*−/− mice exhibited decreased numbers of tumor-infiltrating macrophages in the lung due to both decreased macrophage migration and decreased macrophage proliferation (Balli et al., 2012). The reduced macrophage recruitment to the lung and the reduced pulmonary inflammation in MCA/BHT-treated *macFoxm1*−/− mice resulted in decreased tumor cell proliferation in the lung and thus in decreased lung tumor growth (Balli et al., 2012). This decreased tumor cell proliferation in the lungs of MCA/BHT-treated *macFoxm1*−/− mice might be explained by the given fact that TAMs secrete mitogenic pro-inflammatory cytokines (Allavena et al., 2008a,b; Balli et al., 2012; Bingle et al., 2002; Coffelt et al., 2009; Guruvayoorappan, 2008; Mantovani et al., 2002, 2006, 2009, 2011; Ohno et al., 2002, 2003; Porta et al., 2007; Sica, 2010; Sica et al., 2002; Solinas et al., 2009; Weigert et al., 2009; Yuan et al., 2008).

Adoptive transfer of control monocytes to *macFoxm1*−/− mice restored both macrophage recruitment to the lung and pulmonary inflammation in response to MCA/BHT injury, indicating that FOXM1 plays a cell-autonomous role in macrophages during MCA/BHT-induced lung tumorigenesis (Balli et al., 2012).

Thus, FOXM1 expression in macrophages is important for lung tumor formation and growth after MCA/BHT treatment because FOXM1 is required for macrophage migration and proliferation, for macrophage recruitment to the lung, for pulmonary inflammation, and for tumor cell proliferation (Balli et al., 2012).

In summary, FOXM1 plays a role in recruitment of TAMs to the lung (Fig. 6.3) and in pulmonary inflammation during lung tumorigenesis induced by MCA/BHT treatment (Balli et al., 2012).

In accordance with the decreased inflammation and macrophage infiltration in the lungs of MCA/BHT-treated *macFoxm1*−/− mice (Balli et al.,

2012), the lungs of *FoxM1B*-transgenic mice (with ubiquitous *foxm1b* transgene expression in all cell types) exhibited an increased inflammatory response with macrophage infiltration following MCA/BHT treatment because they displayed persistent pulmonary inflammation whereas the MCA/BHT-induced chronic lung inflammation was completely resolved in wild-type mice (Wang et al., 2008a). Since no inflammation was detected in untreated *FoxM1B*-transgenic lungs, FoxM1B overexpression does not cause lung inflammation, but it merely sustains the MCA/BHT-induced inflammatory response, leading to the observed persistent pulmonary inflammation with infiltration of macrophages (Wang et al., 2008a).

2.9. The transcription factor FOXM1 and its implication in tumorigenesis

2.9.1 Regulators of FOXM1 and their functions in tumors

As outlined above, the typical proliferation-associated transcription factor FOXM1 is intimately involved in tumorigenesis, because it contributes to oncogenic transformation and participates in tumor initiation as well as tumor progression (Fig. 6.3) (Costa et al., 2005a; Gong and Huang, 2012; Kalin et al., 2011; Koo et al., 2011; Laoukili et al., 2007; Myatt and Lam, 2007; Raychaudhuri and Park, 2011; Wierstra and Alves, 2007c). Accordingly, FOXM1 is overexpressed in numerous human cancers (Table 6.1) and its expression is upregulated by proto-oncoproteins, but downregulated by tumor suppressors (Fig. 6.1) (Costa, 2005; Costa et al., 2005a; Gartel, 2008, 2010; Koo et al., 2011; Laoukili et al., 2007; Murakami et al., 2010; Myatt and Lam, 2007; Raychaudhuri and Park, 2011; Wang et al., 2010a; Wierstra and Alves, 2007c).

Since FOXM1 controls the expression of proto-oncogenes and tumor suppressor genes (Fig. 6.6), it represents a valuable target for proto-oncoproteins and tumor suppressors, which interact with FOXM1, modify FOXM1 enzymatically, or/and regulate the transcriptional activity of FOXM1 (Costa, 2005; Costa et al., 2005a; Koo et al., 2011; Laoukili et al., 2007; Murakami et al., 2010; Myatt and Lam, 2007; Raychaudhuri and Park, 2011; Wang et al., 2010a; Wierstra and Alves, 2007c).

First, the tumor suppressor RB (Carr et al., 2012; Major et al., 2004; Wierstra, 2013a; Wierstra and Alves, 2006b), the proto-oncoprotein β-catenin (Zhang et al., 2011b), and the viral oncoprotein HPV16 E7 (Lüscher-Firzlaff et al., 1999) bind directly to FOXM1. In addition, FOXM1 interacts possibly indirectly with the proto-oncoprotein AKT (Mukhopadhyay et al., 2011), with the tumor suppressors ARF (Kalinichenko et al., 2004), p53 (Roy and Tenniswood, 2007), and Cdh1 (Laoukili et al., 2008; Park et al.,

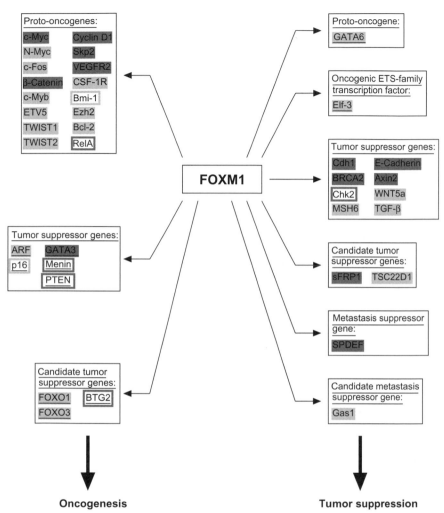

Figure 6.6 The role of FOXM1 target genes in cancer. Shown are direct (dark gray), possibly indirect (light gray), and probably indirect (light gray border) FOXM1 target genes as well as proteins, whose expression is regulated by FOXM1 at the protein level (dark gray border). FOXM1 upregulates the expression of most of its target genes. Those few genes, whose expression is downregulated by FOXM1, are marked (underlined). The FOXM1 target genes are grouped according to their role in cancer. It is indicated whether their regulation by FOXM1 is expected to result in oncogenesis (FOXM1 target genes listed to the left) or in tumor suppression (FOXM1 target genes listed to the right).

2008c), and with the putative tumor suppressors P/CAF (p300/CBP (CREB-binding protein, KAT3A)-associated factor, KAT2B (K-acetyltransferase 2B)) (Wierstra and Alves, 2008), FOXO3a (Madureira et al., 2006) and PP2A (protein phosphatase 2A) (subunits B55α (PPP2R2A, PP2A regulatory subunit Bα) and PP2AC (PP2A catalytic subunit, PPP2C)) (Alvarez-Fernandez et al., 2011).

Second, FOXM1 is phosphorylated by the tumor suppressor Chk2 (Tan et al., 2007) and the proto-oncoproteins cyclin D1/Cdk4 (Anders et al., 2011) as well as dephosphorylated by the potential tumor suppressor PP2A (Alvarez-Fernandez et al., 2011).

Third, as expected, the transcriptional activity of FOXM1 is increased by the proto-oncoproteins cyclin D1/Cdk4 (Anders et al., 2011; Wierstra, 2013a; Wierstra and Alves, 2006b,c, 2008), β-catenin (Zhang et al., 2011b), Ras (Major et al., 2004), Raf-1, and c-Src (Wierstra, 2011b) and by the viral oncoprotein HPV16 E7 (Lüscher-Firzlaff et al., 1999; Wierstra and Alves, 2006d, 2008), whereas it is decreased by the tumor suppressors ARF (Kalinichenko et al., 2004), p16 (Wierstra and Alves, 2006b), and RB (Wierstra, 2013a; Wierstra and Alves, 2006b, 2008) and by the candidate tumor suppressors PP2A (Alvarez-Fernandez et al., 2011), BTG2 (B-cell translocation gene 2, TIS21 (TPA (12-O-tetradecanoylphorbol-13-acetate)-inducible sequence 21), APRO1 (antiproliferative 1)) (Park et al., 2008b) and P/CAF (Wierstra and Alves, 2008).

Surprisingly, the viral oncoprotein adenovirus E1A reduces the transcriptional activity of FOXM1 (Major et al., 2004; Wierstra and Alves, 2008). This apparent contradiction might be reconciled by the given fact that E1A can also exhibit tumor suppressor-like properties in certain contexts (Chinnadurai, 2011; Ferrari et al., 2009; Frisch, 2001; Frisch and Mymryk, 2002; Lavia et al., 2003; Mymryk, 1996; Pelka et al., 2008; Turnell and Mymryk, 2006). For example, E1A can induce p53-dependent apoptosis (Berk, 2005; Roulston et al., 1999; Sang et al., 2002; Weinberg, 1997; White, 2001).

2.9.2 FOXM1 target genes and their roles in cancer

Among the FOXM1 target genes are sixteen proto-oncogenes (*c-myc, n-myc* (*MYCN, Myc oncogene neuroblastoma-derived*), *c-fos, cyclin D1, skp2* (*S-phase kinase-associated protein 2*), *β-catenin, c-myb, etv5, vegfr2* (*VEGF* (*vascular endothelial growth factor*) *receptor type II*), *bcl-2, csf-1r* (*colony-stimulating factor-1 receptor*), *bmi-1, ezh2* (*enhancer of Zeste homolog 2, KMT6* (*K-methyltransferase 6*)), *gata6, twist1, twist2*), ten tumor suppressor genes (*p16, brca2* (*breast cancer-associated gene 2, FANCD1* (*Fanconi anamia complementation group D1*)), *arf, msh6* (*mutS homolog 6*), *axin2* (*axis inhibitor 2, conductin*), *E-cadherin, gata3, tgf-β, wnt5a, cdh1*), and four candidate tumor suppressor genes (*sfrp1* (*secreted Frizzled-related protein 1*), *foxo1, foxo3, tsc22d1* (*TSC22* (*TGF-β-stimulated clone 22*) *domain family, member 1; TGF-*

β-stimulated clone 22 domain 1)) (Fig. 6.6), including eleven direct FOXM1 target genes (*c-myc, cyclin D1, skp2, sfrp1, E-cadherin, β-catenin, axin2, brca2, vegfr2, gata3, cdh1*), seventeen possibly indirect FOXM1 target genes (*n-myc, c-myb, etv5, twist1, twist2, arf, msh6, c-fos, bcl-2, csf-1r, foxo1, foxo3, tsc22d1, gata3, tgf-β, wnt5a, ezh2*), and two indirect FOXM1 target genes (*bmi-1, p16*) (Table 6.3).

In addition, FOXM1 regulates the protein levels of one proto-oncoprotein (RelA), three tumor suppressors (Menin (MEN1 (multiple endocrine neoplasia 1)), PTEN, Chk2) and one candidate tumor suppressor (BTG2) (Fig. 6.6), but an effect of FOXM1 on their mRNA expression has not been reported so far (Table 6.3).

The signaling routes for the FOXM1-mediated regulation of the two indirect FOXM1 target genes are either known (*bmi-1*) or predictable (*p16*):

FOXM1c activates the *bmi-1* promoter indirectly via c-Myc (see Part I of this two-part review, that is, see Wierstra, 2013b; Li et al., 2008a), the protein product of the direct FOXM1c target gene *c-myc* (Fig. 6.3; Table 6.3).

FOXM1, which does not *in vivo* occupy the *p16* promoter (Gemenetzidis et al., 2009), may indirectly repress the *p16* mRNA expression (Teh et al., 2012) through β-catenin/TCF-4, Ezh2, Suz12, Bmi-1 (via c-Myc), or HELLS (helicase, lymphoid-specific; LSH (lymphoid-specific helicase); SMARCA6 (SWI/SNF (switching defective/sucrose nonfermenting)–related, matrix–associated, actin-dependent regulator of chromatin, subfamily A, member 6)) (via E2F-1) (see Part I of this two-part review, that is, see Wierstra, 2013b), which are encoded by direct (*β-catenin, c-myc, e2f-1*), possibly indirect (*ezh2, suz12, tcf-4*), and indirect (*hells, bmi-1*) FOXM1 target genes (Fig. 6.3; Table 6.3).

As anticipated, fifteen proto–oncogenes (*c-myc, n-myc, c-fos, cyclin D1, skp2, β-catenin, c-myb, etv5, vegfr2, bcl-2, csf-1r, bmi-1, ezh2, twist1, twist2*) are activated by FOXM1, whereas three tumor suppressor genes (*p16, arf, gata3*) and two candidate tumor suppressor genes (*foxo1, foxo3*) are repressed by FOXM1 (Fig. 6.6). Also as anticipated, FOXM1 upregulates the protein level of one proto-oncoprotein (RelA), but downregulates those of two tumor suppressors (Menin, PTEN) and one candidate tumor suppressor (BTG2) (Fig. 6.6).

This expected positive and negative control of proto-oncogenes and tumor suppressor genes by FOXM1, respectively, can at least in part account for the promotion of tumor initiation, tumor growth, and tumor progression by FOXM1 (Fig. 6.6).

Surprisingly, FOXM1 represses one proto-oncogene (*gata6*) and activates seven tumor suppressor genes (*E-cadherin*, *msh6*, *tgf-β*, *wnt5a*, *cdh1*, *axin2*, *brca2*) as well as two candidate tumor suppressor genes (*sfrp1*, *tsc22d1*) (Fig. 6.6). Also surprisingly, the protein level of one tumor suppressor (Chk2) is increased by FOXM1 (Fig. 6.6).

This astonishing FOXM1-mediated upregulation of tumor suppressor genes and downregulation of a proto-oncogene is predicted to impair carcinogenesis (Fig. 6.6) so that it contradicts the observed promotion of tumorigenesis by FOXM1 (Costa et al., 2005a; Gong and Huang, 2012; Kalin et al., 2011; Koo et al., 2011; Laoukili et al., 2007; Myatt and Lam, 2007; Raychaudhuri and Park, 2011; Wierstra and Alves, 2007c). Therefore, FOXM1 may also dispose of some tumor-suppressive properties in addition to its many known oncogenic activities (Fig. 6.6) (Kalin et al., 2011; Wierstra, 2011a). Indeed, an unexpected tumor suppressor role of FOXM1 has recently been described (see below) (Balli et al., 2011).

2.10. Some normal functions of FOXM1 might be beneficial for tumor cells

Several functions of FOXM1 in normal cells might also favor carcinogenesis:

The FOXM1-mediated stimulation of cell proliferation (Fig. 6.3) (Costa, 2005; Costa et al., 2003, 2005b; Kalin et al., 2011; Koo et al., 2011; Laoukili et al., 2007; Mackey et al., 2003; Minamino and Komuro, 2006; Raychaudhuri and Park, 2011; Wierstra and Alves, 2007c) is likely to represent one important contribution of FOXM1 to tumorigenesis. In particular, the ability of FOXM1 to promote S-phase entry (Fig. 6.3) (Ackermann Misfeldt et al., 2008; Anders et al., 2011; Brezillon et al., 2007; Calvisi et al., 2009; Chan et al., 2008; Chen et al., 2010, 2012a; Davis et al., 2010; Gusarova et al., 2007; Huang and Zhao, 2012; Kalin et al., 2006; Kalinichenko et al., 2003, 2004; Kim et al., 2005, 2006a; Krupczak-Hollis et al., 2003, 2004; Lefebvre et al., 2010; Liu et al., 2006; Liu et al., 2011; Park et al., 2008c, 2009; Ramakrishna et al., 2007; Wang et al., 2001a,b, 2002a,b, 2005, 2007a, 2008a,b, 2010c; Wonsey and Follettie, 2005; Wu et al., 2010a; Xue et al., 2010; Ye et al., 1999; Yoshida et al., 2007; Zhang et al., 2006a, 2009; Zhao et al., 2006) should be beneficial for tumor cells.

Cancer cells may benefit from FOXM1's capabilities to prevent premature cellular senescence (Anders et al., 2011; Li et al., 2008a; Park et al., 2009; Qu et al., 2013; Rovillain et al., 2011; Wang et al., 2005; Zeng et al., 2009; Zhang et al., 2006a) and to overcome contact inhibition (Fig. 6.3) (Faust et al., 2012).

The positive effect of FOXM1 on the self-renewal capacity of stem cells and cancer cells (Fig. 6.3) (Bao et al., 2011; Wang et al., 2011b; Zhang et al., 2011b) could be favorable for oncogenesis.

FOXM1 plays a role in the maintenance of stem cell pluripotency (Fig. 6.3) (Xie et al., 2010) and the FOXM1 target genes include *oct-4* (*octamer-binding transcription factor 4*, *POU5F1* (*POU* (*Pit-Oct-Unc*) *domain, class 5, transcription factor 1*)), *sox2* (*SRY* (*sex-determing region Y*)-*box 2*, *SRY-related HMG* (*high mobility group*)-*box 2*), and *nanog* (Table 6.3) (Wang et al., 2011b; Xie et al., 2010; Zhang et al., 2011b), which are key factors for pluripotency and self-renewal of ES (embryonic stem) cells (Amabile and Meissner, 2009; Brumbaugh et al., 2011; de Vries et al., 2008; Dejosez and Zwaka, 2012; Geoghegan and Byrnes, 2008; Gonzales and Ng, 2011; Hanna et al., 2010; Hochedlinger and Plath, 2009; Jaenisch and Young, 2008; Li, 2010; Loh et al., 2008; Loh et al., 2011; MacArthur et al., 2009; Ng and Surani, 2011; Niwa, 2007a,b; Ohtsuka and Dalton, 2008; Orkin and Hochedlinger, 2011; Pauklin et al., 2011; Pei, 2009; Ralston and Rossant, 2010; Rossant, 2008; Scheper and Copray, 2009; Silva and Smith, 2008; Yamanaka, 2007, 2008b; Young, 2011; Yu and Thomson, 2008; Zhao and Daley, 2008). These findings are interesting with respect to cancer because similarities exist between ES cells and cancer cells, in particular CSCs (cancer stem cells), and because the direct reprogramming of normal cells into iPS (induced pluripotent stem) cells shows parallels to the transformation of normal cells into tumor cells (Ben-Porath et al., 2008; Deng and Xu, 2009; Geoghegan and Byrnes, 2008; Gupta et al., 2009; Keith and Simon, 2007; Kelleher et al., 2006; Kim et al., 2010; Knoepfler, 2008, 2009; Krizhanovsky and Lowe, 2009; Masip et al., 2010; Morrison and Kimble, 2006; Ramalho-Santos, 2009; Rothenberg et al., 2010; Somervaille et al., 2009; Werbowetzki-Ogilvie and Bhatia, 2008; Wong et al., 2008a,b; Yamanaka, 2007, 2008a; Zhao and Daley, 2008).

It has been suggested that FOXM1 perturbs the balance between pro-genitor cell renewal and commitment to differentiation by favoring clonal expansion of stem/progenitor cells, but disfavoring terminal differentiation (Fig. 6.3) (Carr et al., 2012; Gemenetzidis et al., 2010; Zhang et al., 2011b). This may be advantageous for tumor development.

However, cell sorting based on expression of the oral CSC marker CD44 argues against a CSC-specific role of FOXM1, because it revealed a uniform expression of FOXM1 in the whole population of tumor-derived keratinocytes regardless of their lineage hierarchy within the tumor subpopulations (e.g., CSCs) (Gemenetzidis et al., 2010).

Although FOXM1 seems to play a versatile, cell type-dependent role in the regulation of cellular differentiation (see Part I of this two-part review, that is, see Wierstra, 2013b), the inhibition of differentiation by FOXM1, which was reported in some biological contexts (Carr et al., 2012; Huynh et al., 2010; Wang et al., 2011b; Xie et al., 2010), would favor tumorigenesis.

In fact, FOXM1 inhibits mammary luminal differentiation in virgin mice (Carr et al., 2012). Therefore, FOXM1 overexpression in the mammary gland leads to a loss of differentiated luminal cells with a concomitant expansion of the luminal progenitor pool and the mammary stem cell population (Carr et al., 2012). Accordingly, the *foxm1* expression declines during mammary luminal differentiation so that *foxm1* is highly expressed in luminal progenitors (more abundant than in mammary stem cells), but dramatically downregulated in differentiated luminal cells (Carr et al., 2012).

Carr et al. (2012) point out that these findings implicate another role of FOXM1 in tumor development, namely, promoting the expansion of undifferentiated tumor cells. They note that it is likely that FOXM1 overexpression would drive accumulation of undifferentiated tumor cells in a subset of luminal breast tumors and that it is possible that overexpression of FOXM1 in luminal progenitors, in association with other changes, promotes the development of poorly differentiated mammary tumors (Carr et al., 2012).

FoxM1B overexpression led to global hypomethylation and focal hypermethylation in promoter regions (Fig. 6.3) so that FOXM1 might principally activate proto-oncogenes and repress tumor suppressor genes by triggering their promoter hypomethylation or promoter hypermethylation, respectively (Teh et al., 2012), which would be beneficial for tumor cells. The similarity of the FoxM1B-induced methylation pattern in normal cells (primary NOKs (normal oral human keratinocytes)) with that of a cancer cell line (HNSCC) suggests that FOXM1 may really cause a DNA methylation pattern resembling those found in cancer cells (Teh et al., 2012).

Although a clear-cut role of FOXM1 in either apoptosis or cellular survival has not been defined so far (see Part I of this two-part review, that is see, Wierstra, 2013b), cancer cells may benefit from a negative effect of FOXM1 on apoptosis, which has been described in the absence (Ahmed et al., 2012;

Calvisi et al., 2009; Halasi and Gartel, 2012; Huynh et al., 2010; Lefebvre et al., 2010; Mencalha et al., 2012; Ning et al., 2012; Park et al., 2009; Priller et al., 2011; Ren et al., 2010; Uddin et al., 2011, 2012; Ustiyan et al., 2009; Wan et al., 2012; Wang et al., 2010b; Wonsey and Follettie, 2005; Xia et al., 2012c; Xue et al., 2010; Zhao et al., 2006) as well as in the presence (Ahmed et al., 2012; Bhat et al., 2009a,b; Carr et al., 2010; Halasi and Gartel, 2012; Kwok et al., 2010; McGovern et al., 2009; Millour et al., 2011; Ning et al., 2012; Pandit and Gartel, 2011a,c; Uddin et al., 2011; Xu et al., 2012a) of anticancer drugs. This suggests that a positive effect of FOXM1 on cellular survival may contribute to chemotherapeutic drug resistance of FOXM1-overexpressing cancer cells (Fig. 6.3) (see below).

In summary, the normal functions of FOXM1 (Fig. 6.3) entail several advantages for tumor cells so that they will benefit from the FOXM1 over-expression, which occurs in numerous human cancers (Table 6.1).

2.11. Implication of FOXM1 in carcinogenesis caused by viral oncoproteins

2.11.1 FOXM1 and HPV16 E7

Tumor viruses contribute to 15–20% of human cancers (Butel, 2000; Carillo-Infante et al., 2007; Damania, 2007; Davey et al., 2011; Dayaram and Marriott, 2008; Fernandez and Esteller, 2010; Hoppe-Seyler and Butz, 1995; Javier and Butel, 2008; Liao, 2006; Martin and Gutkind, 2008; McLaughlin-Drubin and Münger, 2008; Moore and Chang, 2010; Parkin, 2006; Ziegler and Bounaguro, 2009; zur Hausen, 2001a,b).

E7 is a transforming oncoprotein of the high-risk human papillomavirus 16 (HPV16), a small DNA tumor virus belonging to the *Papillomaviridae* family (Chakrabarti and Krishna, 2003; Doorbar, 2006; Duensing and Münger, 2002; Duensing and Münger, 2004; Duensing et al., 2009; Fehrmann and Laimins, 2003; Finzer et al., 2002; Galloway and McDougall, 1996; Ganguly and Parihar, 2009; Garnett and Duerksen-Hughes, 2006; Ghittoni et al., 2010; Hamid et al., 2009; Jones and Münger, 1996; Klingelhutz and Roman, 2012; Korzeniewski et al., 2011; Lagunas-Martinez et al., 2010; Ledwaba et al., 2004; Longworth and Laimins, 2004; McCance, 2005; McLaughlin-Drubin and Münger, 2009a,b; Moody and Laimins, 2010; Münger, 2002; Münger and Howley, 2002; Münger et al, 2001, 2004; Narisawa-Saito and Kiyono, 2007; Pim and Banks, 2010; Scheffner and Whitaker, 2003; Tan et al., 2012 ; Wise-Draper and Wells, 2008; Yugawa and Kiyono, 2009; zur Hausen, 1996, 2002; Zwerschke and Jansen-Dürr, 2000). HPV16 infects the basal cell layer of mucosal epithelia in the anogenital tract and causes lesions, which have a propensity for carcinogenic progression, so that HPV16 is the causative agent of about half of all human cervix cancers (Baseman and Koutsky, 2005; Burk et al., 2009; Doorbar, 2006; Fehrmann and Laimins, 2003; Feller et al., 2009; Hebner and Laimins, 2006; Jayshree et al., 2009; Kisseljov et al., 2008; Lehoux et al., 2009; Lizano et al., 2009; Longworth and Laimins, 2004; McLaughlin-Drubin and Münger, 2009b, 2012; Moody and Laimins, 2010; Münger, 2002; Münger et al, 2004; Narisawa-Saito and Kiyono, 2007; Roden and Wu, 2006; Schiffman et al., 2007; Stanley et al., 2007; Stubenrauch and Laimins, 1999; Subramanya and Grivas, 2008; Woodman et al., 2007; zur Hausen, 1996, 2002, 2009).

The viral oncoprotein HPV16 E7 inactivates the tumor suppressor RB by disrupting E2F-RB complexes and by triggering the degradation of RB via the ubiquitin-proteasome pathway (Blanchette and Branton, 2009; Chakrabarti and Krishna, 2003; D'Abramo and Archambault, 2011; Duensing and Münger, 2004; Duensing et al., 2009; Felsani et al., 2006; Ganguly and Parihar, 2009; Ghittoni et al., 2010; Hamid et al., 2009; Helt and Galloway, 2003; Jones and Münger, 1996; Klingelhutz and Roman, 2012; Korzeniewski et al., 2011; Lee and Cho, 2002; Lehoux et al., 2009; Liu and Marmorstein, 2006; Lizano et al., 2009; McLaughlin-Drubin and Münger, 2009a,b; Moody and Laimins, 2010; Münger, 2003; Münger and Howley, 2002; Münger et al., 2001, 2004; Pim and Banks, 2010; Randow and Lehner, 2009; Scheffner and Whitaker, 2003; Tan et al., 2012; Yugawa and Kiyono, 2009; Zwerschke and Jansen-Dürr, 2000). The analysis of a RB mutant, which was selectively defective for binding E7, demonstrated that the inactivation of RB is required for nearly all acute *in vivo* effects of E7 (Balsitis et al., 2005).

E7 binds directly to FOXM1c (Lüscher-Firzlaff et al., 1999) and increases the transcriptional activity of FOXM1c (Lüscher-Firzlaff et al., 1999; Wierstra and Alves, 2006d, 2008). In particular, E7 enhances the transactivation of the human *c-myc* promoter by FOXM1c (Wierstra and Alves, 2006d, 2008). This positive effect of E7 on the FOXM1c-mediated transactivation of the *c-myc* promoter seems to require the binding of E7 to RB (Wierstra and Alves, 2006d, 2008) because the E7 mutant E7ΔRB, which retained the FOXM1c interaction domain (Lüscher-Firzlaff et al., 1999), but did not bind to RB (Münger et al., 1989), failed to increase the transactivation of the *c-myc* promoter by FOXM1c (Wierstra and Alves, 2006d, 2008).

In addition to enhancing the transcriptional activity of FOXM1c (Lüscher-Firzlaff et al., 1999; Wierstra and Alves, 2006d, 2008), E7 appears to upregulate the *foxm1* mRNA expression (Fig. 6.1) because over-expression of E7 increased the *foxm1* mRNA level in senescent HMFs (human mammary fibroblasts), which had first been immortalized with hTERT (human telomerase reverse transcriptase) and SV40 large T antigen and were then induced to senesce by inactivation of SV40 large T (Rovillain et al., 2011). Accordingly, the FOXM1 expression level correlated with the E7 expression level in HPV16-infected invasive cervical carcinoma (Rosty et al., 2005).

The oncoprotein E7 is essential for HPV16-induced cervical carcinogenesis (see above), suggesting that its target FOXM1c could be implicated in the pathology of cervix cancer.

Indeed, FOXM1 is overexpressed in cervical SCC (Table 6.1), namely, in early-stage tumors and in particular in late-stage tumors (Chan et al., 2008; He et al., 2012; Pilarsky et al., 2004; Rosty et al., 2005; Santin et al., 2005). FOXM1 is also highly expressed in cervical cancer cell lines (HeLa, CaSki, SiHa, HCC94, C33A) (Chan et al., 2008; He et al., 2012). With respect to clinicopathological features of cervical SCC, the FOXM1 expression level correlated with tumor stage (Chan et al., 2008; He et al., 2012), recurrence, and poor prognosis (He et al., 2012).

Furthermore, FOXM1 is part of the CCPC (cervical cancer proliferation cluster), which was defined in HPV16/HPV18-positive invasive cervical carcinoma (Rosty et al., 2005). The expression level of the CCPC genes was positively correlated with the E7 mRNA level and to a lower extent with the HPV DNA load (Rosty et al., 2005). The average expression level of all CCPC genes was higher in tumors with an unfavorable outcome (i.e., early relapse) than in tumors with a favorable course so that it may be indicative of poor disease prognosis (Rosty et al., 2005).

In accordance with a possible implication of FOXM1 in cervix cancer, FOXM1 enhanced the anchorage-independent growth of cervical cancer cells (C33A, SiHa) on soft agar (Chan et al., 2008), the tumorigenicity of cervical cancer cells (HCC94, SiHa) in nude mice (measured as volume and weight of xenograft tumors) (He et al., 2012), and the migration and invasion abilities of cervical cancer cells (HCC94, SiHA) (He et al., 2012), which in each case were significantly increased by FOXM1 overexpression (C33A, HCC94), but significantly decreased by shRNA-mediated knockdown of FOXM1 (SiHa) (Chan et al., 2008; He et al., 2012).

2.11.2 FOXM1 and adenovirus E1A

E1A is an oncoprotein of the transforming small DNA tumor virus adenovirus 5 that belongs to the *Adenoviridae* family (Bayley and Mymryk, 1994;
Ben-Israel and Kleinberger, 2002; Berk, 2005; Chinnadurai, 2011; Endter
and Dobner, 2004; Ferrari et al., 2009; Frisch and Mymryk, 2002; Gallimore
and Turnell, 2001; Jones, 1995; Pelka et al., 2008; Sang et al., 2002; Turnell
and Mymryk, 2006; White, 2001; Yousef et al., 2012). Although adenovirus
5 has not been linked to any human cancer, it induces various solid tumors in
rodents and the viral oncoprotein E1A transforms rodent cells in culture in
cooperation with a second oncogene (Chinnadurai, 2011; Endter and
Dobner, 2004; Frisch and Mymryk, 2002; Gallimore and Turnell, 2001;
Pelka et al., 2008; Russell, 2009; Smith et al., 2010a; Turnell and
Mymryk, 2006; Zheng, 2010). Due to alternative splicing of *E1A* transcripts, E1A is expressed in two major isoforms, which are called E1A-
12S (243-aa protein) and E1A-13S (289-aa protein) (Akusjarvi, 2008;
Bayley and Mymryk, 1994; Caron et al., 2002; Chinnadurai, 2011; Flint
and Shenk, 1997; Gallimore and Turnell, 2001; Jansen-Dürr, 1996;
Jones, 1995; Pelka et al., 2008; Zheng, 2010).

Surprisingly, E1A decreases the transcriptional activity of FOXM1
(Major et al., 2004; Wierstra and Alves, 2008). In particular, both E1A-
12S and E1A-13S repress the transactivation of the human *c-myc* promoter
by FOXM1c (Wierstra and Alves, 2008; data not shown).

Since E1A also possesses a tumor-suppressive activity (e.g., induction of
p53-dependent apoptosis) in addition to its dominant oncogenic properties
(Berk, 2005; Chinnadurai, 2011; Ferrari et al., 2009; Frisch, 2001; Frisch and
Mymryk, 2002; Lavia et al., 2003; Mymryk, 1996; Pelka et al., 2008;
Roulston et al., 1999; Sang et al., 2002; Turnell and Mymryk, 2006;
Weinberg, 1997; White, 2001), this surprising repression of FOXM1's tran-
scriptional activity by E1A (Major et al., 2004; Wierstra and Alves, 2008)
might be part of the tumor-suppressive side of E1A.

In contrast to the unexpected negative effect of E1A on the transcriptional activity of FOXM1 (Major et al., 2004; Wierstra and Alves, 2008),
E1A seems to elevate the *foxm1* mRNA expression (Fig. 6.1) because
E1A-12S overexpression raised the *foxm1* mRNA level in senescent HMFs,
which had first been immortalized with hTERT and SV40 large T antigen
and were then induced to senesce by inactivation of SV40 large T (Rovillain
et al., 2011).

2.11.3 FOXM1 and HBX of HBV

HBX is an oncoprotein of the human cancer virus HBV, which belongs to the *Hepadnaviridae* family and infects liver cells leading to chronic hepatitis, cirrhosis, and HCC (Andrisani and Barnabas, 1999; Benhenda et al., 2009; Bouchard and Schneider, 2004; Diao et al., 2001; Kew, 2011; Ma et al., 2011; Madden and Slagle, 2001; Martin-Vilchez et al., 2011; Murakami, 2001; Ng and Lee, 2011; Tang et al., 2006; Wei et al., 2010; Zhang et al., 2006a,b).

The viral oncoprotein HBX activates the *foxm1* promoter and increases the *foxm1* mRNA and protein expression (Fig. 6.1) (Xia et al., 2012b). Accordingly, FOXM1 is overexpressed in HBV-HCC (Table 6.1) (Xia et al., 2012b).

In HBV-HCC, FOXM1 is involved in the promotion of migration and invasion by HBX because shRNA-mediated silencing of FOXM1 markedly inhibited the HBX-enhanced migration and invasion of human HepG2 and Huh7 HCC cells (Xia et al., 2012b). In HBV-HCC, FOXM1 is also involved in the promotion of metastasis by HBX because the incidence of both lung metastasis and intrahepatic metastasis was reduced when the HBX-expressing HepG2 and Huh7 cells, which were transplanted into the livers of nude mice, expressed shRNA against FOXM1 (Xia et al., 2012b).

3. THE UNEXPECTED TUMOR SUPPRESSOR ROLE OF FOXM1

3.1. FOXM1 plays a surprising tumor suppressor role in endothelial lung cells

As described above, FOXM1 is generally regarded as a tumorigenesis-promoting transcription factor with oncogenic properties (Costa et al., 2005a; Gong and Huang, 2012; Kalin et al., 2011; Koo et al., 2011; Laoukili et al., 2007; Myatt and Lam, 2007; Raychaudhuri and Park, 2011; Wierstra and Alves, 2007c).

Yet, more recently, an unexpected tumor suppressor role of FOXM1 was discovered (Fig. 6.3), because Balli et al. (2011) reported that *foxm1* deficiency in endothelial mouse cells increases lung tumorigenesis induced by urethane or MCA/BHT.

In mice, lung tumors can be induced by treatment with either urethane (ethyl carbamate) or MCA/BHT (Forkert, 2010; Malkinson et al., 1997).

Surprisingly, when the *foxm1* gene was specifically deleted from endothelial cells, these *enFoxm1*$^{-/-}$ mice developed more and larger lung tumors than control mice after lung tumor induction with either urethane or MCA/BHT (Balli et al., 2011). Thus, the urethane-induced and MCA/BHT-induced lung tumorigenesis was increased by the endothelial cell-specific *foxm1* deletion, demonstrating an unexpected tumor suppressor role of FOXM1 in endothelial lung cells (Balli et al., 2011).

The urethane-induced lung tumors in *enFoxm1*$^{-/-}$ mice were lung adenocarcinomas, most likely originating from alveolar type II epithelial cells (Balli et al., 2011). Since the endothelial cell-specific deletion of the *foxm1* gene increased the proliferation of epithelial derived tumor cells following urethane exposure, FOXM1 seems to be involved in the cross-talk between endothelial and epithelial cells during urethane-induced lung carcinogenesis (Balli et al., 2011).

Urethane treatment caused severe persistent lung inflammation in *enFoxm1*$^{-/-}$ mice, which was characterized by increased perivascular leukocyte infiltration and increased numbers of inflammatory cells in BALF (bronchoalveolar lavage fluid) (Balli et al., 2011). Moreover, canonical Wnt signaling was activated in epithelial cells of *enFoxm1*$^{-/-}$ lungs after urethane treatment, which was evidenced by the increased nuclear localization of β-catenin and an increase in the activity of the β-catenin/TCF-4 reporter construct TOP-GAL that is driven by TCF-4 binding sites (Balli et al., 2011). Thus, the increased lung tumorigenesis in urethane-treated *enFoxm1*$^{-/-}$ mice seems to result from chronic pulmonary inflammation and from activated Wnt signaling in the respiratory epithelium after urethane treatment (Balli et al., 2011), because both inflammation (Aggarwal and Ghelot, 2009; Aggarwal et al., 2006; Balkwill and Mantovani, 2001, 2010, 2012; Balkwill et al., 2005; Ben-Neriah and Karin, 2011; Bollrath and Greten, 2009; Chow et al., 2012; Colotta et al., 2009; Coussens and Werb, 2002; Coussens et al., 2013; de Visser and Coussens, 2006; de Visser et al., 2006; Del Prete et al., 2011; Demaria et al., 2010; DiDonato et al., 2012; Germano et al., 2011; Grivennikov and Karin, 2010; Grivennikov et al., 2010; Hanahan and Weinberg, 2011; Johansson et al., 2008; Karin, 2008; Karin and Greten, 2005; Karin et al., 2006; Mantovani, 2010; Mantovani et al., 2008; Mantovani et al., 2010; Noonan et al., 2008; Philip et al., 2004; Porta et al., 2009; Sethi et al., 2012; Sica et al., 2008; Solinas et al., 2010; Tan and Coussens, 2007; Wu and Zhou, 2009) and Wnt-induced epithelial proliferation (Barker and Clevers, 2006; Clevers, 2006; de Lau et al., 2007; Fodde and Brabletz, 2007; Karim et al., 2004;

Klaus and Birchmeier, 2008; Lai et al., 2009; Niehrs and Acebron, 2012; Nusse and Varmus, 2012; Polakis, 2007, 2012; Reya and Clevers, 2005) are known to promote tumorigenesis (Balli et al., 2011). Therefore, the endothelial-specific expression of FOXM1 may limit lung inflammation and canonical Wnt signaling in lung epithelial cells, thereby restricting lung tumorigenesis (Balli et al., 2011).

It should be noted that lung inflammation was not detected in untreated $enFoxm1^{-/-}$ mice, which had normal lung morphology (BalLi et al., 2011). Hence, $foxm1$ deficiency does not cause inflammation, but FOXM1 impairs cancer-related inflammation in the context of urethane-induced lung tumorigenesis (BalLi et al., 2011).

The increased canonical Wnt signaling in epithelial cells of $enFoxm1^{-/-}$ lungs after urethane treatment resulted from a decreased $sfrp1$ mRNA expression in $enFoxm1^{-/-}$ lungs (Balli et al., 2011) because sFRP1 is a secreted Wnt inhibitor that represses the canonical Wnt/β-catenin pathway (although it can also exert a positive effect on Wnt signaling at low concentrations) (Bovolenta et al., 2008; Jones and Jomary, 2002; Kawao and Kypta, 2003; Rubin et al., 2006). The candidate tumor suppressor gene $sfrp1$ is a direct FOXM1 target gene (Fig. 6.3; Table 6.3) because the $sfrp1$ promoter was in $vivo$ occupied by FOXM1 in mouse endothelial MFLM-91U cells (Balli et al., 2011). The $sfrp1$ mRNA level was reduced in $enFoxm1^{-/-}$ mouse lungs compared to control lungs before and after lung tumor induction by urethane, indicating that FOXM1 enhances the $sfrp1$ mRNA expression (BalLi et al., 2011). Thus, since FOXM1 activates the $sfrp1$ transcription (Fig. 6.3; Table 6.3), $enFoxm1^{-/-}$ lungs exhibit a decreased $sfrp1$ mRNA expression, which results in increased canonical Wnt signaling in the epithelial cells of urethane-treated $enFoxm1^{-/-}$ lungs (Balli et al., 2011).

In summary, endothelial cell-specific $foxm1$ deficiency increases urethane-induced and MCA/BHT-induced lung tumorigenesis in mice, revealing an unexpected tumor suppressor role of FOXM1 in endothelial lung cells (Fig. 6.3) (BalLi et al., 2011; Kalin et al., 2011), which is in clear contrast to the known oncogenic properties of FOXM1 (see above).

In addition to these surprising findings for chemically induced lung tumorigenesis in $enFoxm1^{-/-}$ mice (BalLi et al., 2011), also individual results of other studies point to a hidden tumor-suppressing activity of FOXM1 (see below), which suggests that FOXM1 may have not only many oncogenic properties but also a few tumor-suppressive properties (Fig. 6.7) (Kalin et al., 2011; Wierstra, 2011a).

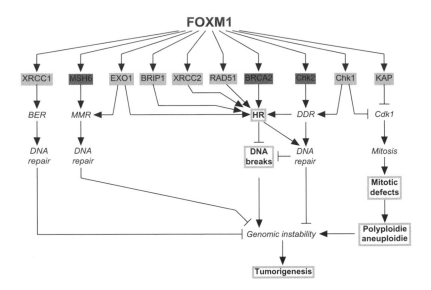

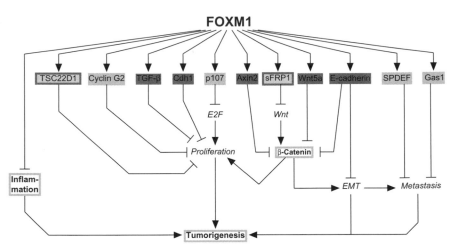

Figure 6.7 Possible tumor-suppressive properties of FOXM1. Surprisingly, FOXM1 exhibits a few tumor-suppressive properties (marked in bold type and with a light gray border) and it upregulates the mRNA or/and protein expression of eight tumor suppressors (dark gray), two candidate tumor suppressors (light gray with a dark gray border), and eleven other proteins (light gray), which counteract tumorigenesis. SPDEF is a metastasis suppressor. Gas1 is a candidate metastasis suppressor. Please note that parts of this figure contradict the many oncogenic properties of FOXM1. The tumor suppressor Wnt5a, a noncanoncial Wnt ligand, inhibits canonical β-catenin/TCF signaling (see figure) through the orphan tyrosine kinase Ror2 (RTK (receptor tyrosine kinase)-like orphan receptor 2), but it can also activate canonical β-catenin/TCF signaling (not shown) through the coreceptors Frz4 (Frizzled 4) and LRP5 (low density lipoprotein receptor-related protein 5). The cytokine TGF-β functions as a tumor suppressor in early stages of carcinogenesis by inducing apoptosis and inhibiting proliferation, that is, causing a cell cycle arrest in G1-phase. In contrast, during later stages of carcinogenesis, TGF-β promotes tumorigenesis by inducing EMT, stimulating metastasis, suppressing the immune reponse, and activating fibroblasts in the tumor stroma. BER, base excision repair; Cdk1, cyclin-dependent kinase 1; DDR, DNA damage response; EMT, epithelial–mesenchymal transition; HR, homologous recombination; MMR, mismatch repair.

3.2. FOXM1 may dispose of a few tumor-suppressive properties in addition to its predominant oncogenic properties

Deletion of the *foxm1* gene from endothelial cells increased the number and size of urethane-induced and MCA/BHT-induced lung tumors, demonstrating an endothelial cell-specific tumor suppressor role of FOXM1 in lung tumorigenesis induced by urethane or MCA/BHT (Fig. 6.3) (Balli et al., 2011; Kalin et al., 2011).

This unexpected result raised the possibility that FOXM1 may dispose of a few tumor-suppressive properties in addition to its predominant oncogenic properties (Kalin et al., 2011; Wierstra, 2011a). This view is supported by the following findings, which all indicate a hidden tumor-suppressing activity of FOXM1 (Fig. 6.7):

FOXM1 promotes HR (Fig. 6.7) (Monteiro et al., 2012; Park et al., 2012), a DNA repair mechanism for DSBs (Amunugama and Fishel, 2012; Bordeianu et al., 2011; Cerbinskaite et al., 2012; Ciaccia and Elledge, 2010; Dever et al., 2012; Evers et al., 2010; Hakem, 2008; Helleday, 2010; Heyer et al., 2010; Hoeijmakers, 2001; Jeggo et al., 2011; Kass and Jasin, 2010; Khanna and Jackson, 2001; Krejci et al., 2012; Moynahan and Jasin, 2010; Peterson and Cote, 2004; Polo and Jackson, 2011; Raassool and Tomkinson, 2010; San Filippo et al., 2008; Scott and Pandita, 2006; Symington and Gautier, 2011; van Gent et al., 2001; Wyman and Kanaar, 2006). Accordingly, FOXM1 deficiency leads to an increase in the number of DNA breaks (Fig. 6.7) (Chetty et al., 2009; Kwok et al., 2010; Monteiro et al., 2012; Tan et al., 2007), whereas overexpression of FOXM1 (Monteiro et al., 2012) and a N-terminally truncated FOXM1 mutant (ΔN-FOXM1) (Kwok et al., 2010) decreased the number of epirubicin-induced and cisplatin-induced DNA breaks in MCF-7 breast cancer cells, respectively.

In accordance with the role of FOXM1 for HR repair of DSBs (Monteiro et al., 2012; Park et al., 2012), FOXM1 activates the transcription of the HR proteins BRCA2 (Kwok et al., 2010; Tan et al., 2007, 2010; Zhang et al., 2012e), XRCC2 (X-ray cross-completing group 2) (Anders et al., 2011), RAD51 (Zhang et al., 2012e), BRIP1 (BRCA1-interacting protein 1, Bach1 (BRAC1-associated C-terminal helicase)) (Monteiro et al., 2012), and EXO1 (exonuclease 1) (Fig. 6.7) (Park et al., 2012), which are encoded by four direct (*brca*, *rad51*, *brip1*, *exo1*) and one possibly indirect (*xrcc2*) FOXM1 target gene (Fig. 6.3; Table 6.3).

FOXM1 seems to be important for maintenance of chromosome integrity and genomic stability (Laoukili et al., 2007; Wierstra and Alves, 2007c; Wilson et al., 2011; Zhao and Lam, 2012), because FOXM1-deficient cells generally display polyploidy and aneuploidy (Fig. 6.7) (Chen et al., 2012a; Fu et al., 2008; Kalinichenko et al., 2004; Kim et al., 2005; Korver et al., 1998; Krupczak-Hollis et al., 2004; Laoukili et al., 2005; Nakamura et al., 2010a; Ramakrishna et al., 2007; Ustiyan et al., 2009; Wan et al., 2012; Wang et al., 2005; Wonsey and Follettie, 2005; Zhao et al., 2006).

Accordingly, FOXM1 plays a role in proper execution of mitosis and thus in preservation of mitotic fidelity so that FOXM1 deficiency leads to pleiotropic mitotic defects, including aneuploidy and polyploidy, a delay in progression through mitosis, a failure to progress beyond the prophase stage of mitosis, aberrant furrow formation, defects in cytokinesis, mitotic spindle defects, misalignment of chromosomes at the metaphase plate, chromosome mis-segregation, a defective spindle assembly checkpoint, a propensity to be binucleated, centrosome amplification, and cell death by mitotic catastrophe (Fig. 6.7) (Chan et al., 2008; Fu et al., 2008; Gusarova et al., 2007; Laoukili et al., 2005; Priller et al., 2011; Ramakrishna et al., 2007; Schüller et al., 2007; Wang et al., 2005; Wonsey and Follettie, 2005; Yoshida et al., 2007).

The target gene spectrum of FOXM1 includes seven tumor suppressor genes (*E-cadherin, msh6, wnt5a, tgf-β, cdh1, axin2, brca2*) and two candidate tumor suppressor genes (*sfrp1, tsc22d1*), which are activated by FOXM1, as well as one proto-oncogene (*gata6*), which is repressed by FOXM1, so that the FOXM1-mediated regulation of these 10 cancer genes is predicted to suppress tumorigenesis (Figs. 6.6 and 6.7):

FOXM1 activates the transcription of the tumor suppressor genes *E-cadherin* (Wierstra, 2011a), *brca2* (Kwok et al., 2010; Tan et al., 2007, 2010a; Zhang et al., 2012e), *axin2* (Zhang et al., 2011b), and *msh6* (Anders et al., 2011) as well as that of the candidate tumor suppressor gene *tsc22d1* (Anders et al., 2011), which are direct (*E-cadherin, brca2, axin2*) or possibly indirect (*msh6, tsc22d1*) FOXM1 target genes (Fig. 6.3; Table 6.3). Moreover, FOXM1 increases the mRNA expression of the tumor suppressors Cdh1 (Chen et al., 2013a), TGF-β (Balli et al., 2013), and Wnt5a (Wang et al., 2012) as well as that of the candidate tumor suppressor sFRP1 (BalLi et al., 2011), which are encoded by direct (*cdh1, sfrp1*) or possibly indirect (*wnt5a, tgf-β*) FOXM1 target genes (Fig. 6.3; Table 6.3).

Conversely, FOXM1 decreases the mRNA expression of the proto-oncogene *gata6* (Wang et al., 2012), a possibly indirect FOXM1 target gene (Fig. 6.3; Table 6.3).

In addition, the protein level of the tumor suppressor Chk2 is upregulated by FOXM1 (Chetty et al., 2009; Tan et al., 2010a), which is anticipated to manifest in tumor suppression (Figs. 6.6 and 6.7).

Also other FOXM1-induced gene expression changes are expected to suppress tumorigenesis, for example (Fig. 6.7):

FOXM1 activates the transcription of the metastasis suppressor SPDEF (Ren et al., 2013), the BER (base excision repair) protein XRCC1 (Chetty et al., 2009; Kwok et al., 2010; Tan et al., 2007, 2010a), the MMR (mismatch repair) protein EXO1 (Park et al., 2012), the checkpoint kinase Chk1 (Chetty et al., 2009; Tan et al., 2010a,b), and the inhibitor of cell cycle progression cyclin G2 (Anders et al., 2011), which all are encoded by direct FOXM1 target genes (Fig. 6.3; Table 6.3). Furthermore, FOXM1 enhances the mRNA expression of the candidate metastasis suppressor Gas1 (growth arrest-specific 1) (Wonsey and Follettie, 2005) and the Cdk-inactivating phosphatase KAP (Cdk-associated protein phosphatase, CDKN3 (Cdk inhibitor 3)) (Laoukili et al., 2005), which both are encoded by possibly indirect FOXM1 target genes (Fig. 6.3; Table 6.3). Additionally, the protein level of the pocket protein p107 (RBL1 (retinoblastoma-like 1)) is elevated by FOXM1 (Xue et al., 2010).

Conversely, FOXM1 reduces the mRNA expression of the oncogenic ETS-family transcription factor Elf-3 (E74-like factor-3) (Ustiyan et al., 2012), which is encoded by a possibly indirect FOXM1 target gene (Fig. 6.3; Table 6.3).

Fig. 6.7 summarizes how FOXM1 might suppress tumorigenesis through the upregulation of eight tumor suppressors, two putative tumor suppressors, and eleven additional proteins with anti-oncogenic activities.

In summary, the above-mentioned examples show that FOXM1 may have not only many oncogenic properties, but also a few tumor-suppressive properties (Figs. 6.6 and 6.7) (Kalin et al., 2011; Wierstra, 2011a).

In fact, the surprising finding that urethane-induced and MCA/BHT-induced lung tumorigenesis are increased in *enFoxm1*$^{-/-}$ mice with an endothelial cell-specific deletion of the *foxm1* gene demonstrated an unexpected tumor suppressor role of FOXM1 in endothelial lung cells (Fig. 6.3) (Balli et al., 2011; Kalin et al., 2011).

4. CONTEXT-DEPENDENT EFFECTS OF FOXM1

4.1. The context dependence of FOXM1 action

More recently, context-dependent functions of FOXM1 are emerging so that FOXM1 is now realized to have not only general biological effects, but also context-dependent effects (Kalin et al., 2011).

Examples for such context-dependent outcomes of FOXM1 action include:

- opposite effects on apoptosis (see Part I of this two-part review, that is see, Wierstra, 2013b);
- opposite effects on cellular differentiation (Fig. 6.3) (see Part I of this two-part review, that is see, Wierstra, 2013b);
- antagonistic regulation of some FOXM1 target genes (e.g., *p21*, *sox2*, *c-myc*, *cyclin D1*, *cyclin B1*, *E-cadherin*, *TCF-4*, *ERα*, *FOXA3*, *GATA3*, *IL-1β* (*interleukin-1β*), *IL-13*, *KRT18*, *MMP-9*, *vimentin*, *fibronectin*, *α-SMA*, *SP-B*, *SP-C*) (Table 6.3);
- opposite effects on MCA/BHT-induced lung tumorigenesis (Balli et al., 2011, 2012);
- opposite effects on cancer-related inflammation in response to urethane or BHT treatment (Fig. 6.3) (Balli et al., 2011, 2012);
- opposite effects on the activity of the canonical Wnt/β-catenin signaling pathway (Fig. 6.3) (Balli et al., 2011; Wang et al., 2012; Zhang et al., 2011b).

4.2. Lung tumor induction by MCA/BHT as an example for context-dependent effects of FOXM1

Exactly opposite results were reported for the influence of FOXM1 on lung tumor formation and growth after lung tumor induction by MCA/BHT (Balli et al., 2011, 2012), which suggests context-dependent effects of FOXM1 on MCA/BHT-induced lung tumorigenesis:

On the one hand, deletion of *foxm1* from endothelial cells in *enFoxm1*$^{-/-}$ mice increased the number and size of lung tumors induced by MCA/BHT, indicating that FOXM1 expression in endothelial cells suppresses lung tumor formation and lung tumor growth after MCA/BHT treatment (Balli et al., 2011). In accordance with this negative effect of FOXM1 on MCA/BHT-induced lung tumorigenesis, the tumor cell proliferation was enhanced in MCA/BHT-treated *enFoxm1*$^{-/-}$ lungs (Balli et al., 2011).

On the other hand, deletion of *foxm1* from macrophages and granulocytes in *macFoxm1$^{-/-}$* mice decreased the number and size of lung tumors induced by MCA/BHT, indicating that FOXM1 expression in macrophages promotes lung tumor formation and lung tumor growth after MCA/BHT treatment (Balli et al., 2012). In accordance with this positive effect of FOXM1 on MCA/BHT-induced lung tumorigenesis, the tumor cell proliferation was reduced in MCA/BHT-treated *macFoxm1$^{-/-}$* lungs (Balli et al., 2012).

Thus, the influence of FOXM1 on MCA/BHT-induced lung tumorigenesis seems to be context-dependent so that FOXM1 can exert either a positive (Balli et al., 2012) or a negative (Balli et al., 2011) effect on lung tumor formation and lung tumor growth induced by MCA/BHT depending on the biological context.

Hence, in response to MCA/BHT lung injury, FOXM1 seems to work as a tumor suppressor in endothelial cells (Balli et al., 2011), but as an oncogene in macrophages (Balli et al., 2012).

4.3. BHT- and urethane-induced lung inflammation as an example for context-dependent effects of FOXM1

Exactly opposite results were reported for the influence of FOXM1 on the lung inflammation caused by BHT (Balli et al., 2012) and urethane (Balli et al., 2011), which suggests context-dependent effects of FOXM1 on pulmonary inflammation:

On the one hand, the endothelial cell-specific deletion of *foxm1* increased the urethane-induced lung inflammation in *enFoxm1$^{-/-}$* mice, including a higher infiltration of inflammatory cells in urethane-treated *enFoxm1$^{-/-}$* lungs and more inflammatory cells in the BAL (brochoalveolar lavage) from these lungs, which indicates that FOXM1 expression in endothelial cells impairs the urethane-induced lung inflammation (Balli et al., 2011).

On the other hand, the macrophage-specific deletion of *foxm1* decreased the BHT-induced lung inflammation in *macFoxm1$^{-/-}$* mice, including a lower infiltration of inflammatory cells in BHT-treated *macFoxm1$^{-/-}$* lungs and less inflammatory cells in the BAL from these lungs, which indicates that FOXM1 expression in macrophages favors the BHT-induced lung inflammation (Balli et al., 2012).

Thus, the influence of FOXM1 on BHT- and urethane-induced pulmonary inflammation seems to be context dependent so that FOXM1 can exert either a positive (Balli et al., 2012) or a negative (Balli et al., 2011) effect on the lung inflammation caused by BHT and urethane, respectively, depending on the biological context.

Hence, with regard to tumor-associated lung inflammation, FOXM1 behaves as an inhibitor of inflammation in endothelial cells (Balli et al., 2011), but as a promoter of inflammation in macrophages, where it plays a role for TAM recruitment to the lung (Balli et al., 2012).

4.4. Canonical Wnt/β-catenin signaling as an example for context-dependent effects of FOXM1

Exactly opposite results were reported for the influence of FOXM1 on the activity of the canonical Wnt/β-catenin signaling pathway (Balli et al., 2011; Wang et al., 2012; Zhang et al., 2011b), which suggests context-dependent effects of FOXM1 on Wnt/β-catenin signaling:

On the one hand, FOXM1 had a positive effect on the canonical Wnt/β-catenin signaling pathway in embryonic kidney cells (293T), NSCs (neural stem cells), GIC (GBM-initiating cell) cells (MD11, MD20s), glioma cells (SW1783), and embryonic fibroblasts because FOXM1 promoted the nuclear translocation of β-catenin in 293T, MD11, and MD20s cells, MEFs (mouse embryonic fibroblasts), and NSCs and because FOXM1 activated the β-catenin/TCF-4 reporter construct TOP-Flash (which is driven by TCF-4 binding sites) in 293T, SW1783, MD11, and MD20s cells (Zhang et al., 2011b).

On the other hand, a negative effect of FOXM1 on the canonical Wnt/β-catenin signaling pathway was reported in epithelial cells of urethane-treated adult mouse lungs (Balli et al., 2011), in the developing respiratory epithelium of mouse embryos, in lung epithelial cells (A549, MLE-12), and in osteosarcoma cells (U2OS) (Wang et al., 2012):

First, Wnt/β-catenin signaling was activated in the epithelial cells of $enFoxm1^{-/-}$ mouse lungs (with an endothelial cell-specific deletion of the foxm1 gene) after urethane treatment as evidenced by the increased nuclear localization of β-catenin and the increased activity of the β-catenin/TCF-4 reporter construct TOP-GAL (which is driven by TCF-4 binding sites), which indicates a negative effect of FOXM1 on the activity of the Wnt/β-catenin signaling pathway in the urethane-treated murine respiratory epithelium (Balli et al., 2011). This increased Wnt/β-catenin signaling in foxm1-deficient cells following urethane exposure results from the decreased sfrp1 mRNA expression in $enFoxm1^{-/-}$ lungs (Wang et al., 2012), because sFRP1 is a secreted Wnt inhibitor (Bovolenta et al., 2008; Jones and Jomary, 2002; Kawao and Kypta, 2003; Rubin et al., 2006).

Second, the conditional deletion of *foxm1* from the lung epithelium in *epFoxm1*$^{-/-}$ mouse embryos activated Wnt/β-catenin signaling as evidenced by the increased TOP-GAL activity (Wang et al., 2012), indicating a negative effect of FOXM1 on the activity of the Wnt/β-catenin signaling pathway in epithelial cells of developing murine lungs. This increased Wnt/β-catenin signaling in *foxm1*-deficient cells may result from the decreased Axin2 and JNK1 (c-Jun (JUN) N-terminal kinase *1*) mRNA and/or protein expression in the distal lung epithelium of *epFoxm1*$^{-/-}$ embryos (Wang et al., 2012), because Axin2 (Kikuchi, 1999a,b; Luo and Lin, 2004; Parveen et al., 2011; Salahshor and Woodgett, 2005) and JNK1 (Amura et al., 2005; Hu et al., 2008; Liao et al., 2006) are two known inhibitors of β-catenin (see below).

Third, the conditional expression of a ca form of FoxM1B (FoxM1B-ΔN lacking the 231 N-terminal aa) in the lung epithelium of *epFoxm1* transgenic mouse embryos inhibited Wnt/β-catenin signaling as evidenced by the decreased TOP-GAL activity (Wang et al., 2012).

Fourth, siRNA-mediated knockdown of FOXM1 in A549 and MLE-12 lung epithelial cells enhanced the nuclear localization of β-catenin, indicating a negative effect of FOXM1 on the Wnt/β-catenin signaling pathway (Wang et al., 2012).

Fifth, FoxM1B inhibited Wnt/β-catenin signaling in U2OS cells because FoxM1B repressed both the basal activity of TOP-Flash and the β-catenin-mediated transactivation of TOP-GAL (Wang et al., 2012).

Thus, the influence of FOXM1 on Wnt/β-catenin signaling seems to be context dependent so that FOXM1 can exert either a positive (Zhang et al., 2011b) or a negative (Balli et al., 2011; Wang et al., 2012) effect on the activity of the canonical Wnt/β-catenin signaling pathway depending on the biological context.

In particular, there seems to be a difference in the outcome of FOXM1 action depending on whether FOXM1 binds directly to β-catenin at Wnt target genes or whether FOXM1 binds to the promoter of a gene, which encodes an inhibitor of the canonical Wnt/β-catenin pathway (i.e., either the Wnt inhibitor sFRP1 or the β-catenin inhibitors Axin2 and JNK1). The direct binding of FoxM1B to β-catenin results in the activation of Wnt/β-catenin signaling because the binding of FoxM1B to β-catenin enhances the nuclear translocation of β-catenin, the interaction of β-catenin with TCF-4, and the transcriptional activity of β-catenin/TCF-4 complexes so that FoxM1B activates β-catenin/TCF-4 target genes in this DNA-binding-independent manner (Zhang et al., 2011b). In contrast, the binding of FOXM1 to the *sFRP1*, *Axin2*, and *JNK1* promoters results in the inhibition of Wnt/β-catenin signaling because FOXM1 activates the transcription of these three inhibitors of the canonical Wnt/β-catenin pathway (Balli et al., 2011; Wang et al., 2012).

However, the ability of FoxM1B to repress TOP-Flash and to inhibit the transactivation of TOP-GAL by β-catenin (Wang et al., 2012) suggests that the direct binding of FoxM1B to β-catenin at (artificial) Wnt target genes can also reduce the transcriptional activity of β-catenin/TCF-4 complexes, implying that also the outcome of the binding of FoxM1B to β-catenin depends on the cellular context.

The β-catenin inhibitors Axin2 and JNK1:

Axin2 is part of the cytoplasmic β-catenin destruction complex, which consists of APC (adenomatous polyposis coli), Axin (Axin1 or Axin2 (conductin)), GSK-3β (glycogen synthase kinase-3β), CK1α, PP2A, and WTX (Wilms tumor gene on the X chromosome) (Cadigan and Peifer, 2009; Clevers, 2006; Clevers and Nusse, 2012; Fu et al., 2011; Giles et al., 2003; Huang and He, 2008; Kimelman and Xu, 2006; Lustig and Behrens, 2003; MacDonald and He, 2012; Macdonald et al., 2007, 2009; Polakis, 2012; Price, 2006; Seidensticker and Behrens, 2000; Stamos and Weis, 2013; Tauriello and Maurice, 2010; Valenta et al., 2012; Verheyen and Gottardi, 2010; Willert and Jones, 2006; Wu and Pan, 2009; Xu and Kimelman, 2007).

JNK1 has been reported to inhibit β-catenin (Hu et al., 2008; Liao et al., 2006): First, JNK1 decreased the β-catenin protein expression because JNK1, which interacted with β-catenin and GSK-3β, promoted the GSK-3β-dependent proteasomal degradation of β-catenin (Hu et al., 2008). Second, JNK1 prevented the nuclear accumulation of β-catenin by stimulating its nuclear export (Liao et al., 2006). Third, JNK1 inhibited the transactivation by β-catenin (Hu et al., 2008; Liao et al., 2006). Fourth, JNK1 downregulated the *Wnt4* and *Wnt6* mRNA levels, which provides another mechanism for repression of Wnt/β-catenin signaling by JNK1 (Amura et al., 2005).

4.5. Control of the *c-myc* and *cyclin D1* genes as an example for context-dependent effects of FOXM1

The direct FOXM1 target genes *c-myc* (Hedge et al., 2011; Wierstra and Alves, 2006d; Zhang et al., 2011b) and *cyclin D1* (Tan et al., 2006) are two examples for antagonistic regulation of the same target gene by FOXM1 in dependence on the cellular context (Table 6.3).

Both *c-myc* and *cyclin D1* are direct target genes of the canonical Wnt/β-catenin signaling pathway (http://www.stanford.edu/~rnusse/wntwindow.html; He et al., 1998; Shtutman et al., 1999; Tetsu and McCormick, 1999), and the secreted Wnt inhibitor sFRP1 is known to repress the *c-myc* and *cyclin D1* mRNA expression (Cheng et al., 2011; Ezan et al., 2004; Gauger et al., 2009; Gumz et al., 2007; Hay et al., 2009; Hu et al., 2009; Jiang et al., 2010; Matsuda et al., 2009).

In general, the promoters of the direct FOXM1 target genes *c-myc* (Wierstra and Alves, 2006d, 2007a,b, 2008) and *cyclin D1* (Tan et al., 2006; Wang et al., 2001a,b) are transactivated by FOXM1 so that FOXM1 increases the *c-myc* and *cyclin D1* mRNA and protein expression directly (Fig. 6.3; Table 6.3) (Bao et al., 2011; Chan et al., 2008; Dai et al., 2010; Green et al., 2011; Li et al., 2008a; Liu et al., 2006; Liu et al., 2013b; Millour et al., 2010; Petrovic et al., 2008; Qu et al., 2013; Ren et al., 2010; Raghavan et al., 2012; Ustiyan et al., 2009; Wang et al., 2001a,b, 2007a, 2010c,e; Wu et al., 2010a; Xia et al., 2009; Xie et al., 2010; Xue et al., 2012; Ye et al., 1999; Zeng et al., 2009; Zhang et al., 2011b, 2012d).

However, there are two exceptional contexts, in which the *c-myc* and *cyclin D1* expression is indirectly decreased by FOXM1 (Table 6.3) because FOXM1 activates the expression of known inhibitors of the Wnt/β-catenin pathway (sFRP1, Axin2, JNK1) so that FOXM1 inhibits Wnt/β-catenin signaling and thereby reduces the transcription of the direct Wnt/β-catenin target genes *c-myc* and *cyclin D1* (Balli et al., 2011; Wang et al., 2012). Thus, in these two special contexts, the indirect FOXM1-mediated downregulation of *c-myc* and *cyclin D1* via the Wnt/β-catenin pathway overwhelms the normal direct upregulation of *c-myc* and *cyclin D1* by FOXM1 itself, resulting in a seeming repression of the *c-myc* and *cyclin D1* genes by FOXM1 (Balli et al., 2011; Wang et al., 2012).

Exceptional context 1:

Following lung tumor induction by urethane, the *c-myc* and *cyclin D1* mRNA expression was increased in the lungs of $enFoxm1^{-/-}$ mice with a endothelial cell-specific deletion of the *foxm1* gene compared to the lungs of control mice, which at first glance suggests a negative effect of FOXM1 on the transcription of *c-myc* and *cyclin D1* (Table 6.3) (BalLi et al., 2011). However, the increased *c-myc* and *cyclin D1* mRNA expression is no direct consequence of the *foxm1* gene deletion from endothelial cells, but instead results indirectly from the increased Wnt/β-catenin signaling in urethane-treated $enFoxm1^{-/-}$ lungs (BalLi et al., 2011) because c-*myc* and *cyclin D1* are direct target genes of the Wnt/β-catenin pathway (http://www.stanford.

edu/~rnusse/wntwindow.html; He et al., 1998; Shtutman et al., 1999; Tetsu and McCormick, 1999).

This increased Wnt/β-catenin signaling (as evidenced by an increase in the nuclear localization of β-catenin and an increase in the activity of the β-catenin/TCF-4 reporter construct TOP-GAL) is caused by the decreased mRNA expression of the secreted Wnt inhibitor sFRP1 in *enFoxm1*$^{-/-}$ lungs (BalLi et al., 2011), which is explained by *sfrp1* being a direct FOXM1 target gene that is activated by FOXM1 (Fig. 6.3; Table 6.3).

Consequently, the increased mRNA levels of the direct β-catenin target genes *c-myc* and *cyclin D1* in urethane-treated *enFoxm1*$^{-/-}$ mouse lungs can be attributed to the decreased *sfrp1* mRNA expression and to the resulting increased canonical Wnt signaling (BalLi et al., 2011).

Thus, following lung tumor induction by urethane, the direct activation of the *sfrp1* transcription by FOXM1 leads via the Wnt inhibitor sFRP1 and via the canonical Wnt/β-catenin signaling pathway to the indirect down-regulation of the *c-myc* and *cyclin D1* transcription by FOXM1 (Table 6.3) so that the apparent FOXM1-mediated repression of the *c-myc* and *cyclin D1* genes results from the function of FOXM1 as a transactivator of the *sfrp1* gene (BalLi et al., 2011).

Exceptional context 2:
siRNA-mediated knockdown of FOXM1 increased the *c-myc* and *cyclin D1* mRNA expression in A549 lung epithelial cells, which at first glance suggests a negative effect of FOXM1 on the transcription of *c-myc* and *cyclin D1* (Table 6.3) (Wang et al., 2012). However, the increased *c-myc* and *cyclin D1* mRNA expression is no direct consequence of the FOXM1 knock-down, but instead results indirectly from the increased Wnt/β-catenin signaling in the FOXM1-depleted A549 cells (Wang et al., 2012) because *c-myc* and *cyclin D1* are direct Wnt/β-catenin target genes (http://www.stanford.edu/~rnusse/wntwindow.html; He et al., 1998; Shtutman et al., 1999; Tetsu and McCormick, 1999).

This increased Wnt/β-catenin signaling (as evidenced by an increased nuclear localization of β-catenin and an increased amount of activated β-catenin (dephosphorylated at S-37 and T-41) in the nuclear fraction) is caused by the decreased mRNA and/or protein expression of the known β-catenin inhibitors Axin2 and JNK1 in FOXM1-depleted A549 cells (Wang et al., 2012), which is explained by *Axin2* and *JNK1* being direct FOXM1 target genes that are activated by FOXM1 (Fig. 6.3; Table 6.3).

Therefore, the increased mRNA levels of the direct β-catenin target genes *c-myc* and *cyclin D1* in FOXM1-depleted A549 cells can be attributed to the decreased Axin2 and JNK1 mRNA and/or protein expression and to the resulting increased canonical Wnt signaling (Wang et al., 2012).

Hence, in A549 cells, the direct activation of the *Axin2* and *JNK1* transcription by FOXM1 leads via the β-catenin inhibitors Axin2 and JNK1 and via the canonical Wnt/β-catenin signaling pathway to the indirect down-regulation of the *c-myc* and *cyclin D1* transcription by FOXM1 (Table 6.3) so that the apparent FOXM1-mediated repression of the *c-myc* and *cyclin D1* genes results from the function of FOXM1 as a transactivator of the *Axin2* and *JNK1* genes (Wang et al., 2012).

In summary, this example shows that a direct effect of FOXM1 on a target gene (i.e., activation of *c-myc* and *cyclin D1*) cannot manifest if it is canceled by an opposite indirect effect of FOXM1 on the same target gene, which is mediated by the protein product of another FOXM1 target gene (i.e., the Wnt inhibitor sFRP1 or the β-catenin inhibitors Axin2 and JNK1).

Such scenarios may also explain why contradictory effects of FOXM1 on the expression of other FOXM1 target genes (*p21, sox2, cyclin B1, E-cadherin, TCF-4, ERα, FOXA3, GATA3, IL-1β, IL-13, KRT18, MMP-9, vimentin, fibronectin, α-SMA, SP-B, SP-C*) were observed in different biological contexts (Table 6.3). Thus, FOXM1 appears to have not only general effects but also context-dependent effects, so that the outcome of FOXM1 action may sometimes depend on the specific biological context (Kalin et al., 2011).

5. FOXM1 AS TARGET FOR ANTICANCER THERAPY

5.1. Antitumor therapy and the attractive target FOXM1

Radiotherapy and classical chemotherapy (e.g., daunorubicin, doxorubicin, epirubicin, cisplatin, oxaliplatin, 5-FU (5-fluorouracil), TMZ (temozolomide)) eliminate cancer cells by causing irradiation-induced and drug-induced DNA damage, respectively, which leads to apoptosis (Brown and Attardi, 2005; Brown and Wouters, 1999; Chaney and Sancar, 1996; Coultas and Strasser, 2000; Espinosa et al., 2003; Havelka et al., 2007; Helleday, 2011; Helleday et al., 2008; Herr and Debatin, 2001; Kim et al., 2006c; Luo et al., 2009; Radford, 1999; Reed, 1999; Waxman and Schwartz, 2003; Xu et al., 2001). Another strategy for clinical tumor treatment is the inhibition of oncogenic RTKs by antibodies (e.g.,

Herceptin) and selective small-molecule compounds (e.g., imatinib, gefitinib, lapatinib, nilotinib, dasatinib, vemurafenib) (Arslan et al., 2006; Arteaga and Baselga, 2004; Baselga, 2006; Baselga and Swain, 2009; Bode and Dong, 2005; Dancey and Sausville, 2003; Dar and Shokat, 2011; Fischer et al., 2003; Grant, 2009; Gschwind et al., 2004; Guitierrez et al., 2009; Haber et al., 2011; Hannah, 2005; Imai and Takaoka, 2006; Krause and Van Etten, 2005; Levitzki, 2004; Logue and Morrison, 2012; Mosesson and Yarden, 2004; Noble et al., 2004; Pytel et al., 2009; Quintas-Cardama et al., 2007, 2009; Sawyers, 2003, 2004, 2005; Sebolt-Leupold and English, 2006; Seliger et al., 2010; Shawver et al., 2002; Stegmeier et al., 2010; Thompson, 2009; Traxler, 2003; Varmus et al., 2005; Zhang et al., 2009). An approach for the treatment of steroid hormone-dependent tumors is endocrine therapy with specific antagonists of the nuclear receptors (e.g., tamoxifen, ICI) (Bamadas et al., 2008; Briest and Wolff, 2007; Chawla et al., 2010; Goss et al., 2008; Higgins and Stearns, 2011; Lumachi et al., 2011; Weinshilboum, 2008). Also the proteasome inhibitor bortezomib is a promising anticancer drug (Adams, 2001, 2002, 2003, 2004a,b; Adams and Kauffman, 2004; Caravita et al., 2006; Chen et al., 2011b; Cusack, 2003; Cvek and Dvorak, 2011; Einsele, 2010; Ludwig et al., 2005; Mitsiades et al., 2005; Montagut et al., 2005, 2006; Mujtaba and Dou, 2011; Nencioni et al., 2007; Orlowski and Kuhn, 2008; Richardson and Anderson, 2003; Richardson and Mitsiades, 2005; Richardson et al., 2005, 2006; Roccaro et al., 2006a,b; Sterz et al., 2009; Voorhees, and Orlowski, 2006; Zvarski et al., 2007). Moreover, cancer immunotherapy exploits the immune system to eradicate tumor cells (Dillman, 2011; Dimberu and Leonhardt, 2011; Dougan and Dranoff, 2009; Freire and Osinaga, 2012; Lake and Robinson, 2005; Lesterhuis et al., 2011; Liu and Zeng, 2012; Mellman et al., 2011; Pandolfi et al., 2011; Pardoll, 2012; Rosenberg et al., 2004, 2008; Schietinger et al., 2008; Sharma et al., 2011; Shiao et al., 2011; Steer et al., 2010; Stewart and Smyth, 2011; Topalian et al., 2011; Vanneman and Dranoff, 2012; Weiner et al., 2012).

One major challenge in antitumor therapy (Pavet et al., 2011) is drug resistance because long-term treatment with anticancer drugs ultimately selects for the worst tumor cells, namely, those which have evolved drug resistance (Baguley, 2010; Bock and Lengauer, 2012; Borst, 2012; Chabner and Roberts, 2005; Daub et al., 2004; Engelman and Settleman, 2008; Garraway and Jänne, 2012; Gillies et al., 2012; Glickman and Sawyers, 2012; Hait and Habley, 2009; Igney and Krammer, 2002; Klein et al., 2005; Livingston and Silver, 2008; Longley and Johnston, 2005; Luqmani, 2005; Martinkova et al., 2009; Milane et al., 2011; Mimeault et al., 2008; O'Connor, 2009; O'Connor et al., 2007; Redmond et al., 2008; Sawyers, 2007, 2009; Sellers, 2011; Sharma et al., 2010; Shekhar, 2011; Singh and Settleman, 2010; Solyanik, 2010; Tan et al., 2010b; Tannock, 2001; Tsuruo et al., 2003; Varmus, 2006; Zhivotovsky and Orrenius, 2003). Therefore, it is important to understand how drug resistance develops and to figure out strategies to overcome this resistance, which is acquired upon prolonged treatment with an initially effective antitumor agent. One possibility to resensitize resistant tumor cells is the observed sensitivity of tumor cells to killing by the combination of two anticancer drugs in spite of the ineffectiveness of each drug as monotherapy (Chan and Giaccia, 2011; Ferrari et al., 2010; Kaelin, 2005, 2009; Mizuarai and Kotani, 2011; Qiao et al., 2010; Weidle et al., 2011).

The manifold implication of FOXM1 in tumorigenesis makes it an attractive target for anticancer therapy (Adami and Ye, 2007; Ahmad et al., 2012; Costa et al., 2005a; Gartel, 2008, 2010, 2012; Halasi and Gartel, 2013; Katoh et al., 2013; Koo et al., 2011; Li et al., 2012b; Myatt and Lam, 2007; Raychaudhuri and Park, 2011; Teh, 2012a,b; Wang et al., 2010a; Wierstra and Alves, 2007c; Wilson et al., 2011; Zhao and Lam, 2012). Accordingly, several trials to interfere with the expression or activity of FOXM1 have been published (see below).

5.2. Anticancer drugs targeting FOXM1
5.2.1 Thiostrepton
The thiazole antibiotic thiostrepton, a natural product isolated from *Streptomyces azureus*, *Streptomyces hawaiiensis*, and *Streptomyces laurentii*, represses the *foxm1* promoter (Kwok et al., 2008). Accordingly, thiostrepton reduces the endogenous *foxm1* mRNA (Hedge et al., 2011; Kwok et al., 2008; Pandit and Gartel, 2011b) and protein (Ahmed et al., 2012; Bellelli et al., 2012; Bhat et al., 2008, 2009a; Chan et al., 2012; Halasi et al., 2010; Hedge et al., 2011; Kwok et al., 2008, 2010; Lok et al., 2011; Newick et al., 2012; Pandit and Gartel, 2010, 2011b,c; Uddin et al., 2011, 2012; Wang and Gartel, 2011; Wang et al., 2013) expression.

Yet, thiostrepton did not decrease the FOXM1 protein level in thiostrepton-resistant untransformed normal MCF-10A breast epithelial cells, whereas it diminished the *foxm1* mRNA level in thiostrepton-sensitive MCF-7 breast cancer cells (Kwok et al., 2008).

Thiostrepton seems to inhibit the FOXM1 protein expression through a redox–dependent and proteasome–dependent mechanism because both the antioxidant NAC (N-acetyl-cysteine) and the proteasome inibitor MG132 prevented the downregulation of the FOXM1 protein level by thiostrepton (Newick et al., 2012).

Since thiostrepton decreased the FOXM1 protein level in OVCA433 ovarian cancer cells (p53 wild type), but not in SKOV3 ovarian cancer cells (p53 deleted), the transcription factor and key tumor suppressor p53 might be involved in the downregulation of the FOXM1 expression by thiostrepton (Lok et al., 2011).

The FOXM1 inhibitor thiostrepton interacts directly with FOXM1, namely, with the splice variants FOXM1c and FoxM1B (Hedge et al., 2011). Circular dichroism (CD) spectroscopy suggested that thiostrepton alters the conformation of FOXM1c through its partial unfolding (Hedge et al., 2011).

Thiostrepton inhibits the binding of FOXM1c and FoxM1B to DNA in vitro and reduces the in vivo occupancy of FOXM1 target genes by FOXM1 (Hedge et al., 2011). Accordingly, thiostrepton represses the transcriptional activity of exogenous FoxM1B (Bhat et al., 2009a) and decreases the mRNA expression of direct FOXM1 target genes (Hedge et al., 2011). Importantly, the negative effects of thiostrepton on the mRNA levels and the FOXM1 occupancy of FOXM1 target genes were already observed after 4 h of thiostrepton treatment when the FOXM1 protein level was still almost unchanged by thiostrepton, ensuring that these early thiostrepton effects are independent of the later thiostrepton–mediated downregulation of the FOXM1 expression (Hedge et al., 2011).

In human HCT-15 CRC cells, thiostrepton reduced the FOXM1-induced anchorage-independent growth on soft agar, the FOXM1-induced invasion and the FOXM1-induced migration, indicating that the FOXM1 inhibitor thiostrepton can suppress some oncogenic activities of FOXM1 in vivo (Uddin et al., 2011).

Thiostrepton induced cell death in TPC-1 PTC cells, and this thiostrepton–induced cell death was decreased by FOXM1 overexpression, but increased by siRNA against FOXM1 (Ahmed et al., 2012). Accordingly, thiostrepton induced the cleavage of caspase-3 and PARP (poly(ADP (adenosine diphosphate)-ribose)polymerase) in TPC-1 cells, and this thiostrepton-induced caspase-3 and PARP cleavage was decreased by FOXM1 overexpression, but increased by siRNA against FOXM1 (Ahmed et al., 2012).

Thiostrepton induced apoptosis in HCT-15 human colon adenocarcinoma cells, and the extent of thiostrepton-induced apoptosis was lower in FOXM1-overexpressing HCT-15 cells than in control HCT-15 cells (Uddin et al., 2011).

In U2OS osteosarcoma cells, thiostrepton induced the cleavage of caspase-3 and exogenous FoxM1B reduced this thiostrepton-induced caspase-3 cleavage, suggesting that FoxM1B might promote cell survival (Bhat et al., 2009a,b).

shRNA-mediated knockdown of FOXM1 sensitized MIA PaCa-2 pancreatic cancer cells, MDA-MB-231 breast cancer cells, and HCT-116 colon cancer cells to thiostrepton-induced apoptosis, because shRNA against FOXM1 increased the extent of the thiostrepton-induced caspase-3 cleavage (Pandit and Gartel, 2011a). However, this FOXM1 knockdown had no effect on apoptosis in the absence of thiostrepton because shRNA against FOXM1 did not cause caspase-3 cleavage in MIA PaCa-2, MDA-MB-231, or HCT-116 cells (Pandit and Gartel, 2011a).

The extent of caspase-3 cleavage induced by the combination of thiostrepton plus bortezomib was larger in $foxm1^{-/-}$ MDA-MB-231 breast cancer cells than in their wild-type counterparts, suggesting that FOXM1 counteracts thiostrepton/bortezomib-induced apoptosis (Pandit and Gartel, 2011c). Yet, this FOXM1 knockout does not cause apoptosis without thiostrepton/bortezomib treatment because no caspase-3 cleavage occurred in the $foxm1^{-/-}$ MDA-MB-231 cells (Pandit and Gartel, 2011c).

Together, these findings suggest that the thiostrepton-mediated repression of FOXM1 is involved in the induction of apoptosis by thiostrepton so that FOXM1 overexpression interferes with the thiostrepton-induced apoptosis of tumor cells, whereas loss of FOXM1 sensitizes them to thiostrepton-induced cell death. Accordingly, the negative effect of thiostrepton on the number of viable cells was antagonized by overexpression of wild-type FOXM1 in human HCT-15 CRC cells (Uddin et al., 2011) and by overexpression of a ca N-terminally truncated FOXM1 mutant (ΔN-FOXM1) in human MCF-7 breast carcinoma cells (Kwok et al., 2008).

5.2.2 Siomycin A

The thiazole antibiotic siomycin A, which differs only in 2 aa from thiostrepton (isoleucine–alanine *versus* valine-dehydroalanine), down-regulates the *foxm1* mRNA (Radhakrishnan et al., 2006) and protein (Bhat et al., 2008, 2009a; Halasi et al., 2010; Li et al., 2013; Pandit and Gartel, 2010; Priller et al., 2011; Radhakrishnan et al., 2006; Wan et al.,

2012) expression. Moreover, siomycin A represses the transcriptional activity of exogenous FoxM1B (Bhat et al., 2009a; Radhakrishnan et al., 2006).

Siomycin A reduced the *in vivo* occupancy of the *uPAR* (uPA (urokinase-type plasminogen activator) receptor) promoter by endogenous FOXM1 in human SW480 and SW620 colon cancer cells, but this siomycin A effect may simply result from the downregulation of the FOXM1 protein expression by siomycin A in these cells (Li et al., 2013).

Siomycin A reduced the FoxMB-induced anchorage-independent growth on soft agar of U2OS osteosarcoma cells, that is, the FOXM1 inhibitor siomycin A is capable of suppressing an oncogenic property of FOXM1 *in vivo* (Radhakrishnan et al., 2006).

In U2OS osteosarcoma cells, siomycin A induced the cleavage of caspase-3 and exogenous FoxM1B reduced this siomycin A-induced caspase-3 cleavage, suggesting that FOXM1 overexpression interferes with the induction of apoptosis by siomycin A because the siomycin A-induced apoptosis involves the siomycin A-mediated repression of FOXM1 (Bhat et al., 2009a,b).

5.2.3 Mithramycin
The antineoplastic antibiotic mithramycin decreased the *foxm1* mRNA expression in human HepG2 hepatoma cells (Petrovich et al., 2010).

5.2.4 Daunorubicin
The DNA-damaging chemotherapeutic drug daunorubicin, a natural anthracyclin isolated from *Streptomyces peucetius*, decreases the *foxm1* mRNA and protein expression in MCF-7 mammary carcinoma cells and HepG2 HCC cells (Barsotti and Prives, 2009).

This daunorubicin-induced *foxm1* downregulation depends on p53 and p21, because it was abolished by knockdown of either p53 or p21 (Barsotti and Prives, 2009): First, daunorubicin upregulated the *foxm1* mRNA and protein levels in MCF-7 cells after silencing of p53 by siRNA or shRNA (Barsotti and Prives, 2009). Second, daunorubicin upregulated the FOXM1 protein level in MCF-7 cells treated with siRNA against p21 and in HepG2 cells treated with siRNA against p53 or p21 (Barsotti and Prives, 2009). Third, the *foxm1* mRNA expression was unaffected by daunorubicin in MCF-7 cells treated with siRNA against p21 and in HepG2 cells treated with siRNA against p53 or p21 (Barsotti and Prives, 2009). Thus, the downregulation of the *foxm1* expression by daunorubicin depends on both p53 and p21 in MCF-7 and HepG2 cells (Barsotti and Prives, 2009).

In accordance with this dependence on p53 (Barsotti and Prives, 2009), p53 binds to (Kurinna et al., 2013) and represses (Chen et al., 2013b; Millour et al., 2011) the *foxm1* promoter so that p53 reduces the *foxm1* mRNA and protein expression (Fig. 6.2; Table 6.2) (Barsotti and Prives, 2009; Bellelli et al., 2012; Bhonde et al., 2006; Chen et al., 2013b; Lok et al., 2011; Millour et al., 2011; Pandit et al., 2009; Qu et al., 2013; Rovillain et al., 2011; Spurgers et al., 2006). In accordance with this dependence on p21 (Barsotti and Prives, 2009), p21 diminishes the *foxm1* mRNA and protein expression (Fig. 6.2) (Barsotti and Prives, 2009; Bellelli et al., 2012; Millour et al., 2011; Stoyanova et al., 2012; Weymann et al., 2009). *p21* is a direct p53 target gene (El-Deiry et al., 1993), but p53 can repress the *foxm1* mRNA expression independently of p21 in MCF-7 cells (Barsotti and Prives, 2009) in accordance with *foxm1* being a direct target gene of p53 (Kurinna et al., 2013). This leaves open the possibilities that p21 might act either downstream of p53 or in parallel with p53 in the repression of the *foxm1* expression by daunorubicin.

Remarkably, the effect of daunorubicin on the FOXM1 expression depends not only on the p53 status and the p21 status of the cell, but also on the cell type, because daunorubicin upregulated the FOXM1 protein level and did not affect the *foxm1* mRNA level in U2OS osteosarcoma cells and HCT116 CRC cells despite the presence of p53 and p21 in both cell types (Barsotti and Prives, 2009).

In MCF-7 cells, pocket proteins seem to play a role in the repression of the *foxm1* mRNA expression by daunorubicin, because siRNA directed against the three pocket proteins (RB, p107, p130 (RBL2 (retinoblastoma-like 2))) partially abrogated the daunorubicin-induced downregulation of the *foxm1* mRNA level (Barsotti and Prives, 2009).

5.2.5 Doxorubicin

The anthracyclin antibiotic doxorubicin (Adriamycin, Rubex), a DNA-damaging chemotherapeutic drug, decreased the *foxm1* mRNA and protein expression in human MCF-7 breast cancer cells, but it increased the *foxm1* mRNA and protein expression after silencing of p53 by shRNA (Pandit et al., 2009). Similarly, in human HCT116 colon cancer cells, the FOXM1 protein level was downregulated by doxorubicin, but left unchanged after shRNA-mediated knockdown of p53 (Halasi and Gartel, 2012; Pandit et al., 2009). Hence, the repression of the FOXM1 expression by doxorubicin depends on p53 (Halasi and Gartel, 2012). Accordingly, doxorubicin upregulated the FOXM1 protein level in three p53-deficient human cancer cell lines, namely, in Hep3B liver cancer cells (p53-null), MDA-MB-231 breast cancer cells (p53 mutant), and MIA PaCa-2 pancreatic cancer cells (p53 mutant) (Halasi and Gartel, 2012).

Doxorubicin treatment causes the cleavage of caspase-3 and PARP-1/2 in HCT116 and MIA PaCa-2 cells (Halasi and Gartel, 2012). Silencing of FOXM1 by shRNA enhanced the doxorubicin-induced cleavage of caspase-3 and PARP-1/2, indicating that FOXM1 deficiency sensitizes these cells to doxorubicin-induced apoptosis (Halasi and Gartel, 2012).

Doxorubicin causes caspase-3 cleavage in Hep3B and MDA-MB-231 cells, too (Halasi and Gartel, 2012). Again, shRNA-mediated depletion of FOXM1 augmented the doxorubicin-induced caspase-3 cleavage, showing that FOXM1 deficiency also sensitizes these cells to doxorubicin-induced apoptosis (Halasi and Gartel, 2012).

However, in all four cell types, FOXM1 had no effect on cellular survial in the absence of daunorubicin-induced DNA damage, because the FOXM1 knockdown did not affect the cleavage of caspase-3 and/or PARP-1/2 without doxorubicin treatment (Halasi and Gartel, 2012).

5.2.6 Epirubicin
5.2.6.1 Downregulation of FOXM1 by epirubicin
The anthracyclin epirubicin, a DNA-damaging anticancer drug, represses the *foxm1* promoter in human MCF-7 breast carcinoma cells (Millour et al., 2011). Accordingly, epirubicin decreases the *foxm1* mRNA and protein expression (Millour et al., 2011; Monteiro et al., 2012).

This epirubicin-induced *foxm1* downregulation depends on p53, but not on p21, because it was abolished by p53 deficiency, but not by p21 deficiency (Millour et al., 2011): First, epirubicin reduces the *foxm1* mRNA and protein expression in MCF-7 control cells (Millour et al., 2011; Monteiro et al., 2012) and if p21 is knocked down with siRNA, but not if p53 is silenced with siRNA (Millour et al., 2011). Second, epirubicin decreases the *foxm1* mRNA level in wild-type MEFs and in $p21^{-/-}$ MEFs, but not in $p53^{-/-}$ MEFs (Millour et al., 2011). Third, the *foxm1* mRNA and protein expression is diminished by epirubicin in parental p53-positive MCF-7 cells, but unaffected in epirubicin-resistant p53-negative MCF-7-EPI^R cells (Millour et al., 2011; Monteiro et al., 2012) and p53-negative MDA-MB-453 breast cancer cells (Millour et al., 2011). Thus, epirubicin downregulates the *foxm1* expression (Millour et al., 2011; Monteiro et al., 2012) dependent on p53, but independent of p21 (Millour et al., 2011).

Yet, the effect of epirubicin on the FOXM1 expression depends not only on the p53 status of the cell but also on the cell type, because the FOXM1 protein level was unaffected by epirubicin in p53-positive U2OS osteosarcoma cells (Millour et al., 2011).

Epirubicin represses the *foxm1* promoter (Millour et al., 2011) and decreases the *foxm1* mRNA and protein expression in parental MCF-7 cells (Millour et al., 2011; Monteiro et al., 2012). In contrast, the *foxm1* mRNA and protein levels are unaffected by epirubicin in epirubicin-resistant MCF-7-EPI^R cells (Millour et al., 2011; Monteiro et al., 2012). The mechanism by which *foxm1* acquires this epirubicin resistance in MCF-7-EPI^R cells might include E2F-1 and RB (at the *foxm1* promoter level) as well as ATM (ataxia telangiectasia mutated) (at the *foxm1* transcription level) (Millour et al., 2011).

In contrast to parental MCF-7 cells, epirubicin-resistant MCF-7-EPI^R cells fail to cease proliferation and to die by apoptosis in response to epirubicin treatment (Millour et al., 2011).

Strikingly, siRNA-mediated depletion of FOXM1 restored at least in part the ability of epirubicin to induce apoptosis and to inhibit cell proliferation in epirubicin-resistant MCF-7-EPI^R cells, indicating that loss of FOXM1 resensitizes MCF-7-EPI^R cells to epirubicin-induced apoptosis and epirubicin-mediated cell growth suppression whereas FOXM1 over-expression seems to interfere with the induction of cell death and the inhibition of proliferation by epirubicin (Millour et al., 2011; Monteiro et al., 2012).

In fact, overexpression of FOXM1 counteracted the epirubicin-induced decrease in the number of viable parental MCF-7 cells, which points to an involvement of FOXM1 in the epirubicin resistance of MCF-7-EPI^R cells (Monteiro et al., 2012).

Yet, in the absence of epirubicin, FOXM1 knockdown by siRNA did not enhance apoptosis in MCF-7-EPI^R cells (Millour et al., 2011; Monteiro et al., 2012) although it could mimic the antiproliferative effect of epirubicin on these cells (Millour et al., 2011).

In accordance with the contribution of FOXM1 to epirubicin resistance, *foxm1*$^{-/-}$ MEFs are more sensitive to epirubicin treatment than wild-type MEFs because epirubicin reduced the cell viability of *foxm1*$^{-/-}$ MEFs more dramatically than that of wild-type MEFs (Monteiro et al., 2012).

Anthracyclin-based chemotherapeutics as epirubicin induce DSBs (Gewirtz, 1999; Kizek et al., 2012; Minotti et al., 2004; Müller et al., 1998; Perego et al., 2001; Rabbani et al., 2005; Taatjes et al., 1998). γH2AX (phosphorylated H2AFX (H2A histone family, member X)) foci are indicative of DSBs (Bekker-Jensen and Mailand, 2010; Fillingham et al., 2006; Firsanov et al., 2011; Greenberg, 2011; Kinner et al., 2008; Lukas et al., 2011; Miller and Jackson, 2012; Pilch et al., 2003; Srivastava et al., 2009).

On epirubicin treatment, epirubicin-resistant MCF-7-EPI^R cells accumulate far fewer γH2AX foci than parental MCF-7 cells (Monteiro et al., 2012). However, siRNA-mediated depletion of FOXM1 led to the accumulation of considerably higher numbers of γH2AX foci in epirubicin-treated MCF-7-EPI^R cells, which indicates that FOXM1 knockdown is sufficient to resensitize epirubicin-resistant MCF-7-EPI^R cells to induction of DSBs by epirubicin (Monteiro et al., 2012). Conversely, overexpression of FOXM1 in MCF-7 cells drastically reduced the level of epirubicin-induced DNA damage (measured as lower tail moments in alkaline comet assays), indicating that FOXM1 overexpression can render parental MCF-7 cells resistant to induction of DNA damage by epirubicin (Monteiro et al., 2012).

This ability of FOXM1 to protect cells against epirubicin-induced DSBs was confirmed with MEFs (Monteiro et al., 2012): First, *foxm1$^{-/-}$* MEFs sustained higher levels of DSBs in response to epirubicin than wild-type MEFs, which was evidenced by the formation of more γH2AX foci in epirubicin-treated *foxm1$^{-/-}$* MEFs and by their longer comet tails in alkaline comet assays (Monteiro et al., 2012). Thus, FOXM1 deficiency results in hypersensitivy to induction of DSBs by epirubicin (Monteiro et al., 2012). Second, reintroduction of exogenous FOXM1 into the hypersensitive *foxm1$^{-/-}$* MEFs abolished the accumulation of γH2AX foci on epirubicin treatment, indicating that FOXM1 reexpression can confer resistance to epirubicin-induced DNA damage (Monteiro et al., 2012).

In summary, FOXM1 overexpression is able to render cells resistant to induction of DNA damage by epirubicin so that epirubicin-resistant MCF-7-EPI^R cells can be resensitized to epirubicin-induced DSBs by silencing of FOXM1 with siRNA (Monteiro et al., 2012).

Time course experiments revealed that the number of γH2AX foci differed between *foxm1$^{-/-}$* MEFs and wild-type MEFs only after prolonged epirubicin treatment (4 and 24 h), but not at earlier time points (30 min and 2 h) (Monteiro et al., 2012). This finding suggests that the epirubicin-hypersensitive *foxm1$^{-/-}$* MEFs are not more susceptible to induction of DNA damage by epirubicin, but instead less effective in repairing the epirubicin-induced DSBs, which results in the observed progressive accumulation of more DSBs in *foxm1$^{-/-}$* MEFs on long-term epirubicin exposure (Monteiro et al., 2012). Thus, FOXM1 deficiency impairs DSB repair

(Monteiro et al., 2012), which might be attributed to the known role of FOXM1 in the HR repair of DSBs (Monteiro et al., 2012; Park et al., 2012).

5.2.6.2 Upregulation of FOXM1 by epirubicin

Confusingly, epirubicin was also reported to upregulate the FOXM1 protein level in human U2OS osteosarcoma cells, MCF-7 breast carcinoma cells, and MEFs (de Olano et al., 2012):

First, epirubicin induced FOXM1 protein expression via the pathway epirubicin $\rightarrow$ p38 $\rightarrow$ MK2 $\dashv$ JNK1/2 $\dashv$ FOXM1 (Fig. 6.2) (de Olano et al., 2012).

Second, epirubicin induced FOXM1 protein expression via the pathway epirubicin $\rightarrow$ p38 $\rightarrow$ MK2 $\rightarrow$ E2F-1 $\rightarrow$ FOXM1 (Fig. 6.2) (de Olano et al., 2012). *foxm1* is a known direct E2F-1 target gene (Chen et al., 2013b; Millour et al., 2011; Wang et al., 2007b, 2008c). The transactivation of the *foxm1* promoter by E2F-1 was repressed by dn (dominant-negative) MK2 (MAPKAPK2 (MAPK-activated protein kinase 2)) and by the p38 inhibitor SB203580, which points to positive effects of MK2 and p38 on the E2F-1-mediated transactivation of the *foxm1* promoter (de Olano et al., 2012). Since the serine/threonine kinase MK2 phosphorylates E2F-1 at S-364 (de Olano et al., 2012) and since MK2 is a known substrate of the serine/threonine kinase p38 (Cargnello and Roux, 2011; Chen et al., 2001; Gaestel, 2006; Keshet and Seger, 2010; Krishna and Narang, 2008; Plotnikov et al., 2011; Roux and Blenis, 2004; Wagner and Nebreda, 2009), epirubicin may increase the FOXM1 protein expression through the pathway epirubicin $\rightarrow$ p38 $\rightarrow$ MK2 $\rightarrow$ E2F-1 $\rightarrow$ *foxm1* promoter (Fig. 6.2) (de Olano et al., 2012).

In the surprising study of de Olano et al. (2012), the FOXM1 protein expression was increased by epirubicin in epirubicin-sensitive MCF-7 cells, but not in epirubicin-resistant MCF-7-EPI[R] cells, which constitutively expressed FOXM1 at a high level.

5.2.7 Cisplatin

The platinum-based chemotherapeutic cisplatin, a DNA-alkylating agent, diminishes the *foxm1* mRNA and protein expression in parental MCF-7 breast cancer cells, but it raises the *foxm1* mRNA and protein levels in cisplatin-resistant MCF-7-CIS[R] cells (Kwok et al., 2010).

Cisplatin induced apoptosis in MCF-7 cells, and overexpression of a ca N-terminally truncated FOXM1 mutant (ΔN-FOXM1) prevented this cisplatin-induced apoptosis (Kwok et al., 2010). The overexpression of

ΔN-FOXM1 also canceled the antiproliferative effect of cisplatin on MCF-7 cells (Kwok et al., 2010). These findings suggest that FOXM1 is implicated in the induction of apoptosis and the inhibition of proliferation by cisplatin so that FOXM1 overexpression prevents cisplatin-induced cell death and cisplatin-mediated cell growth suppression (Kwok et al., 2010).

As mentioned above, cisplatin downregulates the *foxm1* expression in parental MCF-7 cells, but upregulates the *foxm1* expression in cisplatin-resistant MCF-7-CIS^R cells (Kwok et al., 2010).

In contrast to the parental MCF-7 cells, cisplatin fails to reduce the number of viable cisplatin-resistant MCF-7-CIS^R cells (Kwok et al., 2010). Since siRNA-mediated knockdown of FOXM1 restored the negative effect of cisplatin on the viability of MCF-7-CIS^R cells, FOXM1 deficiency seems to resensitize cisplatin-resistant MCF-7-CIS^R cells to cisplatin.

Cisplatin causes the formation of DNA adducts, which can lead to DSBs if DNA repair fails (Nowosielska and Marinus, 2005; Siddik, 2003).

On cisplatin treatment, parental MCF-7 cells accumulate high numbers of γH2AX foci whereas cisplatin-resistant MCF-7-CIS^R do not (Kwok et al., 2010). SiRNA-mediated depletion of FOXM1 increased the level of cisplatin-induced DSBs in MCF-7-CIS^R cells, indicating that FOXM1 knockdown is able to resensitize cisplatin-resistant MCF-7-CIS^R cells to induction of DSBs by cisplatin (Kwok et al., 2010). Conversely, over-expression of a ca FOXM1 mutant in parental MCF-7 cells abolished the formation of γH2AX foci in response to cisplatin, which indicates that ca FOXM1 overexpression can confer resistance to induction of DNA damage by cisplatin (Kwok et al., 2010). Thus, overexpression of ca FOXM1 is suf-ficient to render parental MCF-7 cells resistant to induction of DSBs by cis-platin (Kwok et al., 2010). Accordingly, cisplatin-resistant MCF-7-CIS^R cells can be resensitized to cisplatin-induced DSBs by silencing of FOXM1 with siRNA (Kwok et al., 2010).

This ability of FOXM to protect cells against cisplatin-induced DSBs (Kwok et al., 2010) might be attributed to its known role in the HR repair of DSBs (Monteiro et al., 2012; Park et al., 2012).

A549/DDP cells are the cisplatin-resistant counterparts of human A549 lung adenocarcinoma cells, so that cisplatin decreased the number of viable A549 cells dramatically, but the number of viable A549/DDP cells only modestly (Wang et al., 2013). Since the cisplatin-resistant A549/DDP cells expressed significantly higher FOXM1 mRNA and protein levels than the parental cisplatin-sensitive A549 cells, FOXM1 overexpression might contribute to their cisplatin resistance (Wang et al., 2013). Accordingly, siRNA-mediated silencing of FOXM1 enhanced the negative effect of cisplatin on the viability of both A549 and A549/DDP cells, suggesting that FOXM1 depletion might partially restore chemosensitivity to cisplatin (Wang et al., 2013).

In accordance with the implication of FOXM1 in the anticancer effect of cisplatin (Kwok et al., 2010; Wang et al., 2013), a high FOXM1 expression level was correlated with a poor clinical response to cisplatin-based chemotherapy in advanced NSCLC, because the response rate to cisplatin-based combination chemotherapy (combined with either docetaxel or gemcitabine or vinorelbine) was 52.2% for patients with FOXM1-negative tumors, but only 30.2% for patients with FOXM1-positive tumors (Wang et al., 2013).

5.2.8 Oxaliplatin

The DNA-alkylating agent oxaliplatin (Eloxatin), a third-generation platinum-derived chemotherapeutic drug, decreases the *foxm1* mRNA and protein expression in HepG2 and SMMC-7721 HCC cells (Qu et al., 2013).

This downregulation of the FOXM1 mRNA and protein levels by oxaliplatin depends on p53 because it was abolished by siRNA-mediated knockdown of p53 (Qu et al., 2013). Accordingly, p53 binds to (Kurinna et al., 2013) and represses (Chen et al., 2013b; Millour et al., 2011) the *foxm1* promoter so that p53 reduces the *foxm1* mRNA and protein expression (Barsotti and Prives, 2009; Bellelli et al., 2012; Bhonde et al., 2006; Chen et al., 2013b; Lok et al., 2011; Millour et al., 2011; Pandit et al., 2009; Qu et al., 2013; Rovillain et al., 2011; Spurgers et al., 2006).

In contrast, the oxaliplatin-mediated decline of the FOXM1 mRNA and protein levels is independent of p21 because it was unaffected by siRNA against p21 (Qu et al., 2013).

Oxaliplatin induced senescence in HepG2 and SMMC-7721 HCC cells and overexpression of FOXM1 impaired this oxaliplatin-induced senescence, suggesting that FOXM1 is implicated in the induction of senescence by oxaliplatin so that FOXM1 overexpression prevents oxaliplatin-induced senescence (Qu et al., 2013).

Accordingly, overexpression of FOXM1 in HCC hampered the therapeutic response to oxaliplatin treatment because among oxaliplatin-treated advanced HCC patients the patients with a high FOXM1 expression level exhibited shorter overall survival times and earlier tumor recurrence than the patients with a low FOXM1 expression level (Qu et al., 2013).

5.2.9 Paclitaxel

The microtubule-stabilizing anticancer drug Paclitaxel (Taxol), a taxane isolated from the bark of the Pacific yew tree (*Taxus brevifolia*), modestly reduced the FOXM1 protein level in MEFs (Laoukili et al., 2005).

Overexpression of FOXM1 counteracted the Paclitaxel-induced decrease in the number of viable SKBR3, MDA-MB-453, and BT474 mammary tumor cells (Carr et al., 2010). FOXM1 overexpression also antagonized that decline in the number of viable SKBR3, MDA-MB-453, and BT474 cells, which was induced by treatment with a combination of Paclitaxel and Herceptin (Trastuzumab) (Carr et al., 2010). These findings point to an involvement of FOXM1 in the negative effect of Paclitaxel on cell viability (Carr et al., 2010).

5.2.10 Docetaxel

The taxane docetaxel (taxotere), a microtubule-stabilizing anticancer drug, diminished the *foxm1* mRNA (Ahmad et al., 2011; Li et al., 2005a,b) and protein expression (Ahmad et al., 2011; Wang et al., 2010d) in human LNCaP, PC-3, and/or C4-2B prostate cancer cells and in human MCF-7, SKBR3, MDA-MB-231, and/or MDA-MB-468 breast cancer cells.

In MCF-7 and SKBR3 breast cancer cells (Ahmad et al., 2011) and in MKn45 gastric cancer cells (Okada et al., 2013), FOXM1 overexpression alleviated the docetaxel-induced decline of the number of viable cells, which points to an implication of FOXM1 in the negative effect of docetaxel on cell viability. Conversely, siRNA-mediated silencing of FOXM1 exacerbated the negative effect of docetaxel on the percentage of viable MKn7 gastric cancer cells, which otherwise respond poorly to docetaxel and which overexpress FOXM1 (Okada et al., 2013). Thus, FOXM1 overexpression contributes to the chemoresistance to docetaxel (Ahmad et al., 2011; Okada et al., 2013) so that FOXM1 knockdown recovers the chemosensitivity for docetaxel (Okada et al., 2013).

In accordance with the implication of FOXM1 in the anticancer effects of both docetaxel (Ahmad et al., 2011; Okada et al., 2013) and cisplatin (Kwok et al., 2010; Wang et al., 2013), the FOXM1 expression level is inversely correlated with the clinical response to DFP combination chemotherapy (with docetaxel, cisplatin, and 5-FU) in advanced gastric cancer, because the response rate to DFP therapy was 80% for patients with FOXM1-negative tumors, but only 35% for patients with FOXM1-positive tumors (Okada et al., 2013).

5.2.11 Tamoxifen

The SERM (selective estrogen receptor modulator) tamoxifen (4-OHT, 4-hydroxytamoxifen) diminishes the *foxm1* mRNA and protein expression in ER-positive (MCF-7, ZR-75-1), but not in ER-negative (MDA-MB-231) human breast carcinoma cells (Millour et al., 2010). This selectivity of the ER antagonist 4-OHT is expected because *foxm1* is a direct target gene of ERα, which transactivates the *foxm1* promoter and elevates the *foxm1* mRNA and protein expression (Millour et al., 2010).

The anti-estrogen 4-OHT prevents the transactivation of the *foxm1* promoter by ERα (Millour et al., 2010). This 4-OHT-mediated repression of ERα may involve HDAC1 (histone deacetylase 1) and HDAC2, two known ERα corepressors (Glass and Rosenfeld, 2000; Gurevich et al., 2007; Jepsen and Rosenfeld, 2002; Lonard and O'Malley, 2007; McKenna and O'Malley, 2002; McKenna et al., 1999; Perissi and Rosenfeld, 2005; Rochette-Egly, 2005; Rosenfeld and Glass, 2001; Torchia et al., 1998), because 4-OHT treatment of MCF-7 cells increased the *in vivo* occupancy of the *foxm1* promoter by HDAC1 and HDAC2 whereas the binding of ERα to the *foxm1* promoter remained unchanged (Millour et al., 2010). Accordingly, 4-OHT decreased the acetylation of the histones H3 and H4 at the *foxm1* promoter in MCF-7 cells (Millour et al., 2010).

In ZR-75-1 cells, the binding of ERα to the *foxm1* promoter was enhanced by 4-OHT (Millour et al., 2010).

In MCF-7 cells, 4-OHT causes a G1-phase cell cycle arrest and over-expression of a ca N-terminally truncated FOXM1 mutant prevents this 4-OHT-induced G1-arrest, which suggests that FOXM1 is implicated in the 4-OHT-induced cell cycle arrest in G1-phase so that ca FOXM1 over-expression interferes with the antiproliferative effect of 4-OHT (Millour et al., 2010). However, MCF-7 cells overexpressing full-length FOXM1 underwent the G1-phase arrest in response to 4-OHT treatment (although at slightly slower kinetics) (Millour et al., 2010), indicating that the mere overexpression of intact FOXM1 is insufficient to confer resistance to the antiproliferative effect of 4-OHT.

4-OHT downregulates the *foxm1* mRNA and protein expression in parental MCF-7 cells, but not in tamoxifen-resistant MCF-7-TAM^R cells (Millour et al., 2010).

In contrast to parental MCF-7 cells, tamoxifen-resistant MCF-7-TAM^R fail to cease proliferation (Lykkesfeldt et al., 1994) and to arrest in G1-phase (Millour et al., 2010) after 4-OHT addition.

Estradiol (E2) stimulates the proliferation of tamoxifen-resistant MCF-7-TAM^R cells, which exhibit only a limited reduction of their proliferation rate in response to 4-OHT treatment. siRNA-mediated knockdown of FOXM1 not only decreases the ability of estradiol to induce proliferation of MCF-7-TAM^R cells, but it could also be combined with 4-OHT to cause further decreases in the rate of MCF-7-TAM^R proliferation. Thus, FOXM1 deficiency resensitizes tamoxifen-resistant MCF-7-TAM^R cells to 4-OHT and diminishes their responsiveness to estradiol stimulation (Millour et al., 2010).

5.2.12 ICI

The SERM ICI (ICI182780, fulvestrant, Faslodex) reduces the *foxm1* mRNA and protein expression in ER-positive MCF-7 and ZR-75-1 cells, but not in ER-negative MDA-MB-231 cells (Millour et al., 2010). Again, this selectivity of the pure ER antagonist ICI is expected because *foxm1* is a direct ERα target gene (Millour et al., 2010). In fact, the repression of the *foxm1* expression by ICI involves ERα because ICI reduces the binding of ERα to the *foxm1* promoter in MCF-7 and ZR-75-1 cells (Millour et al., 2010).

5.2.13 Herceptin

Herceptin (Trastuzumab) is a humanized monoclonal antibody directed against the RTK HER2/Neu (Bartsch et al., 2007; Emens, 2005; Hall and Cameron, 2009; Hudis, 2007; Leyland-Jones, 2002; Mariani et al., 2009; Moasser, 2007; Nahta and Esteva, 2006, 2007; Nihira, 2003; Perez et al., 2009; Shepard et al., 2008; Tokunaga et al., 2006; Valabrega et al., 2007).

Herceptin downregulated the FOXM1 protein level in SKBR3, BT474, and MDA-MB-453 mammary tumor cells (Carr et al., 2010). Accordingly, the Herceptin target HER2 activated the *foxm1* promoter and increased the FOXM1 protein expression in SKBR3 and BT474 cells whereas siRNA directed against HER2 decreased the FOXM1 protein level in SKBR3, BT474, and MDA-MB-453 cells (Francis et al., 2009).

The overexpression of FOXM1 counteracted the Herceptin-induced decrease in the number of viable SKBR3, BT474, and MDA-MB-453 cells, which points to an involvement of FOXM1 in the negative effect of Herceptin on cell viability (Carr et al., 2010).

Herceptin induces a G1-phase cell cycle arrest in SKBR3, BT474, and MDA-MB-453 cells (Carr et al., 2010). This Herceptin-induced G1-arrest is mostly antagonized by FOXM1 overexpression because overexpression of FOXM1 prevented the Herceptin-induced increase in the G1-phase cell population (SKBR3, BT474 and MDA-MB-453 cells) and counteracted the Herceptin-induced decrease in the number of BrdU (bromodeoxyuridine)-positive S-phase cells (SKBR3 cells) (Carr et al., 2010). Thus, FOXM1 overexpression overcomes the Herceptin-induced cell cycle arrest in G1-phase, indicating that the overexpression of FOXM1 is capable of rendering tumor cells resistant to the antiproliferative effect of Herceptin treatment (Carr et al., 2010).

Herceptin reduces the FOXM1 protein level in parental BT474 cells, but not in Herceptin-resistant BT474 cells (Carr et al., 2010). Moreover, Herceptin-resistant BT474, SKBR3, and MDA-MB-453 cells display a higher *foxm1* mRNA and protein expression than the parental BT474, SKBR3, and MDA-MB-453 cells (Carr et al., 2010).

In contrast to parental SKBR3 and MDA-MB-453 cells, Herceptin fails to decrease the number of viable Herceptin-resistant SKBR3 and MDA-MB-453 cells (Carr et al., 2010). Since silencing of FOXM1 by siRNA completely restored the strong negative effect of Herceptin on the viability of Herceptin-resistant SKBR3 and MDA-MB-453 cells, FOXM1 deficiency is sufficient to resensitize Herceptin-resistant SKBR3 and MDA-MB-453 cells to Herceptin (Carr et al., 2010).

5.2.14 Gefitinib

Gefitinib (Iressa, ZD1839) is a selective EGFR tyrosine kinase inhibitor (Averbuch, 2003; Blagosklonny, 2004; Cappuzzo et al., 2006; Ciardiello, 2005; Harris, 2010; Herbst and Kies, 2003; Herbst et al., 2004; Hynes and Lane, 2005; Rowinsky, 2004; Sebastian et al., 2006; Siegel-Lakhai et al., 2005; Wakeling, 2002; Yano et al., 2003).

Gefitinib represses the *foxm1* promoter and downregulates the *foxm1* mRNA and protein expression in gefitinib-sensitive BT474 and SKBR3 human breast carcinoma cells, but not in gefitinib-resistant MDA-MB-453 human breast carcinoma cells (McGovern et al., 2009). Accordingly, the FOXM1 protein expression is also unaffected by gefitinib in gefitinib-resistant MCF-7 and MDA-MB-231 human breast carcinoma cells (McGovern et al., 2009).

Similarly, the *foxm1* mRNA and protein levels are decreased by gefitinib in gefitinib-sensitive NCI-H292 human lung mucoepidermoid carcinoma cells, but they are increased by gefitinib in gefitinib-resistant SPC-A-1 human lung adenocarcinoma (NSCLC) cells (Xu et al., 2012a).

In accordance with the *foxm1* downregulation by gefitinib, the gefitinib target EGFR upregulated the FOXM1 protein level in SKBR3 cells (McGovern et al., 2009).

In SKBR3 cells, overexpression of FOXM1c reduced the level of apoptosis induced by gefitinib (McGovern et al., 2009).

Likewise, FOXM1 overexpression alleviated the gefitinib-induced apoptosis in NCI-H292 cells (but it had no effect on the rate of apoptosis in the absence of gefitinib) (Xu et al., 2012a). Accordingly, the negative effect of gefitinib on the number of viable NCI-H292 cells was antagonized by overexpression of FOXM1 (Xu et al., 2012a).

Conversely, siRNA against FOXM1 exacerbated the gefitinib-induced apoptosis in SPC-A-1 cells (but it had no effect on the rate of apoptosis in the absence of gefitinib) (Xu et al., 2012a).

These findings suggest an involvement of FOXM1 in the induction of apoptosis by gefitinib so that FOXM1 overexpression interferes with the gefitinib-induced apoptosis of tumor cells whereas depletion of FOXM1 sensitizes them to gefitinib-induced cell death.

As mentioned above, gefitinib downregulated *foxm1* mRNA and protein levels in gefitinib-sensitive NCI-H292 cells, but it upregulated them in gefitinib-resistant SPC-A-1 cells (Xu et al., 2012a).

In contrast to the pronounced induction of apoptosis and reduction of cell viability by gefitinib in gefitinib-sensitive NCI-H292 cells, gefitinib exerts only a weak positive effect on the apoptosis rate in gefitinib-resistant SPC-A-1 cells and it has only a marginal negative effect on their viability (Xu et al., 2012a).

Strikingly, siRNA-mediated depletion of FOXM1 strongly reinforced the induction of apoptosis by gefitinib in gefitinib-resistant SPC-A-1 cells, demonstrating that loss of FOXM1 is able to resensitize these gefitinib-resistant cells to gefitinib-induced apoptosis (Xu et al., 2012a). Accordingly, FOXM1 knockdown by siRNA considerably strengthened the ability of gefitinib to reduce the number of viable gefitinib-resistant SPC-A-1 cells (Xu et al., 2012a).

5.2.15 Lapatinib
Lapatinib (Tykerb, GW572016) is a small-molecule dual tyrosine kinase inhibitor for HER2/Neu and EGFR (Cameron and Stein, 2008; Carter

et al., 2009; Collins et al., 2009; Harris, 2010; Jo Chien and Rugo, 2009; Kopper, 2008; Lackey, 2006; Montemurro et al., 2007; Mukherjee, et al., 2007; Nelson and Dolder, 2006; Obajimi, 2009; Schneider-Merck and Trepel, 2010).

Lapatinib represses the *foxm1* promoter and decreases *foxm1* mRNA and protein expression in lapatinib-sensitive SKBR3 and BT474 human breast carcinoma cells, but not in lapatinib-resistant MDA-MB-452 human breast carcinoma cells (Francis et al., 2009; Karadedou et al., 2012). Accordingly, lapatinib does not affect the FOXM1 protein level in lapatinib-resistant MCF-7 and MDA-MB-231 human breast carcinoma cells (Francis et al., 2009; Karadedou et al., 2012).

The VEGFR ligand VEGF is a potent pro-angiogenic factor (Coultas et al., 2005; Ferrara, 2004; Ferrara et al., 2003; Giacca, 2010; Gianni-Barrera et al., 2011; Hicklin and Ellis, 2005; Kowanetz and Ferrara, 2006; Lohela et al., 2009; Nagy et al., 2007; Roy et al., 2006; Saharinen et al., 2011; Takahashi and Shibuya, 2005; Zachary, 2003). *vegf* is a direct target gene of both the candidate tumor suppressor FOXO3a (Karadedou et al., 2012) and FOXM1 (Fig. 6.3; Table 6.3), the latter of which contributes to oncogenic transformation and participates in tumor initiation, growth, and progression (Costa et al., 2005a; Gong and Huang, 2012; Kalin et al., 2011; Koo et al., 2011; Laoukili et al., 2007; Myatt and Lam, 2007; Raychaudhuri and Park, 2011; Wierstra and Alves, 2007c).

Lapatinib tips the balance between FOXO3a-mediated *vegf* down-regulation and FOXM1-mediated *vegf* upregulation in favor of the former at the expense of the latter (Karadedou et al., 2012):

Lapatinib treatment leads at the *vegf* promoter to an exchange of FOXM1, which activates the *vegf* promoter, against FOXO3a, which represses the *vegf* promoter, so that lapatinib downregulates the *vegf* mRNA and protein expression (Karadedou et al., 2012). Accordingly, lapatinib decreases the FOXM1 mRNA and protein levels, but increases the FOXO3a mRNA and protein levels (Karadedou et al., 2012). Additionally, lapatinib treatment leads to the redistribution of FOXO3 from the cytoplasm to the nucleus by reducing the inhibitory phosphorylation of FOXO3a at T-32, a known Akt phosphorylation site (Karadedou et al., 2012).

5.2.16 Imatinib

The small-molecule tyrosine kinase inhibitor imatinib (Gleevec, STI571, imatinib mesylate) inhibits ABL, BCR-ABL, PDGFR (PDGF (platelet-derived growth factor) receptor), CSF-1R, and KIT (SCFR (SCF (stem cell factor) receptor)) (Alvarez et al., 2007; Capdeville et al., 2002; Cassier and Blay, 2010; Deininger and Druker, 2003; Deininger et al., 2005; Druker, 2002a,b,c, 2004, 2008, 2009; Duffaud and Le Chesne, 2009; Hernandez-Boluda and Cervantes, 2002; Hunter, 2007; Jones and Judson, 2005; Kaelin, 2004; Lopes and Bacchi, 2010; Lydon, 2009; Nadal and Olavarria, 2004; Peggs, 2004; Piccaluga et al., 2007; Radford, 2002; Redner, 2010; Roskoski, 2003; Sawyers, 2009; Soverini et al., 2008; Waller, 2010).

Imatinib represses the *foxm1* promoter (Mencalha et al., 2012). Accordingly, imatinib downregulated the *foxm1* mRNA expression in CFU-GM, CFU-GEMM, and BFU-E hematopoietic progenitor cells derived from CML patients (Takemura et al., 2010) as well as in the human K562 CML cell line (Mencalha et al., 2012), which was established from a CML patient in blast crisis (Klein et al., 1976).

K562-R cells are imatinib-resistant derivatives of K562 cells (Donato et al., 2003; Nimmanapalli et al., 2003). The *foxm1* mRNA level is higher in imatinib-resistant K562-R cells than in parental K562 cells, suggesting that FOXM1 overexpression might contribute to the imatinib resistance of K562-R cells (Mencalha et al., 2012).

5.2.17 Nilotinib

Nilotinib (Tasigna, AMN107) is a second-generation selective BCR-ABL tyrosine kinase inhibitor that also targets PDGFR and c-Kit (Blay and von Mehren, 2011; Breccia and Alimena, 2010; Deininger, 2008; Deremer et al., 2008; Hantschel et al., 2008; Manley et al., 2010; Quintas-Cardama and Cortes, 2007, 2008; Quintas-Cardama et al., 2010; Weisberg et al., 2006).

Nilotinib reduced the *foxm1* mRNA level in CFU-GM, CFU-GEMM, and BFU-E hematopoietic progenitor cells derived from CML patients (Takemura et al., 2010).

5.2.18 Dasatinib

The small-molecule tyrosine kinase inhibitor Dasatinib (Sprycel, BMS354825) inhibits BCR-ABL, SFKs (SRC family kinases), PDGFR, FGFR (FGF (fibroblast growth factor) receptor), c-KIT, and EPHA (ephrin A) receptor (Aguilera and Tsimberidou, 2009; Araujo and Logothetis, 2010; Chuah and Melo, 2009; Keam, 2008; Lindauer and Hochhaus, 2010; Oliviero and Manzione, 2007; Quintas-Cardama et al., 2006, 2007, 2009; Steinberg, 2007; Verstovsek, 2009).

Dasatinib diminished the *foxm1* mRNA expression in CFU-GM, CFU-GEMM, and BFU-E hematopoietic progenitor cells derived from CML patients (Takemura et al., 2010).

5.2.19 Vemurafenib

Vemurafenib (PLX4032, RG7204, RO5185426) is an ATP-competitive, small-molecule RAF kinase (i.e., B-Raf, C-Raf, A-Raf, B-Raf^{V600E} mutant) inhibitor, which can also inhibit several other protein kinases (SRMS (Src-related kinase lacking C-terminal regulatory tyrosine and N-terminal myristylation sites, SRM), ACK1 (activated p21CDC42 kinase, TNK2 (tyrosine kinase, non-receptor, 2)), MAP4K5 (MAPKKKK5 (MAPK kinase kinase kinase 5)), FGR (Gardner-Rasheed feline sarcoma viral (v-fgr) oncogene homolog), BRK (breast tumor kinase, PTK6 (protein-tyrosine kinase 6))) (Amaria et al., 2012; Bollag et al., 2012; Fisher and Larkin, 2012; Heakal et al., 2012 Jordan and Kelly, 2012; Keating, 2012; Patrawala and Puzanov, 2012; Ravnan and Matalka, 2012).

Vemurafenib decreased the *foxm1* mRNA and protein levels in A375 and 501mel melanoma cells (Bonet et al., 2012).

5.2.20 Cell-penetrating ARF$_{26-44}$ peptide

The tumor suppressor ARF is not only part of the p53 tumor suppressor pathway (Sherr, 1998; Sherr, 2000, 2001, 2002; Sherr and McCormick, 2002; Sherr and Weber, 2000), but has also several p53-independent functions (Carnero et al., 2000; Dominguez-Brauer et al., 2010; Gallagher et al., 2006; Lindström and Zhang, 2006; Lloyd, 2000; Lowe and Sherr, 2003; Pollice et al., 2008; Rocha and Perkins, 2005; Sapotita et al., 2007; Sherr, 2006; Sherr et al., 2005; Weber et al., 2000; Zhang, 2004), one of which is the targeting of FOXM1 (Costa et al., 2005a; Gusarova et al., 2007; Kalinichenko et al., 2004).

The transcriptional activity of exogenous FoxM1B is repressed by a cell-penetrating ARF$_{26-44}$ peptide, which comprises aa 26–44 of ARF and whose cellular uptake is enhanced by nine N-terminal D-arginine residues (Costa et al., 2005a; Kalinichenko et al., 2004). Accordingly, FoxM1B interacts possibly indirectly with aa 26–44 of ARF (Kalinichenko et al., 2004), that is, with that ARF region which is included in the ARF$_{26-44}$ peptide.

Two not mutually exclusive mechanisms could explain the inhibition of FoxM1B by the ARF$_{26-44}$ peptide (Costa et al., 2005a; Gusarova et al., 2007; Kalinichenko et al., 2004). Both rely on the possibly indirect interaction between FoxM1B and ARF (Kalinichenko et al., 2004).

Mechanism 1: In Co-IPs (co-immunopreciptation assays), ARF interacted with full-length FoxM1B, but not with the deletion mutant FoxM1B(1-688) lacking the 60 C-terminal aa of FoxM1B (Kalinichenko et al., 2004), so that ARF probably interacts with the last 60 aa of FoxM1B, which are identical to the last 60 aa of the splice variant FOXM1c (see Part I of this two-part review, that is see, Wierstra, 2013b). For FOXM1c, the TAD (transactivation domain) was mapped to aa 721–762, that is, to the 42 C-terminal aa (see Part I of this two-part review, that is see, Wierstra, 2013b) (Wierstra, 2013a; Wierstra and Alves, 2006a). Thus, ARF seems to interact with the TAD of FOXM1 so that ARF may inhibit the transcriptional activity of FoxM1B (Kalinichenko et al., 2004) by sterically blocking its TAD.

Mechanism 2: Both ARF and the ARF_{26-44} peptide relocalize FOXM1 to the nucleolus (Gusarova et al., 2007; Kalinichenko et al., 2004), which could explain the repression of the transcriptional activity of exogenous FoxM1B by ARF and the ARF_{26-44} peptide (Kalinichenko et al., 2004).

Hence, the ARF_{26-44} peptide could inhibit the transcriptional activity of exogenous FoxM1B either through masking of its TAD or through its relocalization to the nucleolus.

The cell-penetrating ARF_{26-44} peptide abolished the FoxMB-induced anchorage-independent growth on soft agar of U2OS osteosarcoma cells, demonstrating that the FOXM1-inhibiting ARF_{26-44} peptide has the ability to suppress an oncogenic activity of FOXM1 *in vivo* (Kalinichenko et al., 2004).

Injection of HCC cells isolated from *FoxM1B*-transgenic $ARF^{-/-}$ mice into nude mice via the tail vein resulted in considerably more lung nodules than injection of HCC cells derived from $ARF^{-/-}$ mice, which indicates that FoxM1B overexpression enhances the metastatic potential of ARF-deficient HCC cells (Park et al., 2011). The ARF_{26-44} peptide reduced the number of lung nodules in mice injected with HCC cells from *FoxM1B*-transgenic $ARF^{-/-}$ mice, indicating that the ARF_{26-44} peptide inhibits this FoxM1B-driven lung metastasis and thus interferes with an oncogenic activity of FOXM1 *in vivo* (Park et al., 2011).

The cell-penetrating ARF_{26-44} peptide-induced apoptosis in human HepG2 hepatoma cells and siRNA-mediated depletion of FOXM1 mostly prevented this ARF_{26-44} peptide-induced apoptosis, indicating that induction of apoptosis by the ARF_{26-44} peptide requires FOXM1 (Gusarova et al., 2007).

Accordingly, in mice, the ARF_{26-44} peptide caused apoptosis in DEN/PB-induced hepatic adenomas and HCC, but it failed to induce apoptosis in such liver tumors with a *foxm1* knockout (the *foxm1* gene was deleted in preexisting DEN/PB-induced liver tumors) (Gusarova et al., 2007). Hence, *foxm1* deficiency results in resistance of liver tumor cells to ARF_{26-44} peptide-induced cell death, because the induction of apoptosis by the ARF_{26-44} peptide depends on FOXM1 (Gusarova et al., 2007).

5.2.21 Nutlin-3

Nutlin-3 is a specific MDM2 inhibitor, which blocks the p53-binding domain of MDM2 resulting in p53 stabilization and p53 activation (Arkin and Wells, 2004; Chene, 2004a,b; Dudkina and Lindsley, 2007; Fotouhi and Graves, 2005; Fry and Vassilev, 2005; Joseph et al., 2010; Klein and Vassilev, 2004; Secchiero et al., 2008, 2011; Shangary and Wang, 2008, 2009; Shen and Maki, 2011; Vassilev, 2004, 2005, 2007; Vassilev et al., 2004; Wade and Wahl, 2010; Wiman, 2006). For this purpose, Nutlin-3 mimics three hydrophobic residues on p53 required for MDM2 binding, thus working as a competitive inhibitor of the interaction between p53 and MDM2.

Nutlin-3 downregulates the *foxm1* mRNA and protein levels in U2OS osteosarcoma cells and HCT116 colon carcinoma cells (both p53- and p21-positive), but neither in HCT116 $p53^{-/-}$ cells nor in HCT116 $p21^{-/-}$ cells, indicating that Nutlin-3 decreases the *foxm1* mRNA and protein expression dependent on p53 and p21 (Barsotti and Prives, 2009).

This dependence on p53 (Barsotti and Prives, 2009) is expected because Nutlin-3 is a small-molecule activator of p53 and inhibits the interaction between p53 and MDM2 (Arkin and Wells, 2004; Chene, 2004a,b; Dudkina and Lindsley, 2007; Fotouhi and Graves, 2005; Fry and Vassilev, 2005; Joseph et al., 2010; Klein and Vassilev, 2004; Secchiero et al., 2008, 2011; Shangary and Wang, 2008, 2009; Shen and Maki, 2011; Vassilev, 2004, 2005, 2007; Vassilev et al., 2004; Wade and Wahl, 2010; Wiman, 2006).

p21 is a direct p53 target gene (El-Deiry et al., 1993), but p53 can repress the *foxm1* mRNA expression independently of p21 in MCF-7 breast cancer cells (Barsotti and Prives, 2009), in accordance with *foxm1* being a direct target gene of p53 (Kurinna et al., 2013). This leaves open the possibilities that p21 might act either downstream of p53 or in parallel with p53 in the repression of the *foxm1* mRNA and protein expression by nutlin-3.

5.2.22 Bortezomib

Bortezomib (Velcade) is a proteasome inhibitor and shows antitumor activity (Adams, 2001, 2002, 2003, 2004a,b; Adams and Kauffman, 2004; Caravita et al., 2006; Chen et al., 2011b; Cusack, 2003; Cvek and Dvorak, 2011; Einsele, 2010; Ludwig et al., 2005; Mitsiades et al., 2005; Montagut et al.,

2005, 2006; Mujtaba and Dou, 2011; Nencioni et al., 2007; Orlowski and
Kuhn, 2008; Richardson and Anderson, 2003; Richardson and Mitsiades,
2005; Richardson et al., 2005, 2006; Roccaro et al., 2006a,b; Sterz et al.,
2009; Voorhees, and Orlowski, 2006; Zvarski et al., 2007).

Bortezomib decreases the *foxm1* mRNA (Pandit and Gartel, 2011b) and
protein (Bhat et al., 2009b; Pandit and Gartel, 2011b,c) expression. Furthermore, bortezomib represses the transcriptional activity of exogenous
FoxM1B (Bhat et al., 2009b).

In U2OS osteosarcoma cells, bortezomib induced the cleavage of
caspase-3 and exogenous FoxM1B reduced this bortezomib-induced
caspase-3 cleavage, suggesting that FoxM1B might promote cell survival
(Bhat et al., 2009a,b).

shRNA-mediated knockdown of FOXM1 sensitizes MIA PaCa-2 pan-
creatic cancer cells, MDA-MB-231 breast cancer cells, and HCT-116 colon
cancer cells to bortezomib-induced apoptosis because shRNA against
FOXM1 increased the extent of the bortezomib-induced caspase-3 cleavage
(Pandit and Gartel, 2011a). However, this FOXM1 knockdown had no
effect on apoptosis in the absence of bortezomib because shRNA against
FOXM1 did not cause caspase-3 cleavage in MIA PaCa-2, MDA-
MB-231, and HCT-116 cells (Pandit and Gartel, 2011a).

These findings suggest an implication of FOXM1 in the induction of
apoptosis by bortezomib so that FOXM1 overexpression interferes with
bortezomib-induced cell death whereas FOXM1 deficiency sensitizes
tumor cells to bortezomib-induced apoptosis.

However, since bortezomib affects the transcriptional activity of NF-κB
(Abayomi et al., 2007; Adams, 2001, 2004a,b; Aghajanian, 2004; Baud and
Karin, 2009; Berenson et al., 2001; Chaturvedi et al., 2011; Chen et al.,
2011a,b; Cusack, 2003; DeMartino, 2004; Frescas and Pagano, 2008;
Fribley and Wang, 2006; Gartel, 2010; Hideshima et al., 2003; Ishii
et al., 2007; Kitagawa et al., 2009; Lenz, 2003; Li et al., 2008b;
Ludwig et al., 2005; Mattingly et al., 2007; McConkey and Zhu, 2008;
Milano et al., 2007; Montagut et al., 2005, 2006; Moran and Nencioni,
2007; Nakanishi and Toi, 2005; Nakayama and Nakayama, 2006; Olivier
et al., 2006; Orlowski and Kuhn, 2008; Richardson, 2003; Richardson
and Anderson, 2003; Richardson et al., 2005, 2006; Spano et al., 2005;
Sun and Karin, 2008; Voorhees et al., 2003; Zvarski et al., 2005, 2007),
which plays a central role in apoptosis and cellular survival (Aggarwal and
Takada, 2005; Baldwin, 2012; Dutta et al., 2006; Fan et al., 2008; Karin
and Lin, 2002; Kucharczak et al., 2003; Luo et al., 2005a,b; Papa et al.,

2004; Piva et al., 2006; Radhakrishnan and Kamalakaran, 2006; Wang and Cho, 2010), rather NF-κB than FOXM1 might be decisive for the induction of apoptosis by bortezomib.

The same qualification is also valid for the induction of apoptosis by the FOXM1 inhibitors thiostrepton and siomycin A because the TNF-α (tumor necrosis factor-α)-induced transcriptional activity of a NF-κB reporter construct was reduced by thiostrepton and siomycin A (Bhat et al., 2009b).

5.2.23 MG132

The proteasome inhibitor MG132 downregulates the *foxm1* mRNA (Pandit and Gartel, 2011b) and protein (Bhat et al., 2009b; Pandit and Gartel, 2011b) levels. Moreover, MG132 reduces the transcriptional activity of exogenous FoxM1B (Bhat et al., 2009b).

Silencing of FOXM1 by shRNA sensitizes MIA PaCa-2 pancreatic cancer cells, MDA-MB-231 breast cancer cells, and HCT-116 colon cancer cells to MG132-induced apoptosis, because shRNA directed against FOXM1 increased the extent of the MG132-induced caspase-3 cleavage (Pandit and Gartel, 2011a). Yet, this FOXM1 knockdown had no effect on apoptosis in the absence of MG132, because shRNA against FOXM1 did not cause caspase-3 cleavage in MIA PaCa-2, MDA-MB-231, and HCT-116 cells (Pandit and Gartel, 2011a).

Thus, FOXM1 seems to be involved in the induction of apoptosis by MG132 so that depletion of FOXM1 sensitizes tumor cells to MG132-induced apoptosis.

5.2.24 MG115

The proteasome inhibitor MG115 diminishes the FOXM1 protein expression (Bhat et al., 2009b). In addition, MG115 decreases the transcriptional activity of exogenous FOXM1B (Bhat et al., 2009b).

5.2.25 Troglitazone

The antineoplastic thiazolidinedione troglitazone downregulated the *foxm1* mRNA expression in human HepG2 and PLC/PRF/5 hepatoma cells and the FOXM1 protein expression in HepG2 cells (Petrovich et al., 2010).

5.2.26 Pioglitazone

The antineoplastic thiazolidinedione pioglitazone decreased the *foxm1* mRNA level in HepG2 and PLC/PRF/5 cells (Petrovich et al., 2010).

5.2.27 Rosiglitazone

The antineoplastic thiazolidinedione rosiglitazone diminished the *foxm1* mRNA expression in HepG2 cells (Petrovich et al., 2010).

5.2.28 Genistein

The soybean isoflavone genistein, a cancer chemopreventive agent, decreased the *foxm1* mRNA and protein expression in human BxPC-3, HPAC, MIA PaCa-2, and PANC28 pancreatic cancer cells (Wang et al., 2010b), in human PC-3, LNCaP, and C4-2B prostate cancer cells (Wang et al., 2010d), in human SKOV3 and CoC1 ovarian cancer cells (Ning et al., 2012), and in human AGS and SGC-7901 gastric cancer cells (Xiang et al., 2012).

The genistein-induced apoptosis was alleviated by FOXM1 over-expression, but exacerbated by siRNA-mediated silencing of FOXM1 in MIA PaCa-2 (Wang et al., 2010b), SKOV3, and CoC1 cells (Ning et al., 2012). Accordingly, FOXM1 overexpression prevented the genistein-induced decrease in the number of viable MIA PaCa-2 (Wang et al., 2010b) and SKOV3 cells (Ning et al., 2012).

These findings point to an involvement of FOXM1 in the induction of apoptosis by genistein so that the overexpression of FOXM1 antagonizes the genistein-induced apoptosis of tumor cells whereas loss of FOXM1 sensitizes them to genistein-induced cell death.

5.2.29 DFOG

The genistein derivative DFOG (7-difluoromethoxyl-5,4′-di-*n*-octyl-genistein) downregulated the *foxm1* mRNA and protein levels in SKOV3, CoC1 (Ning et al., 2012), AGS, and SGC-7901 cells (Xiang et al., 2012).

The DFOG-induced apoptosis of SKOV3 and CoC1 cells was decreased by FOXM1 overexpression, but increased by siRNA against FOXM1 (Ning et al., 2012). Accordingly, FOXM1 overexpression antagonized the negative effect of DFOG on the viability of SKOV3 (Ning et al., 2012) and AGS cells (Xiang et al., 2012).

These findings suggest that FOXM1 is implicated in the induction of apoptosis by DFOG so that the overexpression of FOXM1 counteracts DFOG-induced cell death whereas FOXM1 deficiency sensitizes tumor cells to DFOG-induced apoptosis.

5.2.30 Ursolic acid

The anticancer agent ursolic acid (3β-hydroxy-12-urs-12-en-28-oic acid), a naturally occurring pentacyclic triterpene, downregulated the *foxm1* mRNA and protein levels in human MCF-7 breast cancer cells (Wang et al., 2011a).

5.2.31 DIM

The anticancer drug DIM (3,3′-diindolylmethane), a bioavailable, nontoxic dietary chemopreventive agent, diminished the FOXM1 mRNA (Ahmad et al., 2011; Rahman et al., 2006) and protein (Ahmad et al., 2011) expression in human MDA-MB-231, MDA-MB-468, SKBR3, and/or MCF-7 breast cancer cells.

FOXM1 overexpression interfered with the DIM-induced decline of the number of viable MCF-7 and SKBR3 cells, which points to an implication of FOXM1 in the negative effect of DIM on cell viability (Ahmad et al., 2011).

5.2.32 Natura-α

The indirubicin derivative Natura-α (NTI-Onco2008-1, N-methyl-Δ3,3′-dihydroindole-2,2′-diketone) downregulated the *foxm1* mRNA and protein expression in human LNCaP (androgen-dependent) and LNCaP-AI (androgen-independent) prostate cancer cells (Li et al., 2011).

Natura-α reduced the FOXM1-induced invasion of LNCaP-AI cells, which shows the ability of Natura-α to suppress an oncogenic property of FOXM1 *in vivo* (Li et al., 2011).

Overexpression of FOXM1 reversed the negative effect of Natura-α on the proliferation of LNCaP-AI cells, suggesting that the Natura-α-mediated cell growth suppression involves FOXM1 so that FOXM1 overexpression hampers the inhibition of tumor cell proliferation by Natura-α (Li et al., 2011).

5.2.33 TMPP

The deoxybromophospha sugar derivative TMPP (2,3,4-tribromo-3-methyl-1-phenylphospholane 1-oxide) decreased the FOXM1 protein level in human U937 and YRK2 leukemia cells (Nakamura et al., 2010b).

5.2.34 BVAN08

BVAN08 (6-bromoisovanillin, 6-bromine-5-hydroxy-4-methoxybenzalde
hyde), a derivative of the natural flowering agent vanillin, diminished the
FOXM1 protein expression in HepG2 cells (Zhang et al., 2011a).

5.2.35 ChemBridge DiverSet library

Ten small molecules (5194403, 5212518, 5217497, 5241816, 5256360,
5316908, 5349968, 5354001, 5356272, 5560509) from the ChemBridge
DiverSet library reduced the *foxm1* mRNA expression in GSC (GBM stem
cell)-enriched GBM cultures (Visnyei et al., 2011).

5.2.36 Temozolomide

The anticancer drug temozolomide (TMZ) (Temodar, Temodal, Temcad),
a DNA-alkylating agent, increases the FOXM1 protein expression dramatically in TMZ-resistant recurrent GBM1 cells, but only marginally in TMZ-
sensitive primary GBM1 cells (Zhang et al., 2011e).

Already prior to TMZ treatment, the FOXM1 protein level is significantly higher in recurrent GBM1 cells than in primary GBM1 cells
(Zhang et al., 2011e). Primary and recurrent GBM1 cells are cultured cell
lines established from paired primary or recurrent GBM tumor samples,
respectively (Zhang et al., 2011e).

The RAD51 protein expression and the percentage of cells with more
than five RAD51 foci are strongly increased by TMZ treatment in
TMZ-resistant recurrent GBM1 cells, whereas they are only weakly
enhanced by TMZ in TMZ-sensitive primary GBM1 cells (Zhang et al.,
2011e). Already prior to TMZ treatment, the RAD51 protein level is higher
in recurrent GBM1 cells than in primary GBM1 cells (Zhang et al., 2011e).

TMZ treatment decreases the proliferation rate of primary GBM1 cells
and arrests human U87 glioma cells in G2-phase of the cell cycle (Zhang
et al., 2011e). Moreover, TMZ reduces the anchorage-independent growth
of primary GBM1 cells on soft agar (Zhang et al., 2011e).

In contrast to TMZ-sensitive primary GBM1 cells, TMZ leaves (almost)
unaffected both the proliferation rate of TMZ-resistant recurrent GBM1
cells and their anchorage-independent growth on soft agar (Zhang et al.,
2011e). In contrast to TMZ-sensitive U87cells, TMZ fails to induce
G2-arrest in TMZ-resistant recurrent GBM1 cells (Zhang et al., 2011e).

The intracranial injection of nude mice with TMZ-resistant recurrent GBM cells results in the development of xenograft tumors (Zhang et al., 2011e). TMZ fails to prolong the survival of these nude mice, which were intracranially injected with TMZ-resistant recurrent GBM cells (Zhang et al., 2011e).

Thus, recurrent GBM1 cells are TMZ-resistant whereas primary GBM1 cells are TMZ-sensitive (Zhang et al., 2011e).

Recurrent GBM1 cells display a higher FOXM1 protein expression than primary GBM1 cells both before and after TMZ treatment so that the TMZ resistance of recurrent GBM1 cells is correlated with a higher FOXM1 protein level (Zhang et al., 2011e).

Silencing of FOXM1 with shRNA resensitizes TMZ-resistant recurrent GBM1 cells to TMZ cytotoxicity, which shows that the strong TMZ-induced upregulation of the FOXM1 protein expression in recurrent GBM1 cells contributes to their TMZ resistance (Zhang et al., 2011e):

First, shRNA against FOXM1 decreased the proliferation rate and the anchorage-independent growth on soft agar of TMZ-treated recurrent GBM1 cells, indicating that FOXM1 knockdown restores the negative effects of TMZ on both cell proliferation and anchorage-independent growth (Zhang et al., 2011e).

Second, shRNA against FOXM1 increased the percentage of G2-arrested TMZ-treated recurrent GBM1 cells, indicating that FOXM1 knockdown restores the TMZ–induced G2-arrest (Zhang et al., 2011e).

Third, shRNA against FOXM1 attenuated the TMZ-mediated upregulation of the RAD51 protein expression and prevented the TMZ-induced rise in the percentage of cells with more than five RAD51 foci (Zhang et al., 2011e).

Fourth, shRNA against FOXM1 reduced the xenograft tumor formation in nude mice, which were intracranially injected with TMZ-resistant recurrent GBM cells, and prolonged their survival (Zhang et al., 2011e). The positive effect of FOXM1 knockdown on the survival time of these nude mice was observed not only in TMZ-treated animals, but also without treatment (Zhang et al., 2011e).

Hence, shRNA-mediated depletion of FOXM1 resensitizes TMZ-resistant recurrent GBM1 cells to TMZ cytotoxicity, which indicates that FOXM1 decreases the TMZ chemosensitivity of recurrent GBM1 cells and suggests that their TMZ resistance is at least partially mediated by

FOXM1 (Zhang et al., 2011e). Thus, the ability of FOXM1 knockdown to increase the TMZ sensitivity of recurrent GBM1 cells points to a causal role of FOXM1 in their TMZ resistance, suggesting that recurrent GBM1 cells attain their TMZ resistance at least in part by strongly upregulating FOXM1 in response to TMZ treatment (Zhang et al., 2011e).

5.2.37 BI2536
The PLK1 inhibitor BI2536, an antineoplastic small-molecule compound, represssed the *foxm1* mRNA expression in human OE33 esophageal adenocarcinoma cells as they progressed from late G2- into M-phase (7 h after release from a double thymidine block) (Dibb et al., 2012).

5.2.38 Casticin
The polymethoxyflavone casticin, a natural flavonoid with anticarcinogenic activities and one of the main components of the fruits of *Vitex rotundifolia* L. (beach vitex, chasteberry, round-leaved or single-leaf chaste tree, Monk's pepper), downregulated the *foxm1* mRNA and protein expression in human HepG2 (p53 wild-type) and PLC/PRF/5 (p53 mutant) HCC cells (He et al., 2013).

5.2.39 5-FU (5-fluorouracil)
The pyrimidine base analog 5-FU (Efudex), a cytostatic antimetabolite for antitumor therapy, decreased the FOXM1 protein expression in parental human MCF-7 breast carcinoma cells, but not in 5-FU-resistant MCF-7-5FuR cells (Monteiro et al., 2012).

5.2.40 Olaparib
The PARP inhibitor olaparib (AZD2281) reduced the viability of $foxm1^{-/-}$ MEFs more dramatically than that of wild-type MEFs, indicating that $foxm1^{-/-}$ MEFs are more sensitive to olaparib treatment than wild-type MEFs (Monteiro et al., 2012). This higher olaparib sensitivity of $foxm1^{-/-}$ MEFs suggests an involvement of FOXM1 in the negative effect of olaparib on cell viability (Monteiro et al., 2012).

5.2.41 Veliparib
The PARP inhibitor veliparib (ABT-888) decreased the viability of $foxm1^{-/-}$ MEFs more strongly than that of wild-type MEFs, which indicates a higher sensitivity of $foxm1^{-/-}$ MEFs to veliparib compared to wild-type MEFs (Monteiro et al., 2012). This suggests that FOXM1 is

implicated in the negative effect of veliparib on cell viability (Monteiro et al., 2012).

PARP inhibitors, like olaparib and veliparib, block the BER pathway and they are used to selectively induce synthetic lethality of HR-deficient tumor cells (Gibson and Kraus, 2012; Krishnakumar and Kraus, 2010; Kruse et al., 2011; Luo and Kraus, 2012).

5.3. Cancer immunotherapy using FOXM1

One aim of cancer immunotherapy is to induce CTLs (cytotoxic T lymphocytes) by a TAA (tumor-associated antigen) and to utilize the CTLs elicited by the TAA for specific attack of tumor cells expressing this target antigen (Bei and Scardino, 2010; Gross and Margalit, 2007; Hamai et al., 2010; Kessler and Melief, 2007; Kuroki et al., 2006; Linley et al., 2011; Liu et al., 2005; Pietersz et al., 2006; Rosenberg et al., 2004; Sato et al., 2009). In this therapeutic approach, a target antigen induces a potent and specific antitumor immune response.

Three FOXM1-derived peptides were identified as HLA (human leukocyte antigen)-A2-restricted antigenic epitopes, suggesting that FOXM1 may be a candidate TAA and a suitable target antigen for cancer immunotherapy (Yokomine et al., 2009).

REFERENCES

Abate-Shen, C., Shen, M.M., 2002. Mouse models of prostate carcinogenesis. Trends Genet. 18, S1–S5.
Abayomi, E.A., Sissolak, G., Jacobs, P., 2007. Use of novel proteosome inhibitors as a therapeutic strategy in lymphomas: current experience and emerging concepts. Transfus. Apher. Sci. 37, 85–92.
Abdulkadir, S.A., Kim, J., 2005. Genetically engineered murine models of prostate cancer: insights into mechanisms of tumorigenesis and potential utility. Future Oncol. 1, 351–360.
Abraham, R.T., 2001. Cell cycle checkpoint signaling through the ATM and ATR kinases. Genes Dev. 15, 2177–2196.
Aburto, M.R., Magarinos, M., Leon, Y., Varela-Nieto, I., Sanchez-Calderon, H., 2012. AKT signaling mediates IGF-I survival actions on otic neural progenitors. PLoS One 7, e30790.
Ackermann Misfeldt, A., Costa, R.H., Gannon, M., 2008. β-Cell proliferation, but not neogenesis, following 60% partial pancreatectomy is impaired in the absence of FoxM1. Diabetes 57, 3069–3077.
Adam, A.P., George, A., Schewe, D., Bragado, P., Iglesias, B.V., Ranganathan, A.C., et al., 2009. Computational identification of a p38^{SAPK}-regulated transcription factor network required for tumor cell quiescence. Cancer Res. 69, 5664–5672.
Adami, G.R., Ye, H., 2007. Future roles for FoxM1 inhibitors in cancer treatments. Future Oncol. 3, 1–3.

Adams, J., 2001. Proteasome inhibition in cancer: development of PS-341. Semin. Oncol. 28, 613–639.

Adams, J., 2002. Proteasome inhibition: a novel approach to cancer therapy. Trends Mol. Med. 8, S49–S54.

Adams, J.M., 2003. Ways of dying: multiple pathways to apoptosis. Genes Dev. 17, 2482–2495.

Adams, J., 2004a. The proteasome: a suitable antineoplastic target. Nat. Rev. Cancer 4, 349–360.

Adams, J., 2004b. The development of proteasome inhibitors as anticancer drugs. Cancer Cell 5, 417–421.

Adams, P.D., 2009. Healing and hurting: molecular mechanisms, functions, and pathologies of cellular senescence. Mol. Cell 36, 2–14.

Adams, J., Kauffman, M., 2004. Development of the proteasome inhibitor Velcade (Bortezomib). Cancer Invest. 22, 304–311.

Aggarwal, B.B., Ghelot, P., 2009. Inflammation and cancer: how friendly is the relationship for cancer patients? Curr. Opin. Pharmacol. 9, 351–369.

Aggarwal, B.B., Takada, Y., 2005. Pro-apoptotic and anti-apoptotic effects of tumor necrosis factor in tumor cells. Role of nuclear transcription factor NF-κB. Cancer Treat. Rev. 126, 103–127.

Aggarwal, B.B., Shishodia, S., Sandur, S., Pandey, M.K., Sethi, G., 2006. Inflammation and cancer: how hot is the link? Biochem. Pharmacol. 72, 1605–1621.

Aggarwal, B.B., Kunnumakkara, A.B., Harikumar, K.B., Gupta, S.R., Tharakan, S.T., Koca, C., et al., 2009a. Signal transducer and activator of transcription-3, inflammation, and cancer: how intimate is the relationship? Ann. N. Y. Acad. Sci. 1171, 59–76.

Aggarwal, B.B., Sethi, G., Ahn, K.S., Sandur, S.K., Pandey, M.K., Kunnumakkara, A.B., et al., 2009b. Targeting signal-transducer-and-activator-of-transcription-3 for prevention and therapy of cancer. Ann. N. Y. Acad. Sci. 1171, 59–76.

Aghajanian, C., 2004. Clinical update: novel targets in gynecologic malignancies. Semin. Oncol. 31, 22–26 (Discussion 33).

Agiostratidou, G., Hulit, J., Phillips, G.R., Hazan, R.B., 2007. Differential cadherin expression: potential markers for epithelial to mesenchymal transformation during tumor progression. J. Mammary Gland Biol. Neoplasia 12, 127–133.

Aguilera, D.G., Tsimberidou, A.M., 2009. Dasatinib in chronic myeloid leukemia: a review. Ther. Clin. Risk Manag. 5, 281–289.

Ahmad, I., Sansom, O.J., Leung, H.Y., 2008. Advances in mouse models of prostate cancer. Expert Rev. Mol. Med. 10, e16.

Ahmad, A., Wang, Z., Kong, D., Ali, S., Li, Y., Banerjee, S., et al., 2010. FoxM1 downregulation leads to inhibition of proliferation, migration and invasion of breast cancer cells through the modulation of extra-cellular matrix degrading factors. Breast Cancer Res. Treat. 122, 337–346.

Ahmad, A., Ali, S., Wang, Z., Ali, A.S., Sethi, S., Sakr, W.A., et al., 2011. 3,3'-Diindolylmethane enhances Taxotere-induced growth inhibition of breast cancer cells through down-regulation of FoxM1. Int. J. Cancer 129, 1781–1791.

Ahmad, A., Sakr, W.A., Rahman, K.W., 2012. Novel targets for detection of cancer and their modulation by chemopreventive natural compounds. Front. Biosci. (Elite Ed.) 4, 410–425.

Ahmed, M., Uddin, S., Hussain, A.R., Alyan, A., Jehan, Z., Al-Dayel, F., et al., 2012. FoxM1 and its association with matrix metalloproteinases (MMP) signaling pathway in papillary thyroid carcinoma. J. Clin. Endocrinol. Metab. 97, E1–E13.

Ahn, J.I., Lee, K.H., Shin, D.M., Shim, J.W., Kim, C.M., Kim, H., et al., 2004a. Temporal expression changes during differentiation of neural stem cells derived from mouse embryonic stem cells. J. Cell. Biochem. 93, 563–578.

Ahn, J., Urist, M., Prives, C., 2004b. The Chk2 protein kinase. DNA Repair 3, 1039–1047.

Ahuja, D., Saenz-Robles, M.T., Pipas, J.M., 2005. SV40 large T antigen targets multiple cellular pathways to elicit cellular transformation. Oncogene 24, 7729–7745.

Akusjarvi, G., 2008. Temporal regulation of adenovirus major late alternative RNA splicing. Front. Biosci. 13, 5006–5015.

Albihn, A., Johnsen, J.I., Henriksson, M.A., 2010. MYC in oncogenesis and as a target for cancer therapy. Adv. Cancer Res. 107, 163–1224.

Ali, S.H., DeCaprio, J.A., 2001. Cellular transformation by SV40 large T antigen: interaction with host proteins. Semin. Cancer Biol. 11, 15–23.

Alitalo, K., Koskinen, P., Mäkelä, T.P., Saksela, K., Sistonen, L., Winqvist, R., 1987. *Myc* oncogenes: activation and amplification. Biochim. Biophys. Acta 907, 1–32.

Allavena, P., Sica, A., Solinas, G., Porta, C., Mantovani, A., 2008a. The inflammatory microenvironment in tumor progression: the role of tumor-associated macrophages. Crit. Rev. Oncol. Hematol. 66, 1–9.

Allavena, P., Sica, A., Garlanda, C., Mantovani, A., 2008b. The Yin-Yang of tumor-associated macrophages in neoplastic progression and immune surveillance. Immunol. Rev. 222, 155–161.

Alpaugh, M.L., Barsky, S.H., 2002. Reversible model of spheroid formation allows for high efficiency of gene delivery *ex vivo* and accurate gene assessment *in vivo*. Hum. Gene Ther. 13, 1245–1258.

Alpaugh, M.L., Tomlinson, J.S., Kasraeian, S., Barsky, S.H., 2002. Cooperative role of E-cadherin and sialyl-Lewis X/A-deficient MUC1 in the passive dissemination of tumor emboli in inflammatory breast carcinoma. Oncogene 21, 3631–3643.

Alvarez, R.H., Kantarijan, H., Cortes, J.E., 2007. The biology of chronic myelogenous leukemia: implications for imatinib therapy. Semin. Hematol. 44, S4–S14.

Alvarez-Fernandez, M., Halim, V.A., Aprelia, M., Mohammed, S., Medema, R.H., 2011. Protein phosphatase 2A (B55 α) prevents premature activation of transcription factor FOXM1 by antagonizing cyclin A/cyclin dependent kinase mediated phosphorylation. J. Biol. Chem. 286, 33029–33036.

Amabile, G., Meissner, A., 2009. Induced pluripotent stem cells: current progress and potential for regenerative medicine. Trends Mol. Med. 15, 59–68.

Amaria, R.N., Lewis, K.D., Jimeno, A., 2012. Vemurafenib: the road to personalized medicine in melanoma. Drugs Today 48, 109–118.

Amatangelo, M.D., Goodyear, S., Varma, D., Stearns, M.E., 2012. c-myc expression and MEK1 induced Erk2 nuclear localization are required for TGF-β induced epithelial-mesenchymal transition and invasion in prostate cancer. Carcinogenesis 33, 1965–1975.

Amunugama, R., Fishel, R., 2012. Homologous recombination in eukaryotes. Prog. Mol. Biol. Transl. Sci. 110, 155–206.

Amura, C.R., Marek, L., Winn, R.A., Heasley, L.E., 2005. Inhibited neurogenesis in JNK1-deficient embryonic stem cells. Mol. Cell. Biol. 25, 10791–10802.

Anders, L., Ke, N., Hydbring, P., Choi, Y.J., Widlund, H.R., Chick, J.M., et al., 2011. A systematic screen for CDK4/6 substrates links FOXM1 phosphorylation to senescence suppression in cancer cells. Cancer Cell 20, 620–634.

Andersson-Sjöland, A., Nihlberg, K., Eriksson, L., Bjermer, L., Westergren-Thorsson, G., 2011. Fibrocytes and the tissue niche in lung repair. Respir. Res. 12, 76.

Andrisani, O.M., 1999. CREB-mediated transcriptional control. Crit. Rev. Eukaryot. Gene Expr. 9, 19–32.

Andrisani, O.M., Barnabas, S., 1999. The transcriptional function of the hepatitis B virus X protein and its role in hepatocarcinogenesis. Int. J. Oncol. 15, 373–379.

Antoni, L., Sodha, N., Collins, I., Garrett, M.D., 2007. CHK2 kinase: cancer susceptibility and cancer therapy—two sides of the same coin? Nat. Rev. Cancer 7, 925–936.

Anzola, M., 2004. Hepatocellular carcinoma: role of hepatitis B and hepatitis C viruses proteins in hepatocarcinogenesis. J. Viral Hepat. 11, 383–393.

Araujo, J., Logothetis, C., 2010. Dasatinib: a potent SRC inhibitor in clinical development for the treatment of solid tumors. Cancer Treat. Rev. 36, 492–500.

Arkin, M.W., Wells, J.A., 2004. Small-molecule inhibitors of protein-protein interactions: progressing toward the dream. Nat. Rev. Drug Discov. 3, 301–317.

Arlinghaus, R., Sun, T., 2004. Signal transduction pathways in Bcr-Abl transformed cells. Cancer Treat. Res. 119, 239–270.

Arslan, M.A., Kutuk, O., Baselga, H., 2006. Protein kinases as drug targets in cancer. Curr. Cancer Drug Targets 6, 623–634.

Arteaga, C.L., Baselga, J., 2004. Tyrosine kinase inhibitors: why does the current process of clinical development no apply to them? Cancer Cell 5, 525–531.

Aster, J.C., Pear, W.S., Blacklow, S.C., 2008. Notch signaling in leukemia. Annu. Rev. Pathol. 3, 587–613.

Auersperg, N., Pan, J., Grove, B.D., Peterson, T., Fisher, J., Maines-Bandiera, S., et al., 1999. E-cadherin induces mesenchymal-to-epithelial transition in human ovarian surface epithelium. Proc. Natl. Acad. Sci. U.S.A. 96, 6249–6254.

Averbuch, S.D., 2003. The impact of gefitinib on epidermal growth factor receptor signaling pathways in cancer. Clin. Lung Cancer 5, S5–S10.

Bae, I., Rih, J.K., Kim, H.J., Kang, H.J., Haddad, B., Kirilyuk, A., et al., 2005. BRCA1 regulates gene expression for orderly mitotic progression. Cell Cycle 4, 1641–1666.

Baguley, B.C., 2010. Multiple drug resistance mechanisms in cancer. Mol. Biotechnol. 46, 308–316.

Baldwin, A.S., 2012. Regulation of cell death and autophagy by IKK and NF-κB: critical mechanisms in immune function and cancer. Immunol. Rev. 246, 327–345.

Balkwill, F.R., Mantovani, A., 2001. Inflammation and cancer: back to virchow? Lancet 357, 539–545.

Balkwill, F., Mantovani, A., 2010. Cancer and inflammation: implications for pharmacology and therapeutics. Clin. Pharmacol. Ther. 87, 401–406.

Balkwill, F.R., Mantovani, A., 2012. Cancer-related inflammation: common themes and therapeutic opportunities. Semin. Cancer Biol. 22, 33–40.

Balkwill, F., Charles, K.A., Mantovani, A., 2005. Smoldering and polarized inflammation in the initiation and promotion of malignant disease. Cancer Cell 7, 211–217.

Balli, D., Zhang, Y., Snyder, J., Kalinichenko, V.V., Kalin, T.V., 2011. Endothelial-specific deletion of transcription factor FoxM1 increases urethane-induced lung carcinogenesis. Cancer Res. 71, 40–50.

Balli, D., Ren, X., Chou, F.S., Cross, E., Zhang, Y., Kalinichenko, V.V., et al., 2012. Foxm1 transcription factor is required for macrophage migration during lung inflammation and tumor formation. Oncogene 31, 3875–3888.

Balli, D., Ustiyan, V., Zhang, Y., Wang, I.C., Masino, A.J., Ren, X., et al., 2013. Foxm1 transcription factor is required for lung fibrosis and epithelial-to mesenchymal transition. EMBO J. 32, 231–244.

Balmain, A., Gray, J., Ponder, B., 2003. The genetics and genomics of cancer. Nat. Genet. 33, 238–244.

Balsitis, S., Dick, F., Lee, D., Farrell, L., Hyde, R.K., Griep, A.E., et al., 2005. Examination of the pRB-dependent and pRB-independent functions of E7 in vivo. J. Virol. 79, 11392–11402.

Bamadas, A., Gil, M., Sanchez-Rovira, P., Llombart, A., Adrover, E., Estevez, L.G., et al., 2008. Neoadjuvant endocrine therapy for breast cancer: past, present and future. Anticancer Drugs 19, 339–347.

Banz, C., Ungethuem, U., Kuban, R.J., Diedrich, K., Lengyel, E., Hornung, D., 2009. The molecular signature of endometriosis-associated endometrioid ovarian cancer differs significantly from endometriosis-independent endometrioid ovarian cancer. Fertil. Steril. 94, 1212–1217.

Bao, B., Wang, Z., Ali, S., Kong, D., Banerjee, S., Ahamd, A., et al., 2011. Over-expression of FoxM1 leads to epithelial-mesenchymal transition and cancer stem cell phenotype in pancreatic cancer cells. J. Cell. Biochem. 112, 2296–2306.

Barakat, M.T., Humke, E.W., Scott, M.P., 2010. Learning from Jekyll to control Hyde: Hedgehog signaling in development and cancer. Trends Mol. Med. 16, 337–348.

Baranwal, S., Alahari, S.K., 2009. Molecular mechanisms controlling E-cadherin expression in breast cancer. Biochem. Biophys. Res. Commun. 384, 6–11.

Bar-Joseph, Z., Siegfried, Z., Brandeis, M., Brors, B., Lu, Y., Eils, R., et al., 2008. Genome-wide transcriptional analysis of the human cell cycle identifies genes differentially regulated in normal and cancer cells. Proc. Natl. Acad. Sci. U.S.A. 105, 955–960.

Barker, N., Clevers, H., 2001. Tumor microenvironment: a potent driving force in colorectal cancer? Trends Mol. Med. 7, 535–537.

Barker, N., Clevers, H., 2006. Mining the Wnt pathway for cancer therapeutics. Nat. Rev. Drug Discov. 5, 997–1014.

Barrallo-Gimeno, A., Nieto, M.A., 2005. The Snail genes as inducers of cell movement and survival: implications in development and cancer. Development 132, 3151–3161.

Barsotti, A.M., Prives, C., 2009. Pro-proliferative FoxM1 is a target of p53-mediated repression. Oncogene 28, 4295–4305.

Bartek, J., Lukas, J., 2003. Chk1 and Chk2 kinases in checkpoint control and cancer. Cancer Cell 3, 421–429.

Bartek, J., Lukas, J., 2007. DNA damage checkpoints: from initiation to recovery or adaptation. Curr. Opin. Cell Biol. 19, 238–245.

Bartek, J., Falck, J., Lukas, J., 2001. CHK2 kinase—a busy manager. Nat. Rev. Mol. Cell Biol. 2, 877–886.

Bartek, J., Lukas, C., Lukas, J., 2004. Checking on DNA damage in S-phase. Nat. Rev. Mol. Cell Biol. 5, 792–804.

Bartsch, R., Wenzel, C., Steger, G.G., 2007. Trastuzumab in the management of early and advanced stage breast cancer. Biologics 1, 19–31.

Baselga, J., 2006. Targeting tyrosine kinases in cancer: the second wave. Science 312, 1175–1178.

Baselga, J., Swain, S.M., 2009. Novel anticancer targets: revisiting ERBB2 and discovering ERBB3. Nat. Rev. Cancer 9, 463–475.

Baseman, J.G., Koutsky, L.A., 2005. The epidemiology of human papillomavirus infections. J. Clin. Virol. 32, S16–S24.

Basseres, D.S., Baldwin, A.S., 2006. Nuclear factor-κB and inhibitor of κB kinase pathways in oncogenic initiation and progression. Oncogene 25, 6817–6830.

Batsche, E., Cremisi, C., 1999. Opposite transcriptional activity between the wild type c-myc gene coding for c-Myc1 and c-Myc2 proteins and c-Myc1 and c-Myc2 separately. Oncogene 18, 5662–5671.

Batsche, E., Muchardt, C., Behrens, J., Hurst, H.C., Cremisi, C., 1998. RB and c-Myc activate expression of the E-cadherin gene in epithelial cells through interaction with transcription factor AP-2. Mol. Cell. Biol. 18, 3647–3658.

Baud, V., Karin, M., 2009. Is NF-κB a good target for cancer therapy? Hopes and pitfalls. Nat. Rev. Drug Discov. 8, 33–40.

Baum, B., Georgiou, M., 2011. Dynamics of adherens junctions in epithelial establishment, maintenance, and remodeling. J. Cell Biol. 192, 907–917.

Baum, B., Settleman, J., Quinlan, M.P., 2008. Transitions between epithelial and mesenchymal states in development and disease. Semin. Cell Dev. Biol. 19, 294–308.

Baumann, P., West, S.C., 1998. Role of the human RAD51 protein in homologous recombination and double-strand-break repair. Trends Biochem. Sci. 23, 247–251.

Bayley, S.T., Mymryk, J.S., 1994. Adenovirus E1A proteins and transformation. Int. J. Oncol. 5, 425–444.

Beavon, I.R.G., 2000. The E-cadherin-catenin complex in tumour metastasis: structure, function and regulation. Eur. J. Cancer 36, 1607–1620.

Bednarz-Knoll, N., Alix-Panabieres, C., Pantel, K., 2012. Plasticity of disseminating cancer cells in patients with epithelial malignancies. Cancer Metastasis Rev. 31, 673–687.

Behren, A., Mühlen, S., Acuna Sanhueza, G.A., Schwager, C., Plinkert, P.K., Huber, P.E., et al., 2010. Phenotype-assisted transcriptome analysis identifies FOXM1 downstream from Ras-MKK3-p38 to regulate in vitro cellular invasion. Oncogene 29, 1519–1530.

Bei, R., Scardino, A., 2010. TAA polypeptide DNA-based vaccines: a potential tool for cancer therapy. J. Biomed. Biotechnol. 2010, 102758.

Beijersbergen, R.L., Bernards, R., 1996. Cell cycle regulation by the retinoblastoma family of growth inhibitory proteins. Biochim. Biophys. Acta 1287, 103–120.

Bekker-Jensen, S., Mailand, N., 2010. assembly and function of DNA double-strand break repair foci in mammalian cells. DNA Repair 9, 1219–1228.

Bektas, N., ten Haaf, A., Veeck, J., Wild, P.J., Lüscher-Firzlaff, J., Hartmann, A., et al., 2008. Tight correlation between expression of the Forkhead transcription factor FOXM1 and HER2 in human breast cancer. BMC Cancer 8, 42.

Bellelli, R., Castellone, M.D., Garcia-Rostan, G., Ugolini, C., Nucera, C., Sadow, P.M., et al., 2012. FOXM1 is a molecular determinant of the mitogenic and invasive phenotype of anaplastic thyroid carcinoma. Endocr. Relat. Cancer 19, 695–710.

Bellini, A., Mattoli, S., 2007. The role of the fibrocyte, a bone marrow-derived mesenchymal progenitor, in reactive and reparative fibroses. Lab. Invest. 87, 858–870.

Benhenda, S., Cougot, D., Buendia, M.A., Neuveut, C., 2009. Hepatitis B virus X protein molecular functions and its role in virus life cycle and pathogenesis. Adv. Cancer Res. 103, 75–109.

Ben-Israel, H., Kleinberger, T., 2002. Adenovirus and cell cycle control. Front. Biosci. 7, d369–d395.

Ben-Neriah, Y., Karin, M., 2011. Inflammation meets cancer, with NF-κB as the matchmaker. Nat. Immunol. 12, 715–723.

Ben-Porath, I., Thomson, M.W., Carey, V.J., Ge, R., Bell, G.W., Regev, A., et al., 2008. An embryonic stem cell-like gene expression signature in poorly differentiated aggressive human tumors. Nat. Genet. 40, 499–507.

Benvenuti, S., Arena, S., Bardelli, A., 2005. Identification of cancer genes by mutational profiling of tumor genomes. FEBS Lett. 579, 1884–1890.

Berenson, J.R., Ma, H.M., Vescio, R., 2001. The role of nuclear factor-kappaB in the biology and treatment of multiple myeloma. Semin. Oncol. 28, 626–633.

Bergamaschi, A., Christensen, B.L., Katzenellenbogen, B.S., 2011. Reversal of endocrine resistance in breast cancer: interrelationships among 14-3-3τ, FOXM1, and a gene signature associated with miosis. Breast Cancer Res. 13, R70.

Berger, E., Rome, S., Vega, N., Ciancia, C., Vidal, H., 2010. Transcriptome profiling in response to adiponectin in human cancer-derived cells. Physiol. Genomics 42A, 61–70.

Berk, A.J., 2005. Recent lessons in gene expression, cell cycle control, and cell biology from adenovirus. Oncogene 24, 7673–7685.

Berman-Booty, L.D., Sargeant, A.M., Rosol, T.J., Rengel, R.C., Clinton, S.K., Chen, C.S., et al., 2012. A review of the existing grading schemes and a proposal for a modified grading scheme for prostatic lesions in TRAMP mice. Toxicol. Pathol. 40, 5–17.

Berx, G., van Roy, F., 2009. Involvement of members of the cadherin superfamily in cancer. Cold Spring Harb. Perspect. Biol. 1, a003129.

Berx, G., Raspe, E., Christofori, G., Thierry, J.P., Sleeman, J.P., 2007. Pre-EMTing metastasis? Recapitulation of morphogenetic processes in cancer. Clin. Exp. Metastasis 24, 587–597.

Bharadwaj, M., Roy, G., Dutta, K., Misbah, M., Husain, M., Husain, S., 2013. Tackling hepatitis B virus-associated hepatocellular carcinoma—the future is now. Cancer Metastasis Rev. 32, 229–268.

Bhat, U.G., Zipfel, P.A., Tyler, D.S., Gartel, A.L., 2008. Novel anticancer compounds induce apoptosis in melanoma cells. Cell Cycle 7, 1851–1855.

Bhat, U.G., Halasi, M., Gartel, A.L., 2009a. Thiazole antibiotics target FoxM1 and induce apoptosis in human cancer cells. PLoS One 4, e5592.

Bhat, U.G., Halasi, M., Gartel, A.L., 2009b. FoxM1 is a general target for proteasome inhibitors. PLoS One 4, e6593.

Bhat, U.G., Jagadeeswaran, R., Halasi, M., Gartel, A.L., 2011. Nucleophosmin interacts with FOXM1 and modulates the level and localization of FOXM1c in human cancer cells. J. Biol. Chem. 286, 41425–41433.

Bhonde, M.R., Hanski, M.L., Budczies, J., Cao, M., Gillissen, B., Moorthy, D., et al., 2006. DNA damage-induced expression of p53 suppresses mitotic checkpoint kinase hMsp1. J. Biol. Chem. 281, 8675–8685.

Biddle, A., Mackenzie, I.C., 2012. Cancer stem cells and EMT in carcinoma. Cancer Metastasis Rev. 31, 285–293.

Bingle, L., Brown, N.J., Lewis, C.E., 2002. The role of tumor-associated macrophages in tumor progression: implications for new anticancer therapies. J. Pathol. 196, 254–265.

Birchmeier, W., 1995. E-cadherin as a tumor (invasion) suppressor gene. Bioessays 17, 97–99.

Bishop, J.M., 1995. Cancer: the rise of the genetic paradigm. Genes Dev. 9, 1309–1315.

Blagosklonny, M.V., 2004. Gefitinib (Iressa) in oncogene-addictive cancers and therapy for common cancers. Cancer Biol. Ther. 3, 436–440.

Blanchette, P., Branton, P.E., 2009. Manipulation of the ubiquitin-proteasome pathway by small DNA tumor viruses. Virology 384, 317–323.

Blanco-Bose, W.E., Murphy, M.J., Ehninger, A., Offner, S., Dubey, C., Huang, W., et al., 2008. c-Myc and its target FoxM1 are critical downstream effectors of TCPOBOP-CAR induced direct liver hyperplasia. Hepatology 48, 1302–1311.

Blay, J.Y., von Mehren, M., 2011. Nilotinib: a novel, selective tyrosine kinase inhibitor. Semin. Oncol. 38, S3–S9.

Blume-Jensen, P., Hunter, T., 2001. Oncogenic kinase signalling. Nature 411, 355–365.

Bocchetta, M., Carbone, M., 2004. Epidemiology and molecular pathology at crossroads to establish causation: molecular mechanisms of malignant transformation. Oncogene 23, 6484–6491.

Bock, C., Lengauer, T., 2012. Managing drug resistance in cancer: lessons from HIV therapy. Nat. Rev. Cancer 12, 494–501.

Bode, A.M., Dong, Z., 2005. Signal transduction pathways in cancer development and as targets for cancer prevention. Prog. Nucleic Acid Res. Mol. Biol. 79, 237–297.

Bollag, G., Tsai, J., Zhang, J., Zhang, C., Ibrahim, P., Nolop, K., et al., 2012. Vemurafenib: the first drug approved for BRAF-mutant cancer. Nat. Rev. Drug Discov. 11, 873–886.

Bollrath, J., Greten, F.R., 2009. IKK/NF-κB and STAT3 pathways: central signalling hubs in inflammation-mediated tumour promotion and metastasis. EMBO Rep. 10, 1314–1319.

Bonet, C., Giuliano, S., Ohanna, M., Bille, K., Allegra, M., Lacour, J.P., et al., 2012. Aurora B is regulated by the MAPK/ERK signaling pathway and is a valuable potential target in melanoma cells. J. Biol. Chem. 287, 29887–29898.

Bonnomet, A., Brysse, A., Tachsidis, A., Waltham, M., Thompson, E.W., Polette, M., et al., 2010. Epithelial-to-mesenchymal transitions and circulating tumor cells. J. Mammary Gland Biol. Neoplasia 15, 261–273.

Bordeianu, G., Zugun-Eloae, F., Rusu, M.G., 2011. The role of DNA repair by homologous recombination in oncogenesis. Rev. Med. Chir. Soc. Med. Nat. Iasi. 115, 1189–1194.

Borghi, N., Nelson, J.W., 2009. Intercellular adhesion in morphogenesis: molecular and bio-physical considerations. Curr. Top. Dev. Biol. 89, 1–32.

Borst, P., 2012. Cancer drug pan-resistance: pumps, cancer stem cells, quiescence, epithelial to mesenchymal transition, blocked cell death pathways, persisters or what? Open Biol. 2, 120066.

Bose, A., Teh, M.T., Hutchison, I.L., Wan, H., Leigh, I.M., Waseem, A., 2012. Two mechanisms regulate keratin K15 expression in keratinocytes: role of PKC/AP-1 and FOXM1 mediated signalling. PLoS One 7, e38599.

Bouchard, M.J., Schneider, R.J., 2004. The enigmatic X gene of hepatitis B virus. J. Virol. 78, 12725–12734.

Boulikas, T., 1995. The phosphorylation connection to cancer. Int. J. Oncol. 6, 271–278.

Bovolenta, P., Esteve, P., Ruiz, J.M., Cisneros, E., Lopez-Rios, J., 2008. Beyond Wnt inhibition: new functions of secreted Frizzled-related proteins in development and disease. J. Cell Sci. 121, 737–746.

Brabletz, T., 2012. To differentiate or not—routes towards metastasis. Nat. Rev. Cancer 12, 425–436.

Brabletz, S., Brabletz, T., 2010. The ZEB/miR-200 feedback loop—a motor of cellular plasticity in development and cancer? EMBO Rep. 11, 670–677.

Brabletz, T., Jung, A., Reu, S., Porzner, M., Hlubek, F., Kunz-Schughart, L.A., et al., 2001. Variable β-catenin expression in colorectal cancers indicates tumor progression driven by the tumor microenvironment. Proc. Natl. Acad. Sci. U.S.A. 98, 10356–10361.

Brabletz, T., Jung, A., Spaderna, S., Hlubek, F., Kirchner, T., 2005a. Migrating cancer stem cells—an integrated concept of malignant tumour progression. Nat. Rev. Cancer 5, 744–749.

Brabletz, T., Hlubek, F., Spaderna, S., Schmalhofer, O., Hiendlmeyer, E., Jung, A., et al., 2005b. Invasion and metastasis in colorectal cancer: epithelial-mesenchymal transition, mesenchymal-epithelial transition, stem cells and β-catenin. Cells Tissues Organs 179, 56–65.

Brahimi-Horn, M.C., Pouyssegur, J., 2005. The hypoxia-inducible factor and tumor progression along the angiogenic pathway. Int. Rev. Cytol. 242, 157–213.

Brahimi-Horn, C., Pouyssegur, J., 2006. The role of the hypoxia-inducible factor in tumor metabolism, growth and invasion. Bull. Cancer 93, E73–E80.

Brahimi-Horn, M.C., Pouyssegur, J., 2009. HIF at a glance. J. Cell Sci. 122, 1055–1057.

Braig, M., Schmitt, C.A., 2006. Oncogene-induced senescence: putting the brakes on tumor development. Cancer Res. 66, 2881–2884.

Branzei, D., Foiani, M., 2008. Regulation of DNA repair throughout the cell cycle. Nat. Rev. Mol. Cell Biol. 9, 297–308.

Bräuning, A., Heubach, Y., Knorpp, T., Kowalik, M.A., Templin, M., Columbano, A., et al., 2011. Gender-specific interplay of signaling through β-catenin and CAR in the regulation of xenobiotic-induced hepatocyte proliferation. Toxicol. Sci. 123, 113–122.

Breccia, M., Alimena, G., 2010. Nilotinib: a second-generation tyrosine kinase inhibitor for chronic myeloid leukemia. Leuk. Res. 34, 129–134.

Brezillon, N., Lambert-Blot, M., Morosan, S., Couton, D., Mitchell, C., Kremsdorf, D., et al., 2007. Transplanted hepatocytes over-expressing FoxM1B efficiently repopulate chronically injured mouse liver independent of donor age. Mol. Ther. 15, 1710–1715.

Briest, S., Wolff, A.C., 2007. Insights on adjuvant endocrine therapy for premenopausal and postmenopausal breast cancer. Expert Rev. Anticancer Ther. 7, 1243–1253.

Brown, J.M., Attardi, L.D., 2005. The role of apoptosis in cancer development and treatment response. Nat. Rev. Cancer 5, 231–237.

Brown, J.M., Wouters, B.G., 1999. Apoptosis, p53, and tumor cell sensitivity to anticancer agents. Cancer Res. 59, 1391–1399.

Brumbaugh, J., Rose, C.M., Phanstiel, D.H., Thomson, J.A., Coon, J.J., 2011. Proteomics and pluripotency. Crit. Rev. Biochem. Mol. Biol. 46, 493–506.

Buda, A., Pignatelli, M., 2011. E-cadherin and the cytoskeleton network in colorectal cancer development and metastasis. Cell Commun. Adhes. 18, 133–143.

Bukholm, I.K., Nesland, J.M., Borresen-Dale, A.L., 2000. Re-expression of E-cadherin, α-catenin and β-catenin, but not of γ-catenin, in metastatic tissue from breast cancer patients. J. Pathol. 190, 15–19.

Bunney, T.D., Katan, M., 2010. Phosphoinositide signalling in cancer: beyond PI3K and PTEN. Nat. Rev. Cancer 10, 342–352.

Burk, R.D., Chen, Z., Van Doorslaer, K., 2009. Human papillomaviruses: genetic basis of carcinogenicity. Public Health Genomics 12, 281–290.

Burke, B.A., Carroll, M., 2010. BCR-ABL: a multi-faceted promoter of DNA mutation in chronic myelogeneous leukemia. Leukemia 24, 1105–1112.

Burkhart, D.L., Sage, J., 2008. Cellular mechanisms of tumour suppression by the retinoblastoma gene. Nat. Rev. Cancer 8, 671–682.

Butel, J.S., 2000. Viral carcinogenesis: revelation of molecular mechanisms and etiology of human disease. Carcinogenesis 21, 405–426.

Cadigan, K.M., Peifer, M., 2009. Wnt signaling from development to disease: insights from model systems. Cold Spring Harb. Perspect. Biol. 1, a002881.

Caestecker, K.W., Van de Walle, G.R., 2013. The role of BRCA1 in DNA double-strand repair: past and present. Exp. Cell Res. 319, 575–587.

Caino, M.C., Meshki, J., Kazanietz, M.G., 2009. Hallmarks of senescence in carcinogenesis: novel signaling players. Apoptosis 14, 392–408.

Caldwell, S.A., Jackson, S.R., Shahriari, K.S., Lynch, T.P., Sethi, G., Walker, S., et al., 2010. Nutrient sensor O-GlcNAc transferase regulates breast cancer tumorigenesis through targeting of the oncogenic transcription factor FoxM1. Oncogene 29, 2831–2842.

Calvisi, D.F., Pinna, F., Ladu, S., Pellegrino, R., Simile, M.M., Frau, M., et al., 2009. Forkhead box M1B is a determinant of rat susceptibility to hepatocarcinogenesis and sustains ERK activity in human HCC. Gut 58, 679–687.

Calvo, F., Aguda-Ibanez, L., Crespo, P., 2010. The Ras-ERK pathway: understanding site-specific signaling provides hope of new anti-tumor therapies. Bioessays 32, 412–421.

Cameron, D.A., Stein, S., 2008. Drug Insight: intracellular inhibitors of HER2—clinical development of lapatinib in breast cancer. Nat. Clin. Pract. Oncol. 5, 512–520.

Campisi, J., 2000. Cancer, aging and cellular senescence. In Vivo 14, 183–188.

Campisi, J., 2001. Cellular senescence as a tumor-suppressor mechanism. Trends Cell Biol. 11, S27–S31.

Campisi, J., 2003. Cancer and aging: rival demons? Nat. Rev. Cancer 3, 339–349.

Campisi, J., 2005a. Senescent cells, tumor suppression, and organismal aging: good citizens, bad neighbours. Cell 120, 513–522.

Campisi, J., 2005b. Aging, tumor suppression and cancer: high wire-act!. Mech. Ageing Dev. 126, 51–58.

Campisi, J., 2011. Cellular senescence: putting paradoxes in perspectives. Curr. Opin. Genet. Dev. 21, 107–112.

Campisi, J., d'Adda di Fagagna, F., 2007. Cellular senescence: when bad things happen to good cells. Nat. Rev. Mol. Cell Biol. 8, 729–740.

Cannito, S., Novo, E., Valfre di Bonzo, L., Busletta, C., Colombatto, S., Parola, M., 2010. Epithelial-mesenchymal transition: from molecular mechanisms, redox regulation to implications in human health and disease. Antioxid. Redox Signal. 12, 1383–1430.

Cano, A., Nieto, M.A., 2008. Non-coding RNAs take centre stage in epithelial-to-mesenchymal transition. Trends Cell Biol. 18, 357–359.

Cano, C., Motoo, Y., Iovanna, J.L., 2010. Epithelial-to-mesenchymal transition in pancreatic adenocarcinoma. Sci. World J. 10, 1947–1957.

Capdeville, R., Buchdunger, E., Zimmermann, J., Matter, A., 2002. Glivec (STI571, imatinib), a rationally developed, targeted anticancer drug. Nat. Rev. Drug Discov. 1, 493–502.

Cappuzzo, F., Finocchiara, G., Metro, G., Bartolini, S., Magrini, E., Cancellieri, A., et al., 2006. Clinical experience with gefitinib: an update. Crit. Rev. Oncol. Hematol. 58, 31–45.

Caracciolo, V., Reiss, K., Khalili, K., De Falco, G., Giordano, A., 2006. Role of the interaction between large T antigen and Rb family members in the oncogenicity of JC virus. Oncogene 25, 5294–5301.

Caravita, T., de Fabritiis, P., Palumbo, A., Amadori, S., Boccadoro, M., 2006. Bortezomib: efficacy comparisons in solid tumors and hematologic malignancies. Nat. Clin. Pract. Oncol. 3, 374–387.

Cargnello, M., Roux, P.P., 2011. Activation and function of MAPKs and their substrates, the MAPK-activated protein kinases. Microbiol. Mol. Biol. Rev. 75, 50–83.

Carillo-Infante, C., Abbadessa, G., Bagella, L., Giordano, A., 2007. Viral infections as a cause of cancer. Int. J. Oncol. 30, 1521–1528.

Carlini, M.J., De Lorenzo, M.S., Puricelli, L., 2011. Cross-talk between tumor cells and the microenvironment at the metastatic niche. Curr. Pharm. Biotechnol. 12, 1900–1908.

Carnero, A., Hudson, J.D., Price, C.M., Beachy, D.H., 2000. p16INK4A and p19ARF act in overlapping pathways in cellular immortalization. Nat. Cell Biol. 2, 148–155.

Caron, C., Col, E., Khochbin, S., 2002. The viral control of cellular acetylation signaling. Bioessays 25, 58–65.

Carr, J.R., Park, H.J., Wang, Z., Kiefer, M.M., Raychaudhuri, P., 2010. FoxM1 mediates resistance to Herceptin and Paclitaxel. Cancer Res. 70, 5054–5063.

Carr, J.R., Kiefer, M.M., Park, H.J., Li, J., Wang, Z., Fontanarosa, J., et al., 2012. FoxM1 regulates mammary luminal cell fate. Cell Rep. 1, 715–729.

Carter, S.L., Eklund, A.C., Kohane, I.S., Harris, L.N., Szallasi, Z., 2006. A signature of chromosomal instability inferred from gene expression profiles predicts clinical outcome in multiple human cancers. Nat. Genet. 38, 1043–1048.

Carter, C.A., Kelly, R.J., Giaccone, G., 2009. Small-molecule inhibitors of the human epidermal receptor family. Expert Opin. Investig. Drugs 18, 1829–1842.

Cassier, P.A., Blay, J.Y., 2010. Imatinib mesylate for the treatment of gastrointestinal stromal tumor. Expert Rev. Anticancer Ther. 10, 623–634.

Cavallaro, U., Christofori, G., 2004. Cell adhesion and signalling by cadherins and Ig-CAMs in cancer. Nat. Rev. Cancer 4, 118–132.

Cavallaro, U., Schaffhauser, B., Christofori, G., 2002. Cadherins and the tumour progression: is it all in a switch? Cancer Lett. 176, 123–128.

Cavey, M., Lecuit, T., 2009. Molecular bases of cell-cell junction stability and dynamics. Cold Spring Harb. Perspect. Biol. 1, a002998.

Celia-Terrassa, T., Meca-Cortes, O., Mateo, F., Martinez de Paz, A., Rubio, N., Arnal-Estape, A., et al., 2012. Epithelial-mesenchymal transition can suppress major attributes of human epithelial tumor-initiating cells. J. Clin. Invest. 122, 1849–1868.

Cerbinskaite, A., Mukhopadhyay, A., Plummer, E.R., Curtin, N.J., Edmondson, R.J., 2012. Defective homologous recombination in human cancers. Cancer Treat. Rev. 38, 89–100.

Chabner, B.A., Roberts, T.G., 2005. Timeline: chemotherapy and the war on cancer. Nat. Rev. Cancer 5, 65–72.

Chaffer, C.L., Thompson, E.W., Williams, E.D., 2007. Mesenchymal to epithelial transition in development and disease. Cells Tissues Organs 185, 7–19.

Chakrabarti, O., Krishna, S., 2003. Molecular interactions of 'high risk' human papillomaviruses E6 and E7 oncoproteins: implications for tumour progression. J. Biosci. 28, 337–348.

Chalhoub, N., Baker, S.J., 2009. PTEN and the PI3-kinase pathway in cancer. Annu. Rev. Pathol. Mech. Dis. 4, 127–150.

Chan, D.A., Giaccia, A.J., 2011. Harnessing synthetic lethal interactions in anticancer drug discovery. Nat. Rev. Drug Discov. 10, 351–364.

Chan, H.M., Shikama, N., La Thangue, N.B., 2001. Control of gene expression and the cell cycle. Essays Biochem. 37, 87–96.

Chan, D.W., Yu, S.Y.M., Chiu, P.M., Yao, K.M., Liu, V.W.S., Cheung, A.N.Y., et al., 2008. Over-expression of FOXM1 transcription factor is associated with cervical cancer progression and pathogenesis. J. Pathol. 215, 245–252.

Chan, D.W., Hui, W.W., Cai, P.C., Liu, M.X., Yung, M.M., Mak, C.S., et al., 2012. Targeting GRB7/ERK/FOXM1 signaling pathway impairs aggressiveness of ovarian cancer cells. PLoS One 7, e52578.

Chandran, U.R., Ma, C., Dhir, R., Bisceglia, M., Lyons-Weiler, M., Liang, W., et al., 2007. Gene expression profiles of prostate cancer reveal involvement of multiple molecular pathways in the metastatic process. BMC Cancer 7, 64.

Chaney, S.G., Sancar, A., 1996. DNA repair: enzymatic mechanisms and relevance to drug response. J. Natl. Cancer Inst. 88, 1346–1360.

Chang, T.C., Wentzel, E.A., Kent, O.A., Ramachandra, K., Mulendore, M., Lee, K.H., et al., 2007. Transactivation of miR-34a by p53 broadly influences gene expression and promotes apoptosis. Mol. Cell 26, 745–752.

Chapman, H.A., 2011. Epithelial-mesenchymal interactions in pulmonary fibrosis. Annu. Rev. Physiol. 3, 413–435.

Charafe-Jauffret, E., Tarpin, C., Bardou, V.J., Bertucci, F., Ginestier, C., Braud, A.C., et al., 2004. Immunophenotypic analysis of inflammatory breast cancers: identification of an "'inflammatory signature" J. Pathol. 202, 265–273.

Charafe-Jauffret, E., Mrad, K., Labidi, S.I., Hamida, A.B., Romdhane, K.B., Abdallah, M.B., et al., 2007. Inflammatory breast cancers in Tunisia and France show similar immunophenotypes. J. Pathol. 202, 265–273.

Chaturvedi, M.M., Sung, B., Yadav, V.R., Kannappan, R., Aggarwal, B.B., 2011. NF-κB addiction and its role in cancer: 'one size does not fit all'. Oncogene 30, 1615–1630.

Chaudhary, J., Mosher, R., Kim, G., Skinner, M.K., 2000. Role of the winged helix transcription factor (WIN) in the regulation of Sertoli cell differentiated functions: WIN acts as an early event gene for follicle-stimulating hormone. Endocrinology 141, 2758–2766.

Chawla, J.S., Ma, C.X., Ellis, M.J., 2010. Neoadjuvant endocrine therapy for breast cancer. Surg. Oncol. Clin. N. Am. 19, 627–638.

Chen, Y., Poon, R.Y., 2008. The multiple checkpoint functions of CHK1 and CHK2 in maintenance of genome stability. Front. Biosci. 13, 5016–5029.

Chen, J., Stubbe, J., 2005. Bleomycins: towards better therapeutics. Nat. Rev. Cancer 5, 102–112.

Chen, Z., Gibson, T.B., Robinson, F., Silvestro, L., Pearson, G., Yu, B.E., et al., 2001. MAP kinases. Chem. Rev. 101, 2449–2476.

Chen, C.H., Chien, C.Y., Huang, C.C., Hwang, C.F., Chuang, H.C., Fang, F.M., et al., 2009. Expression of FLJ10540 is correlated with aggressiveness of oral cavity squamous cell carcinoma by stimulating cell migration and invasion through increased FOXM1 and MMP-2 activity. Oncogene 28, 2723–2737.

Chen, W.D., Wang, Y.D., Zhang, L., Shiah, S., Wang, M., Yang, F., et al., 2010. Farnesoid X receptor alleviates age-related proliferation defects in regenerating mouse livers by activating Forkhead Box m1b transcription. Hepatology 51, 953–962.

Chen, W., Yuan, K., Tao, Z.Z., Xiao, B.K., 2011a. Deletion of Forkhead Box M1 transcription factor reduces malignancy in laryngeal squamous carcinoma cells. Asian Pac. J. Cancer Prev. 12, 1785–1788.

Chen, D., Frezza, M., Schmitt, S., Kanwar, J., Dou, Q.P., 2011b. Bortezomib as the first proteasome inhibitor anticancer drug. Current status and future perspectives. Curr. Cancer Drug Targets 11, 239–253.

Chen, H., Yang, C., Yu, L., Xie, L., Hu, J., Zeng, L., et al., 2012a. Adenovirus-mediated RNA interference targeting FOXM1 transcription factor suppresses cell proliferation and tumor growth of nasopharyngeal carcinoma. J. Gene Med. 14, 231–240.

Chen, C., Zimmermann, M., Tinhofer, I., Kaufmann, A.M., Albers, A.E., 2012b. Epithelial-to-mesenchymal transition and cancer stem(-like) cells in head and neck squamous cell carcinoma. Cancer Lett. Epub Jul 4.

Chen, X., Müller, G.A., Quaas, M., Fischer, M., Han, N., Stutchbury, B., et al., 2013a. The forkhead transcription factor FOXM1 controls cell cycle-dependent gene expression through an atypical chromatin binding mechanism. Mol. Cell. Biol. 33, 227–236.

Chen, P.M., Wu, Y.H., Li, M.C., Cheng, Y.W., Chen, C.Y., Lee, H., 2013b. MnSOD promotes tumor invasion via upregulation of FoxM1-MMP2 axis and related with poor survival and relapse in lung adenocarcinomas. Mol. Cancer Res. 11, 261–271.

Chene, P., 2004a. Inhibition of the p53-MDM2 interaction: targeting a protein-protein interface. Mol. Cancer Res. 2, 20–28.

Chene, P., 2004b. Inhibition of the p53-hdm2 interaction with low molecular weight compounds. Cell Cycle 3, 60–461.

Cheng, J., DeCaprio, J.A., Fluck, M.M., Schaffhausen, B.S., 2009. Cellular transformation by simian virus 40 and murine polyoma virus T antigens. Semin. Cancer Biol. 19, 218–228.

Cheng, C.K., Li, L., Cheng, S.H., Ng, K., Chan, N.P.H., Ip, R.K.L., et al., 2011. Secreted-frizzled related protein 1 is a transcriptional repression target of the t(8;21) fusion protein in acute myeloid leukemia. Blood 118, 6638–6648.

Chetty, C., Bhoopathi, P., Rao, J.A., Lakka, S.S., 2009. Inhibition of matrix metalloproteinase-2 enhances radiosensitivity by abrogating radiation-induced FoxM1-mediated G2/M arrest in A549 lung cancer cells. Int. J. Cancer 124, 2468–2477.

Chibon, F., Lagarde, P., Salas, S., Perot, G., Brouste, V., Tirode, F., et al., 2010. Validated prediction of clinical outcome in sarcomas and multiple types of cancer on the basis of a gene expression signature related to genome complexity. Nat. Med. 16, 781–787.

Chinnadurai, G., 2011. Opposing oncogenic activities of small DNA tumor virus transforming proteins. Trends Microbiol. 19, 174–183.

Chinnam, M., Goodrich, D.W., 2011. RB1, development, and cancer. Curr. Top. Dev. Biol. 94, 129–169.

Chisari, F.V., 2000. Rous-Whipple award lecture. Viruses, immunity and cancer: lessons from hepatitis B. Am. J. Pathol. 156, 1117–1132.

Cho, K.B., Cho, M.K., Lee, W.Y., Kang, K.W., 2010. Overexpression of c-myc induces epithelial mesenchymal transition in mammary epithelial cells. Cancer Lett. 293, 230–239.

Chopra, R., Pu, Q.Q., Elefanty, A.G., 1999. Biology of BCR-ABL. Blood Rev. 13, 211–229.

Chow, M.T., Möller, A., Smyth, M.J., 2012. Inflammation and immune surveillance in cancer. Semin. Cancer Biol. 22, 23–32.

Christiansen, J.J., Rajasekran, A.K., 2006. Reassessing epithelial to mesenchymal transition as a prerequisite for carcinoma invasion and metastasis. Cancer Res. 66, 8319–8326.

Christoffersen, N.R., Shalgi, R., Frankel, L.B., Leucci, E., Lees, M., Klausen, M., et al., 2010. p53-independent upregulation of miR-34a during oncogene-induced senescence represses MYC. Cell Death Differ. 17, 236–245.

Christofori, G., Semb, H., 1999. The role of the cell-adhesion molecule E-cadherin as a tumour-suppressor gene. Trends Biochem. Sci. 24, 73–76.

Chu, X.Y., Zhu, Z.M., Chen, L.B., Wang, J.H., Su, Q.S., Yang, J.R., et al., 2012. FOXM1 expression correlates with tumor invasion and a poor prognosis of colorectal cancer. Acta Histochem. 114, 755–762.

Chua, P.J., Yip, G.W.C., Bay, B.H., 2009. Cell cycle arrest induced by hydrogen peroxide is associated with modulation of oxidative stress related genes in breast cancer cells. Exp. Biol. Med. 234, 1086–1094.

Chuah, C., Melo, J.V., 2009. Targeted treatment of imatinib-resistant chronic myeloid leukemia: focus on dasatinib. Onco Targets Ther. 2, 83–94.

Chung, E., Kondo, M., 2011. Role of Ras/Raf/MEK/ERK signaling in physiological hematopoiesis and leukemia development. Immunol. Res. 49, 248–268.

Chung, C.H., Bernard, P.S., Perou, C.M., 2002. Molecular portraits and the family tree of cancer. Nat. Genet. 32, 533–540.

Ciaccia, A., Elledge, S.J., 2010. The DNA damage response: making it safe to play with knives. Mol. Cell 40, 179–204.

Ciardiello, F., 2005. Epidermal growth factor receptor inhibitors in cancer treatment. Future Oncol. 1, 221–234.

Cicatiello, L., Scafoglio, C., Altucci, L., Cancemi, M., Natoli, G., Facchiano, A., et al., 2004. A genomic view of estrogen actions in human breast cancer cells by expression profiling of the hormone-responsive transcriptome. J. Mol. Endocrinol. 32, 719–775.

Cichowski, K., Hahn, W.C., 2008. Unexpected pieces to the senescence puzzle. Cell 133, 958–961.

Cilloni, D., Saglio, G., 2012. Molecular pathways: BCR-ABL. Clin. Cancer Res. 18, 930–937.

Classon, M., Harlow, E., 2002. The retinoblastoma tumour suppressor protein in development and cancer. Nat. Rev. Cancer 2, 910–917.

Clevers, H., 2006. Wnt/β-catenin signaling in development and disease. Cell 127, 469–480.

Clevers, H., Nusse, R., 2012. Wnt/β-catenin signaling and disease. Cell 149, 1192–1205.

Cobrinik, D., 1996. Regulatory interactions among E2Fs and cell cycle control proteins. Curr. Top. Microbiol. Immunol. 208, 31–61.

Cobrinik, D., 2005. Pocket proteins and cell cycle control. Oncogene 24, 2796–2809.

Coffelt, S.B., Hughes, R., Lewis, C.E., 2009. Tumor-associated macrophages. Effectors of angiogenesis and tumor progression. Biochim. Biophys. Acta 1796, 11–18.

Collado, M., Serrano, M., 2005. The senescent side of tumor suppression. Cell Cycle 4, 1722–1724.

Collado, M., Serrano, M., 2006. The power and promise of oncogene-induced senescence markers. Nat. Rev. Cancer 6, 472–476.

Collado, M., Serrano, M., 2010. Senescence in tumors: evidence from mice and humans. Nat. Rev. Cancer 10, 51–57.

Collado, M., Blasco, M.A., Serrano, M., 2007. Cellular senescence in cancer and ageing. Cell 130, 223–233.

Collins, D., Hill, A.D., Young, L., 2009. Lapatinib. A competitor or companion to trastuzumab? Cancer Treat. Rev. 35, 574–581.

Colotta, F., Allavena, P., Sica, A., Garlanda, C., Mantovani, A., 2009. Cancer-related inflammation, the seventh hallmark of cancer: Links to genetic instability. Carcinogenesis 30, 1073–1081.

Colpaert, C.G., Vermeulen, P.B., Benoy, I., Soubry, A., Van Roy, F., van Beest, P., et al., 2003. Inflammatory breast cancer shows angiogenesis with high endothelial proliferation rate and strong E-cadherin expression. Br. J. Cancer 88, 718–725.

Comen, A., 2012. Tracking the seed and tending the soil: evolving concepts in metastatic breast cancer. Discov. Med. 14, 97–104.

Conacci-Sorrell, M., Zhurinsky, J., Ben-Ze'ev, A., 2002. The cadherin-catenin adhesion system in signaling and cancer. J. Clin. Invest. 109, 987–991.

Coppe, J.P., Desprez, P.Y., Krtolica, A., Campisi, J., 2010. The senescence-associated secretory phenotype: the dark side of tumor suppression. Annu. Rev. Pathol. Mech. Dis. 5, 99–118.

Corvol, H., Flamein, F., Epaud, R., Clement, A., Guillot, L., 2009. Lung alveolar epithelium and interstitial lung disease. Int. J. Biochem. Cell Biol. 41, 1643–1651.

Costa, R.H., 2005. FoxM1 dances with mitosis. Nat. Cell Biol. 7, 108–110.

Costa, R.H., Kalinichenko, V.V., Lim, L., 2001. Transcription factors in lung development and function. Am. J. Physiol. Lung Cell. Mol. Physiol. 280, L823–L838.

Costa, R.H., Kalinichenko, V.V., Holterman, A.X.L., Wang, X., 2003. Transcription factors in liver development, differentiation, and regeneration. Hepatology 38, 1331–1347.

Costa, R.H., Kalinichenko, V.V., Major, M.L., Raychaudhuri, P., 2005a. New and unexpected: forkhead meets ARF. Curr. Opin. Genet. Dev. 15, 42–48.

Costa, R.H., Kalinichenko, V.V., Tan, Y., Wang, I.C., 2005b. The CAR nuclear receptor and hepatocyte proliferation. Hepatology 42, 1004–1008.

Coultas, L., Strasser, A., 2000. The molecular control of DNA damage-induced cell death. Apoptosis 5, 491–507.

Coultas, L., Chawengsaksophak, K., Rossant, J., 2005. Endothelial cells and VEGF in vascular development. Nature 438, 937–945.

Courtois-Cox, S., Jones, S.L., Cichowski, K., 2008. Many roads lead to oncogene-induced senescence. Oncogene 27, 2801–2809.

Coussens, L.M., Werb, Z., 2002. Inflammation and cancer. Nature 420, 860–867.

Coussens, L.M., Zitvogel, L., Palucka, A.K., 2013. Neutralizing tumor-promoting chronic inflammation: a magic bullet? Science 339, 286–291.

Coward, W.R., Saini, G., Jenkins, G., 2010. The pathogenesis of idiopathic pulmonary fibrosis. Ther. Adv. Respir. Dis. 4, 367–388.

Cowling, V.H., Cole, M.D., 2007. E-cadherin repression contributes to c-Myc-induced epithelial cell transformation. Oncogene 26, 3582–3586.

Craig, D.W., O'Shaughnessy, J.A., Kiefer, J.A., Aldrich, J., Sinari, S., Moses, T.M., et al., 2013. Genome and transcriptome sequencing in prospective metastatic negative breast cancer uncovers therapeutic vulnerabilities. Mol. Cancer Ther. 12, 104–116.

Cress, W.D., Nevins, J.R., 1996. Use of the E2F transcription factor by DNA tumor virus regulators proteins. Curr. Opin. Microbiol. Immunol. 208, 63–78.

Cully, M., You, H., Levine, A.J., Mak, T.W., 2006. Beyond PTEN mutations: the PI3K pathway as an integrator of multiple inputs during tumorigenesis. Nat. Rev. Cancer 6, 184–192.

Curtis, C., Shah, S.P., Chin, S.F., Turashvili, G., Rueda, O.M., Dunning, M.J., et al., 2012. The genomic and transcriptomic architecture of 2,000 breast tumours reveals novel subgroups. Nature 486, 346–352.

Cusack, J.C., 2003. Rationale for the treatment of solid tumors with the proteasome inhibitor bortezomib. Cancer Treat. Rev. 29, 21–31.

Cvek, B., Dvorak, Z., 2011. The ubiquitin-proteasome system (UPS) and the mechanism of action of bortezomib. Curr. Pharm. Des. 17, 1483–1499.

Daboussi, F., Dumay, A., Delacote, F., Lopez, B.S., 2002. DNA double-strand break repair signalling: the case of RAD51 post-translational regulation. Cell Cycle 14, 969–975.

D'Abramo, C.M., Archambault, J., 2011. Small molecule inhibitors of human papillomavirus protein–protein interactions. Open Virol. J. 5, 80–95.

d'Adda di Fagagna, F., 2008. Living on a break: cellular senescence as a DNA-damage response. Nat. Rev. Cancer 8, 512–522.

Dai, B.D., Kang, S.H., Gong, W., Liu, M., Aldape, K.D., Sawaya, R., et al., 2007. Aberrant FoxM1B expression increases matrix metalloproteinase-2 transcription and enhances the invasion of glioma cells. Oncogene 26, 6212–6219.

Dai, B., Pieper, R.O., Li, D., Wei, P., Liu, M., Woo, S.Y., et al., 2010. FoxM1B regulates NEDD4-1 expression, leading to cellular transformation and full malignant phenotype in immortalized astrocytes. Cancer Res. 70, 2951–2961.

Dai, B., Gong, A., Jing, Z., Aldape, K.D., Kang, S.H., Sawaya, R., et al., 2013. Forkhead box M1 is regulated by heat shock factor 1 and promotes glioma cells survival under heat shock stress. J. Biol. Chem. 288, 1634–1642.

Damania, B., 2007. DNA tumor viruses and human cancer. Trends Mol. Med. 15, 38–44.

Dancey, J., Sausville, E.A., 2003. Issues and progress with protein kinase inhibitors for cancer treatment. Nat. Rev. Drug Discov. 2, 196–313.

D'Andrea, A.D., 2003. The Fanconi road to cancer. Genes Dev. 17, 1933–1936.

Dar, A.C., Shokat, K.M., 2011. The evolution of protein kinase inhibitors from antagonists to agonists of cellular signaling. Annu. Rev. Biochem. 80, 769–795.

Daub, H., Specht, K., Ullrich, A., 2004. Strategies to overcome resistance to targeted protein kinase inhibitors. Nat. Rev. Drug Discov. 3, 1001–1010.

Davey, N.E., Trave, G.A., Gibson, T.J., 2011. How viruse hijack cell regulation. Trends Biochem. Sci. 36, 159–169.

David, J.M., Rajasekaran, A.K., 2012. Dishonorable discharge: the oncogenic roles of cleaved E-cadherin fragments. Cancer Res. 72, 2917–2923.

Davidson, B., Trope, C.G., Reich, R., 2012. Epithelial-mesenchymal transition in ovarian carcinoma. Front. Oncol. 2, 33.

Davis, D.B., Lavine, J.A., Suhonen, J.I., Krautkramer, K.A., Rabaglia, M.E., Sperger, J.M., et al., 2010. FoxM1 is up-regulated by obesity and stimulates β-cell proliferation. Mol. Endocrinol. 24, 1822–1834.

Dayaram, T., Marriott, S.J., 2008. Effect of transforming viruses on molecular mechanisms associated with cancer. J. Cell. Physiol. 216, 309–314.

Dean, J.L., McClendon, A.K., Stengel, K.R., Knudsen, E.S., 2012. Modification of the DNA damage response by therapeutic CDK4/6 inhibition. Oncogene 29, 68–80.

DeCaprio, J.A., 2009. How the Rb tumor suppressor structure and function was revealed by the study of Adenovirus and SV40. Virology 384, 274–284.

Deckbar, D., Jeggo, P.A., Löbrich, M., 2011. Understanding the limitations of radiation-induced cell cycle checkpoints. Crit. Rev. Biochem. Mol. Biol. 46, 271–283.

Deeb, K.K., Michalowska, A.M., Yoon, C.Y., Krummey, S.M., Hoenerhoff, M.J., Kavanaugh, C., et al., 2007. Identification of an integrated SV40 T/t-antigen cancer signature in aggressive human breast, prostate, and lung carcinomas with poor prognosis. Cancer Res. 67, 8065–8080.

DeGregori, J., 2004. The Rb network. J. Cell Sci. 117, 3411–3413.

Deininger, M.W., 2008. Nilotinib. Clin. Cancer Res. 14, 4027–4031.

Deininger, M., Druker, B.J., 2003. Specific targeted therapy of chronic myeloid leukemia with imatinib. Pharmacol. Rev. 55, 401–423.

Deininger, M., Buchdunger, E., Druker, B.J., 2005. The development of imatinib as a therapeutic agent for chronic myeloid leukemia. Blood 1, 2640–2653.

Dejosez, M., Zwaka, T.P., 2012. Pluripotency and nuclear reprogramming. Annu. Rev. Biochem. 81, 737–765.

de Krijger, I., Mekenkamp, L.J.M., Punt, C.J.A., Nagtegaal, I.D., 2011. MicroRNAs in colorectal cancer metastasis. J. Pathol. 224, 438–447.

de Lau, W., Barker, N., Clevers, H., 2007. WNT signaling in the normal intestine and colorectal cancer. Front. Biosci. 12, 471–491.

Del Prete, A., Allavena, P., Santoro, G., Fumarulo, R., Corci, M.M., Mantovani, A., 2011. Molecular pathways in cancer-related inflammation. Biochem. Med. 21, 264–275.

Delpuech, O., Griffiths, B., East, P., Essafi, A., Lam, E.W.F., Burgering, B., et al., 2007. Induction of Mxi1-SRα by FOXO3a contributes to repression of Myc dependent gene expression. Mol. Cell. Biol. 27, 4917–4930.

De Luca, A., Maiello, M.R., D'Alessio, A., Pergameno, M., Normanno, N., 2012. The RAS/RAF/MEK/ERK and the PI3K/AKT signalling pathways: role in cancer pathogenesis and implications for therapeutic approaches. Expert Opin. Ther. Targets 16, S17–S27.

Demaria, S., Pikarsky, E., Karin, M., Coussens, L.M., Chen, Y.C., El-Omar, E.M., et al., 2010. Cancer and inflammation: promise for biologic therapy. J. Immunother. 33, 335–351.

DeMartino, G.N., 2004. Proteasome inhibition: mechanism of action. J. Natl. Compr. Canc. Netw. 2, S5–S9.

De Marzo, A.M., Knudsen, B., Chan-Tack, K., Epstein, J.I., 1999. E-cadherin expression as a marker of tumor aggressiveness in routinely processed radical prostatectomy specimens. Urology 53, 707–713.

Demirci, C., Ernst, S., Alvarez-Perez, J.C., Rosa, T., Valle, S., Shridhar, V., et al., 2012. Loss of HGF/c-Met signaling in pancreatic β-cells leads to incomplete maternal β-cell adaptation and gestational diabetes. Diabetes 61, 1143–1152.

De Mitri, M.S., Cassino, R., Bernardi, M., 2010. Hepatitis B virus-related hepatocarcinogenesis: molecular oncogenic potential of clear and occult infections. Eur. J. Cancer 46, 2178–2186.

Deng, W., Xu, Y., 2009. Genome integrity: linking pluripotency and tumorigenicity. Trends Genet. 25, 425–427.

de Olano, N., Koo, C.Y., Monteiro, L.J., Pinto, P.H., Gomes, A.R., Aligue, R., et al., 2012. The p38-MAPK-MK2 axis regulates E2F1 and FOXM1 expression after epirubicin treatment. Mol. Cancer Res. 10, 1189–1202.

Deremer, D.L., Ustun, C., Natarajan, K., 2008. Nilotinib: a second-generation tyrosine kinase inhibitor for the treatment of chronic myelogenous leukemia. Clin. Ther. 30, 1956–1975.

Desmouliere, A., Chaponnier, C., Gabbiani, G., 2005. Tissue repair, contraction, and the myofibroblast. Wound Repair Regen. 13, 7–12.

Dever, S.M., White, E.R., Hartman, M.C., Valerie, K., 2012. BRCA1-directed, enhanced and aberrant homologous recombination: mechanism and potential treatment strategies. Cell Cycle 11, 687–694.

de Visser, K.E., Coussens, L.M., 2006. The inflammatory tumor microenvironment and its impact on cancer development. Contrib. Microbiol. 13, 118–137.

de Visser, K.E., Eichten, A., Coussens, L.M., 2006. Paradoxical roles of the immune system during cancer development. Nat. Rev. Cancer 6, 24–37.

de Vries, W.V., Evsikov, A.V., Brogan, L.J., Anderson, C.P., Graber, J.H., Knowles, B.B., et al., 2008. Reprogramming and differentiation in mammals: motifs and mechanisms. Cold Spring Harb. Symp. Quant. Biol. 73, 33–38.

Dhillon, A.S., Hagan, S., Rath, O., Kolch, W., 2007. MAP kinase signalling pathways in cancer. Oncogene 26, 3279–3290.

Diao, J., Garces, R., Richardson, C.D., 2001. X protein of hepatitis B virus modulates cytokine and growth factor related signal transduction pathways during the course of viral infections and hepatocarcinogenesis. Cytokine Growth Factor Rev. 12, 189–205.

Dibb, M., Han, N., Choudhury, J., Hayes, S., Valentine, H., West, C., et al., 2012. The FOXM1-PLK1 axis is commonly upregulated in oesophageal adenocarcinoma. Br. J. Cancer 107, 1766–1775.

DiDonato, J.A., Mercurio, F., Karin, M., 2012. NF-κB and the link between inflammation and cancer. Immunol. Rev. 246, 379–400.

Dillman, R.O., 2011. Cancer immunotherapy. Cancer Biother. Radiopharm. 16, 1–64.

Dimberu, P.M., Leonhardt, R.M., 2011. Cancer immunotherapy takes a multi-faceted approach to kick the immune system into gear. Yale J. Biol. Med. 84, 371–380.

Di Micco, R., Fumagalli, M., d'Adda di Fagagna, F., 2007. Breaking news: high-speed race ends in arrest—how oncogenes induce senescence. Trends Cell Biol. 17, 529–536.

Dominguez-Brauer, C., Brauer, P.M., Chen, Y.J., Pimkina, J., Raychaudhuri, P., 2010. Tumor suppression by ARF: gatekeeper and caretaker. Cell Cycle 9, 86–89.

Donato, N.J., Wu, J.Y., Stapley, J., Gallick, G., Lin, H., Arlinghaus, R., et al., 2003. BCR-ABL independence and LYN kinase overexpression in chronic myelogenous leukemia cells selected for resistance to STI571. Blood 101, 690–698.

Dong, H.M., Liu, G., Hou, Y.F., Wu, J., Lu, J.S., Luo, J.M., et al., 2007. Dominant-negative E-cadherin inhibits the invasiveness of inflammatory breast cancer cells in vitro. J. Cancer Res. Clin. Oncol. 133, 83–92.

Doorbar, J., 2006. Molecular biology of human papillomavirus infection and cervical cancer. Clin. Sci. 110, 525–541.

Dougan, M., Dranoff, G., 2009. Immune therapy for cancer. Annu. Rev. Immunol. 27, 83–117.

Down, C.F., Millour, J., Lam, E.W.F., Watson, R.J., 2012. Binding of FoxM1 to G2/M gene promoters is dependent upon B-Myb. Biochim. Biophys. Acta 1819, 855–862.

Downward, J., 2003. Targeting Ras signalling pathways in cancer therapy. Nat. Rev. Cancer 3, 11–22.

Druker, B.J., 2002a. STI571 (Gleevec) as a paradigm for cancer therapy. Trends Mol. Med. 8, S14–S18.

Druker, B.J., 2002b. Inhibition of Bcr-Abl tyrosine kinase as therapeutic strategy for CML. Oncogene 21, 8541–8546.

Druker, B.J., 2002c. Perspectives on the development of a molecularly targeted agent. Cancer Cell 1, 31–36.

Druker, B.J., 2004. Imatinib as a paradigm of targeted therapies. Adv. Cancer Res. 91, 1–30.

Druker, B.J., 2008. Translation of the Philadelphia chromosome into therapy for CML. Blood 112, 4808–4817.

Druker, B.J., 2009. Perspectives on the development of imatinib and the future of cancer research. Nat. Med. 15, 1149–1152.

Du, W., Pogoriler, J., 2006. Retinoblastoma family genes. Oncogene 25, 5190–5200.

Duda, D.G., Jain, R.K., 2010. Premetastatic lung "niche": is vascular endothelial growth factor receptor 1 activation required? Cancer Res. 70, 5670–5673.

Dudkina, A.S., Lindsley, C.W., 2007. Small molecule protein-protein inhibitors for the p53-MDM2 interaction. Curr. Top. Med. Chem. 7, 952–960.

Duensing, S., Münger, K., 2002. Human papillomaviruses and centrosome duplication errors: modeling the origins of genomic instability. Oncogene 21, 6241–6248.

Duensing, S., Münger, K., 2004. Mechanisms of genomic instability in human cancer: insights from studies with human papillomavirus oncoproteins. Int. J. Cancer 109, 156–162.

Duensing, A., Spardy, N., Chatterjee, P., Zheng, L., Parry, J., Cuevas, R., et al., 2009. Centrosome overduplication, chromosomal instability, and human papillomavirus oncoproteins. Environ. Mol. Mutagen. 50, 741–747.

Duffaud, F., Le Chesne, A., 2009. Imatinib in the treatment of solid tumours. Target. Oncol. 4, 45–56.

Dutta, J., Fan, Y., Gupta, N., Fan, G., Gelinas, C., 2006. Current insights into the regulation of programmed cell death by NF-κB. Oncogene 25, 6800–6816.

Dykxhoorn, D.M., 2010. MicroRNAs and metastasis: litte RNAs go a long way. Cancer Res. 70, 6401–6406.

Dyson, N.B., 1998. The regulation of E2F by pRB-family proteins. Genes Dev. 12, 2245–2262.

Ecker, A., Simma, O., Hoebl, A., Kenner, L., Beug, H., Moriggl, R., et al., 2009. The dark and the bright side of Stat3: proto-oncogene and tumor suppressor. Front. Biosci. 14, 2944–2958.

Einsele, H., 2010. Bortezomib. Recent Results Cancer Res. 184, 173–187.

El-Deiry, W.S., Tokino, T., Velculescu, V.E., Levy, D.B., Parsons, R., Trent, J.M., et al., 1993. *WAF1*, a potential mediator of p53 tumor suppression. Cell 75, 817–825.

Elgaaen, B.V., Olstad, O.K., Sandvik, L., Odegaard, E., Sauer, T., Staff, A.C., et al., 2012. ZNF385B and VEGFA are strongly differentially expressed in serous ovarian carcinomas and correlate with survival. PLoS One 7, e46317.

Emens, L.A., 2005. Trastuzumab: targeted therapy for the management of HER-2/neu-overexpressing metastatic breast cancer. Am. J. Ther. 12, 243–253.

Endter, C., Dobner, T., 2004. Cell transformation by adenoviruses. Curr. Top. Microbiol. Immunol. 273, 163–214.

Engelman, J.A., 2009. Targeting PI3K signalling in cancer: opportunities, challenges and limitations. Nat. Rev. Cancer 9, 550–562.

Engelman, J.A., Settleman, J., 2008. Acquired resistance to tyrosine kinase inhibitors during cancer therapy. Curr. Opin. Genet. Dev. 18, 73–79.

Ertel, A., Dean, J.L., Rui, H., Liu, C., Witkiewicz, A.K., Knudsen, K.E., et al., 2010. RB-pathway disruption in breast cancer. Differential association with disease subtypes, disease-specific prognosis and therapeutic response. Cell Cycle 9, 4153–4163.

Espinosa, E., Zamora, P., Feliu, J., Gonzalez Baron, M., 2003. Classification of anticancer drugs—a new system based on therapeutic targets. Cancer Treat. Rev. 29, 515–523.

Evan, G.I., d'Adda di Fagagna, F., 2009. Cellular senescence: hot or what? Curr. Opin. Genet. Dev. 19, 25–31.

Evan, G.I., Christophorou, M., Lawlor, E.A., Ringshausen, I., Prescott, J., Dansen, T., et al., 2005. Oncogene-dependent tumor suppression: using the dark side of the force for cancer therapy. Cold Spring Harb. Symp. Quant. Biol. 70, 263–273.

Evers, B., Helleday, T., Jonkers, J., 2010. Targeting homologous recombination repair defects in cancer. Trends Pharmacol. Sci. 31, 372–380.

Ewald, J.A., Desotelle, J.A., Wilding, G., Jarrard, D.F., 2010. Therapy-induced senescence in cancer. J. Natl. Cancer Inst. 102, 1536–1546.

Ezan, J., Leroux, L., Barandon, L., Dufourcq, P., Jaspard, B., Moreau, C., et al., 2004. FrzA/sFRP-1, a secreted antagonist of the Wnt-Frizzled pathway, controls vascular cell proliferation in vitro and in vivo. Cardiovasc. Res. 63, 731–738.

Fan, Y., Dutta, J., Gupta, N., Fan, G., Gelinas, C., 2008. Regulation of programmed cell death by NF-κB and its role in tumorigenesis and therapy. Adv. Exp. Med. Biol. 615, 223–250.

Fanciulli, M., 2006. Rb and Tumorigenesis. Landes Bioscience, Austin, TX.

Faust, D., Al-Butmeh, F., Linz, B., Dietrich, C., 2012. Involvement of the transcription factor FoxM1 in contact inhibition. Biochem. Biophys. Res. Commun. 426, 659–663.

Fehrmann, F., Laimins, L.A., 2003. Human papillomaviruses: targeting differentiating epithelial cells for malignant transformation. Oncogene 22, 5201–5207.

Feitelson, M.A., 1999. Hepatitix B virus in hepatocarcinogenesis. J. Cell. Physiol. 181, 188–202.

Feller, L., Wood, N.H., Khammissa, R.A.G., Lemmer, J., 2009. Human papillomavirus-mediated carcinogenesis and HPV-associated oral and oropharyngeal squamous cell carcinoma. Part 1: Human papillomavrius-mediated carcinogenesis. Head Face Med. 6, 14.

Felsani, A., Mileo, A.M., Paggi, M.G., 2006. Retinoblastoma family proteins as key targets of the small DNA virus oncoproteins. Oncogene 25, 5277–5285.

Feng, B., Wang, R., Chen, L.B., 2012. Review of miR-200b and cancer chemosensitivity. Biomed. Pharmacother. 66, 397–402.

Fernandez, I.E., Eickelberg, O., 2012. The impact of TGF-β on lung fibrosis: from targeting to biomarkers. Proc. Am. Thorac. Soc. 9, 111–116.

Fernandez, A.F., Esteller, M., 2010. Viral epigenomes in human tumorigenesis. Oncogene 29, 1405–1420.

Fernandez, P.C., Frank, S.R., Wang, L., Schroeder, M., Liu, S., Greene, J., et al., 2003. Genomic targets of the human c-Myc protein. Genes Dev. 17, 1115–1129.

Ferrara, N., 2004. Vascular endothelial growth factor: basic science and clinical progress. Endocr. Rev. 25, 581–611.

Ferrara, N., Gerber, H.P., LeCouter, J., 2003. The biology of VEGF and its receptors. Nat. Med. 9, 669–676.

Ferrari, R., Berk, A.J., Kurdistani, S.K., 2009. Viral manipulation of the host epigenome for oncogenic transformation. Nat. Rev. Genet. 10, 290–294.

Ferrari, E., Lucca, C., Foiani, M., 2010. Lethal combination for cancer cells: synthetic lethality screenings for drug discovery. Eur. J. Cancer 46, 2889–2895.

Fillingham, J., Keogh, M.C., Krogan, N.J., 2006. γH2AX and its role in DNA double-strand break repair. Biochem. Cell Biol. 84, 568–577.

Finn, K., Lowndes, N.F., Grenon, M., 2012. Eukaryotic DNA damage checkpoint activation in response to double-strand breaks. Cell. Mol. Life Sci. 69, 1447–1473.

Finzer, P., Aguilar-Lemarroy, A., Rösl, F., 2002. The role of human papillomavirus oncoproteins E6 and E7 in apoptosis. Cancer Lett. 188, 15–24.

Fiorentino, F.P., Marchesi, I., Giordano, A., 2013. On the role of retinoblastoma family proteins in the establishment and maintenance of the epigenetic landscape. J. Cell. Physiol. 228, 276–284.

Firsanov, D.V., Solovjeva, L.V., Svetlova, M.P., 2011. H2AX phosphorylation at the sites of DNA double-strand breaks in cultivated mammalian cells and tissues. Clin. Epigenetics 2, 283–297.

Fischer, O.M., Streit, S., Hart, S., Ullrich, A., 2003. Beyond Herceptin and Gleevec. Curr. Opin. Chem. Biol. 7, 490–495.

Fisher, R., Larkin, J., 2012. Vemurafenib: a new treatment for BRAF-V600 mutated advanced melanoma. Cancer Manag. Res. 4, 243–252.

Fisher, G.H., Wellen, S.L., Klimstra, D., Lenczowski, J.M., Tichelaar, J.W., Lizak, M.J., et al., 2001. Induction and apoptotic regression of lung adenocarcinomas by regulation of a K-Ras transgene in the presence and absence of tumor suppressor genes. Genes Dev. 15, 3249–3262.

Flanders, K.C., 2004. Smad 3 as a mediator of the fibrotic response. Int. J. Exp. Pathol. 85, 47–64.

Flint, J., Shenk, T., 1997. Viral transactivating proteins. Annu. Rev. Genet. 31, 177–212.

Floor, S., van Staveren, W.C.G., Larsimont, D., Dumont, J.E., Maenhaut, C., 2011. Cancer cells in epithelial-to-mesenchymal transition and tumor-propagating-cancer stem cells: distinct, overlapping or same populations. Oncogene 30, 4609–4621.

Fodde, R., Brabletz, T., 2007. Wnt/β-catenin signaling in cancer stemness and malignant behaviour. Curr. Opin. Cell Biol. 19, 150–158.

Forget, A.L., Kowalczykowski, S.C., 2010. Single-molecule imaging brings Rad51 nucleoprotein filaments into focus. Trends Cell Biol. 20, 269–276.

Forkert, P.G., 2010. Mechanisms of lung tumorigenesis by ethyl carbamate and vinyl carbamate. Drug Metab. Rev. 42, 355–378.

Fotouhi, N., Graves, B., 2005. Small-molecule inhibitors of p53/MDM2 interaction. Curr. Top. Med. Chem. 5, 159–165.

Foty, R.A., Steinberg, M.S., 2004. Cadherin-mediated cell-cell adhesion and tissue segregation in relation to malignancy. Int. J. Dev. Biol. 48, 397–409.

Fournier, M.V., Martin, K.J., Kenny, P.A., Xhaja, K., Bosch, I., Yaswen, P., et al., 2006. Gene expression signature in organized and growth arrested mammary acini predicts good outcome in breast cancer. Cancer Res. 66, 7095–7102.

Francis, R.E., Myatt, S.S., Krol, J., Hartman, J., Peck, B., McGovern, U.B., et al., 2009. FoxM1 is a downstream target and marker of HER2 overexpression in breast cancer. Int. J. Oncol. 35, 57–68.

Freije, W.A., Castro-Vargas, F.E., Fang, Z., Horvath, S., Cloughesy, T., Liau, L.M., et al., 2004. Gene expression profiling of gliomas strongly predicts survival. Cancer Res. 64, 6503–6510.

Freire, T., Osinaga, E., 2012. The sweet side of tumor immunotherapy. Immunotherapy 4, 719–734.

Frescas, D., Pagano, M., 2008. Deregulated proteolysis by the F-box proteins SKP2 and β-TrCP: tipping the scales of cancer. Nat. Rev. Cancer 8, 438–449.

Fresno Vara, J.A., Casado, E., de Castro, J., Cejas, P., Belda-Iniesta, C., Gonzalez-Baron, M., 2004. PI3K/Akt signalling pathway and cancer. Cancer Treat. Rev. 30, 193–204.

Freund, A., Orjalo, A.V., Desprez, P.Y., Campisi, J., 2010. Inflammatory networks during cellular senescence: causes and consequences. Trends Mol. Med. 16, 238–246.

Fribley, A., Wang, C.Y., 2006. Proteasome inhibitor induces apoptosis through induction of endoplasmic reticulum stress. Cancer Biol. Ther. 5, 745–748.

Fridman, J.S., Tainsky, M.A., 2008. Critical pathways in cellular senescence and immortalization revealed by gene expression profiling. Oncogene 27, 5975–5987.

Frisch, M., 2001. Tumor suppression activity of adenovirus E1a protein: anoikis and the epithelial phenotype. Adv. Cancer Res. 80, 39–49.

Frisch, S.M., Mymryk, J.S., 2002. Adenovirus-5 E1A: paradox and paradigm. Nat. Rev. Mol. Cell Biol. 3, 441–452.

Frolov, M.V., Dyson, N.J., 2004. Molecular mechanisms of E2F-dependent activation and pRB-mediated repression. J. Cell Sci. 117, 2173–2181.

Fry, D.C., Vassilev, L.T., 2005. Targeting protein-protein interactions for cancer therapy. J. Mol. Med. 83, 955–963.

Fu, Z., Malureanu, L., Huang, J., Wang, W., Li, H., van Deursen, J.M., et al., 2008. Plk1-dependent phosphorylation of FoxM1 regulates a transcriptional programme required for mitotic progression. Nat. Cell Biol. 10, 1076–1082.

Fu, Y., Zheng, S., An, N., Athanasopoulos, T., Popplewell, L., Liang, A., et al., 2011. β-catenin as a potential key target for tumor suppression. Int. J. Cancer 129, 1541–1551.

Fujii, T., Ueda, T., Nagata, S., Funaga, R., 2010. Essential role of p400/mDomino chromatin-remodeling ATPase in bone marrow hematopoiesis and cell-cycle progression. J. Biol. Chem. 285, 30214–30223.

Fung, J., Lai, C.L., Yuen, M.F., 2009. Hepatitis B and C virus-related carcinogenesis. Clin. Microbiol. Infect. 15, 964–970.

Futreal, P.A., Coin, L., Marshall, M., Down, T., Hubbard, T., Wooster, R., et al., 2004. A census of human cancer genes. Nat. Rev. Cancer 4, 177–183.

Fuxe, J., Vincent, T., Garcia de Herreros, A., 2010. Transcriptional crosstalk between TGF-β and stem cell pathways in tumor cell invasion. Role of EMT promoting Smad complexes. Cell Cycle 9, 2363–2374.

Gaestel, M., 2006. MAPKAP kinases—MKs—two's company, three's a crowd. Nat. Rev. Mol. Cell Biol. 7, 120–130.

Gallagher, S.J., Kefford, R.F., Rizos, H., 2006. The ARF tumor suppressor. Int. J. Biochem. Cell Biol. 28, 1637–1641.

Gallimore, P.H., Turnell, A.S., 2001. Adenovirus E1A: remodelling the host cell, a life and death experience. Oncogene 20, 7824–7835.

Gallo, D., Ferlini, C., Scambia, G., 2010. The epithelial-mesenchymal transition and the estrogen-signaling in ovarian cancer. Curr. Drug Targets 11, 474–481.

Galloway, D.A., McDougall, J.K., 1996. The disruption of cell cycle checkpoints by papillomavirus oncoproteins contributes to anogenital neoplasia. Semin. Cancer Biol. 7, 309–315.

Ganguly, N., Parihar, S., 2009. Human papillomavirus E6 and E7 oncoproteins as risk factors for tumorigenesis. J. Biosci. 34, 113–123.

Gao, D., Vahdat, L.T., Wong, S., Chang, J.C., Mittal, V., 2012. Microenvironmental regulation of epithelial-mesenchymal transitions in cancer. Cancer Res. 72, 4883–4889.

Garber, M.E., Troyanskaya, O.G., Schluens, K., Petersen, S., Thaesler, Z., Pacyna-Gengelbach, M., et al., 2001. Diversity of gene expression in adenocarcinoma of the lung. Proc. Natl. Acad. Sci. U.S.A. 98, 13784–13789.

Garcia, S., Dales, J.P., Jacquemier, J., Charafe-Jauffret, E., Birnbaum, D., Andrac-Meyer, L., et al., 2007. c-Met overexpression in inflammatory breast carcinomas: automated quantification on tissue microarrays. Br. J. Cancer 96, 329–335.

Garnett, T.O., Duerksen-Hughes, P.J., 2006. Modulation of apoptosis by human papillomavirus (HPV) oncoproteins. Arch. Virol. 151, 2321–2335.

Garraway, L.A., Jänne, P.A., 2012. Circumventing cancer drug resistance in the era of personalized medicine. Cancer Discov. 2, 214–226.

Gartel, A.L., 2008. FoxM1 inhibitors as potential anticancer drugs. Expert Opin. Ther. Targets 12, 663–665.

Gartel, A.L., 2010. A new target for proteasome inhibitors: FoxM1. Expert Opin. Investig. Drugs 19, 235–242.

Gartel, A.L., 2012. The oncogenic transcription factor FOXM1 and anticancer therapy. Cell Cycle 10, 1–2.

Gates, J., Peifer, M., 2005. Can 1000 reviews be wrong? Actin, α-catenin, and adherens junctions. Cell 123, 769–772.

Gauger, K.J., Hugh, J.M., Troester, M.A., Schneider, S.S., 2009. Down-regulation of sfrp1 in a mammary epithelial cell line promotes the development of a cd44high/cd24low population which is invasive and resistant to anoikis. Cancer Cell Int. 9, 11.

Gavert, N., Ben-Ze'ev, A., 2010. Coordinating changes in cell adhesion and phenotype during EMT-like processes in cancer. Biol. Rep. 2, 86.

Gavert, N., Ben-Ze'ev, A., 2007. β-Catenin signaling in biological control and cancer. J. Cell. Biochem. 102, 820–828.

Gavert, N., Ben-Ze'ev, A., 2008. Epithelial-mesenchymal transition and the invasive potential of tumors. Trends Mol. Med. 14, 199–209.

Gemenetzidis, E., Bose, A., Riaz, A.M., Chaplin, T., Young, B.D., Ali, M., et al., 2009. FOXM1 upregulation is an early event in human squamous cell carcinoma and it is enhanced by nicotine during malignant transformation. PLoS One 4, e4849.

Gemenetzidis, E., Elena-Costea, D., Parkinson, E.K., Waseem, A., Wan, H., Teh, M.T., 2010. Induction of epithelial stem/progenitor expansion by FOXM1. Cancer Res. 70, 9515–9526.

Geoghegan, E., Byrnes, L., 2008. Mouse induced pluripotent stem cells. Int. J. Dev. Biol. 52, 1015–1022.

Germano, G., Mantovani, A., Allavena, P., 2011. Targeting of the innate immunity/inflammation as complementary anti-tumor therapies. Ann. Med. 43, 581–593.

Gewirtz, D.A., 1999. A critical evaluation of the mechanisms of action proposed for the antitumor effects of the anthracycline antibiotics adriamycin and daunorubicin. Biochem. Pharmacol. 57, 727–741.

Gharaee-Khermani, M., Hu, B., Phan, S.H., Gyetko, M.R., 2009. Recent advances in molecular targets and treatment of idiopathic pulmonary fibrosis: focus on TGFβ signaling and the myofibroblast. Curr. Med. Chem. 16, 1400–1417.

Ghittoni, R., Accardi, R., Hasan, U., Gheit, T., Sylla, B., Tommasino, M., 2010. The biological properties of E6 and E7 oncoproteins from human papillomaviruses. Virus Genes 40, 1–13.

Giacca, M., 2010. Non-redundant functions of the protein isoforms arising from alternative splicing of the VEGF-A pre-mRNA. Transcription 1, 149–153.

Giacinti, C., Giordano, A., 2006. RB and cell cycle progression. Oncogene 25, 5220–5227.

Gialmanidis, I.P., Bravou, V., Amanetopoulou, S.G., Varakis, J., Kourea, H., Papadaki, H., 2009. Overexpression of hedgehog pathway molecules and FOXM1 in non-small cell lung carcinomas. Lung Cancer 66, 64–74.

Gianni-Barrera, R., Trani, M., Reginato, S., Banfi, A., 2011. To sprout o to split? VEGF, Notch and vascular morphogenesis. Biochem. Soc. Trans. 39, 1644–1648.

Giannoni, E., Parri, M., Chiarugi, P., 2012. EMT and oxidative stress: a bidirectional interplay affecting tumor malignancy. Antioxid. Redox Signal. 16, 1248–1263.

Gibson, B.A., Kraus, W.L., 2012. New insights into the molecular and cellular function of poly(ADP-ribose) and PARPs. Nat. Rev. Mol. Cell Biol. 13, 411–424.

Giehl, K., Menke, A., 2008. Microenvironmental regulation of E-cadherin-mediated adherens junctions. Front. Biosci. 13, 3975–3985.

Gieling, R.G., Elsharkawy, A.M., Caamano, J.H., Cowie, D.E., Wright, M.C., Ebrahimkhani, M.R., et al., 2010. The c-Rel subunit of nuclear factor-κB regulates murine liver inflammation, wound-healing, and hepatocyte proliferation. Hepatology 51, 922–931.

Giles, R.H., van Es, J.H., Clevers, H., 2003. Caught up in a Wnt storm: Wnt signaling in cancer. Biochim. Biophys. Acta 1653, 1–24.

Gillies, R.J., Verduzco, D., Gatenby, R.A., 2012. Evolutionary dynamics of carcinogenesis and why targeted therapy does not work. Nat. Rev. Cancer 12, 487–493.

Gingrich, J.R., Barrios, R.J., Morton, R.A., Boyce, B.F., DeMayo, F.J., Finegold, M.J., et al., 1996. Metastatic prostate cancer in a transgenic mouse. Cancer Res. 56, 4096–4102.

Glass, C.K., Rosenfeld, M.G., 2000. The coregulator exchange in transcriptional functions of nuclear receptors. Genes Dev. 14, 121–141.

Glickman, M.S., Sawyers, C.L., 2012. Converting cancer therapies into cures: lessons from infectious diseases. Cell 148, 1089–1098.

Goldman, J.M., Melo, J.V., 2008. BCR-ABL in chronic myelogenous leukemia—how does it work? Acta Haematol. 119, 212–217.

Gong, A., Huang, S., 2012. FoxM1 and Wnt/β-catenin signaling in glioma stem cells. Cancer Res. 72, 5658–5662.

Gonzales, K.A.U., Ng, H.H., 2011. Choreographing pluripotency and cell fate with transcription factors. Biochim. Biophys. Acta 1809, 337–349.

Gooding, J.M., Yap, K.L., Ikura, M., 2004. The cadherin-catenin complex as focal point of cell adhesion and signalling: new insights from three-dimensional structures. Bioessays 26, 497–511.

Goodwin, M., Yap, A.S., 2004. Classical cadherin adhesion molecules: coordinating cell adhesion, signaling and the cytoskeleton. J. Mol. Histol. 35, 839–844.

Gordon, G.M., Du, W., 2011. Conserved RB functions in development and tumor suppression. Protein Cell 2, 864–878.

Goss, P.E., Muss, H.B., Ingle, J.N., Whelan, T.J., Wu, M., 2008. Extended adjuvant endocrine therapy in breast cancer: current status and future directions. Clin. Breast Cancer 8, 411–417.

Gotzmann, J., Mikula, M., Eger, A., Schulte-Hermann, R., Foisner, R., Beug, H., et al., 2004. Molecular aspects of epithelial cell plasticity: implications for local tumor invasion. Mutat. Res. 566, 9–20.

Grabowska, M.M., Day, M.L., 2012. Soluble E-cadherin: more than a symptom of disease. Front. Biosci. 17, 1948–1964.

Grana, X., Garriga, J., Mayol, X., 1998. Role of the retinoblastoma family, pRB, p107, and p130 in the negative control of cell growth. Oncogene 17, 3365–3383.

Granovsky, A.E., Rosner, M.R., 2008. Raf kinase inhibitory protein: a signal transduction modulator and metastasis suppressor. Cell Res. 18, 452–457.

Grant, S.K., 2009. Therapeutic protein kinase inhibitors. Cell. Mol. Life Sci. 66, 1163–1177.

Green, K.J., Getsios, S., Troyanovsky, S., Godsel, L.M., 2010. Intercellular junction assembly, dynamics, and homeostasis. Cold Spring Harb. Symp. Quant. Biol. 2, a000125.

Green, M.R., Aya-Bonilla, C., Gandhi, M.K., Lea, R.A., Wellwood, J., Wood, P., et al., 2011. Integrative genomic profiling reveals conserved genetic mechanisms for tumorigenesis in common entities of Non-Hodgkin's lymphoma. Genes Chromosomes Cancer 50, 313–326.

Greenberg, R.A., 2011. Histone tails: directing the chromatin response to DNA damage. FEBS Lett. 585, 2883–2890.

Greenberg, N.M., DeMayo, F., Finegold, M.J., Medina, D., Tilley, W.D., Asspinall, J.O., et al., 1995. Prostate cancer in a transgenic mouse. Proc. Natl. Acad. Sci. U.S.A. 92, 3439–3443.

Gregory, P.A., Bracken, C.P., Bert, A.G., Goodall, G.J., 2008. MicroRNAs as regulators of epithelial-mesenchymal transition. Cell Cycle 7, 3112–3118.

Grivennikov, S.I., Karin, M., 2010. Inflammation and oncogenesis: a vicious connection. Curr. Opin. Genet. Dev. 20, 65–71.

Grivennikov, S.I., Greten, F.R., Karin, M., 2010. Immunity, inflammation, and cancer: the good, the bad and the ugly. Cell 140, 883–899.

Gross, G., Margalit, A., 2007. Targeting tumor-associated antigens to the MHC class I presentation pathway. Endocr. Metab. Immune Disord. Drug Targets 7, 99–109.

Gschwind, A., Fischer, O.M., Ullrich, A., 2004. The discovery of receptor tyrosine kinases: targets for cancer therapy. Nat. Rev. Cancer 4, 361–370.

Guarino, M., Rubino, B., Ballabio, G., 2007. The role of epithelial-mesenchymal transition in cancer pathology. Pathology 39, 305–318.

Guerra, E., Trerotola, M., Aloisi, A.L., Tripaldi, R., Vacca, G., La Sodda, R., et al., 2013. The Trop-2 signalling network in cancer growth. Oncogene 32, 1594–1600.

Guitierrez, M.E., Kummar, S., Giaccone, G., 2009. Next generation oncology drug development: opportunities and challenges. Nat. Rev. Clin. Oncol. 6, 259–265.

Gumbiner, B.M., 2005. Regulation of cadherin-mediated adhesion in morphogenesis. Nat. Rev. Mol. Cell Biol. 6, 622–634.

Gumz, M.L., Zou, H., Kreinest, P.A., Childs, A.C., Belmonte, L.S., LeGrand, S.N., et al., 2007. Secreted frizzled-related protein 1 loss contributes to tumor phenotype of clear renal cell carcinoma. Clin. Cancer Res. 13, 4740–4749.

Gunasinghe, N.P.A.D., Wells, A., Thompson, E.W., Hugo, H.J., 2012. Mesenchymal-epithelial transition (MET) as a mechanism for metastatic colonisation in breast cancer. Cancer Metastasis Rev. 31, 469–478.

Günther, A., Korfei, M., Mahavadi, P., von der Beck, D., Ruppert, C., Markart, P., 2012. Unravelling the progressive pathophysiology of idiopathic pulmonary fibrosis. Eur. Respir. Rev. 21, 152–160.

Guo, S., Liu, M., Gonzalez-Perez, R.R., 2011. Role of Notch and its oncogenic signaling crosstalk in breast cancer. Biochim. Biophys. Acta 1815, 197–213.

Gupta, P.B., Chaffer, C.L., Weinberg, R.A., 2009. Cancer stem cells: mirage or reality? Nat. Med. 15, 1010–1012.

Gurevich, I., Flores, A.M., Aneskievich, B.J., 2007. Corepressors of agonist-bound nuclear receptors. Toxicol. Appl. Pharmacol. 223, 288–298.

Guruvayoorappan, C., 2008. Tumor versus tumor-associated macrophages: how hot is the link? Integr. Cancer Ther. 7, 90–95.

Gusarova, G.A., Wang, I.C., Major, M.L., Kalinichenko, V.V., Ackerson, T., Petrovic, V., et al., 2007. A cell-penetrating ARF peptide inhibitor of FoxM1 in mouse hepatocellular carcinoma treatment. J. Clin. Invest. 117, 99–111.

Ha, S.Y., Lee, C.H., Chang, H.K., Chang, S., Kwon, K.Y., Lee, E.H., et al., 2012. Differential expression of forkhead box M1 and its downstream cyclin-dependent kinase inhibitors p27^{kip1} and p21$^{waf1/cip1}$ in the diagnosis of pulmonary neuroendocrine tumours. Histopathology 60, 731–739.

Haber, D.A., Gray, N.S., Baselga, J., 2011. The evolving war on cancer. Cell 145, 19–24.

Hagan, S., Garcia, R., Dhillon, A., Kolch, W., 2006. Raf kinase inhibitor protein regulation of raf and MAPK signaling. Methods Enzymol. 407, 248–259.

Hait, W.N., Habley, T.W., 2009. Targeted cancer therapeutics. Cancer Res. 69, 1263–1267.

Hajra, K.M., Fearon, E.R., 2002. Cadherin and catenin alterations in human cancer. Genes Chromosomes Cancer 34, 255–268.

Hakem, R., 2008. DNA-damage repair; the good, the bad, and the ugly. EMBO J. 27, 589–605.

Halaban, R., 2005. Rb/E2F: a two-edged sword in the melanocytic system. Cancer Metastasis Rev. 24, 339–356.

Halasi, M., Gartel, A.L., 2009. A novel mode of FoxM1 regulation: positive auto-regulatory loop. Cell Cycle 8, 1966–1967.

Halasi, M., Gartel, A.L., 2012. Suppression of FOXM1 sensitizes human cancer cells to cell death induced by DNA-damage. PLoS One 7, e31761.

Halasi, M., Gartel, A., 2013. Targeting FOXM1 in cancer. Biochem. Pharmacol. 85, 644–652.

Halasi, M., Dahari, H., Bhat, U.G., Gonzalez, E.B., Lyubimov, A.V., Tonetti, D.A., et al., 2010. Thiazole antibiotics against breast cancer. Cell Cycle 9, 1214–1217.

Halbleib, J.M., Nelson, W.J., 2006. Cadherins in development: cell adhesion, sorting, and tissue morphogenesis. Genes Dev. 20, 3199–3324.

Hall, P.S., Cameron, D.A., 2009. Current perspective—trastuzumab. Eur. J. Cancer 45, 12–18.

Hall, M., Peters, G., 1996. Genetic alterations of cyclins, cyclin-dependent kinases, and cdk inhibitors in human cancer. Adv. Cancer Res. 68, 67–108.

Hallstrom, T.C., Mori, S., Nevins, J.R., 2008. An E2F1-dependent gene expression program that determines the balance between proliferation and cell death. Cancer Cell 13, 11–22.

Hamai, A., Benlalam, H., Meslin, F., Hasmim, M., Carre, T., Akalay, I., et al., 2010. Immune surveillance of human cancer: if the cytotoxic T-lymphocytes play the music, does the tumoral system call the tune? Tissue Antigens 75, 1–8.

Hamid, N.A., Brown, C., Gaston, K., 2009. The regulation of cell proliferation by the papillomavirus early proteins. Cell. Mol. Life Sci. 66, 1700–1717.

Hanahan, D., Weinberg, R.A., 2000. The hallmarks of cancer. Cell 100, 57–70.

Hanahan, D., Weinberg, R.A., 2011. Hallmarks of cancer: the next generation. Cell 144, 646–674.

Hanna, J.H., Saha, K., Jaenisch, R., 2010. Pluripotency and cellular reprogramming: facts, hypotheses, unresolved issues. Cell 143, 508–525.

Hannah, A.L., 2005. Kinases as drug discovery targets in hematologic malignancies. Curr. Mol. Med. 5, 625–642.

Hantschel, O., Superti-Furga, G., 2004. Regulation of the c-Abl and Bcr-Abl tyrosine kinases. Nat. Rev. Mol. Cell Biol. 5, 33–44.

Hantschel, O., Rix, U., Superti-Furga, G., 2008. Target spectrum of the BCR-ABL inhibitors imatinib, nilotinib and dasatinib. Leuk. Lymphoma 49, 615–619.

Harder, J.L., Margolis, B., 2008. SnapShot: tight and adherens junction signaling. Cell 133, 1118 1118.e2.

Hardie, W.G., Hagood, J.S., Dave, V., Perl, A.K., Whitsett, J.A., Korfhagen, T.R., et al., 2010. Signaling pathways in the epithelial origins of pulmonary fibrosis. Cell Cycle 9, 1769–1776.

Harper, J.W., Elledge, S.J., 2007. The DNA damage response: ten years after. Mol. Cell 28, 739–745.

Harris, T., 2010. Gene and drug matrix for personalized cancer therapy. Nat. Rev. Drug Discov. 7, 660.

Harris, T., McCormick, F., 2010. Gene and drug matrix for personalized cancer therapy. The molecular pathology of cancer. Nat. Rev. Drug Discov. 7, 251–265.

Harris, T.J., Tepass, U., 2010. Adherens junctions: from molecules to morphogenesis. Nat. Rev. Mol. Cell Biol. 11, 502–514.

Harris, L.G., Samant, R.S., Shevde, L.A., 2011. Hedgehog signaling: networking to nurture a promalignant tumor microenvironment. Mol. Cancer Res. 9, 1165–1174.

Harrison, J.C., Haber, J.E., 2006. Surviving the break: the DNA damage checkpoint. Annu. Rev. Genet. 40, 209–235.

Hartsock, A., Nelson, W.J., 2008. Adherens and tight junctions: structure, function and connections to the actin cytoskeleton. Biochim. Biophys. Acta 1778, 660–669.

Haura, E.B., Turkson, J., Jove, R., 2005. Mechanisms of disease: insights into the emerging role of signal transducers and activators of transcription in cancer. Nat. Clin. Pract. Oncol. 2, 315–324.

Havelka, A.M., Berndtsson, M., Olofsson, M.H., Shoshan, M.C., Linder, S., 2007. Mechanisms of action of DNA-damaging anticancer drugs in treatment of carcinomas: is acute apoptosis an "off-target" effect? Mini Rev. Med. Chem. 7, 1035–1039.

Hay, N., 2005. The Akt-mTOR tango and its relevance to cancer. Cancer Cell 8, 179–183.

Hay, E., Nouraud, A., Marie, P.J., 2009. N-Cadherin negatively regulates osteoblast proliferation and survival by antagonizing Wnt, ERK and PI3K/Akt signalling. PLoS One 4, e8284.

Hayashida, T., Jinno, H., Kitagawa, Y., Kitajima, M., 2011. Cooperation of cancer stem cell properties and epithelial-mesenchymal transition in the establishment of breast cancer metastasis. J. Oncol. 2011, 591427.

Hazlehurst, L.A., Bewry, N.N., Nair, R.R., Pinilla-Ibarz, J., 2009. Signaling networks associated with BCR-ABL-dependent transformation. Cancer Control 16, 100–107.

He, T.C., Sparks, A.B., Rago, C., Hermeking, H., Zawel, L., da Costa, L.T., et al., 1998. Identification of c-MYC as a target of the APC pathway. Science 281, 1509–1512.

He, L., He, X., Lim, L.P., de Stanchina, E., Xuan, Z., Liang, Y., et al., 2007. A microRNA component of the p53 tumour suppressor network. Nature 447, 1130–1134.

He, S.Y., Shen, H.W., Xu, L., Zhao, X.H., Yuan, L., Niu, G., et al., 2012. FOXM1 promotes tumor cell invasion and correlates with poor prognosis in early-stage cervical cancer. Gynecol. Oncol. 127, 601–610.

He, L., Yang, X., Cao, X., Liu, F., Quan, M., Cao, J., 2013. Casticin induces growth suppression and cell cycle arrest through activation of FOXO3a in hepatocellular carcinoma. Oncol. Rep. 29, 103–108.

Heakal, Y., Kester, M., Savage, S., 2012. Vemurafenib (PLX4032): an orally available inhibitor of mutated BRAF for the treatment of metastatic melanoma. Ann. Pharmacother. 45, 1399–1405.

Hebner, C.M., Laimins, L.A., 2006. Human papillomaviruses: basic mechanisms of pathogenesis and oncogenicity. Rev. Med. Virol. 16, 83–97.

Hedge, N.S., Sanders, D.A., Rodriguez, R., Balasubramanian, S., 2011. The transcription factor FOXM1 is a cellular target of the natural product thiostrepton. Nat. Chem. 3, 725–731.

Helleday, T., 2010. Homologous recombination in cancer development, treatment and development of drug resistance. Carcinogenesis 31, 955–960.

Helleday, T., 2011. DNA repair as treatment target. Eur. J. Cancer 47 (Suppl. 3), S333–S335.

Helleday, T., Petermann, E., Lundin, C., Hodgson, B., Sharma, R.A., 2008. DNA repair pathways as targets for cancer therapy. Nat. Rev. Cancer 8, 193–204.

Helt, A.M., Galloway, D.A., 2003. Mechanisms by which DNA tumor virus oncoproteins target the Rb family of pocket proteins. Carcinogenesis 24, 159–169.

Hemann, M.T., Narita, M., 2007. Oncogenes and senescence: breaking down the fast lane. Genes Dev. 21, 1–5.

Henley, S.A., Dick, F.A., 2012. The retinoblastoma family of proteins and their regulatory functions in the mammalian cell division cycle. Cell Div. 7, 10.

Henning, W., Stürzbecher, H.W., 2003. Homologous recombination and cell cycle checkpoints; Rad51 in tumour progression and therapy resistance. Toxicology 193, 91–109.

Hensley, P.J., Kyprianou, N., 2012. Modeling prostate cancer in mice: limitations and opportunities. J. Androl. 33, 133–144.

Herbst, R.S., Kies, M.S., 2003. Gefitinib: current and future status in cancer therapy. Clin. Adv. Hematol. Oncol. 1, 466–472.

Herbst, R.S., Fukuoka, M., Baselga, J., 2004. Gefitinib—a novel targeted approach to treating cancer. Nat. Rev. Cancer 4, 956–965.

Hernandez-Boluda, J.C., Cervantes, F., 2002. Imatinib mesylate (Gleevec/Glivec): a new therapy for chronic myeloid leukemia and other malignancies. Drugs Today 38, 601–613.

Herr, I., Debatin, K.M., 2001. Cellular stress response and apoptosis in cancer therapy. Blood 98, 2603–2614.

Herwig, S., Strauss, M., 1997. The retinoblastoma protein: a master regulator of cell cycle, differentiation and apoptosis. Eur. J. Biochem. 246, 581–601.

Herzog, E.L., Bucala, R., 2010. Fibrocytes in health and disease. Exp. Hematol. 38, 548–556.

Hesketh, R., 1997. The Oncogene and Tumor Suppressor Gene Facts Book. Academic Press, San Diego, CA.

Heuberger, J., Birchmeier, W., 2010. Interplay of Cadherin-mediated cell adhesion and canonical Wnt signaling. Cold Spring Harb. Perspect. Biol. 2, a002915.

Heyer, W.D., Ehmsen, K.T., Liu, J., 2010. Regulation of homologous recombination in eukaryotes. Annu. Rev. Genet. 44, 113–139.

Hicklin, D.J., Ellis, L.M., 2005. Role of the vascular endothelial growth factor pathway in tumor growth and angiogenesis. J. Clin. Oncol. 5, 1011–1027.

Hideshima, T., Richardson, P.G., Anderson, K.C., 2003. Targeting proteasome inhibition in hematologic malignancies. Rev. Clin. Exp. Hematol. 7, 191–204.

Higgins, M.J., Stearns, V., 2011. Pharmacogenetics of endocrine therapy for breast cancer. Annu. Rev. Med. 62, 281–293.

Hill, L., Browne, G., Tulchinsky, E., 2013. ZEB/miR-200 feedback loop: at the crossroads of signal transduction in cancer. Int. J. Cancer 132, 745–754.

Hinz, B., Phan, S.H., Thannickal, V.J., Galli, A., Bochaton-Piallat, M.L., Gabbiani, G., 2007. The myofibroblast: one function, multiple origins. Am. J. Pathol. 170, 1807–1816.

Ho, C., Wang, C., Mattu, S., Destefanis, G., Ladu, S., Delogu, S., et al., 2012. AKT and N-Ras co-activation in the mouse liver promotes rapid carcinogenesis via mTORC1, FOXM1/SKP2, and c-Myc pathways. Hepatology 55, 833–845.

Hochedlinger, K., Plath, K., 2009. Epigenetic reprogramming and induced pluripotency. Development 136, 509–523.

Hodgson, G., Yeh, R.F., Ray, A., Wang, N., Smirnov, I., Yu, M., et al., 2009. Comparative analyses of gene copy number and mRNA expression in GBM tumors and GBM xeno-grafts. Neuro Oncol. 11, 477–487.

Hoeijmakers, J.H.J., 2001. Genome maintenance mechanisms for preventing cancer. Nature 411, 366–374.

Hoppe-Seyler, F., Butz, K., 1995. Molecular mechanisms of virus-induced carcinogenesis: the interaction of viral factors with cellular tumor suppressor proteins. J. Mol. Med. 73, 529–538.

Horimoto, Y., Hartman, J., Millour, J., Pollock, S., Olmos, Y., Ho, K.K., et al., 2011. ERβ1 represses FOXM1 expression through targeting ERα to control cell proliferation in breast cancer. Am. J. Pathol. 179, 1148–1156.

Howard, E.W., Camm, K.D., Wong, Y.C., Wang, X.H., 2008. E-cadherin upregulation as a therapeutic goal in cancer treatment. Mini Rev. Med. Chem. 8, 496–518.

Howe, E.N., Cochrane, D.R., Richer, J.K., 2012. The miR-200 and miR-221/222 microRNA families: opposing effects on epithelial identity. J. Mammary Gland Biol. Neoplasia 17, 65–77.

Howley, P.M., Livingston, D.M., 2009. Small DNA tumor viruses: large contributors to biomedical sciences. Virology 384, 256–259.

Hu, D., Fang, W., Han, A., Gallagher, L., Davis, R.J., Xiong, B., et al., 2008. c-Jun N-terminal kinase 1 interacts with and negatively regulates Wnt/β-catenin signaling through GSK3β pathway. Carcinogenesis 29, 2317–2324.

Hu, J., Dong, A., Fernandez-Ruiz, V., Shan, J., Kawa, M., Martinez-Anso, E., et al., 2009. Blockade of Wnt signaling inhibits angiogenesis and tumor growth in hepatocellular carcinoma. Cancer Res. 69, 6951–6959.

Huang, H., He, X., 2008. Wnt/β-catenin signaling: new (and old) players and new insights. Curr. Opin. Cell Biol. 20, 119–125.

Huang, X., Zhao, Y.Y., 2012. Transgenic expression of FoxM1 promotes endothelial repair following lung injury induced by polymicrobial sepsis in mice. PLoS One 7, e50094.

Huang, Y.H., Li, D., Winoto, A., Robey, E.A., 2004. Distinct transcriptional programs in thymocytes responding to T cell receptor, Notch, and positive selection signals. Proc. Natl. Acad. Sci. U.S.A. 101, 4936–4941.

Huang, W., Ma, K., Zhang, J., Qatanani, M., Cuvillier, J., Liu, J., et al., 2006. Nuclear receptor-dependent bile acid signaling is required for normal liver regeneration. Science 312, 233–236.

Huber, M.A., Kraut, N., Beug, H., 2005. Molecular requirements for epithelial-mesenchymal transition during tumor progression. Curr. Opin. Cell Biol. 17, 548–558.

Hudis, C.A., 2007. Trastuzumab—mechanism of action and use in clinical practice. N. Engl. J. Med. 357, 39–51.

Huen, M.S., Sy, S.M., Chen, J., 2010. BRCA1 and its toolbox for the maintenance of genomic integrity. Nat. Rev. Mol. Cell Biol. 11, 138–148.

Hugo, H., Ackland, M.L., Blick, T., Lawrence, M.G., Clements, J.A., Williams, E.D., et al., 2007. Epithelial-mesenchymal and mesenchymal-epithelial transitions in carcinoma progression. J. Cell. Physiol. 213, 374–383.

Hugo, H.J., Kokkinos, M.I., Blick, T., Ackland, M.L., Thompson, E.W., Newgreen, D.F., 2011. Defining the E-cadherin repressor interactome in epithelial-mesenchymal transition: the PMC43 model as a case study. Cells Tissues Organs 193, 23–40.

Hui, M.K.C., Chan, K.W., Luk, J.M., Lee, N.P., Chung, Y., Cheung, L.C.M., et al., 2012. Cytoplasmic Forkhead Box M1 (FoxM1) in esophageal squamous cell carcinoma significantly correlates with pathological disease stage. World J. Surg. 36, 90–97.

Hunter, T., 1998. The Croonian Lecture 1997. The phosphorylation of proteins on tyrosine: its role in cell growth and disease. Philos. Trans. R. Soc. Lond. B Biol. Sci. 353, 583–605.

Hunter, T., 1999. The role of tyrosine phosphorylation in cell growth and disease. Harvey Lect. 94, 81–119.

Hunter, T., 2007. Treatment for chronic myelogenous leukemia: the long road to imatinib. J. Clin. Invest. 117, 2036–2043.

Hunter, T., 2009. Tyrosine phosphorylation: thirty years and counting. Curr. Opin. Cell Biol. 21, 140–146.

Huss, W.J., Maddison, L.A., Greenberg, N.M., 2001. Autochthonous mouse models for prostate cancer: past, present and future. Semin. Cancer Biol. 11, 245–260.

Huynh, K.M., Kim, G., Kim, D.J., Yang, S.J., Park, S.M., Yeom, Y.I., et al., 2009. Gene expression analysis of terminal differentiation of human melanoma cells highlights global reductions in cell cycle-associated genes. Gene 433, 32–39.

Huynh, K.M., Soh, J.W., Dash, R., Sarkar, D., Fisher, P.B., Kang, D., 2010. FOXM1 expression mediates growth suppression during terminal differentiation of HO-1 human metastatic melanoma cells. J. Cell. Physiol. 226, 194–204.

Hynes, N.E., Lane, H.A., 2005. ERBB receptors and cancer: the complexity of targeted inhibitors. Nat. Rev. Cancer 5, 341–354.

Iacobuzio-Donahue, C.A., Maitra, A., Olsen, M., Lowe, A.W., Van Heek, N.T., Rosty, C., et al., 2003. Exploration of global gene expression patterns in pancreatic adenocarcinoma using cDNA microarrays. Am. J. Pathol. 162, 1151–1162.

Igney, F.H., Krammer, P.H., 2002. Death and anti-death: tumour resistance to apoptosis. Nat. Rev. Cancer 2, 277–288.

Imai, K., Takaoka, A., 2006. Comparing antibody and small-molecule therapies for cancer. Nat. Rev. Cancer 6, 714–727.

Imai, T., Horiuchi, A., Shiozawa, T., Osada, R., Kiuchi, N., Ohira, S., et al., 2004. Elevated expression of E-cadherin and α-, β-, γ-catenins in metastatic lesions compared with primary epithelial ovarian carcinomas. Hum. Pathol. 35, 1469–1476.

Inoue, J.I., Gohda, J., Akiyama, T., Semba, K., 2007. NF-κB activation in development and progression of cancer. Cancer Sci. 98, 268–274.

Inui, M., Martello, G., Piccolo, S., 2010. MicroRNA control of signal transduction. Nat. Rev. Mol. Cell Biol. 11, 252–263.

Ishii, Y., Waxman, S., Germain, D., 2007. Targeting the ubiquitin-proteasome pathway in cancer. Anticancer Agents Med Chem. 7, 359–365.

Ishikawa, K., Ishii, H., Saito, T., 2006. DNA damage-dependent cell cycle checkpoints and genomic stability. DNA Cell Biol. 25, 406–411.

Iwatsuki, M., Mimori, K., Yokobori, T., Ishi, H., Beppu, T., Nakamori, S., et al., 2010. Epithelial-mesenchymal transition in cancer development and its clinical significance. Cancer Sci. 101, 293–299.

Jackson, S.P., 2009. The DNA-damage response: new molecular insights and new approaches to cancer therapy. Biochem. Soc. Trans. 37, 483–494.

Jackson, S.P., Bartek, J., 2009. The DNA damage response in human biology and disease. Nature 461, 1071–1078.

Jaenisch, R., Young, R., 2008. Stem cells, the molecular circuitry of pluripotency and nuclear reprogramming. Cell 132, 567–582.

Jankowski, J.A., Bedford, F.K., Kim, Y.S., 1997. Changes in gene structure and regulation of E-cadherin during epithelial development, differentiation, and disease. Prog. Nucleic Acid Res. Mol. Biol. 57, 187–215.

Jansen-Dürr, P., 1996. How viral oncogenes make the cell cycle. Trends Genet. 12, 270–275.

Janus, J.R., Laborde, R.R., Greenberg, A.J., Wang, V.W., Wei, W., Trier, A., et al., 2011. Linking expression of FOXM1, CEP55 and HELLS to tumorigenesis in oropharyngeal squamous cell carcinoma. Laryngoscope 121, 2598–2603.

Järvinen, P.M., Laiho, M., 2012. LIM-domain proteins in transforming growth factor β-induced epithelial-to-mesenchymal transition and myofibroblast differentiation. Cell. Signal. 24, 819–825.

Jatiani, S.S., Baker, S.J., Silverman, L.R., Reddy, P., 2011. JAK/STAT pathways in cytokine signaling and myeloproliferative disorders: approaches for targeted therapy. Genes Cancer 1, 979–993.

Javelaud, D., Pierrat, M.J., Mauviel, A., 2012. Crosstalk between TGF-β and hedgehog signaling in cancer. FEBS Lett. 586, 2016–2025.

Javier, R.T., Butel, J.S., 2008. The history of tumor virology. Cancer Res. 68, 7693–7706.

Jayshree, R.S., Sreevinas, A., Tessy, M., Krishna, S., 2009. Cell intrinsic & extrinsic factors in cervical carcinogenesis. Indian J. Med. Res. 130, 286–295.

Jeanes, A., Gottardi, C.J., Yap, A.S., 2008. Cadherins and cancer: how does cadherin dysfunction promote tumor progression? Oncogene 27, 6920–6929.

Jechinger, M., Grünert, S., Beug, H., 2003. Mechanisms in epithelial plasticity and metastasis: insights from 3D cultures and expression profiling. J. Mammary Gland Biol. Neoplasia 7, 415–432.

Jeggo, P.A., Geuting, V., Löbrich, M., 2011. The role of homologous recombination in radiation-induced double-strand break repair. Radiother. Oncol. 101, 7–12.

Jepsen, K., Rosenfeld, M.G., 2002. Biological roles and mechanistic actions of co-repressor complexes. J. Cell Sci. 115, 689–698.

Ji, Z., Flaherty, K.T., Tsao, H., 2011. Targeting the RAS pathway in melanoma. Trends Mol. Med. 18, 27–35.

Jia, R., Li, C., McCoy, J.P., Deng, C.X., Zheng, Z.M., 2010. SRp20 is a proto-oncogene critical for cell proliferation and tumor induction and maintenance. Int. J. Biol. Sci. 6, 806–826.

Jiang, B.H., Liu, L.Z., 2008. PI3K/PTEN signaling in tumorigenesis and angiogenesis. Biochim. Biophys. Acta 1784, 150–158.

Jiang, G.X., Liu, W., Cui, Y.F., Zhong, X.Y., Tai, S., Wang, Z.D., et al., 2010. Reconstitution of secreted frizzled-related protein 1 suppresses tumor growth and lung metastasis in an orthotopic model of hepatocellular carcinoma. Dig. Dis. Sci. 55, 2838–2843.

Jiang, L.Z., Wang, P., Deng, B., Huang, C., Tang, W.X., Lu, H.Y., et al., 2011. Overexpression of Forkhead Box M1 transcription factor and nuclear factor-κB in laryngeal squamous cell carcinoma: a potential indicator for poor prognosis. Hum. Pathol. 42, 1185–1193.

Jin, J., Wang, G.L., Iakova, P., Shi, X., Haefliger, S., Finegold, M., et al., 2010. Epigenetic changes play critical role in age-associated dysfunctions of the liver. Aging Cell 9, 895–910.

Jing, Y., Han, Z., Zhang, S., Liu, Y., Wei, L., 2011. Epithelial-mesenchymal transition in tumor microenvironment. Cell Biosci. 1, 29.

Jo Chien, A., Rugo, H.S., 2009. Lapatinib: new directions in HER2 directed therapy for early stage breast cancer. Cancer Treat. Res. 151, 197–215.

Johannessen, M., Moens, U., 2007. Multisite phosphorylation of the cAMP response element-binding protein (CREB) by a diversity of protein kinases. Front. Biosci. 12, 1814–1832.

Johannessen, M., Pedersen Delghandi, M., Moens, U., 2004. What turns CREB on? Cell. Signal. 16, 1211–1227.

Johansson, M., DeNardo, D.G., Coussens, L.M., 2008. Polarized immune responses differentially regulate cancer development. Immunol. Rev. 222, 145–154.

Johnson, D.G., Schneider-Broussard, R., 1998. Role of E2F in cell cycle control and cancer. Front. Biosci. 3, d447–d458.

Johnson, L., Mercer, K., Greenbaum, D., Bronson, R.T., Crowley, D., Tuveson, D.A., et al., 2001. Somatic activation of the K-ras oncogene causes onset of lung cancer in mice. Nature 410, 1111–1116.

Jones, N., 1995. Transcriptional modulation by the adenovirus E1A gene. Curr. Top. Microbiol. Immunol. 1999, 59–80.

Jones, S.E., Jomary, C., 2002. Secreted Frizzled-related proteins: searching for relationships and patterns. Bioessays 24, 811–820.

Jones, R.L., Judson, I.R., 2005. The development and application of imatinib. Expert Opin. Drug Saf. 4, 183–191.

Jones, D.L., Münger, K., 1996. Interactions of the human papillomavirus E7 protein with cell cycle regulators. Semin. Cancer Biol. 7, 327–337.

Jordan, E.J., Kelly, C.M., 2012. Vemurafenib for the treatment of melanoma. Expert Opin. Pharmacother. 13, 2533–2543.

Joseph, T.L., Madhumalar, A., Brown, C.J., Lane, D.P., Verma, C.S., 2010. Differential binding of p53 and nutlin to MDM2 and MDMX: computational studies. Cell Cycle 9, 1167–1181.

Junttila, M.R., Westermarck, J., 2008. Mechanisms of MYC stabilization in human malignancies. Cell Cycle 7, 592–596.

Kaelin, W.G., 1999. Functions of the retinoblastoma protein. Bioessays 21, 950–958.

Kaelin, W.G., 2004. Gleevec: prototype or outlier? Sci. STKE 2004, PE12.

Kaelin, W.G., 2005. The concept of synthetic lethality in the context of anticancer therapy. Nat. Rev. Cancer 5, 689–698.

Kaelin, W.G., 2007. von Hippel-Lindau disease. Annu. Rev. Pathol. 2, 145–173.

Kaelin, W.G., 2008. The von Hippel-Lindau tumour suppressor protein: O2 sensing and cancer. Nat. Rev. Cancer 8, 865–873.

Kaelin, W.G., 2009. Synthetic lethality: a framework for the development of wiser cancer therapeutics. Genome Med. 1, 99.

Kaelin, W.G., Radcliffe, P.J., 2008. Oxygen sensing by metazoans: the central role of the HIF hydroxylase pathway. Mol. Cell 30, 393–402.

Kalin, T.V., Wang, I.C., Ackerson, T.J., Major, M.L., Detrisac, C.J., Kalinichenko, V.V., et al., 2006. Increased levels of the FoxM1 transcription factor accelerate development and progression of prostate carcinomas in both TRAMP and LADY transgenic mice. Cancer Res. 66, 1712–1720.

Kalin, T.V., Ustiyan, V., Kalinichenko, V.V., 2011. Multiple faces of FoxM1 transcription factor. Lessons from transgenic mouse models. Cell Cycle 10, 396–405.

Kalinichenko, V.V., Lim, L., Shin, B., Costa, R.H., 2001. Differential expression of forkhead box transcription factors following butylated hydroxytoluene lung injury. Am. J. Physiol. Lung Cell. Mol. Physiol. 280, L695–L704.

Kalinichenko, V.V., Gusarova, G.A., Tan, Y., Wang, I.C., Major, M.L., Wang, X., et al., 2003. Ubiquitous expression of the Forkhead Box M1B transgene accelerates proliferation of distinct pulmonary cell types following lung injury. J. Biol. Chem. 39, 37888–37894.

Kalinichenko, V.V., Major, M.L., Wang, X., Petrovic, V., Kuechle, J., Yoder, H.M., et al., 2004. Foxm1b transcription factor is essential for development of hepatocellular carcinomas and is negatively regulated by the p19ARF tumor suppressor. Genes Dev. 18, 830–850.

Kalinina, O.A., Kalinin, S.A., Polack, E.W., Mikaelian, I., Panda, S., Costa, R.H., et al., 2003. Sustained hepatic expression of FoxM1B in transgenic mice has minimal effects on hepatocellular carcinoma development but increases cell proliferation rates in preneoplastic and early neoplastic lesions. Oncogene 22, 6266–6276.

Kalluri, R., Neilson, E.G., 2003. Epithelial-mesenchymal transition and its implications for fibrosis. J. Clin. Invest. 112, 1776–1784.

Kalluri, R., Weinberg, R.A., 2009. The basics of epithelial-mesenchymal transition. J. Clin. Invest. 119, 1420–1428.

Kanemitsu, S., Hisaoka, M., Shimajiri, S., Matsuyama, A., Hashimoto, H., 2007. Molecular detection of SS18-SSX fusion gene transcripts by cRNA in situ hybridization in synovial sarcoma using formalin-fixed, paraffin-embedded tumor tissue specimens. Diagn. Mol. Pathol. 16, 9–17.

Kang, Y., Massagué, J., 2004. Epithelial-mesenchymal transitions: twist in development and metastasis. Cell 118, 277–279.

Kantak, S.S., Kramer, R.H., 1998. E-cadherin regulates anchorage-independent growth and survival in oral squamous cell carcinoma cells. J. Biol. Chem. 273, 16953–19961.

Kaplan, R.N., Rafii, S., Lyden, D., 2006. Preparing the "soil": the premetastatic niche. Cancer Res. 66, 11089–11093.

Karadedou, C.T., 2006. Regulation of the FOXM1 transcription factor by the estrogen receptor α at the protein level in breast cancer. Hippokratia 10, 128–132.

Karadedou, C.T., Gomes, A.R., Chen, J., Petkovic, M., Ho, K.K., Zwolinska, A.K., et al., 2012. FOXO3a represses VEGF expression through FOXM1-dependent and -independent mechanisms in breast cancer. Oncogene 31, 1845–1858.

Karge, H., Borok, Z., 2012. EMT and interstitial lung disease: a mysterious relationship. Curr. Opin. Pulm. Med. 18, 517–523.

Karim, R., Tse, G., Putti, T., Scolyer, R., Lee, S., 2004. The significance of the Wnt pathway in the pathology of human cancers. Pathology 36, 120–128.

Karin, M., 2006a. Nuclear factor-κB in cancer development and progression. Nature 441, 431–436.

Karin, M., 2006b. NF-κB and cancer: mechanisms and targets. Mol. Carcinog. 45, 355–361.

Karin, M., 2008. The IκB kinase—a bridge between inflammation and cancer. Cell Res. 18, 334–342.

Karin, M., Greten, F.R., 2005. NF-κB: linking inflammation and immunity to cancer development and progression. Nat. Rev. Immunol. 5, 749–759.

Karin, M., Lin, A., 2002. NF-κB at the crossroads of life and death. Nat. Immunol. 2, 221–227.

Karin, M., Lawrence, T., Nizet, V., 2006. Innate immunity gone awry: linking microbial infections to chronic inflammation and cancer. Cell 124, 823–835.

Kasai, H., Allen, J.T., Mason, R.M., Kamimura, T., Zhang, Z., 2005. TGF-β1 induces human alveolar epithelial to mesenchymal cell transition (EMT). Respir. Res. 6, 56.

Kasper, S., 2005. Survey of genetically engineered mouse models for prostate cancer: analyzing the molecular basis of prostate cancer development, progression and metastasis. J. Cell. Biochem. 94, 279–297.

Kasper, S., Sheppard, P.C., Yan, Y., Pettigrew, N., Borowsky, A.D., Prins, G.S., et al., 1998. Development, progression, and androgen-dependence prostate tumors in probasin-large T antigen transgenic mice: a model for prostate cancer. Lab. Invest. 78, 319–333.

Kass, E.M., Jasin, M., 2010. Collaboration and competition between DNA double-strand break repair pathways. FEBS Lett. 584, 3703–3708.

Kastan, M.B., 2008. DNA damage responses: mechanisms and roles in human disease. Mol. Cancer Res. 6, 517–524.

Kastan, M.B., Bartek, J., 2004. Cell-cycle checkpoints and cancer. Nature 432, 316–323.

Katoh, M., Igarashi, M., Fukuda, H., Nakagama, H., Katoh, M., 2013. Cancer genetics and genomics of human FOX family genes. Cancer Lett. 328, 198–206.

Kawabata, M., Kawabata, T., Nishibori, M., 2005. Role of recA/RAD51 family proteins in mammals. Acta Med. Okayama 59, 1–9.

Kawao, Y., Kypta, R., 2003. Secreted antagonists of the Wnt signalling pathway. J. Cell Sci. 116, 2627–2634.

Keam, S.J., 2008. Dasatinib: in chronic myeloid leukemia and Philadelphia chromosome-positive acute lymphoblastic leukemia. BioDrugs 22, 59–69.

Keating, G.M., 2012. Vemurafenib: in unresectable or metastatic melanoma. BioDrugs 26, 325–334.

Keith, B., Simon, M.C., 2007. Hypoxia-inducible factors, stem cells, and cancer. Cell 129, 465–472.

Keith, B., Johnson, R.S., Simon, M.C., 2012. HIF1α and HIF2α: sibling rivalry in hypoxic tumour growth and progression. Mol. Cell 40, 294–309.

Kelleher, F.C., Fennelly, D., Rafferty, M., 2006. Common critical pathways in embryogenesis and cancer. Acta Oncol. 45, 375–388.

Keller, E.T., 2004. Metastasis suppressor genes: a role for raf kinase inhibitor protein (RKIP). Anticancer Drugs 15, 663–669.

Keller, E.T., Fu, Z., Brennan, M., 2004a. The role of Raf kinase inhibitor protein (RKIP) in health and disease. Biochem. Pharmacol. 68, 1049–1053.

Keller, E.T., Fu, Z., Brennan, M., 2004b. Raf kinase inhibitor protein: a prostate cancer metastasis suppressor gene. Cancer Lett. 207, 131–137.

Keller, E.T., Fu, Z., Brennan, M., 2005. The biology of a prostate cancer metastasis suppressor protein: Raf kinase inhibitor protein. J. Cell. Biochem. 94, 273–278.

Keller, U.B., Old, J.B., Dorsey, F.C., Nilsson, J.A., Nilsson, L., MacLean, K.H., et al., 2007. Myc targets Cks1 to provoke the suppression of p27KIp1, proliferation and lymphomagenesis. EMBO J. 26, 2562–2574.

Kennedy, R.D., D'Andrea, A.D., 2005. The Fanconi Anaemia/BRCA pathway: new faces in the crowd. Genes Dev. 19, 2925–2940.

Keshet, Y., Seger, R., 2010. The MAP kinase signaling cascades: a system of hundreds of components regulates a diverse array of physiological functions. Methods Mol. Biol. 661, 3–38.

Kessler, J.H., Melief, C.J., 2007. Identification of T-cell epitopes for cancer immunotherapy. Leukemia 21, 1859–1874.

Kew, M.C., 2011. Hepatitis B virus X protein in the pathogenesis of hepatitis B virus-induced hepatocellular carcinoma. J. Gastroenterol. Hepatol. 26 (Suppl. 1), 144–152.

Khanna, K.K., Jackson, S.P., 2001. DNA double-strand breaks: signaling, repair and the cancer connection. Nat. Genet. 27, 247–254.

Khidr, L., Chen, P.L., 2006. RB, the conductor that orchestrates life, death and differentiation. Oncogene 25, 5210–5219.

Kikuchi, A., 1999a. Roles of Axin in the Wnt signalling pathway. Cell. Signal. 11, 777–788.

Kikuchi, A., 1999b. Modulation of Wnt signaling by Axin and Axil. Cytokine Growth Factor Rev. 10, 255–265.

Kim, I.M., Ramakrishna, S., Gusarova, G.A., Yoder, H.M., Costa, R.H., Kalinichenko, V.V., 2005. The Forkhead box m1 transcription factor is essential for embryonic development of pulmonary vasculature. J. Biol. Chem. 280, 22278–22286.

Kim, I.M., Ackerson, T., Ramakrishna, S., Tretiakova, M., Wang, I.C., Kalin, T.V., et al., 2006a. The Forkhead Box M1 transcription factor stimulates the proliferation of tumor cells during development of lung cancer. Cancer Res. 66, 2153–2161.

Kim, H.J., Hawke, N., Baldwin, A.S., 2006b. NF-κB and IKK as therapeutic targets in cancer. Cell Death Differ. 13, 738–747.

Kim, R., Emi, M., Tanabe, K., 2006c. The role of apoptosis in cancer cell survival and therapeutic outcome. Cancer Biol. Ther. 5, 1429–1442.

Kim, J., Woo, A.J., Chu, J., Sow, J.W., Fujiwara, Y., Kim, C.G., et al., 2010. A Myc network accounts for similarities between embryonic stem and cancer cell transcription programs. Cell 143, 313–324.

Kim, Y.J., Cha, H.J., Nam, K.H., Yoon, Y., Lee, H., An, S., 2011. *Centella asiatica* extracts modulate hydrogen peroxide-induced senescence in human dermal fibroblasts. Exp. Dermatol. 20, 998–1003.

Kimelman, D., Xu, W., 2006. β-Catenin destruction complex: insights and questions from a structural perspective. Oncogene 25, 7482–7491.

King, T.E., Pardo, A., Selman, M., 2011. Idiopathic pulmonary fibrosis. Lancet 378, 1949–1961.

Kinner, A., Wu, W., Staudt, C., Iliakis, G., 2008. γ-H2AX in recognition and signaling of DNA double-strand breaks in the context of chromatin. Nucleic Acids Res. 36, 5678–5694.

Kis, K., Liu, X., Hagood, J.S., 2011. Myofibroblast differentiation and survival in fibrotic disease. Expert Rev. Mol. Med. 13, e27.

Kisseleva, T., Brenner, D.A., 2008a. Fibrogenesis of parenchymal organs. Proc. Am. Thorac. Soc. 5, 338–342.

Kisseleva, T., Brenner, D.A., 2008b. Mechanisms of fibrogenesis. Exp. Biol. Med. 233, 109–122.

Kisseljov, F., Sakharova, O., Kondratjeva, T., 2008. Cellular and molecular biological aspects of cervical intraepithelial neoplasia. Int. Rev. Cell Mol. Biol. 271, 35–95.

Kitagawa, K., Kotake, Y., Kitagawa, M., 2009. Ubiquitin-mediated control of oncogene and tumor suppressor gene products. Cancer Sci. 100, 1374–1381.

Kizek, R., Adam, V., Hrabeta, J., Eckschlager, T., Smutny, S., Burda, J.V., et al., 2012. Anthracyclines and ellipticines as DNA-damaging anticancer drugs: recent advances. Pharmacol. Ther. 133, 26–39.

Klaus, A., Birchmeier, W., 2008. Wnt signalling and its impact on development and cancer. Nat. Rev. Cancer 8, 387–399.

Kleer, C.G., van Golen, K.L., Braun, T., Merajver, S.D., 2001. Persistent E-cadherin expression in inflammatory breast cancer. Mod. Pathol. 14, 458–464.

Klein, G., 2002. Perspectives in studies of human tumor viruses. Front. Biosci. 7, d268–d274.

Klein, C., Vassilev, L.T., 2004. Targeting the p53-MDM2 interaction to treat cancer. Br. J. Cancer 91, 1415–1419.

Klein, E., Ben-Bassat, H., Neumann, H., Ralph, P., Zeuthen, J., Polliack, A., et al., 1976. Properties of the K562 cell line, derived from a patient with chronic myeloid leukemia. Int. J. Cancer 18, 421–432.

Klein, S., McCormick, F., Levitzki, A., 2005. Killing time for cancer cells. Nat. Rev. Cancer 5, 573–580.

Klingelhutz, A.J., Roman, A., 2012. Cellular transformation by human papillomaviruses: lessons learned by comparing high- and low-risk viruses. Virology 424, 77–98.

Klymkowsky, M.W., Savagner, P., 2009. Epithelial-mesenchymal transition. A cancer researcher's conceptual friend and foe. Am. J. Pathol. 174, 1588–1593.

Klysik, J., Theroux, S.J., Sedivy, J.M., Moffit, J.S., Boekelheide, K., 2008. Signaling crossroads: the function of Raf kinase inhibitory protein in cancer, the central nervous system and reproduction. Cell. Signal. 20, 1–9.

Knight, A.S., Notaridou, M., Watson, R.J., 2009. A Lin-9 complex is recruited by B-Myb to activate transcription of G_2/M genes in undifferentiated embryonal carcinoma cells. Oncogene 28, 1737–1747.

Knoepfler, P.S., 2008. Why myc? An unexpected ingredient in the stem cell cocktail. Cell Stem Cell 2, 18–21.

Knoepfler, P.S., 2009. Deconstructing stem cell tumorigenicity: a roadmap to safe regenerative medicine. Stem Cells 27, 1050–1056.

Knudsen, E.S., Knudsen, K.E., 2006. Retinoblastoma tumor suppressor: where cancer meets the cell cycle. Exp. Biol. Med. 231, 1271–1281.

Koch, U., Radtke, F., 2007. Notch and cancer: a double-edged sword. Cell. Mol. Life Sci. 64, 2746–2762.

Koch, A.W., Manzur, K.L., Shan, W., 2004. Structure-based models of cadherin-mediated cell adhesion: the evolution continues. Cell. Mol. Life Sci. 61, 1884–1895.

Kolibaba, K.S., Druker, B.J., 1997. Protein tyrosine kinases and cancer. Biochim. Biophys. Acta 1333, F217–F248.

Koo, C.Y., Muir, K.W., Lam, E.W.F., 2011. FOXM1: from cancer initiation to progression and treatment. Biochim. Biophys. Acta 1819, 28–37.

Kopper, L., 2008. Lapatinib: a sword with two edges. Pathol. Oncol. Res. 14, 1–8.

Korpal, M., Kang, Y., 2008. The emerging role of miR-200 family of microRNAs in epithelial-mesenchymal transition and cancer metastasis. RNA Biol. 5, 115–119.

Korver, W., Roose, J., Clevers, H., 1997a. The winged-helix transcription factor Trident is expressed in cycling cells. Nucleic Acids Res. 25, 1715–1719.

Korver, W., Roose, J., Heinen, K., Weghuis, D.O., de Bruijn, D., van Kessel, A.G., et al., 1997b. The human TRIDENT/HFH-11/FKHL16 gene: structure, localization, and promoter characterization. Genomics 46, 435–442.

Korver, W., Roose, J., Wilson, A., Clevers, H., 1997c. The winged-helix transcription factor Trident is expressed in actively dividing lymphocytes. Immunobiology 198, 157–161.

Korver, W., Schilham, M.W., Moerer, P., van den Hoff, M.J., Lamers, W.H., Medema, R.H., et al., 1998. Uncoupling of S-phase and mitosis in cardiomyocytes and hepatocytes lacking the winged-helix transcription factor Trident. Curr. Biol. 8, 1327–1330.

Korzeniewski, N., Spardy, N., Duensing, A., Duensing, S., 2011. Genomic instability and cancer: lessons learned from human papillomaviruses. Cancer Lett. 305, 113–122.

Kowalczyk, A.P., Nanes, B.A., 2012. Adherens junction turnover: regulating adhesion through cadherin endocytosis, degradation, and recycling. Subcell. Biochem. 60, 197–222.

Kowalski, P.J., Rubin, M.A., Kleer, C.G., 2003. E-cadherin expression in primary carcinomas of the breast and its distant metastases. Breast Cancer Res. 5, R217–R222.

Kowanetz, M., Ferrara, N., 2006. Vascular endothelial growth factor signaling pathways: therapeutic perspective. Clin. Cancer Res. 12, 5018–5022.

Krause, D.S., Van Etten, R.A., 2005. Tyrosine kinases as targets for cancer therapy. N. Engl. J. Med. 353, 172–187.

Krejci, L., Altmannova, V., Spirek, M., Zhao, X., 2012. Homologous recombination and its regulation. Nucleic Acids Res. 40, 5795–5818.

Kretschmer, C., Sterner-Kock, A., Siedentopf, F., Schlag, P.M., Kemmner, W., 2011. Identification of early molecular markers for breast cancer. Mol. Cancer 10, 15.

Krishna, M., Narang, H., 2008. The complexity of mitogen-activated protein kinases (MAPKs) made simple. Cell. Mol. Life Sci. 65, 3525–3544.

Krishnakumar, R., Kraus, W.L., 2010. The PARP side of the nucleus: molecular actions, physiological outcomes, and clinical targets. Mol. Cell 39, 8–24.

Krizhanovsky, V., Lowe, S.W., 2009. The promises and perils of p53. Nature 460, 1085–1086.

Krupczak-Hollis, K., Wang, X., Dennewitz, M.B., Costa, R.H., 2003. Growth hormone stimulates proliferation of old-aged regenerating liver through Forkhead Box m1b. Hepatology 38, 1552–1562.

Krupczak-Hollis, K., Wang, X., Kalinichenko, V.V., Gusarova, G.A., Wang, I.C., Dennewitz, M.B., et al., 2004. The mouse Forkhead Box m1 transcription factor is essential for hepatoblast mitosis and development of intrahepatic bile ducts and vessels during liver morphogenesis. Dev. Biol. 276, 74–88.

Kruse, V., Rottey, S., De Backer, O., Van Belle, S., Cocquyt, V., Denys, H., 2011. PARP inhibitors in oncology: a new synthetic lethal approach to cancer therapy. Acta Clin. Belg. 66, 2–9.

Kucharczak, J., Simmons, M.J., Fan, Y., Gelinas, C., 2003. To be, or not to be: NF-κB is the answer—role of Rel/NF-κB in the regulation of apoptosis. Oncogene 22, 8961–8982.

Kuilman, T., Michloglou, C., Mooi, W.J., Peeper, D.S., 2010. The essence of senescence. Genes Dev. 24, 2463–2479.

Kurinna, S., Stratton, S.A., Coban, Z., Schumacher, J.M., Grompe, M., Duncan, A.W., et al., 2013. p53 regulates a mitotic transcription program and determines ploidy in normal mouse liver. Hepatology 57, 2004–2013.

Kuroki, M., Huang, J., Shibaguchi, T., Tanaka, T., Zhao, J., Luo, N., et al., 2006. Possible applications of antibodies or their genes in cancer therapy. Anticancer Res. 26, 4019–4025.

Kwok, J.M.M., Myatt, S.S., Marson, C.M., Coombes, R.C., Constantinidou, D., Lam, E.W.F., 2008. Thiostrepton selectively targets breast cancer cells through inhibition of forkhead box M1 expression. Mol. Cancer Ther. 7, 2022–2032.

Kwok, J.M.M., Peck, B., Monteiro, L.J., Schwenen, H.D.C., Millour, J., Coombes, R.C., et al., 2010. FOXM1 confers acquired cisplatin resistance in breast cancer cells. Mol. Cancer Res. 8, 24–34.

Lackey, K.E., 2006. Lessons from the drug discovery of lapatinib, a dual ErbB1/2 tyrosine kinase inhibitor. Curr. Top. Med. Chem. 6, 435–460.

Lagunas-Martinez, A., Madrid-Marina, V., Gariglio, P., 2010. Modulation of apoptosis by early human papillomavirus proteins in cervical cancer. Biochim. Biophys. Acta 1805, 6–16.

Lai, S.L., Chien, A.J., Moon, R.T., 2009. Wnt/Fz signaling and the cytoskeleton: potential roles in tumorigenesis. Cell Res. 19, 532–545.

Lake, R.A., Robinson, B.W., 2005. Immunotherapy and chemotherapy—practical partnership. Nat. Rev. Cancer 5, 397–405.

Lange, A.W., Keiser, A.R., Wells, J.M., Zorn, A.M., Whitsett, J.A., 2009. Sox17 promotes cell cycle progression and inhibits TGF-β/Smad3 signaling to initiate progenitor cell behaviour in the respiratory epithelium. PLoS One 4, e5711.

Lanigan, F., Geraghty, J.G., Bracken, A.P., 2011. Transcriptional regulation of cellular senescence. Oncogene 30, 2901–2911.

Laoukili, J., Kooistra, M.R.H., Brás, A., Kauw, J., Kerkhoven, R.M., Morrison, A., et al., 2005. FoxM1 is required for execution of the mitotic programme and chromosome stability. Nat. Cell Biol. 7, 126–136.

Laoukili, J., Stahl, M., Medema, R.H., 2007. FoxM1: at the crossroads of ageing and cancer. Biochim. Biophys. Acta 1775, 92–102.

Laoukili, J., Alvarez-Fernandez, M., Stahl, M., Medema, R.H., 2008. FoxM1 is degraded at mitotic exit in a Cdh1-dependent manner. Cell Cycle 7, 2720–2726.

Larue, L., Bellacosa, A., 2005. Epithelial-mesenchymal transition in development and cancer: role of phosphatidylinositol 3' kinase/AKT pathways. Oncogene 24, 7443–7454.

Laskin, W.B., Miettinen, M., 2002. Epithelial-type and neural-type cadherin expression in malignant noncarcinomatous neoplasms with epithelioid features that involve the soft tissues. Arch. Pathol. Lab. Med. 126, 425–431.

LaTulippe, E., Satagopan, J., Smith, A., Scher, H., Scardino, P., Reuter, V., et al., 2002. Comprehensive gene expression analysis of prostate cancer reveals distinct transcriptional programs associated with metastatic disease. Cancer Res. 62, 4499–4506.

Laurendeau, I., Ferrer, M., Garrido, D., D'Haene, N., Ciavarelli, P., Basso, A., et al., 2010. Gene expression profiling of the Hedgehog signaling pathway in human meningiomas. Mol. Med. 16, 262–270.

Lavia, P., Mileo, A.M., Giordano, A., Paggi, M.G., 2003. Emerging roles of DNA tumor viruses in cell proliferation: new insights into genomic instability. Oncogene 22, 6508–6516.

Le Bras, G.F., Taubenslag, K.J., Andl, C.D., 2012. The regulation of cell-cell adhesion during epithelial-mesenchymal-transition, motility and tumor progression. Cell Adh. Migr. 6, 365–373.

Leckband, D., Prakasam, A., 2006. Mechanism and dynamics of cadherin adhesion. Annu. Rev. Biomed. Eng. 8, 259–287.

Leckband, D., Sivasankar, S., 2012. Cadherin recognition and adhesion. Curr. Opin. Cell Biol. 24, 620–627.

Ledda-Columbano, G.M., Pibiri, M., Cossu, C., Molotzu, F., Locker, J., Columbano, A., 2004. Aging does not reduce the hepatocyte proliferative response of mice to the primary mitogen TCPOBOP. Hepatology 40, 981–988.

Ledwaba, T., Dlamini, Z., Naicker, S., Bhoola, K., 2004. Molecular genetics of human cervical cancer: role of papillomavirus and the apoptotic cascade. Biol. Chem. 385, 671–682.

Lee, G.H., 2000. Paradoxical effects of phenobarbital on mouse hepatocarcinogenesis. Toxicol. Pathol. 28, 215–225.

Lee, C., Cho, Y., 2002. Interactions of SV40 large T antigen and other viral proteins with retinoblastoma tumour suppressor. Rev. Med. Virol. 12, 81–92.

Lee, A.T., Lee, C.G., 2007. Oncogenesis and transforming viruses: the hepatitis B virus and hepatocellular carcinoma—the etiopathogenic link. Front. Biosci. 12, 234–245.

Lee, E.Y., Muller, W.J., 2010. Oncogenes and tumor suppressor genes. Cold Spring Harb. Perspect. Biol. 2, a003236.

Lee, J.M., Dedhar, S., Kalluri, R., Thompson, E.W., 2006. The epithelial-mesenchymal transition: new insights in signaling, development, and disease. J. Cell Biol. 172, 973–981.

Lee, J., Jeong, D.J., Kim, J., Lee, S., Park, J.H., Chang, B., et al., 2012. Epithelial-mesenchymal transition in cervical carcinoma. Am. J. Transl. Res. 4, 1–13.

Lefebvre, C., Rajbhandari, P., Alvarez, M.J., Bandaru, P., Lim, W.K., Sato, M., et al., 2010. A human B-cell interactome identifies MYB and FOXM1 as master regulators of proliferation in germinal centers. Mol. Syst. Biol. 6, 377.

Lehmann, K., Tschuor, C., Rickenbacher, A., Jang, J.H., Oberkofler, C.E., Tschopp, O., et al., 2012. Liver failure after extended hepatectomy in mice is mediated by a p21-dependent barrier to liver regeneration. Gastroenterology 143, 1609–1619.

Lehoux, M., D'Abramo, C.M., Archambault, J., 2009. Molecular mechanisms of human papillomavirus-induced carcinogenesis. Public Health Genomics 12, 268–280.

Lenz, H.J., 2003. Clinical update: proteasome inhibitors in solid tumors. Cancer Treat. Rev. 29 (Suppl. 1), 41–48.

Lesterhuis, W.J., Haanen, J.B., Punt, C.J., 2011. Cancer immunotherapy—revisited. Nat. Rev. Drug Discov. 10, 591–600.

Leung, T.W.C., Lin, S.S.W., Tsang, A.C.C., Tong, C.S.W., Ching, J.C.Y., Leung, W.Y., et al., 2001. Overexpression of FoxM1 stimulates cyclin B1 expression. FEBS Lett. 507, 59–66.

Levayer, R., Lecuit, T., 2008. Breaking down EMT. Nat. Cell Biol. 10, 757–759.

Levine, A.J., 1994. The origins of the small DNA tumor viruses. Adv. Cancer Res. 65, 141–168.

Levine, A.J., 2009. The common mechanisms of transformation by the small DNA tumor viruses: the inactivation of tumor suppressor gene products: p53. Virology 384, 285–293.

Levitzki, A., 2004. Introduction: signal transduction therapy—10 years later. Semin. Cancer Biol. 14, 219–221.

Leyland-Jones, B., 2002. Trastuzumab: hopes and realities. Lancet Oncol. 3, 137–144.

Li, Y.Q., 2010. Master stem cell transcription factors and signaling regulation. Cell. Reprogram. 12, 3–13.

Li, M.L., Greenberg, R.A., 2012. Links between genome integrity and BRCA1 tumor suppression. Trends Biochem. Sci. 37, 418–424.

Li, L., Zou, L., 2005. Sensing, signaling, and responding to DNA damage: organization of the checkpoint pathways in mammalian cells. J. Cell. Biochem. 94, 298–306.

Li, X., Hong, X., Hussain, M., Sarkar, S.H., Li, R., Sarkar, F.H., 2005a. Gene expression profiling revealed novel molecular targets of docetaxel and estramustine combination treatment in prostate cancer cells. Mol. Cancer Ther. 4, 389–398.

Li, X., Hussain, M., Sarkar, S.H., Eliason, J., Li, R., Sarkar, F.H., 2005b. Gene expression profiling revealed novel mechanism of action of Taxotere and Furtulon in prostate cancer cells. BMC Cancer 5, 7.

Li, S.K.M., Smith, D., Leung, W.Y., Cheung, A.M.S., Lam, E.W.F., Dimri, G.P., et al., 2008a. FOXM1c counteracts oxidative stress-induced senescence and stimulates Bmi-1 expression. J. Biol. Chem. 283, 16545–16553.

Li, Z.W., Chen, H., Campbell, R.A., Bonavida, B., Berenson, J.R., 2008b. NF-κB in the pathogenesis and treatment of multiple myeloma. Curr. Opin. Hematol. 15, 391–399.

Li, Q., Zhang, N., Jia, Z., Le, X., Dai, B., Wie, D., et al., 2009. Critical role and regulation of transcription factor FoxM1 in human gastric cancer angiogenesis and progression. Cancer Res. 69, 3501–3509.

Li, Y., Ligr, M., McCarron, J.P., Daniels, G., Zhang, D., Zhao, X., et al., 2011. Natura-α targets Forkhead box M1 and inhibits androgen-dependent and -independent prostate cancer growth and invasion. Clin. Cancer Res. 17, 4414–4424.

Li, Q., Jia, Z., Wang, L., Kong, X., Li, Q., Guo, K., et al., 2012a. Disruption of Klf4 in Villin-positive gastric progenitor cells promotes formation and progression of tumors of the antrum in mice. Gastroenterology 142, 531–542.

Li, Y., Zhang, S., Huang, S., 2012b. FoxM1: a potential drug target for glioma. Future Oncol. 8, 223–226.

Li, J., Wang, Y., Luo, J., Fu, Z., Ying, J., Yu, Y., et al., 2012c. miR-134 inhibits epithelial to mesenchymal transition by targeting FOXM1 in non-small cell lung cancer cells. FEBS Lett. 586, 3761–3765.

Li, Q., Peng, Z., Hung, C., Tang, H., Jia, Z., Cui, J., et al., 2013. The critical role of dys-regulated FoxM1-uPAR signaling in human colon cancer progression and metastasis. Clin. Cancer Res. 19, 62–72.

Liang, Y., Lin, S.Y., Brunicardi, C., Goss, J., Li, K., 2009. DNA damage response pathways in tumor suppression and cancer treatment. World J. Surg. 33, 661–666.

Liao, J.B., 2006. Viruses and human cancer. Yale J. Biol. Med. 79, 115–122.

Liao, G., Tao, Q., Kofron, M., Chen, J.S., Schloemer, A., Davis, R.J., et al., 2006. Jun NH_2-terminal kinase (JNK) prevents nuclear β-catenin accumulation and regulates axis forma-tion in Xenpopus embryos. J. Biol. Chem. 103, 16313–16318.

Lien, W.H., Klezovitch, O., Vasloukhin, V., 2006. Cadherin-catenin proteins in vertebrate development. Curr. Opin. Cell Biol. 18, 499–506.

Lin, B., Madan, A., Yoon, J.G., Fang, X., Yan, X., Kim, T.K., et al., 2010a. Massively par-allel signature sequencing and bioinformatics analysis identifies up-regulation of TGFBI and SOX4 in human glioblastoma. PLoS One 5, e10210.

Lin, M., Guo, L.M., Liu, H., Du, J., Yang, J., Zhang, L.J., et al., 2010b. Nuclear accumu-lation of glioma-associated oncogene 2 protein and enhanced expression of forkhead-box transcription factor M1 protein in human hepatocellular carcinoma. Histol. Histopathol. 25, 1269–1275.

Lin, P.C., Chiu, Y.L., Banerjee, S., Park, K., Mosquera, J.M., Giannopoulou, E., et al., 2013. Epigenetic repression of miR-31 disrupts androgen receptor homeostasis and contributes to prostate cancer progression. Cancer Res. 73, 1232–1244.

Lindauer, M., Hochhaus, A., 2010. Dasatinib. Recent Results Cancer Res. 184, 83–102.

Lindström, M.S., Zhang, Y., 2006. B23 and ARF: friends or foes? Cell Biochem. Biophys. 46, 79–90.

Linley, A.J., Ahmad, M., Rees, R.C., 2011. Tumour-associated antigens: considerations for their use in tumor immunotherapy. Int. J. Hematol. 93, 427–431.

Liu, X., Marmorstein, R., 2006. When viral oncoprotein meets tumor suppressor: a struc-tural view. Genes Dev. 20, 2332–2337.

Liu, Y., Zeng, G., 2012. Cancer and innate immune system interactions: translational poten-tials for cancer immunotherapy. J. Immunother. 35, 299–308.

Liu, S.H., Zhang, M., Zhang, W.G., 2005. Strategies of antigen-specific T-cell-based immu-notherapy for cancer. Cancer Biother. Radiopharm. 20, 491–501.

Liu, M., Dai, B., Kang, S.H., Ban, K., Huang, F.J., Lang, F.F., et al., 2006. FoxM1B is over-expressed in human glioblastomas and critically regulates the tumorigenicity of glioma cells. Cancer Res. 66, 3593–3602.

Liu, P., Cheng, H., Roberts, T.M., Zhao, J.J., 2009a. Targeting the phosphoinositide 3-kinase pathway in cancer. Nat. Rev. Drug Discov. 8, 627–644.

Liu, M., Casimiro, M.C., Wang, C., Shirley, L.A., Jiao, X., Katiyar, S., et al., 2009b. p21[CIP1] attenuates Ras- and c-Myc-dependent breast tumor epithelial mesenchymal transition and cancer stem cell-like gene expression in vivo. Proc. Natl. Acad. Sci. U.S.A. 106, 19035–19039.

Liu, L., Guo, X., Rao, J.N., Zou, T., Xiao, L., Yu, T., 2009c. Polyamines regulate E-cadherin transcription through c-Myc modulating intestinal epithelial barrier func-tion. Am. J. Physiol. Cell Physiol. 296, C801–C810.

Liu, Y., Sadikot, R.T., Adami, G.R., Kalinichenko, V.V., Pendyala, S., Natarajan, V., et al., 2011. FoxM1 mediates the progenitor function of type II epithelial cells in repairing alveolar injury induced by *Pseudomonas aeruginosa*. J. Exp. Med. 208, 1473–1484.

Liu, J., Guo, S., Li, Q., Yang, L., Xia, Z., Zhang, L., et al., 2013a. Phosphoglycerate dehydrogenase induces glioma cells proliferation and invasion by stabilizing forkhead box M1. J. Neurooncol 111, 245–255.

Liu, Y., Hock, J.M., Van Beneden, R.J., Li, X., 2013b. Aberrant overexpression of FOXM1 transcription factor plays a critical role in lung carcinogenesis induced by low doses of arsenic. Mol. Carcinog. Epub Dec 19.

Livingston, D.M., Silver, D.P., 2008. Crossing over to drug resistance. Nature 451, 1066–1067.

Lizano, M., Berumen, J., Garcia-Carranca, A., 2009. HPV-related carcinogenesis: basic concepts, viral types and variants. Arch. Med. Res. 40, 428–434.

Llaurado, M., Majem, B., Casellvi, J., Cabrera, S., Gil-Moreno, A., Reventos, J., et al., 2012. Analysis of gene expression regulated by the ETV5 transcription factor in OV90 ovarian cancer cells identifies FoxM1 over-expression in ovarian cancer. Mol. Cancer Res. 10, 914–924.

Lleonart, M.E., Artero-Castro, A., Kondoh, H., 2009. Senescence induction; a possible cancer therapy. Mol. Cancer 8, 3.

Lloyd, A.C., 2000. p53: only ARF the story. Nat. Cell Biol. 2, E48–E50.

Logue, J.S., Morrison, D.K., 2012. Complexity in the signaling network: insights from the use of targeted inhibitors in cancer therapy. Genes Dev. 26, 641–650.

Loh, Y.H., Ng, J.H., Ng, H.H., 2008. Molecular framework underlying pluripotency. Cell Cycle 7, 885–891.

Loh, Y.H., Yang, L., Yang, J.C., Li, H., Collins, J.J., Daley, G.Q., 2011. Genomic approaches to deconstruct pluripotency. Annu. Rev. Genomics Hum. Genet. 12, 165–185.

Lohela, M., Bry, M., Tammela, T., Alitalo, K., 2009. VEGFs and receptors involved in angiogenesis versus lymphangiogenesis. Curr. Opin. Cell Biol. 21, 154–165.

Lok, G.T.M., Chan, D.W., Liu, V.W.S., Hui, W.W.Y., Leung, T.H.Y., Yao, K.M., et al., 2011. Aberrant activation of ERK/FOXM1 signaling cascade triggers the cell migration/invasion in ovarian cancer cells. PLoS One 6, e23790.

Lonard, D.M., O'Malley, B.W., 2007. Nuclear receptor coregulators: judges, juries, and executioners of cellular regulation. Mol. Cell 27, 691–700.

Longley, D.B., Johnston, P.G., 2005. Molecular mechanisms of drug resistance. J. Pathol. 205, 275–292.

Longworth, M.S., Laimins, L.A., 2004. Pathogenesis of human papillomaviruses in differentiating epithelia. Microbiol. Mol. Biol. Rev. 68, 362–372.

Lonze, B.E., Ginty, D.D., 2002. Function and regulation of CREB family transcription factors in the nervous system. Neuron 35, 605–623.

Lopes, L.F., Bacchi, C.E., 2010. Imatinib treatment for gastrointestinal stromal tumour (GIST). J. Cell. Mol. Med. 14, 42–50.

Lopez-Bergami, P., 2011. The role of mitogen- and stress-activated protein kinase pathways in melanoma. Pigment Cell Melanoma Res. 24, 902–9121.

Lopez-Novoa, J.M., Nieto, M.A., 2009. Inflammation and EMT: and alliance towards organ fibrosis and cancer progression. EMBO Mol. Med. 1, 303–314.

Lorvellec, M., Dumon, S., Maya-Mendoza, A., Jackson, D., Frampton, J., Garcia, P., 2010. B-Myb is critical for proper DNA duplication during an unperturbed S phase in mouse embryonic stem cells. Stem Cells 28, 1751–1759.

Lowe, S.W., Sherr, C.J., 2003. Tumor suppression by Ink4a-Arf: progress and puzzles. Curr. Opin. Genet. Dev. 13, 77–83.

Lowe, S.W., Cepero, E., Evan, G., 2004. Intrinsic tumour suppression. Nature 432, 307–315.

Ludwig, H., Khayat, D., Giaccone, G., Facon, T., 2005. Proteasome inhibition and its clinical prospects in the treatment of hematologic and solid malignancies. Cancer 104, 1794–1807.

Lukas, J., Lukas, C., Bartek, J., 2004. Mammalian cell cycle checkpoints: signalling pathways and their organization in space and time. DNA Repair 3, 997–1007.

Lukas, J., Lukas, C., Bartek, J., 2011. More than just a focus: the chromatin response to DNA damage and its role in genome integrity. Nat. Cell Biol. 13, 1161–1169.

Lumachi, F., Luisetto, G., Basso, S.M., Basso, U., Brunello, A., Camozzi, V., 2011. Endocrine therapy of breast cancer. Curr. Med. Chem. 18, 513–522.

Luo, X., Kraus, W.L., 2012. On PAR with PARP: cellular stress signaling through poly (ADP-ribose) and PARP-1. Genes Dev. 26, 417–432.

Luo, W., Lin, S.C., 2004. Axin: a master scaffold for multiple signaling pathways. Neurosignals 13, 99–113.

Luo, J.L., Kamata, H., Karin, M., 2005a. The anti-death machinery in IKK/NF-κB signaling. J. Clin. Immunol. 25, 541–550.

Luo, J.L., Kamata, H., Karin, M., 2005b. IKK/NF-κB signaling: balancing life and death—a new approach to cancer therapy. J. Clin. Invest. 115, 2625–2632.

Luo, J., Solimini, N.L., Elledge, S.J., 2009. Principles of cancer therapy: oncogene and non-oncogene addiction. Cell 136, 823–837.

Lupberger, J., Hildt, E., 2007. Hepatitis B virus-induced oncogenesis. World J. Gastroenterol. 13, 74–81.

Luqmani, Y.A., 2005. Mechanisms of drug resistance in cancer chemotherapy. Med. Princ. Pract. 14, 35–48.

Lüscher-Firzlaff, J.M., Westendorf, J.M., Zwicker, J., Burkhardt, H., Henriksson, M., Müller, R., et al., 1999. Interaction of the fork head domain transcription factor MPP2 with the human papilloma virus 16 E7 protein: enhancement of transformation and transactivation. Oncogene 18, 5620–5630.

Lustig, B., Behrens, J., 2003. The Wnt signaling pathway and its role in tumor development. J. Cancer Res. Clin. Oncol. 129, 199–221.

Ly, D.H., Lockhart, D.J., Lerner, R.A., Schultz, P.G., 2000. Mitotic misregulation and human ageing. Science 287, 2486–2492.

Lydon, N., 2009. Attacking cancer at its foundation. Nat. Med. 15, 1153–1157.

Lykkesfeldt, A.E., Madsen, M.W., Briand, P., 1994. Altered expression of estrogen-regulated genes in tamoxifen-resistant and ICI 164384 and ICI 182780 sensitive human breast cancer cell line MCF-7/TAMR-1. Cancer Res. 54, 1587–1595.

Lynch, T.P., Ferrer, C.M., Jackson, R., Shahriari, K.S., Vosseller, K., Reginato, M.J., 2012. Critical role of O-GlcNAc transferase in prostate cancer invasion, angiogenesis and metastasis. J. Biol. Chem. 287, 11070–11081.

Ma, L., Weinberg, R.A., 2008a. Micromanagers of malignancy: role of microRNAs in regulating metastasis. Trends Genet. 24, 448–456.

Ma, L., Weinberg, R.A., 2008b. MicroRNAs in malignant progression. Cell Cycle 7, 570–572.

Ma, L., Young, J., Prabhala, H., Pan, E., Mestdagh, P., Muth, D., et al., 2010. miR-9, a MYC/MYCN-activated microRNA, regulates E-cadherin and cancer metastasis. Nat. Cell Biol. 12, 247–256.

Ma, J., Sun, T., Park, S., Shen, G., Liu, J., 2011. The role of hepatitis B virus X protein is related to its differential intracellular localization. Acta Biochim. Biophys. Sin. 43, 583–588.

Macaluso, M., Montanari, M., Giordano, A., 2006. Rb family proteins as modulators of gene expression and new aspects regarding the interaction with chromatin remodeling enzymes. Oncogene 25, 5263–5267.

MacArthur, B.D., Ma'ayan, A., Lemischka, I.R., 2009. Systems biology of stem cell fate and cellular reprogramming. Nat. Rev. Mol. Cell Biol. 10, 672–681.

MacDonald, B.T., He, X., 2012. Frizzled and LRP5/6 receptors for Wnt/β-catenin signaling. Cold Spring Harb. Perspect. Biol. 4, a007880.

MacDonald, B.T., Semenov, M.V., He, X., 2007. SnapShot: Wnt/β-catenin signaling. Cell 131, 1204 1204.e1.

MacDonald, B.T., Tamai, K., He, X., 2009. Wnt/β-catenin signaling: components, mechanisms, and disease. Dev. Cell 17, 9–26.

Mackey, S., Singh, P., Darlington, G.J., 2003. Making the liver young again. Hepatology 38, 1349–1352.

Madden, C.R., Slagle, B.L., 2001. Stimulation of cellular proliferation by hepatitis B virus X protein. Dis. Markers 17, 153–157.

Madureira, P.A., Varshochi, R., Constantinidou, D., Francis, R.E., Coombes, R.C., Yao, K.M., et al., 2006. The Forkhead box M1 protein regulates the transcription of the estrogen receptor α in breast cancer cells. J. Biol. Chem. 281, 25167–25176.

Major, M.L., Lepe, R., Costa, R.H., 2004. Forkhead Box M1B transcriptional activity requires binding of Cdk-cyclin complexes for phosphorylation-dependent recruitment of p300/CBP coactivators. Mol. Cell. Biol. 24, 2649–2661.

Majumdar, A.J., Wong, W.J., Simon, M.C., 2010. Hypoxia-inducible factors and the response to hypoxic stress. Mol. Cell 40, 294–309.

Malkinson, A.M., Koski, K.M., Evans, W.A., Festing, M.F., 1997. Butylated hydroxytoluene exposure is necessary to induce lung tumors in BALB mice treated with 3-methylcholanthrene. Cancer Res. 57, 2832–2834.

Manley, P.W., Drueckes, P., Fendrich, G., Furet, P., Liebetanz, J., Martiny-Baron, G., et al., 2010. Extended kinase profile and properties of the protein kinase inhibitor nilotinib. Biochim. Biophys. Acta 1804, 445–453.

Mantovani, A., 2010. Molecular pathways linking inflammation and cancer. Curr. Mol. Med. 10, 369–373.

Mantovani, A., Sozzani, S., Locati, M., Allavena, P., Sica, A., 2002. Macrophage polarization: tumor-associated macrophages as a paradigm for polarized M2 mononuclear phagocytes. Trends Immunol. 23, 549–555.

Mantovani, A., Schioppa, T., Porta, C., Allavena, P., Sica, A., 2006. Role of tumor-associated macrophages in tumor progression and invasion. Cancer Metastasis Rev. 25, 315–322.

Mantovani, A., Allavena, P., Sica, A., Balkwill, F., 2008. Cancer-related inflammation. Nature 454, 436–444.

Mantovani, A., Sica, A., Allavena, P., Garlanda, C., Locati, M., 2009. Tumor-associated macrophages and the related myeloid-derived suppressor cells as paradigm of the diversity of macrophage activation. Hum. Immunol. 70, 325–330.

Mantovani, A., Garlanda, C., Allavena, P., 2010. Molecular pathways and targets in cancer-related inflammation. Ann. Med. 42, 161–170.

Mantovani, A., Germano, G., Marchesi, F., Locatelli, M., Biswas, S.K., 2011. Cancer-promoting tumor-associated macrophages: new vistas and open questions. Eur. J. Immunol. 41, 2522–2525.

Marcu, K.B., Bossone, S.A., Patel, A.J., 1992. myc function and regulation. Annu. Rev. Biochem. 61, 809–860.

Margineanu, E., Cotrutz, C.E., Cotrutz, C., 2008. Correlation between E-cadherin abnormal expressions in different types of cancer and the process of metastasis. Rev. Med. Chir. Soc. Med. Nat. Iasi. 112, 432–436.

Mariani, G., Fasolo, A., De Benedictis, E., Gianni, L., 2009. Trastuzumab as adjuvant systemic therapy for HER2-positive breast cancer. Nat. Clin. Pract. Oncol. 6, 93–104.

Markey, M.P., Bergseid, J., Bosco, E.E., Stengel, K., Xu, H., Mayhew, C.N., et al., 2007. Loss of the retinoblastoma tumor suppressor: differential action on transcriptional programs related to cell cycle control and immune function. Oncogene 26, 6307–6318.

Markman, B., Dienstmann, R., Tabernero, J., 2010. Targeting the PI3K/Akt/mTOR pathway—beyond rapalogs. Oncotarget 1, 530–543.

Martin, D., Gutkind, J.S., 2008. Human tumor-associated viruses and new insights into the molecular mechanisms of cancer. Oncogene 27, S31–S42.

Martin, K.J., Patrick, D.R., Bissell, M.J., Fournier, M.V., 2008. Prognostic breast cancer signature identified from 3D culture model accurately predicts clinical outcome across independent datasets. PLoS One 3, e2994.

Martinez Arias, A., 2001. Epithelial mesenchymal interactions in cancer and development. Cell 105, 425–431.

Martinkova, J., Gadher, S.J., Hajduch, M., Kovarova, H., 2009. Challenges in cancer research and multifaceted approaches for cancer biomarker quest. FEBS Lett. 583, 1772–1784.

Martin-Vilchez, S., Lara-Pezzi, E., Trapero-Marugan, M., Moreno-Otero, R., Sanz-Cameno, P., 2011. The molecular and pathophysiological implications of hepatitis B X antigen in chronic hepatitis B virus infection. Rev. Med. Virol. Epub Jul 14.

Masip, M., Veiga, A., Izpisua, J.C., Simon, C., 2010. Reprogramming with defined factors: from induced pluripotency to induced transdifferentiation. Mol. Hum. Reprod. 16, 856–868.

Massagué, J., 2004. G1 cell-cycle control and cancer. Nature 432, 298–306.

Masson, J.Y., West, S.C., 2001. The Rad51 and Dmc1 recombinases: a non-identical twin relationship. Trends Biochem. Sci. 26, 131–136.

Masterson, J., O'Dea, S., 2007. Posttranslational truncation of E-cadherin and significance for tumor progression. Cells Tissues Organs 185, 175–179.

Masumori, N., Thoams, T.Z., Chaurand, P., Case, T., Paul, M., Kasper, S., et al., 2001. A probasin-large T antigen transgenic mouse line develops prostate adenocarcinoma and neuroendocrine carcinoma with metastatic potential. Cancer Res. 61, 2239–2249.

Masumoto, N., Tateno, C., Tachibana, A., Utoh, R., Morikawa, Y., Shimada, T., et al., 2007. GH enhances proliferation of human hepatocytes grafted into immunodeficient mice with damaged liver. J. Endocrinol. 194, 529–537.

Mathias, R.A., Simpson, R.J., 2009. Towards understanding epithelial-mesenchymal transition: a proteomics perspective. Biochim. Biophys. Acta 1794, 1325–1331.

Mathon, N.F., Lloyd, A.C., 2001. Cell senescence and cancer. Nat. Rev. Cancer 1, 203–213.

Matsuda, Y., Ichida, T., 2009. Impact of hepatitis B virus X protein on DNA damage response during hepatocarcinogenesis. Med. Mol. Morphol. 42, 138–142.

Matsuda, Y., Schlange, T., Oakeley, E.J., Boulay, A., Hynes, N.E., 2009. WNT signaling enhances breast cancer cell motility and blockade of the WNT pathway by sFRP1 suppresses MDA-MB-231 xenograft growth. Breast Cancer Res. 11, R32.

Matsuo, T., Yamaguchi, S., Mitsui, S., Emi, A., Shimoda, F., Okamura, H., 2003. Control mechanism of the circadian clock for timing of cell division in vivo. Science 302, 255–259.

Matsushima-Nishiu, M., Unoki, M., Ono, K., Tsunoda, T., Minaguchi, T., Kuramoto, H., et al., 2001. Growth and gene expression profile analyses of endometrial cancer cells expressing exogenous PTEN. Cancer Res. 61, 3741–3749.

Mattingly, L.H., Gault, R.A., Murhy, W.J., 2007. Use of systemic proteasome inhibition as an immune-modulating agent in disease. Endocr. Metab. Immune Disord. Drug Targets 7, 29–34.

Mattoli, S., Bellini, A., Schmidt, M., 2009. The role of a human hematopoietic mesenchymal progenitor in wound healing and fibrotic diseases and implications for therapy. Curr. Stem Cell Res. Ther. 4, 266–280.

Maurer, G., Tarkowski, B., Baccarini, M., 2011. Raf kinases in cancer—role and therapeutic opportunities. Oncogene 30, 3477–3488.

May, C.D., Sphyris, N., Evans, K.W., Werden, S.J., Guo, W., Mani, S.A., 2011. Epithelial-mesenchymal transition and cancer stem cells: a dangerously dynamic duo in breast cancer progression. Breast Cancer Res. 13, 202.

Mayr, B., Montminy, M., 2001. Transcriptional regulation by the phosphorylation-dependent factor CREB. Nat. Rev. Mol. Cell Biol. 2, 599–609.

McCance, D.J., 2005. Transcriptional regulation by human papillomaviruses. Curr. Opin. Genet. Dev. 15, 515–519.

McConkey, D.J., Zhu, K., 2008. Mechanisms of proteasome inhibitor action and resistance in cancer. Drug Resist. Updat. 11, 164–179.

McCormick, F., 2011. Cancer therapy based on oncogene addiction. J. Surg. Oncol. 103, 464–467.

McCubrey, J.A., Steelman, L.S., Chappell, W.H., Abrams, S.L., Wong, E.W.T., Chang, F., et al., 2007. Roles of the Raf/MEK/ERK pathway in cell growth, malignant transformation and drug resistance. Biochim. Biophys. Acta 1773, 1263–1284.

McCubrey, J.A., Steelman, L.S., Kempf, C.R., Chappell, W.H., Abrams, S.L., Stivala, F., et al., 2011. Therapeutic resistance resulting from mutations in Raf/MEK/ERK and PI3K/PTEN/Akt/mTOR signaling pathways. J. Cell. Physiol. 226, 2762–2781.

McDuff, F.K.E., Turner, S.D., 2011. Jailbreak: oncogene-induced senescence and its evasion. Cell. Signal. 23, 6–13.

McGovern, U.B., Francis, R.E., Peck, B., Guest, S.K., Wang, J., Myatt, S.S., et al., 2009. Gefitinib (Iressa) represses FOXM1 expression via FOXO3a in breast cancer. Mol. Cancer Ther. 8, 582–591.

McGowan, C.H., 2002. Checking on Cds1 (Chk2): a checkpoint kinase and tumor suppressor. Bioessays 24, 502–511.

McKenna, N.J., O'Malley, B.W., 2002. Combinatorial control of gene expression by nuclear receptors and coregulators. Cell 108, 465–474.

McKenna, N.J., Lanz, R.B., O'Malley, B.W., 1999. Nuclear receptor coregulators: cellular and molecular biology. Endocr. Rev. 20, 321–344.

McKinnon, P.J., Caldecott, K.W., 2007. DNA strand break repair and human genetic disease. Annu. Rev. Genomics Hum. Genet. 8, 37–55.

McLaughlin-Drubin, M.E., Münger, K., 2008. Viruses associated with human cancer. Biochim. Biophys. Acta 1782, 127–150.

McLaughlin-Drubin, M.E., Münger, K., 2009a. The human papillomavirus E7 oncoprotein. Virology 384, 335–344.

McLaughlin-Drubin, M.E., Münger, K., 2009b. Oncogenic activities of human papillomaviruses. Virus Res. 143, 195–208.

McLaughlin-Drubin, M.E., Münger, K., 2012. Cancer-associated human papillomaviruses. Curr. Opin. Virol. 2, 459–466.

Mellman, I., Coukos, G., Dranoff, G., 2011. Cancer immunotherapy comes of age. Nature 480, 480–489.

Mencalha, A.L., Binato, R., Ferreira, G.M., Du Rocher, B., Abdelhay, E., 2012. Forkhead box M1 (FoxM1) gene is a new STAT3 transcriptional factor target and is essential for proliferation, survival and DNA repair of K562 cell line. PLoS One 7, e48160.

Meng, W., Takeichi, M., 2009. Adherens junction: molecular architecture and regulation. Cold Spring Harb. Perspect. Biol. 1, a002899.

Meng, Z., Wang, Y., Wang, L., Jin, W., Liu, N., Pan, H., et al., 2010. FXR regulates liver repair after CCl_4-induced toxic injury. Mol. Endocrinol. 24, 886–897.

Meng, Z., Liu, N., Fu, X., Wang, X., Wang, Y.D., Chen, W.D., et al., 2011. Insufficient bile acids signaling impairs liver repair in CYP27-/- mice. J. Hepatol. 55, 885–895.

Meyer, N., Penn, L.Z., 2008. Reflecting on 25 years with MYC. Nat. Rev. Cancer 8, 976–990.

Micalizzi, D.S., Farabaugh, S.M., Ford, H.L., 2010. Epithelial-mesenchymal transition in cancer: parallels between normal development and tumor progression. J. Mammary Gland Biol. Neoplasia 15, 117–134.

Milane, L., Ganesh, S., Shah, S., Duan, Z.F., Amiji, M., 2011. Multi-modal strategies for overcoming tumor drug resistance: hypoxia, the Warburg effect, stem cells, and multifunctional nanotechnology. J. Control. Release 155, 237–247.

Milano, A., Iaffaioli, R.V., Caponigro, F., 2007. The proteasome: a worthwhile target for the treatment of solid tumors? Eur. J. Cancer 43, 1125–1133.

Miller, K.M., Jackson, S.P., 2012. Histone marks: repairing DNA breaks within the context of chromatin. Biochem. Soc. Trans. 40, 370–376.

Millour, J., Constantinidou, D., Stavropoulou, A.V., Wilson, M.S.C., Myatt, S.S., Kwok, J.M.M., et al., 2010. FOXM1 is a transcriptional target of ERα and has a critical role in breast cancer endocrine sensitivity and resistance. Oncogene 29, 2983–2995.

Millour, J., de Olano, N., Horimoto, Y., Monteiro, L.J., Langer, J.K., Aligue, R., et al., 2011. ATM and p53 regulate FOXM1 expression in breast cancer epirubicin treatment and resistance. Mol. Cancer Ther. 10, 1046–1058.

Mimeault, M., Batra, S.K., 2007. Interplay of distinct growth factors during epithelial-mesenchymal transition of cncer progenitor cells and molecular targeting as novel cancer therapies. Ann. Oncol. 18, 1605–1619.

Mimeault, M., Hauke, R., Batra, S.K., 2008. Recent advances on the molecular mechanisms involved in the drug resistance of cancer cells and novel targeting therapies. Clin. Pharmacol. Ther. 83, 673–691.

Minamino, T., Komuro, I., 2006. Regeneration of the endothelium as a novel therapeutic strategy for acute lung injury. J. Clin. Invest. 116, 2316–23319.

Minotti, G., Menna, P., Salvatorelli, E., Cairo, G., Gianni, L., 2004. Anthracyclines: molecular advances and pharmacologic developments in antitumor activity and cardiotoxicity. Pharmacol. Rev. 56, 185–229.

Mirza, M.K., Sun, Y., Zhao, Y.D., Potula, H.H.S.K., Frey, R.S., Vogel, S.M., et al., 2010. FoxM1 regulates re-annealing of endothelial adherens junctions through transcriptional control of β-catenin expression. J. Exp. Med. 207, 1675–1685.

Mito, J.K., Riedel, R.F., Dodd, L., Lahat, G., Lazar, A.J., Dodd, R.D., et al., 2009. Cross species genomic analysis identifies a mouse model as undifferentiated pleomorphic sarcoma/malignant fibrous histiocytoma. PLoS One 4, e8075.

Mitra, M., Kandalam, M., Sundaram, C.S., Shenkar Verma, R., Maheswari, U.K., Swaminathan, S., et al., 2011. Reversal of stathmin-mediated microtubule destabilization sensitizes retinoblastoma cells to a low dose of antimicrotubule agents: a novel synergistic therapeutic intervention. Invest. Ophthalmol. Vis. Sci. 52, 5441–5448.

Mitsiades, C.S., Mitsiades, N., Koutsilieris, M., 2004. The Akt pathway: molecular targets for anti-cancer drug development. Curr. Cancer Drug Targets 4, 235–256.

Mitsiades, C.S., Mitsiades, N., Hideshima, T., Richardson, P.G., Anderson, K.C., 2005. Proteasome inhibitors as therapeutics. Essays Biochem. 45, 1–28.

Mizuarai, S., Kotani, H., 2011. Synthetic lethal interactions for the development of cancer therapeutics: biological and methodological advancements. Hum. Genet. 128, 567–575.

Mizuno, T., Murakami, H., Fujii, M., Ishiguro, F., Tanaka, I., Kondo, Y., et al., 2012. YAP induces malignant mesothelioma cell proliferation by upregulating transcription of cell cycle-promoting genes. Oncogene 31, 5117–5122.

Moasser, M.M., 2007. Targeting the function of HER2 oncogene in human cancer therapeutics. Oncogene 26, 6577–6592.

Moeller, A., Ask, K., Warburton, D., Gauldie, J., Kolb, M., 2008. The bleomycin animal model: a useful tool to investigate treatment options for idiopathic pulmonary fibrosis? Int. J. Biochem. Cell Biol. 40, 362–382.

Moens, U., Van Ghelue, M., Johannessen, M., 2007. Oncogenic potentials of the human polyomavirus regulatory proteins. Cell. Mol. Life Sci. 64, 1656–1678.

Mongroo, P.S., Rustgi, A.K., 2010. The role of the miR-200 family in epithelial-mesenchymal transition. Cancer Biol. Ther. 10, 219–222.

Montagut, C., Rovira, A., Mellado, B., Gascon, P., Ross, J.S., Albanell, J., 2005. Preclinical and clinical development of the proteasome inhibitor bortzomib in cancer treatment. Drugs Today 41, 299–315.

Montagut, C., Rovira, A., Albanell, J., 2006. The proteasome: a novel target for anticancer therapy. Clin. Transl. Oncol. 8, 313–317.

Monteiro, L.J., Khongkow, P., Kongsema, M., Morris, J.R., Man, C., Weekes, D., et al., 2012. The Forkhead Box M1 protein regulates BRIP1 expression and DNA damage repair in epirubicin treatment. Oncogene. Epub Oct 29.

Montemurro, F., Valabrega, G., Aglietta, M., 2007. Lapatinib: a dual inhibitor of EGFR and HER2 tyrosine kinase activity. Expert Opin. Biol. Ther. 7, 257–268.

Montminy, M., 1997. Transcriptional regulation by cyclic AMP. Annu. Rev. Biochem. 66, 807–822.

Moody, C.A., Laimins, L.A., 2010. Human papillomavirus oncoproteins: pathways to transformation. Nat. Rev. Cancer 10, 550–560.

Mooi, W.J., Peeper, D.S., 2006. Oncogene-induced cell senescence—halting on the road to cancer. N. Engl. J. Med. 355, 1037–1046.

Moore, P.S., Chang, Y., 2010. Why do viruses cause cancer? Highlights of the first century of human tumour virology. Nat. Rev. Cancer 10, 878–889.

Moore, B.B., Hogaboam, C.M., 2008. Murine models of pulmonary fibrosis. Am. J. Physiol. Lung Cell. Mol. Physiol. 294, L152–L160.

Moran, E., Nencioni, A., 2007. The role of proteasome in malignant diseases. J. BUON 12, S95–S99.

Moreno-Bueno, G., Portillo, F., Cano, A., 2008. Transcriptional regulation of cell polarity in EMT and cancer. Oncogene 27, 6958–6969.

Morris, J.R., 2010. More modifiers move on DNA damage. Cancer Res. 70, 3861–3863.

Morrison, S.J., Kimble, J., 2006. Asymmetric ad symmetric stem-cell divisions in development and cancer. Nature 441, 1068–1074.

Mosesson, Y., Yarden, Y., 2004. Oncogenic growth factor receptors: implications for signal transduction therapy. Semin. Cancer Biol. 14, 262–270.

Mouratis, M.A., Aidini, V., 2011. Modeling pulmonary fibrosis with bleomycin. Curr. Opin. Pulm. Med. 17, 355–361.

Moustakas, A., Heldin, C.H., 2007. Signaling networks guiding epithelial-mesenchymal transitions during embryogenesis and cancer progression. Cancer Sci. 98, 1512–1520.

Moynahan, M.E., Jasin, M., 2010. Mitotic homologous recombination maintains genomic stability and suppresses tumorigenesis. Nat. Rev. Mol. Cell Biol. 11, 196–207.

Mujtaba, T., Dou, Q.P., 2011. Advances in the understanding of mechanisms and therapeutic use of bortezomib. Discov. Med. 12, 471–480.

Mukherjee, A., Dhadda, A.S., Shehata, M., Chan, S., 2007. Lapatinib: a tyrosine kinase inhibitor with a clinical role in breast cancer. Expert Opin. Pharmacother. 8, 2189–2204.

Mukhopadhyay, B., Cinar, R., Yin, S., Liu, J., Tam, J., Godlewski, G., et al., 2011. Hyperactivation of anandamide synthesis and regulation of cell-cycle progression via cannabinoid tape 1 (CB_1) receptors in the regenerating liver. Proc. Natl. Acad. Sci. U.S.A. 108, 6323–6328.

Mullan, P.B., Quinn, J.E., Harkin, D.P., 2006. The role of BRCA1 in transcriptional regulation and cell cycle control. Oncogene 25, 5854–5863.

Müller, H., Helin, K., 2000. The E2F transcription factors: key regulators of cell proliferation. Biochim. Biophys. Acta 1470, M1–M12.

Müller, I., Niethammer, D., Bruchelt, G., 1998. Anthracycline-derived chemotherapeutics in apoptosis and free radical cytotoxicity. Int. J. Mol. Med. 1, 491–494.

Münger, K., 2002. The role of human papillomaviruses in human cancers. Front. Biosci. 7, d641–d649.

Münger, K., 2003. Clefts, grooves, and (small) pockets: the structure of retinoblastoma tumor suppressor in complex with its cellular target E2F unveiled. Proc. Natl. Acad. Sci. U.S.A. 100, 2165–2167.

Münger, K., Howley, P.M., 2002. Human papillomavirus immortalization and transformation. Virus Res. 89, 213–228.

Münger, K., Werness, B.A., Dyson, N., Phelps, W.C., Harlow, E., Howley, P.M., 1989. Complex formation of human papillomavirus E7 proteins with the retinoblastoma tumor suppressor gene product. EMBO J. 8, 4099–4105.

Münger, K., Basile, J.R., Duensing, S., Eichten, A., Gonzalez, S.L., Grace, M., et al., 2001. Biological activities and molecular targets of the human papillomavirus E7 oncoprotein. Oncogene 20, 7888–7898.

Münger, K., Baldwin, A., Edwards, K.M., Hayakawa, H., Nguyen, C.L., Owens, M., et al., 2004. Mechanisms of human papillomavirus-induced oncogenesis. J. Virol. 78, 11451–11460.

Munro, S., Carr, S.M., La Thangue, N.B., 2012. Diversity within the pRB pathway: is there a code of conduct? Oncogene 31, 4343–4352.

Murakami, S., 2001. Hepatitis B virus X protein: a multifunctional viral regulator. J. Gastroenterol. 36, 651–660.

Murakami, H., Aiba, H., Nakanishi, M., Murakami-Tonami, Y., 2010. Regulation of yeast forkhead transcription factors and FoxM1 by cyclin-dependent and polo-like kinases. Cell Cycle 9, 3233–3242.

Murray, M.M., Mullan, P.B., Harkin, D.P., 2007. Role played by BRCA1 in transcriptional regulation in response to therapy. Biochem. Soc. Trans. 35, 1342–1346.

Myatt, S.S., Lam, E.W.F., 2007. The emerging roles of forkhead box (Fox) proteins in cancer. Nat. Rev. Cancer 7, 847–859.

Mymryk, J.S., 1996. Tumour suppressive properties of the adenovirus 5 E1A oncogene. Oncogene 13, 1581–1589.

Nadal, E., Olavarria, E., 2004. Imatinib mesylate (Gleevec/Glivec) a molecular-targeted therapy for chronic myeloid leukaemia and other malignancies. Int. J. Clin. Pract. 58, 511–516.

Nagy, J.A., Dvorak, A.M., Dvorak, H.F., 2007. VEGF-A and the induction of pathological angiogenesis. Annu. Rev. Pathol. Mech. Dis. 2, 25–275.

Nahta, R., Esteva, F.J., 2006. Herceptin: mechanisms of action and resistance. Cancer Lett. 232, 123–138.

Nahta, R., Esteva, F.J., 2007. Trastuzumab: triumphs and tribulations. Oncogene 26, 3637–3643.

Nakamura, T., Furukawa, Y., Nakagawa, H., Tsunoda, T., Ohigashi, H., Murata, K., et al., 2004. Genome-wide cDNA microarray analysis of gene expression profiles in pancreatic cancers using populations of tumor cells and nomal ductal epithelial cells selected for purity by laser microdissection. Oncogene 23, 2385–2400.

Nakamura, S., Hirano, I., Okinaka, K., Takemura, T., Yokota, D., Ono, T., et al., 2010a. The FOXM1 transcriptional factor promotes the proliferation of leukemia cells through modulation of cell cycle progression in acute myeloid leukemia. Carcinogenesis 31, 2012–2021.

Nakamura, S., Yamashita, M., Yokota, D., Hirano, I., Ono, T., Fujie, M., et al., 2010b. Development and pharmacologic characterization of deoxybromophospha sugar derivatives with antileukemic activity. Invest. New Drugs 28, 381–391.

Nakanishi, C., Toi, M., 2005. Nuclear factor-κB inhibitors as sensitizers to anticancer drugs. Nat. Rev. Cancer 5, 297–309.

Nakanishi, M., Shimada, M., Niida, H., 2006. Genetic instability in cancer cells by impaired cell cycle checkpoints. Cancer Sci. 97, 984–989.

Nakayama, K.I., Nakayama, K., 2006. Ubiquitin ligases: cell-cycle control and cancer. Nat. Rev. Cancer 6, 369–381.

Nardella, C., Clohessy, J.G., Alimonti, A., Pandolfi, P.P., 2011. Pro-senescence therapy for cancer treatment. Nat. Rev. Cancer 11, 503–511.

Narisawa-Saito, M., Kiyono, T., 2007. Basic mechanisms of high-risk human papillomavirus-induced carcinogenesis: role of E6 and E7 proteins. Cancer Sci. 98, 1505–1511.

Naugler, W.E., Karin, M., 2008. NF-κB and cancer—identifying targets and mechanisms. Curr. Opin. Genet. Dev. 18, 19–26.

Navone, N.M., Logothetis, C.J., von Eschenbach, A.C., Tronsoco, P., 1999. Model systems of prostate cancer: uses and limitations. Cancer Metastasis Rev. 17, 361–371.

Nelson, W.J., 2008. Regulation of cell adhesion by the cadherin-catenin complex. Biochem. Soc. Trans. 36, 149–155.

Nelson, M.H., Dolder, C.R., 2006. Lapatinib: a novel dual tyrosine kinase inhibitor with activity in solid tumors. Ann. Pharmacother. 40, 261–269.

Nelson, W.J., Nusse, R., 2004. Convergence of Wnt, β-catenin, and cadherin pathways. Science 303, 1483–1487.

Nencioni, A., Grünebach, F., Patrone, F., Ballestrero, A., Brossart, P., 2007. Proteasome inhibitors: antitumor effects and beyond. Leukemia 21, 30–36.

Nesbit, C.E., Tersak, J.M., Prochownik, E.V., 1999. MYC oncogenes and human neoplastic disease. Oncogene 18, 3004–3016.

Neuveut, C., Wei, Y., Buendia, M.A., 2010. Mechanisms of HBV-related hepatocarcinogenesis. J. Hepatol. 52, 594–6904.

Nevins, J.R., 1998. Toward an understanding of the functional complexity of the E2F and retinoblastoma families. Cell Growth Differ. 9, 585–593.

Newick, K., Cunniff, B., Preston, K., Held, P., Arbiser, J., Pass, H., et al., 2012. Peroxiredoxin 3 is a redox-dependent target of thiostrepton in malignant mesothelioma cells. PLoS One 7, e39404.

Ng, J.M.Y., Curran, T., 2011. The Hedgehog's tale: developing strategies for targeting cancer. Nat. Rev. Cancer 11, 493–501.

Ng, S.A., Lee, C., 2011. Hepatitis B virus X gene and hepatocarcinogenesis. J. Gastroenterol. 46, 974–990.

Ng, H.H., Surani, M.A., 2011. The transcriptional and signalling networks of pluripotency. Nat. Cell Biol. 13, 490–496.

Nguyen, D.M., Sam, K., Tsimelzon, A., Li, X., Wong, H., Mohsin, S., et al., 2006. Molecular heterogeneity of inflammatory breast cancer: a hyperproliferative phenotype. Clin. Cancer Res. 12, 5047–5054.

Nicoloso, M.S., Spizzo, R., Shimizu, M., Rossi, S., Calin, G.A., 2009. MicroRNAs—the micro steering wheel of tumour metastasis. Nat. Rev. Cancer 9, 293–302.

Niehrs, C., Acebron, S.P., 2012. Mitotic and mitogenic Wnt signalling. EMBO J. 31, 2705–2713.

Niessen, C.M., Gottardi, C.J., 2008. Molecular components of the adherens junction. Biochim. Biophys. Acta 1778, 562–571.

Nieto, M.A., 2002. The Snail superfamily of zinc-finger transcription factors. Nat. Rev. Mol. Cell Biol. 3, 155–166.

Nieto, M.A., 2011. The ins and outs of the epithelial to mesenchymal transition in health and disease. Annu. Rev. Cell Dev. Biol. 27, 347–376.

Nihira, S., 2003. Development of HER2-specific humanized antibody Herceptin (trastuzumab). Nippon Yakurigaku Zasshi 122, 504–514.

Nimmanapalli, R., O'Bryan, E., Huang, M., Bali, P., Burnette, P., Loughran, T., et al., 2003. Molecular characterization and sensitivity of STI-571 (imatinib mesylate, Gleevec)-resistant, Bcr-Abl-positive, human acute leukemia cells to SRC kinase inhibitor PD180970 and 17-allylamino-17-demethoxygeldanamycin. Cancer Res. 62, 5761–5769.

Ning, Y., Li, Q., Xiang, H., Liu, F., Cao, J., 2012. Apoptosis induced by 7-difluoromethoxyl-5,4′-di-n-octyl genistein via the inactivation of FoxM1 in ovarian cancer cells. Oncol. Rep. 27, 1857–1864.

Nishidate, T., Katagiri, T., Lin, M.L., Mano, Y., Miki, Y., Kasumi, F., et al., 2004. Genome-wide gene-expression profiles of breast-cancer cells purified with laser microbeam microdissection: identification of genes associated with progression and metastasis. Int. J. Oncol. 25, 797–819.

Nishikawa, T., Yamashita, T., Yamada, T., Kobayashi, H., Ohkawara, A., Fujinaga, K., 1991. Tumorigenic transformation of primary rat embryonal fibroblasts by human papillomavirus type 8 E7 gene in collaboration with the activated H-ras gene. Jpn. J. Cancer Res. 82, 1340–1343.

Niwa, H., 2007a. How is pluripotency determined and maintained? Development 134, 635–646.

Niwa, H., 2007b. Open conformation chromatin and pluripotency? Genes Dev. 21, 2671–2676.

Noble, M.E., Endicott, J.A., Johnson, L.N., 2004. Protein kinase inhibitors: insights into drug design from structure. Science 303, 1800–1805.

Noonan, D.M., de Lerma Barbaro, A., Vannini, N., Mortara, L., Albini, A., 2008. Inflammation, inflammatory cells and angiogenesis: decisions and indecisions. Cancer Metastasis Rev. 27, 31–40.

Nowosielska, A., Marinus, M.G., 2005. Cisplatin induces DNA double-strand break formation in Escherichia coli dam mutants. DNA Repair 4, 773–781.

Nusse, R., Varmus, H., 2012. Three decades of Wnts: a personal perspective on how a scientific field developed. EMBO J. 31, 2670–2684.

Nyberg, K.A., Michelson, R.J., Putnam, C.W., Weinert, T.A., 2002. Toward maintaining the genome: DNA damage and replication checkpoints. Annu. Rev. Genet. 36, 617–656.

Obajimi, O., 2009. Lapatinib as a chemotherapeutic drug. Recent Pat. Anticancer Drug Discov. 4, 216–226.

Obama, K., Ura, K., Li, M., Katagiri, T., Tsunoda, T., Nomura, A., et al., 2005. Genome-wide analysis of gene expression in human intrahepatic cholangiocarcinoma. Hepatology 41, 1339–1348.

O'Connor, R., 2009. A review of mechanisms of circumvention and modulation of chemotherapeutic drug resistance. Curr. Cancer Drug Targets 9, 273–280.

O'Connor, R., Clynes, M., Dowling, P., O'Donnovan, N., O'Driscoll, L., 2007. Drug resistance in cancer—searching for mechanisms, markers and therapeutic agents. Curr. Expert Opin. Drug Metab. Toxicol. 3, 805–817.

Oda, H., Takeichi, M., 2011. Structural and functional diversity of cadherin at the adherens junction. J. Cell Biol. 193, 1137–1146.

Odabaei, G., Chatterjee, G., Jazirehi, A.R., Goodglick, L., Yeung, K., Bonavida, B., 2004. Raf-1 kinase inhibitor protein: structure, function, regulation of cell signaling, and pivotal role in apoptosis. Adv. Cancer Res. 91, 169–200.

Ogrunc, M., d'Adda di Fagana, F., 2011. Never-aging cellular senescence. Eur. J. Cancer 47, 1616–1622.

Ohno, S., Inagawa, H., Soma, G., Nagasue, N., 2002. Role of tumor-associated macrophages in malignant tumors: should the location of the infiltrated macrophages be taken into account during evaluation? Anticancer Res. 22, 4269–4275.

Ohno, S., Suzuki, N., Ohno, Y., Inagawa, H., Inoue, M., 2003. Tumor-associated macrophages: foe or accomplice of tumors? Anticancer Res. 23, 4395–4409.

Ohta, T., Sato, K., Wu, W., 2011. The BRCA1 ubiquitin ligase and homologous recombination repair. FEBS Lett. 585, 2836–2844.

Ohtani, N., Mann, D.J., Hara, E., 2009. Cellular senescence: its role in tumor suppression and aging. Cancer Sci. 100, 792–797.

Ohtsuka, S., Dalton, S., 2008. Molecular and biological properties of pluripotent stem cells. Gene Ther. 15, 74–81.

Okabe, H., Satoh, S., Kato, T., Kitahara, O., Yanagawa, R., Yamaoka, Y., et al., 2001. Genome-wide analysis of gene expression in human hepatocellular carcinomas using cDNA microarray: identification of genes involved in viral carcinogenesis and tumor progression. Cancer Res. 61, 2129–2137.

Okada, K., Fujiwara, Y., Takahashi, T., Nakamura, Y., Takiguchi, S., Nakajima, K., et al., 2013. Overexpression of Forkhead box M1 transcription factor (FOXM1) is a potential prognostic marker and enhances chemoresistance for docetaxel in gastric cancer. Ann. Surg. Oncol. 20, 1035–1043.

Olivier, S., Robe, P., Bours, V., 2006. Can NF-κB be a target for novel and efficient anti-cancer agents? Biochem. Pharmacol. 72, 1054–1068.

Oliviero, A., Manzione, L., 2007. Dasatinib: a new step in molecular target therapy. Ann. Oncol. 18, vi42–vi46.

Oltean, S., Sorg, B.S., Albrecht, T., Bonano, V.I., Brazas, R.M., Dewhirst, M.W., et al., 2006. Alternative inclusion of fibroblast growth factor receptor 2 exon IIIc in Dunning prostate tumors reveals unexpected epithelial mesenchymal plasticity. Proc. Natl. Acad. Sci. U.S.A. 103, 14116–14121.

Ong, A., Maines-Bandiera, S.L., Roskelley, C.D., Auersperg, N., 2000. An ovarian adeno-carcinoma line derived from SV40/E-cadherin-transfected normal human ovarian sur-face epithelium. Int. J. Cancer 85, 430–437.

Onishi, H., Katano, M., 2011. Hedgehog signaling pathway as a therapeutic target in various types of cancer. Cancer Sci. 102, 1756–1760.

Orkin, S.H., Hochedlinger, K., 2011. Chromatin connections to pluripotency and cellular reprogramming. Cell 145, 835–849.

Orlowski, R.Z., Kuhn, D.J., 2008. Proteasome inhibitors in cancer therapy: lessons from the first decade. Clin. Cancer Res. 14, 1649–1657.

Osborn, S.L., Sohn, S.J., Winoto, A., 2007. Constitutive phosphorylation mutation in FADD results in early cell cycle defects. J. Biol. Chem. 282, 22786–22792.

O'Shea, C.C., 2005a. DNA tumor viruses—the spies who lyse us. Curr. Opin. Genet. Dev. 15, 18–26.

O'Shea, C.C., 2005b. Viruses—seeking and destroying the tumor program. Oncogene 24, 7640–7655.

O'Shea, C.C., Fried, M., 2005. Modulation of the ARF-p53 pathway by the small DNA tumor viruses. Cell Cycle 4, 449–452.

Oster, S.K., Ho, C.S.W., Soucle, E.L., Penn, L.Z., 2002. The myc oncogene: MarvelouslY Complex. Adv. Cancer Res. 84, 81–154.

Ouyang, G., Wang, Z., Fang, X., Liu, J., Yang, C.J., 2010. Molecular signaling of the epi-thelial to mesenchymal transition in generating and maintaining cancer stem cells. Cell. Mol. Life Sci. 67, 2605–2618.

Paggi, M.G., Felsani, A., Giordano, A., 2003. Growth control by the retinoblastoma gene family. Methods Mol. Biol. 222, 3–19.

Pan, D., 2010. The Hippo pathway in development and cancer. Dev. Cell 19, 491–505.

Pandit, B., Gartel, A.L., 2010. New potential anti-cancer agents synergize with bortezomib and ABT-737 against prostate cancer. Prostate 70, 825–833.

Pandit, B., Gartel, A.L., 2011a. FoxM1 knockdown sensitizes human cancer calls to proteasome inhibitor-induced apoptosis but not autophagy. Cell Cycle 10, 3269–3273.

Pandit, B., Gartel, A.L., 2011b. Proteasome inhibitors suppress expression of NPM and ARF proteins. Cell Cycle 10, 3827–3829.

Pandit, B., Gartel, A.L., 2011c. Thiazole antibiotic thiostrepton synergizes with bortezomib to induce apoptosis in cancer cells. PLoS One 6, e17110.

Pandit, B., Halasi, M., Gartel, A.L., 2009. p53 negatively regulates expression of FoxM1. Cell Cycle 8, 3425–3427.

Pandolfi, F., Cianci, R., Pagliari, D., Casciano, F., Bagala, C., Astone, A., et al., 2011. The immune response to tumors as a tool toward immunotherapy. Clin. Dev. Immunol. 2011, 894704.

Papa, S., Zazzeroni, F., Pham, C.G., Bubici, C., Franzoso, G., 2004. Linking JNK signaling to NF-κB: a key to survival. J. Cell Sci. 117, 5197–5208.

Pardoll, D.M., 2012. The blockade of immune checkpoints in cancer immunotherapy. Nat. Rev. Cancer 12, 252–264.

Paredes, J., Figueiredo, J., Albergaria, A., Oliveira, P., Carvalho, J., Ribeiro, A.S., et al., 2012. Epithelial E- and P-cadherins: role and clinical significance in cancer. Biochim. Biophys. Acta 1826, 297–311.

Park, N.H., Song, I.H., Chung, Y.H., 2007. Molecular pathogenesis of hepatitis-B-virus-associated hepatocellular carcinoma. Gut Liver 1, 101–117.

Park, H.J., Wang, Z., Costa, R.H., Tyner, A., Lau, L.F., Raychaudhuri, P., 2008a. An N-terminal inhibitory domain modulates activity of FoxM1 during cell cycle. Oncogene 27, 1696–1704.

Park, T.J., Kim, J.Y., Oh, P., Kang, S.Y., Kim, B.W., Wang, H.J., et al., 2008b. TIS21 negatively regulates hepatocarcinogenesis by disruption of cyclin B1-Forkhead Box M1 regulation loop. Hepatology 47, 1533–1543.

Park, H.J., Cosat, R.H., Lau, L.F., Tyner, A.L., Raychaudhuri, P., 2008c. APC/C-Cdh1 mediated proteolysis of the Forkhead Box M1 transcription factor is critical for regulated entry into S phase. Mol. Cell. Biol. 28, 5162–5171.

Park, H.J., Carr, J.R., Wang, Z., Nogueira, V., Hay, N., Tyner, A.L., et al., 2009. FoxM1, a critical regulator of oxidative stress during oncogenesis. EMBO J. 28, 2908–2918.

Park, H.J., Gusarova, G., Wang, Z., Carr, J.R., Li, J., Kim, K.H., et al., 2011. Deregulation of FoxM1b leads to tumour metastasis. EMBO Mol. Med. 2, 21–34.

Park, Y.Y., Jung, S.Y., Jennings, N.B., Rodriguez-Aguayo, C., Peng, G., Lee, S.R., et al., 2012. FOXM1 mediates Dox resistance in breast cancer by enhancing DNA repair. Carcinogenesis 33, 1843–1853.

Parkin, D.M., 2006. The global health burden of infection-associated cancers in the year 2002. Int. J. Cancer 118, 3030–3044.

Parveen, N., Hussain, M.U., Pandith, A.A., Mudassar, S., 2011. Diversity of axin in signaling pathways and its relation to colorectal cancer. Med. Oncol. 28, S259–S267.

Patel, S.D., Chen, C.P., Bahna, F., Honig, B., Shapiro, L., 2003. Cadherin-mediated cell-cell adhesion: sticking together as a family. Curr. Opin. Struct. Biol. 13, 690–698.

Patrawala, S., Puzanov, I., 2012. Vemurafenib (RG67204, PLX4032): a potent, selective BRAF kinase inhibitor. Future Oncol. 8, 509–523.

Pauklin, S., Pedersn, R.A., Vallier, L., 2011. Mouse pluripotent stem cells at a glance. J. Cell Sci. 124, 3727–3732.

Pazolli, E., Stewart, S.A., 2008. Senescence: the good the bad and the dysfunctional. Curr. Opin. Genet. Dev. 18, 42–47.

Peggs, K., 2004. Imatinib mesylate—gold standards and silver linings. Clin. Exp. Med. 4, 1–9.

Pei, D., 2009. Regulation of pluripotency and reprogramming by transcription factors. J. Biol. Chem. 284, 3365–3369.

Peinado, H., Portillo, F., Cano, A., 2004. Transcriptional regulation of cadherins during development and carcinogenesis. Int. J. Dev. Biol. 48, 365–375.

Peinado, H., Portillo, F., Cano, A., 2005. Switching on-off Snail. Cell Cycle 4, 1749–1752.

Peinado, H., Olmeda, D., Cano, A., 2007. Snail, Zeb and bHLH factors in tumor progression: an alliance against the epithelial phenotype? Nat. Rev. Cancer 7, 415–428.

Pelka, P., Ablack, J.N.G., Fonseca, G.J., Yousef, A.F., Mymryk, J.S., 2008. Intrinsic structural disorder in adenovirus E1A: a viral molecular hub linking multiple diverse processes. J. Virol. 82, 7252–7263.

Pellegrino, R., Calvisi, D.F., Ladu, S., Ehemann, V., Staniscia, T., Evert, M., et al., 2010. Oncogenic and tumor suppressive roles of Polo-like kinases in human hepatocellular carcinoma. Hepatology 51, 857–868.

Penzo, M., Massa, P.E., Olivotto, E., Bianchi, F., Borzi, R.M., Hanidu, A., et al., 2009. Sustained NF-κB activation produces a short-term cell proliferation block in conjunction with repressing effectors of cell cycle progression controlled by E2F and FoxM1. J. Cell. Physiol. 218, 215–227.

Perego, P., Corna, E., De Cesare, M., Gatti, L., Palizzi, D., Pratesi, G., et al., 2001. Role of apoptosis and apoptosis-related genes in cellular response and antitumor efficacy of anthracyclines. Curr. Med. Chem. 8, 31–37.

Perez, E.A., Palmieri, F.M., Brock, S.M., 2009. Trastuzumab. Cancer Treat. Res. 151, 181–196.

Perez-Moreno, M., Jamora, C., Fuchs, E., 2003. Sticky business: orchestrating cellular signals at adherens junctions. Cell 112, 535–548.

Perissi, V., Rosenfeld, M.G., 2005. Controlling nuclear receptors: the circular logic of cofactor cycles. Nat. Rev. Mol. Cell Biol. 6, 542–554.

Perkins, N.D., 2012. The diverse and complex roles of NF-κB subunits in cancer. Nat. Rev. Cancer 12, 121–132.

Perona, R., Moncho-Amor, V., Machado-Pinilla, R., Belad-Iniesta, C., Sanchez Perez, I., 2008. Role of CHK2 in cancer development. Clin. Transl. Oncol. 10, 538–542.

Perou, C.M., Sorlie, T., Eisen, M.B., van de Rijn, M., Jeffrey, S.S., Rees, C.A., et al., 2000. Molecular portraits of human breast tumors. Nature 406, 747–752.

Peter, M.E., 2009. Let-7 and miR-200 microRNAs. Guardians against pluripotency and cancer progression. Cell Cycle 8, 83–852.

Peterson, C.L., Cote, J., 2004. Cellular machineries for chromosomal DNA repair. Genes Dev. 18, 602–616.

Petrocca, F., Lieberman, J., 2009. Micromanipulating cancer: microRNA-based therapeutics? RNA Biol. 6, 335–340.

Petrovic, V., Costa, R.H., Lau, L.F., Raychaudhuri, P., Tyner, A.L., 2008. FOXM1 regulates growth factor induced expression of the KIS kinase to promote cell cycle progression. J. Biol. Chem. 283, 453–460.

Petrovich, V., Costa, R.H., Lau, L.H., Raychaudhuri, P., Tyner, A.L., 2010. Negative regulation of the oncogenic transcription factor FoxM1 by thiazolidinediones and mithramycin. Cancer Biol. Ther. 9, 1008–1016.

Phan, S.H., 2002. The myofibroblast in pulmonary fibrosis. Chest 122, 286S–289S.

Phan, S.H., 2012. Genesis of the myofibroblast in lung injury and fibrosis. Proc. Am. Thorac. Soc. 9, 148–152.

Philip, M., Rowley, D.A., Schreiber, H., 2004. Inflammation as a tumor promoter in cancer induction. Semin. Cancer Biol. 14, 433–439.

Piccaluga, P.P., Rondoni, M., Paolini, S., Rosti, G., Martinelli, G., Baccarani, M., 2007. Imatinib mesylate in the treatment of hematologic malignancies. Expert Opin. Biol. Ther. 7, 1597–1611.

Piera-Velazquez, S., Li, Z., Jimenez, S.A., 2011. Role of endothelial-mesenchymal transition (EndoMT) in the pathogenesis of fibrotic disorders. Am. J. Pathol. 179, 1074–1080.

Pietersz, G.A., Pouniotis, D.S., Apostolopoulos, V., 2006. Design of peptide-based vaccines for cancer. Curr. Med. Chem. 13, 1591–1607.

Pignot, G., Vieillefond, A., Vacher, S., Zerbib, M., Debre, B., Lidereau, R., et al., 2012. Hedgehog pathway activation in human translational cell carcinoma of the bladder. Br. J. Cancer 106, 1177–1186.

Pilarsky, C., Wenzig, M., Specht, T., Saeger, H.D., Grutzmann, R., 2004. Identification and validation of commonly overexpressed genes in solid tumors by comparison of microarray data. Neoplasia 6, 744–750.

Pilch, D.R., Sedelnikova, O.A., Redon, C., Celeste, A., Nussenzweig, A., Bonner, W.M., 2003. Characteristics of γ-H2Ax foci ast DNA double-strand break sites. Biochem. Cell Biol. 81, 123–129.

Pim, D., Banks, L., 2010. Interaction of viral oncoproteins with cellular target molecules: infection with high-risk vs low-risk human papillomaviruses. APMIS 118, 471–493.

Pipas, J.M., 2009. SV40: cell transformation and tumorigenesis. Virology 384, 294–303.

Pipas, J.M., Levine, A.J., 2001. Role of T antigen interactions with p53 in tumorigenesis. Semin. Cancer Biol. 11, 23–30.

Pitot, H.C., Dragan, Y.P., Teeguarden, J., Hsia, S., Campbell, H., 1996. Quantitation of multistage carcinogenesis in rat liver. Toxicol. Pathol. 24, 119–128.

Piva, R., Belardo, G., Santoro, M.G., 2006. NF-κB: a stress-regulated switch for cell survival. Antioxid. Redox Signal. 8, 478–486.

Plank, J.L., Frist, A.Y., LeGrone, A.W., Magnuson, M.A., Labosky, P.A., 2011. Loss of Foxd3 results in decreased β-cell proliferation and glucose intolerance during pregnancy. Endocrinology 152, 4589–4600.

Platanias, L.C., 2003. Map kinase signaling pathways and hematologic malignancies. Blood 101, 4667–4679.

Plotnikov, A., Zehoraj, E., Procaccia, S., Seger, R., 2011. The MAPK cascades: signaling components, nuclear roles and mechanisms of nuclear translocation. Biochim. Biophys. Acta 1813, 1619–1633.

Pokutta, S., Weis, W.I., 2007. Structure and mechanism of cadherins and catenins in cell-cell contacts. Annu. Rev. Cell Dev. Biol. 23, 237–262.

Polakis, P., 2007. The many ways of Wnt in cancer. Curr. Opin. Genet. Dev. 17, 45–51.

Polakis, P., 2012. Drugging Wnt signalling in cancer. EMBO J. 31, 2737–2746.

Pollice, A., Vivo, M., La Mantia, G., 2008. The promiscuity of ARF interactions with the proteasome. FEBS Lett. 582, 3257–3262.

Polo, S.E., Jackson, S.O.P., 2011. Dynamics of DNA damage response proteins at DNA breaks: a focus on protein modifications. Genes Dev. 25, 409–433.

Polsky, D., Cordon-Cardo, C., 2003. Oncogenes in melanoma. Oncogene 22, 3087–3091.

Polyak, K., Weinberg, R.A., 2009. Transitions between epithelial and mesenchymal states: acquisition of malignant and stem cell traits. Nat. Rev. Cancer 9, 265–273.

Ponder, B.A., 2001. Cancer genetics. Nature 411, 336–341.

Ponzielli, R., Katz, S., Barsyte-Lovejoy, D., Penn, L.Z., 2005. Cancer therapeutics: targeting the dark side of Myc. Eur. J. Cancer 41, 2485–2501.

Popescu, N.C., Zimonjic, D.B., 2002. Chromosome-mediated alterations of the MYC gene in human cancer. J. Cell. Mol. Med. 6, 151–159.

Porta, C., Subhra Kumar, B., Larghi, P., Rubino, L., Mancino, A., Sica, A., 2007. Tumor promotion by tumor-associated macrophages. Adv. Exp. Med. Biol. 604, 67–86.

Porta, C., Larghi, P., Rimoldi, M., Totaro, M.G., Allavena, P., Mantovani, A., et al., 2009. Cellular and molecular pathways linking inflammation and cancer. Immunobiology 214, 761–777.

Postic, C., Magnuson, M.A., 2000. DNA excision in liver by an albumin-Cre transgene occurs progressively with age. Genesis 26, 149–150.

Poznic, M., 2009. Retinoblastoma protein: a central processing unit. J. Biosci. 34, 305–312.

Pratilas, C.A., Solit, D.B., 2010. Targeting the mitogen-activated protein kinase pathway: physiological feedback and drug response. Clin. Cancer Res. 16, 3329–3334.

Price, M.A., 2006. CKI, there's more than one: casein kinase I family members in Wnt and Hedgehog signalling. Genes Dev. 20, 399–410.

Prieur, A., Peeper, D.S., 2008. Cellular senescence in vivo: a barrier to tumorigenesis. Curr. Opin. Cell Biol. 20, 150–155.

Priller, M., Pöschel, J., Abrao, L., von Bueren, A.O., Cho, Y.J., Rutkowski, S., et al., 2011. Expression of FoxM1 is required for the proliferation of medulloblastoma cells and indicates worse survival of patients. Clin. Cancer Res. 17, 6791–6801.

Prots, I., Skapenko, A., Lipsky, P.E., Schulze-Koops, H., 2011. Analysis of the transcriptional program of developing induced regulatory T cells. PLoS One 6, e16913.

Pylayeva-Gupta, Y., Grabocka, E., Bar-Sagi, D., 2011. RAS oncogenes: weaving a tumorigenic web. Nat. Rev. Cancer 11, 761–774.

Pytel, D., Sliwinski, T., Poplawski, T., Ferriola, D., Maisterek, I., 2009. Tyrosine kinase blockers: new hope for successful cancer therapy. Anticancer Agents Med Chem. 9, 66–76.

Qiao, M., Shi, Q., Pardee, A.B., 2010. The pursuit of oncotargets through understanding defective cell regulation. Oncotarget 1, 544–551.

Qu, K., Xu, X., Li, C., Wu, Q., Wei, J., Meng, F., et al., 2013. Negative regulation of transcription factor FoxM1 by p53 enhances oxaliplatin-induced senescence in hepatocellular carcinoma. Cancer Lett. 331, 105–114.

Quan, T.E., Cowper, S.E., Bucala, R., 2006. The role of circulating fibrocytes in fibrosis. Curr. Rheumatol. Rep. 8, 145–150.

Quintas-Cardama, A., Cortes, J., 2007. Nilotinib therapy in chronic myeloid leukemia. Drugs Today 43, 691–702.

Quintas-Cardama, A., Cortes, J., 2008. Nilotinib: a phenylamino-pyrimidine derivative with activity against BCR-ABL, KIT and PDGFR kinases. Future Oncol. 4, 611–621.

Quintas-Cardama, A., Kantarijan, H., Cortes, J., 2006. Targeting ABL and SRC kinases in chronic myeloid leukemia: experience with dasatinib. Future Oncol. 2, 655–665.

Quintas-Cardama, A., Kantarijan, H., Cortes, J., 2007. Flying under the radar: the new wave of BCR-ABL inhibitors. Nat. Rev. Drug Discov. 6, 834–848.

Quintas-Cardama, A., Kantarijan, H., Cortes, J., 2009. Imatinib and beyond—exploring the full potential of targeted therapy for CML. Nat. Rev. Clin. Oncol. 6, 535–543.

Quintas-Cardama, A., Kim, T.D., Cataldo, V., Cortes, J., 2010. Nilotinib. Recent Results Cancer Res. 184, 103–117.

Raassool, F.V., Tomkinson, A.E., 2010. Targeting abnormal DNA double strand break repair in cancer. Cell. Mol. Life Sci. 67, 3699–3710.

Rabbani, A., Finn, R.M., Ausio, J., 2005. The anthracycline antibiotics: antitumor drugs that alter chromatin structure. Bioessays 27, 50–56.

Radford, I.R., 1999. Initiation of ionizing radiation-induced apoptosis: DNA damage-mediated or does ceramide have a role? Int. J. Radiat. Biol. 75, 521–528.

Radford, I.R., 2002. Imatinib. Novartis. Curr. Opin. Investig. Drugs 3, 492–499.

Radhakrishnan, S.K., Kamalakaran, S., 2006. Pro-apoptotic role of NF-κB: implications for cancer therapy. Biochim. Biophys. Acta 1766, 53–62.

Radhakrishnan, S.K., Bhat, U.G., Hughes, D.E., Wang, I.C., Costa, R.H., Gartel, A.L., 2006. Identification of a chemical inhibitor of the oncogenic transcription factor Forkhead box M1. Cancer Res. 66, 9731–9735.

Raghavan, A., Zhou, G., Zhou, Q., Ibe, J.C.F., Ramchandran, R., Yang, Q., et al., 2012. Hypoxia induced pulmonary arterial smooth muscle cell proliferation is controlled by FOXM1. Am. J. Respir. Cell Mol. Biol. 46, 431–436.

Rahman, K.M.W., Li, Y., Wang, Z., Sarkar, S.H., Sarkar, F.H., 2006. Gene expression profiling revealed survivin as a target of 3,3′-diindolyl-methane-induced cell growth inhibition and apoptosis in breast cancer cells. Cancer Res. 66, 4952–4960.

Ralston, A., Rossant, J., 2010. The genetics of induced pluripotency. Reproduction 139, 35–44.

Ramakrishna, S., Kim, I.M., Petrovic, V., Malin, D., Wang, I.C., Kalin, T.V., et al., 2007. Myocardium defects and ventricular hypoplasia in mice homozygous null for the Forkhead Box M1 transcription factor. Dev. Dyn. 236, 1000–1013.

Ramalho-Santos, M., 2009. iPS cells: insights into basic biology. Cell 138, 616–618.

Randow, F., Lehner, P.J., 2009. Viral avoidance and exploitation of the ubiquitin system. Nat. Cell Biol. 11, 527–534.

Ranganathan, P., Weaver, K.L., Capobianco, A.J., 2011. Notch signalling in solid tumors: a little bit of everything but not all the time. Nat. Rev. Cancer 11, 338–351.

Rangarajan, A., Weinberg, R.A., 2003. Comparative biology of mouse versus human cells: modelling human cancer in mice. Nat. Rev. Cancer 3, 952–959.

Ravnan, M.C., Matalka, M.S., 2012. Vemurafenib in patients with BRAF V600E mutation-positive advanced melanoma. Clin. Ther. 34, 1474–1486.

Raychaudhuri, P., Park, H.J., 2011. FoxM1 a master regulator of tumor metastasis. Cancer Res. 71, 4329–4333.

Reddy, P., Liu, L., Ren, C., Lindgren, P., Boman, K., Shen, Y., et al., 2005. Formation of E-Cadherin-mediated cell-cell adhesion activates Akt and mitogen activated protein kinase via phosphatidylinositol 3 kinase and ligand-independent activation of epidermal growth factor receptor in ovarian cancer cells. Mol. Endocrinol. 19, 2564–2578.

Redmond, K.M., Wilson, T.R., Johnston, P.G., Longley, D.B., 2008. Resistance mechanisms to cancer chemotherapy. Front. Biosci. 13, 5138–5154.

Redner, R.L., 2010. Why doesn't imatinib cure chronic myeloid leukemia? Oncologist 15, 182–186.

Reed, J.C., 1999. Mechanisms of apoptosis in avoidance of cancer. Curr. Opin. Oncol. 11, 68–75.

Reinhardt, H.C., Yaffe, M.B., 2009. Kinases that control the cell cycle in response to DNA damage: Chk1, Chk2, and MK2. Curr. Opin. Cell Biol. 21, 245–255.

Ren, R., 2005. Mechanisms of BCR-ABL in the pathogenesis of chronic myelogenous leukemia. Nat. Rev. Cancer 5, 172–183.

Ren, X., Zhang, Y., Snyder, J., Cross, E.R., Shah, T.A., Kalin, T.V., et al., 2010. Forkhead Box M1 transcription factor is required for macrophage recruitment during liver repair. Mol. Cell. Biol. 30, 5381–5393.

Ren, X., Shah, T.A., Ustiyan, V., Zhang, Y., Shinn, J., Chen, G., et al., 2013. FOXM1 promotes allergen-induced goblet cell metaplasia and pulmonary inflammation. Mol. Cell. Biol. 33, 371–386.

Reshmi, G., Sona, C., Pillai, M.R., 2011. Comprehensive patterns in microRNA regulation of transcription factors during tumor metastasis. J. Cell. Biochem. 112, 2210–2217.

Reya, T., Clevers, H., 2005. Wnt signalling in stem cells and cancer. Nature 434, 843–850.

Rhodes, D.R., Yu, J., Shanker, K., Deshpande, N., Varambally, R., Ghosh, D., et al., 2004. Large-scale meta-analysis of cancer microarray data identifies common transcriptional profiles of neoplastic transformation and progression. Proc. Natl. Acad. Sci. U.S.A. 101, 9309–9314.

Richardson, P., 2003. Clinical update: proteasome inhibitors in hematologic malignancies. Cancer Treat. Rev. 29 (Suppl. 1), 33–39.

Richardson, C., 2005. RAD51, genomic stability, and tumorigenesis. Cancer Lett. 218, 127–139.

Richardson, P.G., Anderson, K.C., 2003. Bortezomib: a novel therapy approved for multiple myeloma. Clin. Adv. Hematol. Oncol. 1, 596–600.

Richardson, P.G., Mitsiades, C., 2005. Bortezomib: proteasome inhibition as an effective anticancer therapy. Future Oncol. 1, 161–171.

Richardson, P.G., Mitsiades, C., Hideshima, T., Anderson, K.C., 2005. Proteasome inhibition in the treatment of cancer. Cell Cycle 4, 290–296.

Richardson, P.G., Mitsiades, C., Hideshima, T., Anderson, K.C., 2006. Bortezomib: proteasome inhibition as an effective anticancer therapy. Annu. Rev. Med. 57, 33–47.

Rickman, D.S., Bobek, M.P., Misek, D.E., Kuick, R., Blaivas, M., Kurnit, D.M., et al., 2001. Distinctive molecular profiles of high-grade and low-grade gliomas based on oligonucleotide microarray analysis. Cancer Res. 61, 6885–6891.

Roberts, P.J., Der, C.J., 2007. Targeting the Raf-MEK-ERK mitogen-activated protein kinase cascade for the treatment of cancer. Oncogene 26, 3291–3310.

Robinson, W.S., 1994. Molecular events in the pathogenesis of hepadnavirus-associated hepatocellular carcinoma. Annu. Rev. Med. 45, 297–323.

Roccaro, A.M., Hideshima, T., Richardson, P.G., Russo, D., Ribatti, D., Vacca, A., et al., 2006a. Bortezomib as an antitumor agent. Curr. Pharm. Biotechnol. 7, 441–448.

Roccaro, A.M., Vacca, A., Ribatti, D., 2006b. Bortezomib in the treatment of cancer. Recent Pat. Anticancer Drug Discov. 1, 397–403.

Rocha, S., Perkins, N.D., 2005. ARF the integrator. Linking NF-κB, p53 and checkpoint kinases. Cell Cycle 4, 756–759.

Rochette-Egly, C., 2005. Dynamic combinatorial networks in nuclear receptor-mediated transcription. J. Biol. Chem. 280, 32565–32568.

Roden, R., Wu, T.C., 2006. How will HPV vaccines affect cervical cancer? Nat. Rev. Cancer 6, 753–763.

Rodriguez, J.A., Li, M., Yao, Q., Chen, C., Fisher, W.E., 2005. Gene overexpression in pancreatic adenocarcinoma: diagnostic and therapeutic implications. World J. Surg. 29, 297–305.

Rodriguez, F.J., Lewis-Tuffin, L.J., Anastasiadis, P.Z., 2012. E-cadherin's dark side: possible role in tumor progression. Biochim. Biophys. Acta 1826, 23–31.

Romagnoli, S., Fasoli, E., Vaira, V., Falleni, M., Pellegrini, C., Catania, A., et al., 2009. Identification of potential therapeutic targets in malignant mesothelioma using cell-cycle gene expression profiling. Am. J. Pathol. 174, 762–770.

Roninson, I.B., 2003. Tumor cell senescence in cancer treatment. Cancer Res. 63, 2705–2715.

Rosenberg, S.A., Yang, J.C., Restifo, N.P., 2004. Cancer immunotherapy: moving beyond current vaccines. Nat. Med. 10, 909–915.

Rosenberg, S.A., Restifo, N.P., Yang, J.C., Morgan, R.A., Dudley, M.E., 2008. Adoptive cell transfer: a clinical path to effective cancer immunotherapy. Nat. Rev. Cancer 8, 299–308.

Rosenfeld, M.G., Glass, C.K., 2001. Coregulator codes of transcriptional regulation by nuclear receptors. J. Biol. Chem. 276, 36865–36868.

Roskoski, R., 2003. STI-571: an anticancer protein-tyrosine kinase inhibitor. Biochem. Biophys. Res. Commun. 309, 709–717.

Rossant, J., 2008. Stem cells and early lineage development. Cell 132, 527–531.

Rosty, C., Sheffer, M., Tsfrir, D., Stransky, N., Tsafrir, I., Peter, M., et al., 2005. Identification of a proliferation gene cluster associated with HPV E6/E7 expression level and viral DNA load in invasive cervical carcinoma. Oncogene 24, 7094–7104.

Rothenberg, M.E., Clarke, M.F., Diehn, M., 2010. The Myc connection: ES cells and cancer. Cell 143, 184–186.

Roulston, A., Marcellus, R.C., Branton, P.E., 1999. Viruses and apoptosis. Annu. Rev. Microbiol. 53, 577–628.

Rouse, J., Jackson, S.P., 2002. Interfaces between the detection, signalling, and repair of DNA damage. Science 297, 547–551.

Roussos, E.T., Keckesova, Z., Haley, J.D., Epstein, D.M., Weinberg, R.A., Condeelis, J.S., 2010. AACR special conference on epithelial-mesenchymal transition and cancer progression and treatment. Cancer Res. 70, 7360–7364.

Roux, P.P., Blenis, J., 2004. ERK and p38 MAPK-activated protein kinases: a family of protein kinases with diverse biological functions. Microbiol. Mol. Biol. Rev. 68, 320–344.

Rovillain, E., Mansfeild, L., Caetano, C., Alvarez-Fernandez, M., Caballero, O.L., Medema, R.H., et al., 2011. Activation of nuclear factor-kappa B signalling promotes cellular senescence. Oncogene 30, 2356–2366.

Rowinsky, E.K., 2004. The ERBB family: targets for therapeutic development against cancer and therapeutic strategies using monoclonal antibodies and tyrosine kinase inhibitors. Annu. Rev. Med. 55, 433–457.

Roy, S., Tenniswood, M., 2007. Site-specific acetylation of p53 directs selective transcription complex assembly. J. Biol. Chem. 282, 4765–4771.

Roy, H., Bhardwaj, S., Ylä-Herttuala, S., 2006. Biology of vascular endothelial growth factors. FEBS Lett. 580, 2879–2887.

Roy, R., Chun, J., Powell, S.N., 2011. BRCA1 and BRCA2: different roles in a common pathway of genome protection. Nat. Rev. Cancer 12, 68–78.

Rubin, M.A., Mucci, N.R., Figurski, J., Fecko, A., Pienta, K.J., Day, M.L., 2001. E-cadherin expression in prostate cancer: a broad survey using high-density tissue microarray technology. Hum. Pathol. 32, 690–697.

Rubin, J.S., Barshishat-Kupper, M., Feroze-Merzoug, F., Xi, Z.F., 2006. Secreted WNT antagonists as tumor suppressors: pro and con. Front. Biosci. 11, 2093–2105.

Rucci, N., Sanita, P., Angelucci, A., 2011. Role of metalloproteases in the metastatic niche. Curr. Mol. Med. 11, 609–622.

Rudini, N., Dejana, E., 2008. Adherens junctions. Curr. Biol. 18, R1080–R1082.

Rumpold, H., Webersinke, G., 2011. Molecular pathogenesis of Philadelphia-positive chronic myeloid leukemia—is it all BCR-ABL? Curr. Cancer Drug Targets 11, 3–19.

Russell, W.C., 2009. Adenoviruses: update on structure and function. J. Gen. Virol. 90, 1–20.

Sadavisam, S., Duan, S., DeCarpio, J.A., 2012. The MuvB complex sequentially recruits B-Myb and FoxM1 to promote mitotic gene expression. Genes Dev. 26, 474–489.

Saenz-Robles, M., Pipas, J.M., 2009. T antigen transgenic mouse models. Semin. Cancer Biol. 19, 229–235.

Saenz-Robles, M., Sullivan, C.S., Pipas, J.M., 2001. Transforming functions of simian virus 40. Oncogene 20, 7899–7907.

Sage, J., 2005. Making young tumors old: a new weapon against cancer? Sci. Aging Knowledge Environ. 2005, pe25.

Saharinen, P., Eklund, L., Pulkki, K., Bono, P., Alitalo, K., 2011. VEGF and angiopoietin signaling in tumor angiogenesis and metastasis. Trends Mol. Med. 17, 347–362.

Said, N.A.B.M., Williams, E.D., 2011. Growth factors in induction of epithelial-mesenchymal transition and metastasis. Cells Tissues Organs 193, 85–97.

Saito, T., Oda, Y., Sakamoto, A., Tamiya, S., Kinukawa, N., Hayashi, K., et al., 2000. Prognostic value of the preserved expression of the E-cadherin and catenin families of adhesion molecules and of β-catenin mutations in synovial sarcoma. J. Pathol. 192, 342–350.

Saito, M., Tucker, D.K., Kohlhorst, D., Niessen, C.M., Kowalczyk, A.P., 2012. Classical and desmosomal cadherins at a glance. J. Cell Sci. 125, 2547–2552.

Salahshor, S., Woodgett, J.R., 2005. The links between axin and carcinogenesis. J. Clin. Pathol. 58, 225–236.

Salesse, S., Verfaillie, C.M., 2002. BCR/ABL: from molecular mechanisms of leukemia induction to treatment of chronic myelogenous leukemia. Oncogene 21, 8547–8559.

Salon, C., Lantuejoul, S., Eymin, B., Gazzeri, S., Brambilla, C., Brambilla, E., 2005. The E-cadherin-β-catenin complex and its implication in cancer progression and prognosis. Future Oncol. 1, 649–660.

Salvatore, G., Nappi, T.C., Salerno, P., Jiang, Y., Garbi, C., Ugolini, C., et al., 2007. A cell proliferation and chromosomal instability signature in anaplastic thyroid carcinoma. Cancer Res. 67, 10148–10158.

Samuel, T., Weber, H.O., Funk, J.O., 2002. Linking DNA damage to cell cycle checkpoints. Cell Cycle 1, 162–168.

Sancar, A., Lindsey-Boltz, L.A., Ünsal-Kacmaz, K., Linn, S., 2004. Molecular mechanisms of mammalian DNA repair and the DNA damage checkpoints. Annu. Rev. Biochem. 73, 39–85.

Sanchez-Calderon, H., Rodriguez-de la Rosa, L., Milo, M., Pichel, J.G., Holley, M., Varela-Nieto, I., 2010. RNA microarray analysis in prenatal mouse cochlea reveals novel IGF-I target genes: implication of MEF2 and FOXM1 transcription factors. PLoS One 5, e8699.

San Filippo, J., Sung, P., Klein, H., 2008. Mechanisms of eukaryotic homologous recombination. Annu. Rev. Biochem. 77, 229–257.

Sang, N., Caro, J., Giordano, A., 2002. Adenovirus E1A: everlasting tool, versatile applications, continuous contributions and new hypotheses. Front. Biosci. 7, d407–d413.

Sansal, I., Sellers, W.R., 2004. The biology and clinical relevance of the PTEN tumor suppressor pathway. J. Clin. Oncol. 22, 2954–2963.

Santin, A.D., Zhan, F., Bignotti, E., Siegel, E.R., Cane, S., Bellone, S., et al., 2005. Gene expression profiles of primary HPV-16- and HPV18-infected early stage cervical cancers and normal cervical epithelium: identification of novel candidate molecular markers for cervical cancer diagnosis and therapy. Virology 331, 269–291.

Sapotita, A.J., Maggi, L.B., Apicelli, A.J., Weber, J.D., 2007. Therapeutic targets in the ARF tumor suppressor pathway. Curr. Med. Chem. 14, 1815–1827.

Sargent, L., Dragan, Y., Xu, Y.H., Sattler, G., Wiley, J., Pitot, H.C., 1996. Karyotypic changes in a multistage model of chemical hepatocarcinogenesis in the rat. Cancer Res. 56, 2985–2991.

Sato, H., Hasegawam, T., Abe, Y., Sakai, H., Hirohashi, S., 1999. Expression of E-cadherin in bone and soft tissue sarcomas: a possible role in epithelial differentiation. Hum. Pathol. 30, 1344–1349.

Sato, N., Hirohashi, Y., Tsukahara, T., Kikuchi, T., Sahara, H., Kamiguchi, K., et al., 2009. Molecular pathological approaches to human tumor immunology. Pathol. Int. 59, 205–217.

Sato, M., Shames, D.S., Hasegawa, Y., 2012. Emerging evidence of epithelial-to-mesenchymal transition in lung carcinogenesis. Respirology 17, 1045–1059.

Sattler, M., Griffin, J.D., 2001. Mechanisms of transformation by the BCR/ABL oncogene. Int. J. Hematol. 73, 278–291.

Sattler, M., Griffin, J.D., 2003. Molecular mechanisms of transformation by the BCR-ABL oncogene. Semin. Hematol. 40, 4–10.

Savagner, P., 2001. Leaving the neighbourhood: molecular mechanisms involved during epithelial-mesenchymal transition. Bioessays 23, 912–923.

Savagner, P., 2010. The epithelial-mesenchymal transition (EMT) phenomenon. Ann. Oncol. 21 (Suppl. 7), vii89–vii92.

Sawyers, C.L., 2003. Opportunities and challenges in the development of kinase inhibitor therapy for cancer. Genes Dev. 17, 2998–3010.

Sawyers, C., 2004. Targeted cancer therapy. Nature 432, 294–297.

Sawyers, C.L., 2005. Making progress through molecular attacks on cancer. Cold Spring Harb. Symp. Quant. Biol. 70, 479–482.

Sawyers, C., 2007. Mixing cocktails. Nature 449, 993–996.

Sawyers, C., 2009. Shifting paradigms: the seeds of oncogene addiction. Nat. Med. 15, 1158–1161.

Scales, S.J., de Sauvage, F.J., 2009. Mechanisms of Hedgehog pathway activation in cancer and implications for therapy. Trends Pharmacol. Sci. 30, 303–312.

Scheffner, M., Whitaker, N.J., 2003. Human papillomavirus-induced carcinogenesis and the ubiquitin-proteasome system. Semin. Cancer Biol. 13, 59–67.

Scheijen, B., Griffin, J.D., 2002. Tyrosine kinase oncogenes in normal hematopoiesis and hematological disease. Oncogene 21, 3314–3333.

Scheper, W., Copray, S., 2009. The molecular mechanism of induced pluripotency: a two-stage switch. Stem Cell Rev. Rep. 5, 204–223.

Schietinger, A., Philip, M., Schreiber, H., 2008. Specificity in cancer immunotherapy. Semin. Immunol. 20, 276–285.

Schiffman, M., Castle, P.E., Jeronimo, J., Rodriguez, A.C., Wacholder, S., 2007. Human papillomavirus and cervical cancer. Lancet 370, 890–907.

Schmalhofer, O., Brabletz, S., Brabletz, T., 2009. E-cadherin, β-catenin, and ZEB1 in malignant progression of cancer. Cancer Metastasis Rev. 28, 151–166.

Schmitt, C.A., 2003. Senescence, apoptosis and therapy—cutting the lifelines of cancer. Nat. Rev. Genet. 3, 286–295.

Schmitt, C.A., 2007. Cellular senescence and cancer treatment. Biochim. Biophys. Acta 1775, 5–20.

Schneider-Merck, T., Trepel, M., 2010. Lapatinib. Recent Results Cancer Res. 184, 45–59.

Schubbert, S., Shannon, K., Bollag, G., 2007. Hyperactive Ras in developmental disorders and cancer. Nat. Rev. Cancer 7, 295–308.

Schüller, U., Kho, A.T., Zhao, Q., Ma, Q., Rowitch, D.H., 2006. Cerebellar "transcriptome" reveals cell-type and stage-specific expression during postnatal development and tumorigenesis. Mol. Cell. Neurosci. 33, 247–259.

Schüller, U., Zhao, Q., Godinho, S.A., Heine, V.M., Medema, R.H., Pellman, D., et al., 2007. Forkhead transcription factor FoxM1 regulates mitotic entry and prevents spindle defects in cerebellar granule neuron precursors. Mol. Cell. Biol. 27, 8259–8270.

Scott, S.P., Pandita, T.K., 2006. The cellular control of DNA double-strand breaks. J. Cell. Biochem. 99, 1463–1475.

Scotton, C.J., Chambers, R.C., 2007. Molecular targets in pulmonary fibrosis: the myofibroblast in focus. Chest 132, 1311–1321.

Sebastian, S., Settlemen, J., Reshkin, S.J., Azzariti, A., Bellizzi, A., Paradisio, A., 2006. The complexity of targeting EGFR signalling in cancer: from expression to turnover. Biochim. Biophys. Acta 1766, 120–139.

Sebolt-Leupold, J.S., English, J.M., 2006. Mechanisms of drug inhibition of signalling molecules. Nature 441, 457–462.

Sebolt-Leupold, J.S., Herrera, R., 2004. Targeting the mitogen-activated protein kinase cascade to treat cancer. Nat. Rev. Cancer 4, 937–947.

Secchiero, P., Bosco, R., Celeghini, C., Zauli, G., 2008. The MDM2 inhibitor Nutlins as an innovative therapeutic tool for the treatment of haematological malignancies. Curr. Pharm. Des. 14, 2100–2110.

Secchiero, P., Bosco, R., Celeghini, C., Zauli, G., 2011. Recent advances in the therapeutic perspectives of Nutlin-3. Curr. Pharm. Des. 17, 569–577.

Seidensticker, M.J., Behrens, J., 2000. Biochemical interactions in the wnt pathway. Biochim. Biophys. Acta 1495, 168–182.

Seliger, B., Massa, C., Rini, B., Ko, J., Finke, J., 2010. Antitumor and immune-adjuvant activities of protein-kinase inhibitors. Trends Mol. Med. 16, 184–192.

Sellers, W.R., 2011. A blueprint for advancing genetics-based cancer therapy. Cell 147, 26–31.

Sellers, W.R., Kaelin, W.G., 1996. RB as a modulator of transcription. Biochim. Biophys. Acta 1288, M1–M5.

Semenza, G.L., 2010. Defining the role of hypoxia-inducible factor 1 in cancer biology and therapeutics. Oncogene 29, 625–634.

Semenza, G.L., 2011. Hypoxia-inducible factor 1: regulator of mitochondrial metabolism and mediator of ischemic preconditioning. Biochim. Biophys. Acta 1813, 1263–1268.

Semenza, G.L., 2012. Hypoxia-inducible factors in physiology and medicine. Cell 148, 399–407.

Sengupta, A., Kalinichenko, V.V., Yutzey, K.E., 2013. FoxO and FoxM1 transcription factors have antagonistic functions in neonatal cardiomyocyte cell cycle withdrawal and IGF1 gene regulation. Circ. Res. 112, 267–277.

Sethi, G., Shanmugam, M.K., Ramachandran, L., Kumar, A.P., Tergaonkar, V., 2012. Multifaceted link between cancer and inflammation. Biosci. Rep. 32, 1–15.

Seville, L.L., Shah, N., Westwell, A.D., Chan, W.C., 2005. Modulation of pRB/E2F functions in the regulation of cell cycle and in cancer. Curr. Cancer Drug Targets 5, 159–170.

Shangary, A., Wang, S., 2008. Targeting the MDM2-p53 interaction for cancer therapy. Clin. Cancer Res. 14, 5318–5324.

Shangary, A., Wang, S., 2009. Small-molecule inhibitors of the MDM2-p53 protein-protein interaction to reactivate p53 function: a novel approach for cancer therapy. Annu. Rev. Pharmacol. Toxicol. 49, 223–241.

Shapiro, L., Weis, W.I., 2009. Structure and biochemistry of cadherins and catenins. Cold Spring Harb. Perspect. Biol. 1, a003053.

Sharma, S.V., Haber, D.A., Settleman, J., 2010. Cell line-based platforms to evaluate the therapeutic efficacy of candidate anticancer agents. Nat. Rev. Cancer 10, 241–253.

Sharma, P., Wagner, K., Wolchok, J.D., Allison, J.P., 2011. Novel cancer immunotherapy agents with survival benefit: recent successes and next steps. Nat. Rev. Cancer 11, 805–812.

Sharpless, N.E., DePinho, R.A., 2004. Telomeres, stem cells, senescence and cancer. J. Clin. Invest. 113, 160–168.

Shawver, L.K., Slamon, D., Ullrich, A., 2002. Smart drugs: tyrosine kinase inhibitors in cancer therapy. Cancer Cell 1, 117–123.

Shay, J.W., Roninson, I.B., 2004. Hallmarks of senescence in carcinogenesis and tumor therapy. Oncogene 23, 2919–2933.

Shaywitz, A.J., Greenberg, M.E., 1999. CREB: a stimulus-induced transcription factor activated by a diverse array of extracellular signals. Annu. Rev. Biochem. 68, 821–861.

Shekhar, M.P., 2011. Drug resistance: challenges to effective therapy. Curr. Cancer Drug Targets 11, 613–623.

Shen, X., Kramer, R.H., 2004. Adhesion-mediated squamous cell carcinoma survival through ligand-independent activation of epidermal growth factor receptor. Am. J. Pathol. 165, 1315–1329.

Shen, H., Maki, C.G., 2011. Pharmacological activation of p53 by small-molecule MDM2 antagonists. Curr. Pharm. Des. 17, 560–568.

Shepard, H.M., Jin, P., Slamon, D.J., Pirot, Z., Maneval, D.C., 2008. Herceptin. Handb. Exp. Pharmacol. 181, 183–219.

Sherr, C.J., 1998. Tumor surveillance via the ARF-p53 pathway. Genes Dev. 12, 2984–2991.

Sherr, C.J., 2000. The Pezcoller Lecture: cancer cell cycles revisited. Cancer Res. 60, 3689–3695.

Sherr, C.J., 2001. The INK4a/ARF network in tumour suppression. Nat. Rev. Mol. Cell Biol. 2, 731–737.

Sherr, C.J., 2002. Cell cycle control and cancer. Harvey Lect. 96, 73–92.

Sherr, C.J., 2004. Principles of tumor suppression. Cell 116, 235–246.

Sherr, C.J., 2006. Divorcing ARF and p53: an unsettled case. Nat. Rev. Cancer 6, 663–673.

Sherr, C.J., McCormick, F., 2002. The RB and p53 pathways in cancer. Cancer Cell 2, 103–112.

Sherr, C.J., Weber, J.D., 2000. The ARF/p53 pathway. Curr. Opin. Genet. Dev. 10, 94–99.

Sherr, C.J., Bertwistle, D., den Besten, W., Kuo, M.L., Sugimoto, M., Tago, K., et al., 2005. p53-dependent and -independent functions of the Arf tumor suppressor. Cold Spring Harb. Symp. Quant. Biol. 70, 129–137.

Shiao, S.L., Ganesan, A.P., Rugo, H.S., Coussens, L.M., 2011. Immune microenvironments in solid tumors: new targets for therapy. Genes Dev. 25, 2559–2572.

Shinohara, A., Ogawa, T., 1999. Rad51/RecA protein families and the associated proteins in eukaryotes. Mutat. Res. 435, 13–21.

Shtutman, M., Zhurinsky, J., Simcha, I., Albanese, C., D'Amico, M., Pestell, R., et al., 1999. The cylin D1 gene is a target of the β-catenin /LEF-1 pathway. Proc. Natl. Acad. Sci. U.S.A. 96, 5522–5527.

Sica, A., 2010. Role of tumor-associated macrophages in cancer-related inflammation. Exp. Oncol. 32, 153–158.

Sica, A., Saccani, A., Mantovani, A., 2002. Tumor-associated macrophages: a molecular perspective. Int. Immunopharmacol. 2, 1045–1054.

Sica, A., Allavena, P., Mantovani, A., 2008. Cancer related inflammation: the macrophage connection. Cancer Lett. 267, 204–215.

Siddik, Z.H., 2003. Cisplatin: mode of cytotoxic action and molecular basis of resistance. Oncogene 22, 7265–7279.

Siegel-Lakhai, W.S., Beijnen, J.H., Schellens, J.H., 2005. Current knowledge and future directions of the selective epidermal growth factor receptor inhibitors erlotinib (Tarceva) and gefitinib (Iressa). Oncologist 10, 579–589.

Silva, J., Smith, A., 2008. Capturing pluripotency. Cell 132, 532–536.

Silver, D.P., Livingston, D.M., 2012. Mechanisms of BRCA1 tumor suppression. Cancer Discov. 2, 679–684.

Singh, A., Settleman, J., 2010. EMT, cancer stem cells and drug resistance: an emerging axis of evil in the war on cancer. Oncogene 29, 4741–4751.

Singh, S., Johnson, J., Chellappan, S., 2010. Small molecule regulators of Rb-E2F pathway as modulators of transcription. Biochim. Biophys. Acta 1799, 788–794.

Smith, A.P., Verrecchia, A., Faga, G., Doni, M., Perma, D., Martinato, F., et al., 2009. A positive role for Myc in TGFβ-induced Snail transcription and epithelial-to-mesenchymal transition. Oncogene 28, 422–430.

Smith, J.G., Wiethoff, C.M., Stewart, P.L., Nemerow, G.R., 2010a. Adenovirus. Curr. Top. Microbiol. Immunol. 343, 195–224.

Smith, J., Tho, L.M., Xu, N., Gillespie, D.A., 2010b. The ATM-Chk2 and ATR-Chk1 pathways in DNA damage signaling and cancer. Adv. Cancer Res. 108, 73–112.

Sobrinho-Simeos, M., Oliveira, C., 2002. Different types of epithelial cadherin alterations play different roles in human carcinogenesis. Adv. Anat. Pathol. 9, 329–337.

Solinas, G., Germano, G., Mantovani, A., Allavena, P., 2009. Tumor-associated macrophages (TAM) as major players of the cancer-related inflammation. J. Leukoc. Biol. 86, 1065–1073.

Solinas, G., Marchesi, F., Garlanda, C., Mantovani, A., Allavena, P., 2010. Inflammation-mediated promotion of invasion and metastasis. Cancer Metastasis Rev. 29, 243–248.

Solyanik, G.I., 2010. Multifactorial nature of tumor drug resistance. Exp. Oncol. 32, 181–185.

Somervaille, T.C.P., Matheny, C.J., Spencer, G.J., Iwasaki, M., Rinn, J.L., Witten, D.M., et al., 2009. Hierarchical maintenance of MLL myeloid leukemia stem cells employs a transcriptional program shared with embryonic rather than adult stem cells. Cell Stem Cell 4, 129–140.

Sorlie, T., Perou, C.M., Tibshirani, R., Aas, T., Geisler, S., Johnsen, H., et al., 2001. Gene expression patterns of breast carcinomas distinguish tumor subclasses with clinical implications. Proc. Natl. Acad. Sci. U.S.A. 98, 10869–10874.

Sotiriou, C., Wirapati, P., Loi, S., Harris, A., Fox, S., Smeds, J., et al., 2006. Gene expression profiling in breast cancer: understanding the molecular basis of histologic grade to improve prognosis. J. Natl. Cancer Inst. 98, 262–272.

Soverini, S., Martinelli, G., Iacobucci, I., Baccarani, M., 2008. Imatinib mesylate for the treatment of chronic myeloid leukemia. Expert Rev. Anticancer Ther. 8, 853–864.

Spano, J.P., Bay, J.O., Blay, J.Y., Rixe, O., 2005. Proteasome inhibition: a new approach for the treatment of malignancies. Bull. Cancer 92, E61–E66, 945–952.

Spencer, C.A., Groudine, M., 1991. Control of c-myc regulation in normal and neoplastic cells. Adv. Cancer Res. 56, 1–48.

Spizzo, R., Nicoloso, M.S., Croce, C.M., Calin, G.A., 2009. SnapShot: microRNAs in cancer. Cell 137, 586, 586.e1.

Spurgers, K.B., Gold, D.L., Coombes, K.R., Bohnenstiel, N.L., Mullins, B., Mey, R.E., et al., 2006. Identification of cell cycle regulatory genes as principal targets of p53-mediated transcriptional repression. J. Biol. Chem. 281, 25134–25142.

Sreekumar, R., Sayan, B.S., Mirnezami, A.H., Sayan, A.E., 2011. MicroRNA control of invasion and metastasis pathways. Front. Genet. 2, 58.

Srivastava, N., Gochhait, S., de Boer, P., Bamezai, R.N., 2009. Role of H2AX in DNA damage response and human cancers. Mutat. Res. 681, 180–188.

Stamos, J.L., Weis, W.I., 2013. The β-catenin destruction complex. Cold Spring Harb. Perspect. Biol. 5, a007898.

Stanger, B.Z., 2012. Quit your YAPing: a new target for cancer therapy. Genes Dev. 26, 1263–1267.

Stanley, M.A., Pett, M.R., Coleman, N., 2007. HPV: from infection to cancer. Biochem. Soc. Trans. 35, 1456–1460.

Staudt, L.M., 2010. Oncogenc activation of NF-κB. Cold Spring Harb. Perspect. Biol. 2, a000109.

Steelman, L.S., Pohnert, S.C., Shelton, J.G., Franklin, R.A., Bertrand, F.E., McCubrey, J.A., 2004. JAK/STAT, Raf/MEK/ERK, PI3K/Akt and BCR-ABL in cell cycle progression and leukemogenesis. Leukemia 18, 189–218.

Steelman, L.S., Franklin, R.A., Abrams, S.L., Chappell, W., Kempf, C.R., Bäsecke, J., et al., 2011a. Roles of the Ras/Raf/MEK/ERK pathway in leukemia therapy. Leukemia 25, 1080–1094.

Steelman, L.S., Chappell, W.H., Abrams, S.L., Kempf, C.R., Long, J., Laidler, P., et al., 2011b. Roles of the Raf/MEK/ERK and PI3K/PTEN/Akt/mTOR pathways in controlling growth and sensitivity to therapy—implications for cancer and aging. Aging 3, 192–222.

Steer, H.J., Lake, R.A., Nowak, A.K., Robinson, B.W.S., 2010. Harnessing the immune response to treat cancer. Oncogene 29, 6301–6313.

Stegmeier, F., Warmuth, M., Sellers, W.R., Dorsch, M., 2010. Targeted cancer therapies in the twenty-first century: lessons from imatinib. Clin. Pharmacol. Ther. 87, 543–552.

Stein, G.S., Pardee, A.B., 2004. Cell Cycle and Growth Control. Wiley-Liss, Hoboken.

Steinberg, M., 2007. Dasatinib: a tyrosine kinase inhibitor for the treatment of chronic myelogenous leukemia and philadelphia chromosome-positive acute lymphoblastic leukemia. Clin. Ther. 29, 2289–2308.

Stemmler, M.P., 2008. Cadherins in development and cancer. Mol. Biosyst. 4, 835–850.

Stepniak, E., Radice, G.L., Vasioukhin, V., 2009. Adhesive and signaling functions of cadherins and catenins in vertebrate development. Cold Spring Harb. Perspect. Biol. 1, a002949.

Sterz, J., von Metzler, I., Hahne, J.C., Lamottke, B., Rademacher, J., Heider, U., et al., 2009. The potential of proteasome inhibitors in cancer therapy. Expert Opin. Investig. Drugs 17, 879–895.

Stewart, T.J., Smyth, M.J., 2011. Improving cancer immunotherapy by targeting tumor-induced immune suppression. Cancer Metastasis Rev. 30, 125–140.

Stewart, S.A., Weinberg, R.A., 2002. Senescence: does it all happen at the ends? Oncogene 21, 627–630.

Stewart, S.A., Weinberg, R.A., 2006. Telomeres: cancer to human ageing. Annu. Rev. Cell Dev. Biol. 22, 531–557.

Stoyanova, T., Roy, N., Bhattacharjee, S., Kopanja, D., Valli, T., Bagchi, S., et al., 2012. p21 cooperates with DDB2 in suppression of UV-induced skin malignancies. J. Biol. Chem. 287, 3019–3028.

Stracker, T.H., Usui, T., Petrini, J.H., 2009. Taking the time to make important decisions: the checkpoint effector kinases Chk1 and Chk2 and the DNA damage response. DNA Repair 8, 1047–1054.

Strathdee, G., 2002. Epigenetic *versus* genetic alteration in the inactivation of *E-cadherin*. Semin. Cancer Biol. 12, 373–379.

Strieter, R.M., Keeley, E.C., Burdick, M.D., Mehrad, B., 2009a. The role of circulating mesenchymal progenitor cells, fibrocytes, in promoting pulmonary fibrosis. Trans. Am. Clin. Climatol. Assoc. 120, 49–59.

Strieter, R.M., Keeley, E.C., Hughes, M.A., Burdick, M.D., Mehrad, B., 2009b. The role of circulating mesenchymal progenitor cells (fibrocytes) in the pathogenesis of pulmonary fibrosis. J. Leukoc. Biol. 86, 1111–1118.

Stubenrauch, F., Laimins, L.A., 1999. Human papillomavirus life cycle: active and latent phases. Semin. Cancer Biol. 9, 379–386.

Su, T.T., 2006. Cellular responses to DNA damage: one signal, multiple choices. Annu. Rev. Genet. 40, 187–208.

Subramanya, D., Grivas, P.D., 2008. HPV and cervical cancer: updates on an established relationship. Postgrad. Med. 120, 7–13.

Sullivan, C.A., Pipas, J.M., 2002. T antigens of SV40: molecular chaperones for viral replication and tumorigenesis. Microbiol. Mol. Biol. Rev. 66, 179–202.

Sun, B., Karin, M., 2008. NF-κB signaling, liver disease and hepatoprotective agents. Oncogene 27, 6228–6244.

Sun, A., Bagella, L., Tutton, S., Romano, G., Giordano, A., 2007. From G0 to S phase: a view of the roles played by the retinoblastoma (Rb) family members in the Rb-E2F pathway. J. Cell. Physiol. 102, 1400–1404.

Sun, H.C., Li, M., Lu, J.L., Yan, D.W., Zhou, C.Z., Fan, J.W., et al., 2011a. Overexpression of Forkhead box M1 protein associates with aggressive tumor features and poor prognosis of hepatocellular carcinoma. Oncol. Rep. 25, 1533–1539.

Sun, H., Teng, M., Liu, J., Jin, D., Wu, J., Yan, D., et al., 2011b. FOXM1 expression predicts the prognosis in hepatocellular carcinoma patients after orthotopic liver transplantation combined with the Milan criteria. Cancer Lett. 306, 214–222.

Sundfeldt, K., 2003. Cell-cell adhesion in the normal ovary and ovarian tumors of epithelial origin: an exception to the rule. Mol. Cell. Endocrinol. 202, 89–96.

Sung, P., Krejci, L., Van Komen, S., Sehorn, M.G., 2003. Rad51 recombinase and recombination mediators. J. Biol. Chem. 278, 42729–42732.

Symington, L.S., Gautier, J., 2011. Double-strand break end resection and repair pathway choice. Annu. Rev. Genet. 45, 247–271.

Taatjes, D.J., Fenick, D.J., Gaudiana, G., Koch, T.H., 1998. A redox pathway leading to the alkylation of nucleic acids by doxorubicin and related anthracyclines: application to the design of antitumor drugs for resistant cancer. Curr. Pharm. Des. 4, 203–218.

Taipale, J., Beachy, P.A., 2001. The Hedgehog and Wnt signalling pathways in cancer. Nature 411, 349–354.

Takahashi, H., Shibuya, M., 2005. The vascular endothelial growth factor (VEGF)/VEGF receptor system and its role under physiological and pathological conditions. Clin. Sci. 109, 227–241.

Takahashi, Y., Li, L., Kamiryo, M., Asteriou, T., Moustakas, A., Yamashita, H., et al., 2005. Hyaluronan fragments induce endothelial cell differentiation in a CD44- and CXCL1/GRO1-dependent manner. J. Biol. Chem. 280, 24195–24204.

Takahashi, K., Furukawa, C., Takano, A., Ishikawa, N., Kato, T., Hayama, S., et al., 2006. The neuromedin U-growth hormone secretagogue receptor 1b/neurotensin receptor 1 oncogenic signaling pathway as a therapeutic target for lung cancer. Cancer Res. 66, 9408–9419.

Takebe, N., Warren, R.Q., Ivy, S.P., 2011. Breast cancer growth and metastasis: interplay between cancer stem cells, embryonic signaling pathways and epithelial-to-mesenchymal transition. Breast Cancer Res. 13, 211.

Takemura, T., Nakamura, S., Yokota, D., Hirano, I., Ono, T., Shigeno, K., et al., 2010. Reduction of RAF kinase inhibitor protein expression by BCR-ABL contributes to chronic myelogenous leukemia proliferation. J. Biol. Chem. 285, 6585–6594.

Takizawa, D., Kakizaki, S., Horiguchi, N., Yamazaki, Y., Tojima, H., Mori, M., 2011. Constitutive active/androstane receptor promotes hepatocarcinogenesis in a mouse model of non-alcoholic steatohepatitis. Carcinogenesis 32, 576–583.

Talbot, S.J., Crawford, D.H., 2004. Viruses and tumours—an update. Eur. J. Cancer 40, 1998–2005.

Talluri, S., Dick, F.A., 2012. Regulation of transcription and chromatin structure by RB. Here, there and everywhere. Cell Cycle 11, 3189–3198.

Talora, C., Campese, A.F., Bellavia, D., Felli, M.P., Vacca, A., Gulino, A., et al., 2008. Notch signaling and diseases: an evolutionary journey from a simple beginning to complex outcomes. Biochim. Biophys. Acta 1782, 489–497.

Tamano, S., Merlino, G., Ward, J.M., 1994. Rapid development of hepatic tumors in transforming growth factor a transgenic mice associated with increased cell proliferation in precancerous hepatocellular lesions initiated by N-nitrosodiethylamine and promoted by phenobarbital. Carcinogenesis 15, 1791–1798.

Tan, T.T., Coussens, L.M., 2007. Humoral immunity, inflammation and cancer. Curr. Opin. Immunol. 19, 209–216.

Tan, Y., Yoshida, Y., Hughes, D.E., Costa, R.H., 2006. Increased expression of hepatocyte nuclear factor 6 stimulates hepatocyte proliferation during mouse liver regeneration. Gastroenterology 130, 1283–1300.

Tan, Y., Raychaudhuri, P., Costa, R.H., 2007. Chk2 mediates stabilization of the FoxM1 transcription factor to stimulate expression of DNA repair genes. Mol. Cell. Biol. 27, 1007–1016.

Tan, Y., Chen, L., Yu, H., Zhu, X., Meng, X., Huang, L., et al., 2010a. Two-fold elevation of expression of FoxM1 transcription factor in mouse embryonic fibroblasts enhances cell cycle checkpoint activity by stimulating *p21* and *Chk1* transcription. Cell Prolif. 43, 494–504.

Tan, D.S., Gerlinger, M., Teh, B.T., Swanton, C., 2010b. Anti-cancer drug resistance: understanding the mechanisms through the use of integrative genomics and functional RNA interference. Eur. J. Cancer 46, 2166–2177.

Tan, S., de Vries, E.G., van der Zee, A.G., de Jong, S., 2012. Anticancer drugs aimed at E6 and E7 activity in HPV-positive cervical cancer. Curr. Cancer Drug Targets 12, 170–184.

Tang, H., Oishi, N., Kaneko, S., Murakami, S., 2006. Molecular functions and biological roles of hepatitis B virus X protein. Cancer Sci. 97, 977–983.

Tannock, I.F., 2001. Tumor physiology and drug resistance. Cancer Metastasis Rev. 20, 123–132.

Tarin, D., 2005. The fallacy of epithelial mesenchymal transition in neoplasia. Cancer Res. 65, 5996–6000.

Tauriello, D.V.F., Maurice, M.M., 2010. The various roles of ubiquitin in Wnt pathway regulation. Cell Cycle 9, 3700–3709.

Teglund, S., Toftgard, R., 2010. Hedgehog beyond medulloblastoma and basal cell carcinoma. Biochim. Biophys. Acta 1805, 181–208.

Teh, M.T., 2012a. Cell brainwashed by FOXM1: do they have potential as biomarkers of cancer? Biomark. Med. 6, 499–501.

Teh, M.T., 2012b. FOXM1 coming of age: time for translation into clinical benefits? Front. Oncol. 2, 146.

Teh, M.T., Wong, S.T., Neill, G.W., Ghali, L.R., Philpott, M.P., Quinn, A.G., 2002. FOXM1 is a downstream target of Gli1 in basal cell carcinomas. Cancer Res. 62, 4773–4780.

Teh, M.T., Gemenetzidis, E., Patel, D., Tariq, R., Nadir, A., Bahta, A.W., et al., 2012. FOXM1 induces a global methylation signature that mimics the cancer epigenome in head and neck squamous cell carcinoma. PLoS One 7, e34329.

Tepass, U., 2002. Adherens junctions: new insight into assembly, modulation and function. Bioessays 24, 690–695.

Tetsu, O., McCormick, F., 1999. β-catenin regulates expression of cyclin D1 in colon carcinoma cells. Nature 398, 422–426.

Thacker, J., 2005. The RAD51 gene family, genetic instability and cancer. Cancer Lett. 219, 125–135.

Thannickal, V.J., Toews, G.B., White, E.S., Lynch, J.P., Martinez, F.J., 2004. Mechanisms of pulmonary fibrosis. Annu. Rev. Med. 55, 395–417.

The Cancer Genome Atlas Research Network, 2011. Integrated genomic analyses of ovarian carcinoma. Nature 474, 609–615.

Thierry, F., Benotmane, M.A., Demeret, C., Mori, M., Teissier, S., Desaintes, C., 2004. A genomic approach reveals a novel mitotic pathway in papillomavirus carcinogenesis. Cancer Res. 64, 895–903.

Thiery, J.P., 2002. Epithelial-mesenchymal transitions in tumour progression. Nat. Rev. Cancer 2, 442–454.

Thiery, J.P., 2003. Epithelial-mesenchymal transitions in development and pathologies. Curr. Opin. Cell Biol. 15, 740–746.

Thiery, J.P., Sleeman, J.P., 2006. Complex networks orchestrate epithelial-mesenchymal transitions. Nat. Rev. Mol. Cell Biol. 7, 131–142.

Thiery, J.P., Acloque, H., Huang, R.Y., Nieto, M.A., 2009. Epithelia-mesenchymal transitions in development and disease. Cell 139, 871–890.

Thompson, C.B., 2009. Attacking cancer at its root. Cell 138, 1051–1054.

Thompson, E.W., Newgreen, D.F., 2005. Carcinoma invasion and metastasis: a role for epithelial-mesenchymal transition? Cancer Res. 65, 5991–5995.

Tian, X., Liu, Z., Niu, B., Zhang, J., Tan, T.K., Lee, S.R., et al., 2011. E-cadherin/β-catenin complex and the epithelial barrier. J. Biomed. Biotechnol. 2011, 567305.

Tokunaga, E., Oki, E., Nishida, K., Koga, T., Egashira, A., Morita, M., et al., 2006. Trastuzumab and breast cancer: developments and current status. Int. J. Clin. Oncol. 11, 199–208.

Tomasek, J.J., Gabbiani, G., Hinz, B., Chaponnier, C., Brown, R.A., 2002. Myofibroblasts and mechano-regulation of connective tissue remodelling. Nat. Rev. Mol. Cell Biol. 3, 349–363.

Tomlinson, J.S., Alpaugh, M.L., Barsky, S.H., 2001. An intact overexpressed E-cadherin/α, β-catenin axis characterizes the lymphovascular emboli of inflammatory breast carcinoma. Cancer Res. 61, 5231–5241.

Tompkins, D.H., Besnard, V., Lang, A.W., Keiser, A.R., Wert, S.E., Bruno, M.D., et al., 2011. Sox2 activates cell proliferation and differentiation in the respiratory epithelium. Am. J. Respir. Cell Mol. Biol. 45, 101–110.

Topalian, S.L., Weiner, G.J., Pardoll, D.M., 2011. Cancer immunotherapy comes of age. J. Clin. Oncol. 29, 4828–4836.

Torchia, J., Glass, C., Rosenfeld, M.G., 1998. Co-activators and co-repressors in the integration of transcriptional responses. Curr. Opin. Cell Biol. 10, 373–383.

Trakul, N., Rosner, M.R., 2005. Modulation of the MAP kinase signaling cascade by Raf kinase inhibitory protein. Cell Res. 15, 19–23.

Traxler, P., 2003. Tyrosine kinases as targets in cancer therapy—successes and failures. Expert Opin. Ther. Targets 7, 215–234.

Trimboli, A.J., Fukino, K., de Bruin, A., Wei, G., Shen, L., Tanner, S.M., et al., 2008. Direct evidence for epithelial-mesenchymal transitions in breast cancer. Cancer Res. 68, 937–945.

Troyanovsky, S., 2005. Cadherin dimers in cell-cell adhesion. Eur. J. Cell Biol. 84, 225–233.

Tse, J.C., Kalluri, R., 2007. Mechanisms of metastasis: epithelial-to-mesenchymal transition and contribution of tumor microenvironment. J. Cell. Biochem. 101, 816–829.

Tsuji, T., Ibaragi, S., Hu, G.F., 2009. Epithelial-mesenchymal transition and cell cooperativity in metastasis. Cancer Res. 69, 7135–7139.

Tsuruo, T., Naito, M., Tomida, A., Fujita, N., Mashima, T., Sakamoto, H., et al., 2003. Molecular targeting therapy of cancer: drug resistance, apoptosis and survival signal. Cancer Sci. 94, 15–21.

Turkson, J., 2004. STAT proteins as novel targets for cancer drug discovery. Expert Opin. Ther. Targets 8, 409–422.

Turley, E.A., Veiseh, M., Radisky, D.C., Bissell, M.J., 2008. Mechanisms of disease: epithelial-mesenchymal transition—does cellular plasticity fuel neoplastic progression? Nat. Clin. Pract. Oncol. 5, 280–290.

Turnell, A.S., Mymryk, J.S., 2006. Roles for the coactivators CBP and p300 and the APC/C E3 ubiquitin ligase in E1A-dependent cell transformation. Br. J. Cancer 95, 555–560.

Tutt, A., Ashworth, A., 2002. The relationship between the roles of BRCA genes in DNA repair and cancer predisposition. Trends Mol. Med. 8, 571–576.

Uddin, S., Ahmed, M., Hussain, A., Akubaker, J., Al-Sanea, N., AbdulJabbar, A., et al., 2011. Genome-wide expression analysis of Middle Eastern colorectal cancer reveals FOXM1 as a novel target for cancer therapy. Am. J. Pathol. 178, 537–547.

Uddin, S., Hussain, A.R., Ahmed, M., Siddiqui, K., Al-Dayel, F., Bavi, P., et al., 2012. Over-expression of FoxM1 offers a promising therapeutic target in diffuse large B-cell lymphoma. Haematologica 97, 1092–1100.

Ueno, H., Nakajo, N., Watanabe, M., Isoda, M., Sagata, N., 2008. FoxM1-driven cell division is required for neuronal differentiation in early Xenopus embryos. Development 135, 2023–2030.

Untergasser, G., Gander, R., Lilg, C., Lepperdinger, G., Plas, E., Berger, P., 2005. Profiling molecular targets of TGF-β1 in prostate fibroblast-to-myofibroblast transdifferentiation. Mech. Ageing Dev. 126, 59–69.

Ustiyan, V., Wang, I.C., Ren, X., Zhang, Y., Snyder, J., Xu, Y., et al., 2009. Forkhead box M1 transcriptional factor is required for smooth muscle cells during embryonic development of blood vessels and esophagus. Dev. Biol. 336, 266–279.

Ustiyan, V., Wert, S.E., Ikegami, M., Wang, I.C., Kalin, T.V., Whitsett, J.A., et al., 2012. Foxm1 transcription factor is critical for proliferation and differentiation of Clara cells during development of conducting airways. Dev. Biol. 370, 198–212.

Vakiani, E., Solit, D.B., 2011. KRAS and BRAF: drug targets and predictive biomarkers. J. Pathol. 223, 219–229.

Valabrega, G., Montemurro, F., Aglietta, M., 2007. Trastuzumab: mechanism of action, resistance and future perspectives in HER2-overexpressing breast cancer Ann. Oncol. 18, 977–984.

Valastyan, S., Weinberg, R.A., 2011. Roles for microRNAs in theregulation of cell adhesion molecules. J. Cell Sci. 124, 999–1006.

Valastyn, S., Weinberg, R.A., 2009. MicroRNAs: crucial multi-tasking components in the complex circuitry of tumor metastasis. Cell Cycle 8, 3506–3512.

Valenta, T., Hausmann, G., Basler, K., 2012. The many faces and functions of β-catenin. EMBO J. 31, 2714–2736.

Valkenburg, K.C., Williams, B.O., 2011. Mouse models of prostate cancer. Prostate Cancer 2011, 95238.

van den Boom, J., Wolter, M., Kuick, R., Misek, D.E., Youkilis, A.S., Wechsler, D.S., et al., 2003. Characterization of gene expression profiles associated with glioma progression using oligonucleotide-based microarray analysis and real-time transcription-polymerase chain reaction. Am. J. Pathol. 163, 1033–1043.

van der Horst, G., Bos, L., van der Pluijm, G., 2012. Epithelial plasticity, cancer stem cells, and the tumor-supportive stroma in bladder carcinoma. Mol. Cancer Res. 10, 995–1009.

van der Pluijm, G., 2011. Epithelial plasticity, cancer stem cells, and bone metastasis formation. Bone 48, 37–43.

van der Wurff, A.A.M., Arends, J.W., van der Linden, E.P.M., ten Kate, J., Bosman, F.T., 1994. L-CAM expression in lymp node and liver metastases of colorectal carcinomas. J. Pathol. 172, 177–182.

Van Etten, R.A., 2004. Mechanisms of transformation by the BCR-ABL oncogene: new perspectives in the post-imatinib era. Leuk. Res. 28, S21–S28.

van Gent, D.C., Hoeijmakers, J.H., Kanaar, R., 2001. Chromosomal stability and the DNA double-strand break connection. Nat. Rev. Genet. 2, 196–206.

Vanneman, M., Dranoff, G., 2012. Combining immunotherapy and targeted therapies in cancer treatment. Nat. Rev. Cancer 12, 237–251.

van Roy, F., Berx, G., 2008. The cell-cell adhesion molecule E-cadherin. Cell. Mol. Life Sci. 65, 3756–3788.

Varmus, H., 2006. The new era in cancer research. Science 312, 1162–1165.

Varmus, H., Pao, W., Politi, K., Podsypanina, K., Du, Y.C., 2005. Oncogenes come of age. Cold Spring Harb. Symp. Quant. Biol. 70, 1–9.

Vassilev, L.T., 2004. Small-molecule antagonists of p53-MDM2 binding. Research tools and potential therapeutics. Cell Cycle 3, 419–421.

Vassilev, L.T., 2005. p53 activation by small-molecules: application ion oncology. J. Med. Chem. 48, 4491–4499.

Vassilev, L.T., 2007. MDM2 inhibitors for cancer therapy. Trends Mol. Med. 13, 23–31.

Vassilev, L.T., Vu, B.T., Graves, B., Carjaval, D., Podlaski, F., Filipovic, Z., et al., 2004. In vivo activation of the p53 pathway by small-molecule antagonists of mdm2. Science 303, 844–848.

Venkitaraman, A.R., 2004. Tracing the network connecting BRCA and Fanconi Anaemia proteins. Nat. Rev. Cancer 4, 266–276.

Ventura, A., Jacks, T., 2009. MicroRNAs and cancer: short RNAs go a long way. Cell 136, 586–591.

Verheyen, E.M., Gottardi, C.J., 2010. Regulation of Wnt/β-catenin signaling by protein kinases. Dev. Dyn. 239, 34–44.

Verstovsek, S., 2009. Preclinical and clinical experience with dasatinib in Philadelphia chromosme-negative leukemias and myeloid disorders. Leuk. Res. 33, 617–623.

Visnyei, K., Onodera, H., Damoiseaux, R., Saigusa, K., Petosyan, S., De Vries, D., et al., 2011. A molecular screening approach to identify and characterize inhibitors of glioblastoma stem cells. Mol. Cancer Ther. 10, 1818–1829.

Vispe, S., Defais, M., 1997. Mammalian Rad51 protein: a RecA homologue with pleiotropic functions. Biochimie 79, 587–592.

Vita, M., Henriksson, M., 2006. The Myc oncoprotein as a therapeutic target for human cancer. Semin. Cancer Biol. 16, 318–330.

Vogelstein, B., Kinzler, K.W., 2004. Cancer genes and the pathways they control. Nat. Med. 10, 789–799.

Voorhees, P.M., Orlowski, R.Z., 2006. The proteasome and proteasome inhibitors in cancer therapy. Annu. Rev. Pharmacol. Toxicol. 46, 189–213.

Voorhees, P.M., Dees, E.C., O'Neil, B., Orlowski, R.Z., 2003. The proteasome as a target for cancer therapy. Clin. Cancer Res. 9, 6316–6325.

Voulgari, A., Pintzas, A., 2009. Epithelial-mesenchymal transition in cancer metastasis: mechanisms, markers, and strategies to overcome drug resistance in the clinic. Biochim. Biophys. Acta 1796, 75–90.

Vousden, K.H., 1995. Regulation of the cell cycle by viral oncoproteins. Semin. Cancer Biol. 6, 109–116.

Voutsadakis, I.A., 2012a. Ubiquitination and the ubiquitin-proteasome system as regulators of transcription and transcription factors in epithelial mesenchymal transition of cancer. Tumour Biol. 33, 897–910.

Voutsadakis, I.A., 2012b. The ubiquitin-proteasome system and signal transduction pathways regulating epithelial mesenchymal transition of cancer. J. Biomed. Sci. 19, 67.

Wade, M., Wahl, G.M., 2010. Targeting Mdm2 and Mdmx in cancer therapy: better living through medical chemistry? Mol. Cancer Res. 7, 1–11.

Wagner, E.F., Nebreda, A.R., 2009. Signal integration by JNK and p38 MAPK pathways in cancer development. Nat. Rev. Cancer 9, 537–549.

Wakeling, A.E., 2002. Epidermal growth factor receptor tyrosine kinase inhbitors. Curr. Opin. Pharmacol. 2, 382–387.

Waller, C.F., 2010. Imatinib mesylate. Recent Results Cancer Res. 184, 3–20.

Walworth, N.C., 2000. Cell-cycle checkpoint kinases: checking in on the cell cycle. Curr. Opin. Cell Biol. 12, 697–704.

Wan, X., Yeung, C., Young Kim, S., Dolan, J.G., Ngo, V.N., Burkett, S., et al., 2012. Identification of FoxM1/Bub1b signaling pathway as a required component for growth and survival of rhabdomyosarcoma. Cancer Res. 72, 5889–5899.

Wang, W., 2007. Emergence of a DNA-damage response network consisting of Fanconi anaemia and BRCA proteins. Nat. Rev. Genet. 8, 735–748.

Wang, F., 2011. Modeling human prostate cancer in genetically engineered mice. Prog. Mol. Biol. Transl. Sci. 100, 1–49.

Wang, H., Cho, C.H., 2010. Effect of NF-κB signaling on apoptosis in chronic inflammation-associated carcinogenesis. Curr. Cancer Drug Targets 10, 593–599.

Wang, M., Gartel, A.L., 2011. Micelle-encapsulated thiostrepton as an effective nanomedicine for inhibiting tumor growth and for suppressing FOXM1 in human xenografts. Mol. Cancer Ther. 10, 2287–2297.

Wang, L., Wang, J., 2012. MicroRNA-mediated breast cancer metastasis: from primary site to distant organs. Oncogene 31, 2499–2511.

Wang, Y., Zhou, B.P., 2011. Epithelial-mesenchymal transition in breast cancer progression and metastasis. Chin. J. Cancer 30, 603–611.

Wang, X., Hung, N.J., Costa, R.H., 2001a. Earlier expression of the transcription factor HFH-11B diminishes induction of p21CIP1/WAF1 levels and accelerates mouse hepatocyte entry into S-phase following carbon tetrachloride liver injury. Hepatology 33, 1404–1414.

Wang, X., Quail, E., Hung, N.J., Tan, Y., Ye, H., Costa, R.H., 2001b. Increased levels of forkhead box M1B transcription factor in transgenic mouse hepatocytes prevent age-related proliferation defects in regenerating liver. Proc. Natl. Acad. Sci. U.S.A. 98, 11468–11473.

Wang, X., Krupczak-Hollis, K., Tan, Y., Dennewitz, M.B., Adami, G.R., Costa, R.H., 2002a. Increased hepatic Forkhead Box M1B (FoxM1B) levels in old-aged mice stimulated liver regeneration through diminished p27Kip1 protein levels and increased Cdc25B expression. J. Biol. Chem. 277, 44310–44316.

Wang, X., Kiyokawa, H., Dennewitz, M.B., Costa, R.H., 2002b. The Forkhead Box m1b transcription factor is essential for hepatocyte DNA replication and mitosis during mouse liver regeneration. Proc. Natl. Acad. Sci. U.S.A. 99, 16881–16886.

Wang, I.C., Chen, Y.J., Hughes, D., Petrovic, V., Major, M.L., Park, H.J., et al., 2005. Forkhead Box M1 regulates the transcriptional network of genes essential for mitotic progression and genes encoding the SCF (Skp2-Cks1) ubiquitin ligase. Mol. Cell. Biol. 25, 10875–10894.

Wang, Z., Banerjee, S., Kong, D., Li, Y., Sarkar, F.H., 2007a. Down-regulation of Forkhead Box M1 transcription factor leads to the inhibition of invasion and angiogenesis of pancreatic cancer cells. Cancer Res. 67, 8293–8300.

Wang, G.L., Shi, X., Salisbury, E., Sun, Y., Albrecht, J.H., Smith, R.G., et al., 2007b. Growth hormone corrects proliferation and transcription of phosphoenolpyruvate carboxykinase in liver of old mice via elimination of CCAAT/enhancer-binding protein α-Brm complex. J. Biol. Chem. 282, 1468–1478.

Wang, I.C., Meliton, L., Tretiakova, M., Costa, R.H., Kalinichenko, V.V., Kalin, T.V., 2008a. Transgenic expression of the forkhead box M1 transcription factor induces formation of lung tumors. Oncogene 27, 4137–4149.

Wang, I.C., Chen, Y.J., Hughes, D.E., Ackerson, T., Major, M.L., Kalinichenko, V.V., et al., 2008b. FOXM1 regulates transcription of $JNK1$ to promote the G_1/S transition and tumor cell invasiveness. J. Biol. Chem. 283, 20770–20778.

Wang, G.L., Salisbury, E., Shi, X., Timchenko, L., Medrano, E.E., Timchenko, N.A., 2008c. HDAC1 cooperates with C/EBPα in the inhibition of liver proliferation in old mice. J. Biol. Chem. 283, 26169–26178.

Wang, I.C., Meliton, L., Ren, X., Zhang, Y., Balli, D., Snyder, J., et al., 2009a. Deletion of Forkhead Box M1 transcription factor from respiratory epithelial cells inhibits pulmonary tumorigenesis. PLoS One 4, e6609.

Wang, Y., Ji, P., Liu, J., Broaddus, R.R., Xue, F., Zhang, W., 2009b. Centrosome-associated regulators of the G_2/M checkpoint as targets for cancer therapy. Mol. Cancer 8, 8.

Wang, Z., Ahmad, A., Li, Y., Banerjee, S., Kong, D., Sarkar, F.H., 2010a. Forkhead box M1 transcription factor: a novel target for cancer therapy. Cancer Treat. Rev. 36, 151–156.

Wang, Z., Ahmad, A., Banerjee, S., Azmi, A., Kong, D., Li, Y., et al., 2010b. FoxM1 is a novel target of a natural agent in pancreatic cancer. Pharm. Res. 27, 1159–1168.

Wang, H., Teh, M.T., Ji, Y., Patel, V., Firouzabadian, S., Patel, A.A., et al., 2010c. EPS8 upregulates FOXM1 expression, enhancing cell growth and motility. Carcinogenesis 31, 1132–1141.

Wang, Z., Li, Y., Ahmad, A., Banerjee, S., Azmi, A.S., Kong, D., et al., 2010d. Down-regulation of Notch-1 is associated with Akt and FoxM1 In inducing cell growth inhibition and apoptosis in prostate cancer cells. J. Cell. Biochem. 112, 78–88.

Wang, I.C., Zhang, Y., Snyder, J., Sutherland, M.J., Burhans, M.S., Shannon, J.M., et al., 2010e. Increased expression of FoxM1 transcription factor in respiratory epithelium inhibits lung sacculation and causes Clara cell hyperplasia. Dev. Biol. 347, 301–314.

Wang, J.S., Ren, N., Xi, T., 2011a. Ursolic acid induces apoptosis by suppressing the expression of FoxM1 in MCF-7 human breast cancer cells. Med. Oncol. 29, 10–15.

Wang, Z., Park, H.J., Carr, J.R., Chen, Y.J., Zheng, Y., Li, J., et al., 2011b. FoxM1 in tumorigenicity of the neuroblastoma cells and renewal of the neural progenitors. Cancer Res. 71, 4292–4302.

Wang, J.L., Lin, Y.W., Chen, H.M., Kong, X., Xiong, H., Shen, N., et al., 2011c. Calcium prevents tumorigenesis in a mouse model of colorectal cancer. PLoS One 6, e22566.

Wang, I.C., Snyder, J., Zhang, Y., Landar, J., Nakafuku, Y., Lin, J., et al., 2012. Foxm1 mediates a cross-talk between Kras/MAPK and canonical Wnt pathways during development of respiratory epithelium. Mol. Cell. Biol. 32, 3838–3850.

Wang, Y., Wen, L., Zhao, S.H., Ai, Z.H., Guo, J.Z., Liu, W.C., 2013. FoxM1 expression is significantly associated with cisplatin-based chemotherapy resistance and poor prognosis in advanced non-small cell lung cancer patients. Lung Cancer 79, 173–179.

Waseem, A., Ali, M., Odell, E.W., Fortune, F., Teh, M.T., 2010. Downstream targets of FOXM1: CEP55 and HELLS are cancer progression markers of head and neck squamous cell carcinoma. Oral Oncol. 46, 536–542.

Watanabe, T., Sato, K., Kaibuchi, K., 2009. Cadherin-mediated intercellular adhesion and signaling cascades involving small GTPases. Cold Spring Harb. Perspect. Biol. 1, a003020.

Waxman, D.J., Schwartz, P.S., 2003. Harnessing apoptosis for improved anticancer gene therapy. Cancer Res. 63, 8563–8572.

Weber, J.D., Jeffers, J.R., Rehg, J.E., Randle, D.H., Lozano, G., Roussel, M.F., et al., 2000. p53-independent functions of the p19(ARF) tumor suppressor. Genes Dev. 14, 2358–2365.

Wei, Y., Neuveut, C., Tiollais, P., Buendia, M.A., 2010. Molecular biology of the hepatitis B virus and role of the X gene. Pathol. Biol. 58, 267–272.

Weidle, U.H., Maisel, D., Eick, D., 2011. Synthetic lethality-based targets for discovery of new cancer therapeutics. Cancer Genomics Proteomics 8, 159–172.

Weigert, A., Sekar, D., Brüne, B., 2009. Tumor-associated macrophages a targets for tumor immunotherapy. Immunotherapy 1, 83–95.

Weinberg, R.A., 1997. The cat and mouse games that genes, viruses and cells play. Cell 88, 573–575.

Weinberg, R.A., 2006. The Biology of Cancer. Garland Science, New York.

Weiner, L.M., Murray, J.C., Shuptrine, C.W., 2012. Antibody-based immunotherapy of cancer. Cell 148, 1081–1084.

Weinshilboum, R., 2008. Pharmacogenomics of endocrine therapy in breast cancer. Adv. Exp. Med. Biol. 630, 220–231.

Weis, W.I., Nelson, W.J., 2006. Re-solving the cadherin-catenin conundrum. J. Biol. Chem. 281, 35593–35597.

Weisberg, E., Manley, P., Mestan, J., Cowan-Jacob, S., Ray, A., Griffin, J.D., 2006. AMN107 (nilotinib): a novel and selective inhibitor of BCR-ABL. Br. J. Cancer 94, 1765–1769.

Wells, A., Yates, C., Shepard, C.R., 2008. E-cadherin as an indicator of mesenchymal to epithelial reverting transitions during metastatic seeding of disseminated carcinomas. Clin. Exp. Metastasis 25, 621–628.

Wendt, M.K., Tian, M., Schiemann, W.P., 2012. Deconstructing the mechanisms and consequences of TGF-β-induced EMT during cancer progression. Cell Tissue Res. 347, 8–101.

Werbowetzki-Ogilvie, T.E., Bhatia, M., 2008. Pluripotent human stem cell lines: what we can learn about cancer initiation. Trends Mol. Med. 14, 323–332.

Weymann, A., Hartman, E., Gazit, V., Wang, C., Glauber, M., Turmelle, Y., et al., 2009. p21 is required for dextrose-mediated inhibition of mouse liver regeneration. Hepatology 50, 207–215.

Wheelock, M.J., Johnson, K.R., 2003. Cadherins as modulators of cellular phenotypes. Annu. Rev. Cell Dev. Biol. 19, 207–235.

Wheelock, M.J., Shintani, Y., Maeda, M., Fukumoto, Y., Johnson, K.R., 2007. Cadherin switching. J. Cell Sci. 121, 727–735.

White, E., 2001. Regulation of the cell cycle and apoptosis by the oncogenes of adenovirus. Oncogene 20, 7836–7846.

White, M.K., Khalili, K., 2004. Polyomaviruses and human cancer: molecular mechanisms underlying patterns of tumorigenesis. Virology 324, 1–16.

White, M.K., Khalili, K., 2006. Interaction of retinoblastoma protein family members with large T-antigen of primate polyomaviruses. Oncogene 25, 5286–5293.

Whitfield, M.L., Sherlock, G., Saldanha, A.L., Murray, J.I., Ball, C.A., Alexander, K.E., et al., 2002. Identification of genes periodically expressed in the human cell cycle and their expression in tumors. Mol. Biol. Cell 13, 1977–2000.

Whitfield, M.L., George, L.K., Grant, G.D., Perou, C.M., 2006. Common markers of proliferation. Nat. Rev. Cancer 6, 99–106.

Wierstra, I., 2011a. The transcription factor FOXM1c binds to and transactivates the promoter of the tumor suppressor gene E-cadherin. Cell Cycle 10, 760–766.

Wierstra, I., 2011b. The transcription factor FOXM1c is activated by protein kinase CK2, protein kinase A (PKA), c-Src and Raf-1. Biochem. Biophys. Res. Commun. 413, 230–235.

Wierstra, I., 2013a. Cyclin D1/Cdk4 increases the transcriptional activity of FOXM1c without phosphorylating FOXM1c. Biochem. Biophys. Res. Commun. 431, 753–759.

Wierstra, I., 2013b. The transcription factor FOXM1 (Forkhead box M1): proliferation-specific expression, transcription factor function, target genes, mouse models, and normal biological roles. Adv. Cancer Res. 118, 97–398.

Wierstra, I., Alves, J., 2006a. Despite its strong transactivation domain transcription factor FOXM1c is kept almost inactive by two different inhibitory domains. Biol. Chem. 387, 963–976.

Wierstra, I., Alves, J., 2006b. Transcription factor FOXM1c is repressed by RB and activated by cyclin D1/Cdk4. Biol. Chem. 387, 949–962.

Wierstra, I., Alves, J., 2006c. FOXM1c is activated by cyclin E/Cdk2, cyclin A/Cdk2 and cyclin A/Cdk1, but repressed by GSK-3α. Biochem. Biophys. Res. Commun. 348, 99–108.

Wierstra, I., Alves, J., 2006d. FOXM1c transactivates the human c-myc promoter directly via the two TATA-boxes P1 and P2. FEBS J. 273, 4645–4667.

Wierstra, I., Alves, J., 2007a. FOXM1c and Sp1 transactivate the P1 and P2 promoters of human c-myc synergistically. Biochem. Biophys. Res. Commun. 352, 61–68.

Wierstra, I., Alves, J., 2007b. The central domain of transcription factor FOXM1c directly interacts with itself in vivo and switches from an essential to an inhibitory domain depending on the FOXM1c binding site. Biol. Chem. 388, 805–818.

Wierstra, I., Alves, A., 2007c. FOXM1, a typical proliferation-associated transcription factor. Biol. Chem. 388, 1257–1274.

Wierstra, I., Alves, J., 2008. Cyclin E/Cdk2, P/CAF and E1A regulate the transactivation of the c-myc promoter by FOXM1. Biochem. Biophys. Res. Commun. 368, 107–115.

Willert, K., Jones, K.A., 2006. Wnt signaling: is the party in the nucleus? Genes Dev. 20, 1394–1404.

Willis, B.C., Borok, Z., 2007. TGF-β-induced EMT: mechanisms and implications for fibrotic lung disease. Am. J. Physiol. Lung Cell. Mol. Physiol. 293, L525–L534.

Willis, B.C., duBois, R.M., Borok, Z., 2006. Epithelial origin of myofibroblasts during fibrosis in the lung. Proc. Am. Thorac. Soc. 3, 377–382.

Wilson, M.S.C., Brosens, J.J., Schwenen, H.D.C., Lam, E.W.F., 2011. FOXO and FOXM1 in cancer: the FOXO-FOXM1 axis shapes the outcome of cancer therapy. Curr. Drug Targets 12, 1256–1266.

Wiman, K.G., 2006. Strategies for therapeutic targeting of the p53 pathway in cancer. Cell Death Differ. 13, 921–926.

Wise-Draper, T.M., Wells, S.I., 2008. Papillomavirus E6 and E7 proteins and their cellular targets. Front. Biosci. 13, 1003–1017.

Wong, S., Witte, O.N., 2004. The BCR-ABL story: bench to bedside and back. Annu. Rev. Immunol. 22, 247–306.

Wong, D.J., Liu, H., Ridky, T.W., Cassarino, D., Segal, E., Chang, H.Y., 2008a. Module map of stem cell genes guides creation of epithelial cancer stem cells. Cell Stem Cell 2, 333–344.

Wong, D.J., Segal, E., Chang, H.Y., 2008b. Stemness, cancer and cancer stem cells. Cell Cycle 7, 3622–3624.

Wonsey, D.R., Follettie, M.T., 2005. Loss of the forkhead transcription factor FoxM1 causes centrosome amplification and mitotic catastrophe. Cancer Res. 65, 5181–5189.

Wood, B., Leong, A., 2003. The biology and diagnostic applications of cadherins in neoplasia: a review. Pathology 35, 101–105.

Woodfield, G.W., Horan, A.D., Chen, Y., Weigel, R.J., 2007. TFAP2C controls hormone response in breast cancer cells through multiple pathways of estrogen signaling. Cancer Res. 67, 8439–8443.

Woodman, C.B.J., Collins, S.I., Young, L.S., 2007. The natural history of cervical HPV infection: unresolved issues. Nat. Rev. Cancer 7, 11–22.

Wright, J.A., Richer, J.K., Goodall, G.J., 2010. microRNAs and EMT in mammary cells and breast cancer. J. Mammary Gland Biol. Neoplasia 15, 213–223.

Wu, D., Pan, W., 2009. GSK3: a multifaceted kinase in Wnt signaling. Trends Biochem. Sci. 35, 161–168.

Wu, Y., Zhou, B.P., 2009. Inflammation: a driving force speeds cancer metastasis. Cell Cycle 8, 3267–3273.

Wu, Y., Zhou, B.P., 2010. Snail. More than EMT. Cell Adh. Migr. 4, 199–203.

Wu, Q.F., Liu, C., Tai, M.H., Liu, D., Lei, L., Wang, R.T., et al., 2010a. Knockdown of FoxM1 by siRNA interference decreases cell proliferation, induces cell cycle arrest and inhibits cell invasion in MHCC-97H cells in vitro. Acta Pharmacol. Sin. 31, 361–366.

Wu, J., Lu, L.Y., Yu, X., 2010b. The role of BRCA1 in DNA damage response. Protein Cell 1, 117–123.

Wu, C.Y., Tsai, Y.P., Wu, M.Z., Teng, S.C., Wu, K.J., 2012. Epigenetic reprogramming and post-transcriptional regulation during the epithelial-mesenchymal transition. Trends Genet. 28, 454–463.

Wu, X.R., Chen, Y.H., Liu, D.M., Sha, J.J., Xuan, H.Q., Bo, J.J., et al., 2013. Increased expression of forkhead box M1 protein is associated with poor prognosis in clear cell renal cell carcinoma. Med. Oncol. 30, 346.

Wyman, C., Kanaar, R., 2006. DNA double-strand break repair: all's well the ends well. Annu. Rev. Genet. 40, 363–383.

Wynn, T.A., 2004. Fibrotic disease and the T(H)1/T(H)2 paradigm. Nat. Rev. Immunol. 4, 583–594.

Wynn, T.A., 2007. Common and unique mechanisms regulate fibrosis in various fibroproliferative diseases. J. Clin. Invest. 117, 524–529.

Wynn, T.A., 2008. Cellular and molecular mechanisms of fibrosis. J. Pathol. 214, 199–210.

Wynn, T.A., 2011. Integrating mechanisms of pulmonary fibrosis. J. Exp. Med. 208, 1339–1350.

Wynn, T.A., Barron, L., 2010. Macrophages: master regulators of inflammation and fibrosis. Semin. Liver Dis. 30, 245–257.

Xia, L.M., Huang, W.J., Wang, B., Liu, M., Zhang, Q., Yan, W., et al., 2009. Transcriptional up-regulation of FoxM1 in response to hypoxia is mediated by HIF-1. J. Cell. Biochem. 106, 247–256.

Xia, J.T., Wang, H., Liang, L.J., Peng, B.G., Wu, Z.F., Chen, L.Z., et al., 2012a. Overexpression of FOXM1 is associated with poor prognosis and clinicopathologic stage of pancreatic ductal adenocarcinoma. Pancreas 41, 629–635.

Xia, L., Huang, W., Tian, D., Zhu, H., Zhang, Y., Hu, H., et al., 2012b. Upregulated FoxM1 expression induced by hepatitis B virus X protein promotes tumor metastasis and indicates poor prognosis in hepatitis B virus-related hepatocellular carcinoma. J. Hepatol. 57, 600–612.

Xia, L., Mo, P., Huang, W., Zhang, L., Wang, Y., Zhu, H., et al., 2012c. The TNF-α/ROS/HIF-1-induced upregulation of FoxM1 expression promotes HCC proliferation and resistance to apoptosis. Carcinogenesis 33, 2250–2259.

Xiang, H.L., Liu, F., Quan, M.F., Cao, J.G., Lv, Y., 2012. 7-difluoromethoxyl-5,4'-di-n-octylgenistein inhibits growth of gastric cancer cells through downregulating forkhead box M1. World J. Gastroenterol. 18, 4618–4626.

Xiao, D., He, J., 2010. Epithelial mesenchymal transition and lung cancer. J. Thorac. Dis. 2, 154–159.

Xie, Z., Tan, G., Ding, M., Dong, D., Chen, T., Meng, X., et al., 2010. Foxm1 transcription factor is required for maintenance of pluripotency of P19 embryonal carcinoma cells. Nucleic Acids Res. 38, 8027–8038.

Xu, W., Kimelman, D., 2007. Mechanistic insights from structural studies of β-catenin and its binding partners. J. Cell Sci. 120, 3337–3344.

Xu, L., Pirollo, K.F., Chang, E.H., 2001. Tumor-targeted p53-therapy enhances efficacy of conventional chemo/radiotherapy. J. Control. Release 74, 115–128.

Xu, N., Zhang, X., Wang, X., Ge, H., Wang, X.Y., Garfield, D., et al., 2012a. FoxM1-mediated resistance of non-small-cell lung cancer cells to gefitinib. Acta Pharmacol. Sin. 33, 675–681.

Xu, Y., Wang, Y., Besnard, V., Ikegami, M., Wert, S.E., Heffner, C., et al., 2012b. Transcriptional programs controlling perinatal lung maturation. PLoS One 7, e37046.

Xu, N., Wu, S.D., Wang, H., Wang, Q., Bai, C.X., 2012c. Involvement of FoxM1 in non-small cell lung cancer recurrence. Asian Pac. J. Cancer Prev. 13, 4739–4743.

Xue, L., Chiang, L., He, B., Zhao, Y.Y., Winoto, A., 2010. FoxM1, a Forkhead transcription factor is a master cell cycle regulator for mouse mature T cells but not double positive thymocytes. PLoS One 5, e9229.

Xue, X.J., Xiao, R.H., Long, D.Z., Zou, X.F., Zhang, G.X., Yuan, Y.H., et al., 2012. Overexpression of FoxM1 is associated with tumor progression in patients with clear cell renal cell carcinoma. J. Transl. Med. 10, 200.

Yamada, S., Pokutta, S., Drees, F., Weis, W.I., Nelson, W.J., 2005. Deconstructing the cadherin-catenin-actin complex. Cell 123, 889–901.

Yamanaka, S., 2007. Strategies and new developments in the generation of patient-specific pluripotent stem cells. Cell Stem Cell 1, 39–49.

Yamanaka, S., 2008a. Induction of pluripotent stem cells from mouse fibroblasts by four transcription factors. Cell Prolif. 41, 51–56.

Yamanaka, S., 2008b. Pluripotency and nuclear reprogramming. Philos. Trans. R. Soc. Lond. B Biol. Sci. 363, 2079–2087.

Yamashita, T., Segawa, K., Fujinaga, Y., Nishikawa, T., Fujinaga, K., 1993. Biological and biochemical activity of E7 genes of the cutaneous human papillomavirus type 5 and 8. Oncogene 8, 2433–2441.

Yamashita, T., Segawa, K., Jimbow, K., Fujinaga, K., 2001. Both of the N-terminal and C-terminal regions of human papillomavirus type 16 E7 are essential for immortalization of primary rat cells. J. Investig. Dermatol. Symp. Proc. 6, 69–75.

Yang, J., Weinberg, R.A., 2008. Epithelial-mesenchymal transition: at the crossroads of development and tumor metastasis. Dev. Cell 14, 818–829.

Yang, E.S., Xia, F., 2010. BRCA1 16 years later: DNA damage-induced BRCA1 shuttling. FEBS J. 277, 3079–3085.

Yang, X.H., Zou, L., 2006. Checkpoint and coordinated cellular responses to DNA damage. Results Probl. Cell Differ. 42, 65–92.

Yang, D.K., Son, C.H., Lee, S.K., Choi, P.J., Lee, K.E., Roh, M.S., 2009. Forkhead box M1 expression in pulmonary squamous cell carcinoma: correlation with clinicopathologic features and its prognostic significance. Hum. Pathol. 40, 464–470.

Yang, L., Xie, G., Fan, Q., Xie, J., 2010. Activation of the hedgehog-signaling pathway in human cancer and the clinical implications. Oncogene 29, 469–481.

Yang, P., Huang, S., Liu, D., Zhou, Q., Wen, Y.A., Xiang, Y., et al., 2011. Hepatic expression profile of forkhead transcription factor genes in normal Balb/c mice and their dynamic changes after bile duct ligation. Mol. Biol. Rep. 38, 2665–2671.

Yano, S., Yamaguchi, M., Dong, R.P., 2003. EGFR receptor kinase inhibitor "gefitinib (Iressa)" for cancer therapy. Nippon Yakurigaku Zasshi 122, 491–497.

Yao, D., Dai, C., Peng, S., 2011. Mechanism of the mesenchymal–epithelial transition and its relationship with metastatic tumor formation. Mol. Cancer Res. 9, 1608–1620.

Yap, J.L., Worlikar, S., MacKerell, A.D., Shapiro, P., Fletcher, S., 2011. Small-molecule inhibitors of the ERK signaling pathway: towards novel anticancer therapeutics. ChemMedChem 6, 38–48.

Yarden, R.I., Papa, M.Z., 2006. BRCA1 at the crossroads of multiple cellular pathways: approaches for therapeutic interventions. Mol. Cancer Ther. 5, 1396–1404.

Yau, C., Wang, Y., Zhang, Y., Foekens, J.A., Benz, C.C., 2011. Young age, increased tumor proliferation and FOXM1 expression predict early metastatic relapse only for endocrine-dependent breast cancer. Breast Cancer Res. Treat. 126, 803–810.

Ye, H., Kelly, T.F., Samadani, U., Lim, L., Rubio, S., Overdier, D.G., et al., 1997. Hepatocyte nuclear factor3/fork head homolog 11 is expressed in proliferating epithelial and mesenchymal cells of embryonic and adult tissues. Mol. Cell. Biol. 17, 1626–1641.

Ye, H., Holterman, A.X., Yoo, K.W., Franks, R.R., Costa, R.H., 1999. Premature expression of the winged helix transcription factor HFH-11B in regenerating mouse liver accelerates hepatocyte entry into S-phase. Mol. Cell. Biol. 19, 8570–8580.

Yeang, C.H., McCormick, F., Levine, A., 2008. Combinatorial patterns of somatic mutations in cancer. FASEB J. 22, 2605–2622.

Yilmaz, M., Christofori, G., 2009. EMT, the cytoskeleton, and cancer cell invasion. Cancer Metastasis Rev. 28, 15–33.

Yilmaz, M., Christofori, G., 2010. Mechanisms of motility in metastasizing cells. Mol. Cancer Res. 8, 629–642.

Yokomine, K., Senju, S., Nakatsura, T., Irie, A., Hayashida, Y., Ikuta, Y., et al., 2009. The Forkhead Box M1 transcription factor as a candidate target for anti-cancer immunotherapy. Int. J. Cancer 126, 2153–2163.

Yoshida, K., Miki, Y., 2004. Role of BRCA1 and BRCA2 as regulators of DNA repair, transcription, and cell cycle in response to DNA damage. Cancer Sci. 95, 866–871.

Yoshida, Y., Wang, I.C., Yoder, H.M., Davidson, N.O., Costa, R.H., 2007. The Forkhead Box M1 transcription factor contributes to the development and growth of mouse colorectal cancer. Gastroenterology 132, 1420–1431.

Young, R.A., 2011. Control of the embryonic stem cell state. Cell 144, 940–954.

Yousef, A.F., Fonseca, G.J., Cohen, M.J., Mymryk, J.S., 2012. The C-terminal region of E1A: a molecular toolbox for cellular cartography. Biochem. Cell Biol. 90, 153–163.

Yu, H., Jove, R., 2004. The STATs of cancer—new molecular targets come of age. Nat. Rev. Cancer 4, 97–105.

Yu, J., Thomson, J.A., 2008. Pluripotent stem cell lines. Genes Dev. 22, 1987–1997.

Yu, H., Pardoll, D., Jove, R., 2009. STAT3 in cancer inflammation and immunity: a leading role for STAT3. Nat. Rev. Cancer 9, 798–809.

Yu, J., Deshmukh, H., Payton, J.E., Dunham, C., Scheithauer, B.W., Tihan, T., et al., 2011. Array-based comparative genomic hybridization identifies CDK4 and FOXM1 alterations as independent predictors of survival in malignant peripheral nerve sheath tumor. Clin. Cancer Res. 17, 1924–1934.

Yuan, T.L., Cantley, L.C., 2008. PI3K pathway alterations in cancer: variations on a theme. Oncogene 27, 5497–5510.

Yuan, A., Chen, J.J., Yang, P.C., 2008. Pathophysiology of tumor-associated macrophages. Adv. Clin. Chem. 45, 199–223.

Yugawa, T., Kiyono, T., 2009. Molecular mechanisms of cervical carcinogenesis by high-risk human papillomaviruses: novel functions of E6 and E7 oncoproteins. Rev. Med. Virol. 19, 97–113.

Zachary, I., 2003. VEGF signalling: integration and multi-tasking in endothelial cell biology. Biochem. Soc. Trans. 31, 1171–1177.

Zavadil, J., Böttinger, E.P., 2005. TGF-β and epithelial-to-mesenchymal transitions. Oncogene 24, 5764–5774.

Zeng, Q., Hong, W., 2008. The emerging role of the Hippo pathway in cell contact inhibition, organ size control, and cancer development in mammals. Cancer Cell 13, 188–192.

Zeng, L., Imamoto, A., Rosner, M.R., 2008. Raf kinase inhibitory protein (RKIP): a physiological regulator and future therapeutic target. Expert Opin. Ther. Targets 12, 1275–1287.

Zeng, J., Wang, L., Li, Q., Li, W., Björkholm, M., Jia, J., et al., 2009. FoxM1 is up-regulated in gastric cancer and its inhibition leads to cellular senescence, partially dependent on p27kip1. J. Pathol. 218, 419–427.

Zhang, Y., 2004. The ARF-B23 connection. Implications for growth control and cancer treatment. Cell Cycle 3, 259–2632.

Zhang, J., Ma, L., 2012. MicroRNA control of epithelial-mesenchymal transition and metastasis. Cancer Metastasis Rev. 31, 653–662.

Zhang, H., Ackermann, A.M., Gusarova, G.A., Lowe, D., Feng, X., Kopsombut, U.G., et al., 2006a. The Foxm1 transcription factor is required to maintain pancreatic beta cell mass. Mol. Endocrinol. 20, 1853–1866.

Zhang, X., Zhang, H., Ye, L., 2006b. Effects of hepatitis B virus X protein on the development of liver cancer. J. Lab. Clin. Med. 147, 58–66.

Zhang, Y., Zhang, N., Dai, B., Liu, M., Sawaya, R., Xie, K., et al., 2008. FoxM1B transcriptionally regulates vascular endothelial growth factor expression and promotes angiogenesis and growth of glioma cells. Cancer Res. 68, 8733–8742.

Zhang, J., Yang, P.L., Gray, N.S., 2009. Targeting cancer with small molecule kinase inhibitors. Nat. Rev. Cancer 9, 28–39.

Zhang, H., Zhang, J., Pope, C.F., Crawford, L.A., Vasavada, R.C., Jagasia, S.M., et al., 2010a. Gestational diabetes resulting from impaired β-cell compensation in the absence of FoxM1, a novel downstream effector of placental lactogen. Diabetes 59, 143–152.

Zhang, H., Li, Y., Lai, M., 2010b. The microRNA network and tumor metastasis. Oncogene 29, 937–948.

Zhang, B., Huang, B., Guan, H., Zhang, S.M., Xu, Q.Z., He, X.P., et al., 2011a. Proteomic profiling revealed the functional networks associated with mitotic catastrophe of HepG2 hepatoma cells induced by 6-bromine-5-hydroxy-4-methoxybenzaldehyde. Toxicol. Appl. Pharmacol. 252, 307–317.

Zhang, N., Wei, P., Gong, A., Chiu, W.T., Lee, H.T., Colman, H., et al., 2011b. FoxM1 promotes β-catenin nuclear localization and controls Wnt target-gene expression and glioma tumorigenesis. Cancer Cell 20, 427–442.

Zhang, L., Li, T., Yu, D., Forman, B.M., Huang, W., 2012a. FXR protects lung from lipopolysaccharide-induced acute injury. Mol. Endocrinol. 26, 27–36.

Zhang, L., Wang, Y.D., Chen, W.D., Wang, X., Lou, G., Liu, N., et al., 2012b. Promotion of liver regeneration/repair by farnesoid X receptor in both liver and intestine. Hepatology 56, 2336–2343.

Zhang, X., Bai, Q., Kakiyama, G., Xu, L., Kin, J.K., Pandak, W.M., et al., 2012c. Cholesterol metabolite, 5-cholesten-3b, 25-diol 3-sulfate, promotes hepatic proliferation in mice. J. Steroid Biochem. Mol. Biol. 132, 262–270.

Zhang, X., Zeng, J., Zhu, M., Li, B., Zhang, Y., Huang, T., et al., 2012d. The tumor suppressor role of miR-370 by targeting FoxM1 in acute myeloid leukemia. Mol. Cancer 11, 56.

Zhang, N., Wu, X., Yang, L., Xiao, F., Zhang, H., Zhou, A., et al., 2012e. FoxM1 inhibition sensitizes resistant glioblastoma cells to temozolomide by downregulating the expression of DNA repair gene Rad51. Clin. Cancer Res. 18, 5961–5971.

Zhao, R., Daley, G.Q., 2008. From fibroblasts to iPS cells: induced pluripotency by defined factors. J. Cell. Biochem. 105, 949–955.

Zhao, F., Lam, E.W.F., 2012. Role of the forkhead transcription factor FOXO-FOXM1 axis in cancer and drug resistance. Front. Med. 6, 376–380.

Zhao, Y.Y., Gao, X.P., Zhao, Y.D., Mirza, M.K., Frey, R.S., Kalinichenko, V.V., et al., 2006. Endothelial cell-restricted disruption of FoxM1 impairs endothelial repair following LPS-induced vascular injury. J. Clin. Invest. 116, 2333–2343.

Zhao, B., Li, L., Lei, Q., Guan, K.L., 2010. The Hippo-YAP pathway in organ size control and tumorigenesis: an updated version. Genes Dev. 24, 862–874.

Zheng, Z.M., 2010. Viral oncogenes, noncoding RNAs, and RNA splicing in human tumor viruses. Int. J. Biol. Sci. 6, 730–755.

Zheng, L., Lee, W.H., 2001. The retinoblastoma gene: a prototypic and multifunctional tumor suppressor. Exp. Cell Res. 264, 2–18.

Zhivotovsky, B., Orrenius, S., 2003. Defects in the apoptotic machinery of cancer cells: role in drug resistance. Semin. Cancer Biol. 13, 125–134.

Zhou, B.B.S., Elledge, S.J., 2000. The DNA damage response: putting checkpoints in perspective. Nature 408, 433–439.

Zhou, J., Wang, C., Wang, Z., Dampier, W., Wu, K., Casimiro, M.C., et al., 2010a. Attenuation of Forkhead signaling by the retinal determination factor DACH1. Proc. Natl. Acad. Sci. U.S.A. 107, 6864–6869.

Zhou, J., Liu, Y., Zhang, W., Popov, V.M., Wang, M., Pattabiraman, N., et al., 2010b. Transcription elongation regulator 1 is a co-integrator of the cell fate determination factor Dachshund homolog 1. J. Biol. Chem. 285, 40342–40350.

Zhu, L., 2005. Tumour suppressor retinoblastoma protein Rb: a transcriptional regulator. Eur. J. Cancer 41, 2415–2427.

Zhu, W., Cai, M.Y., Tong, Z.T., Dong, S.S., Mai, S.J., Liao, Y.J., et al., 2012. Overexpression of EIF5A2 promotes colorectal carcinoma cell aggressiveness by upregulating MTA1 through C-myc to induce epithelial-mesenchymal transition. Gut 61, 562–575.

Ziegler, J.L., Bounaguro, F.M., 2009. Infectious agents and human malignancies. Front. Biosci. 14, 3455–3464.

Ziober, B.L., Silverman, S.S., Kramer, R.H., 2001. Adhesive mechanisms regulating invasion and metastasis in oral cancer. Crit. Rev. Oral Biol. Med. 12, 499–510.

Zoccoli, A., Iuliani, M., Pantano, F., Imperatori, M., Intagliata, S., Vincenzi, B., et al., 2012. Premetastatic niche: ready for new therapeutic interventions? Expert Opin. Ther. Targets 16 (Suppl. 2), S119–S129.

Zou, Y., Tsai, W.B., Cheng, C.J., Hsu, C., Chung, Y.M., Li, P.C., et al., 2008. Forkhead box transcription factor FOXO3a suppresses estrogen-dependent breast cancer cell proliferation and tumorigenesis. Breast Cancer Res. 10, R21.

zur Hausen, H., 2001a. Oncogenic DNA viruses. Oncogene 20, 7820–7823.

zur Hausen, H., 2001b. Viruses in human cancers. Curr. Sci. 81, 523–527.

zur Hausen, H., 2002. Papillomaviruses and cancer: from basic studies to clinical application. Nat. Rev. Cancer 2, 342–350.

zur Hausen, H., 2009. Papillomavirus in the causation of human cancers—a brief historical account. Virology 384, 342–350.

zur Hausen, H., 1996. Papillomavirus infections: a major cause of human cancers. Biochim. Biophys. Acta 1288, F55–F78.

Zvarski, I., Jakob, C., Schmid, P., Krebbel, H., Kaiser, M., Fleissner, C., et al., 2005. Proteasome: an emerging target for cancer therapy. Anticancer Drugs 16, 475–481.

Zvarski, I., Kleeberg, L., Kaiser, M., Fleissner, C., Heider, U., Sterz, J., et al., 2007. Proteasome as an emerging therapeutic target in cancer. Curr. Pharm. Des. 13, 471–485.

Zwerschke, W., Jansen-Dürr, P., 2000. Cell transformation by the E7 oncoprotein of human papillomavirus type 16: interactions with nuclear and cytoplasmic target proteins. Adv. Cancer Res. 78, 1–29.

CHAPTER SEVEN

Therapeutic Cancer Vaccines: Past, Present, and Future

Chunqing Guo[*,†,‡], Masoud H. Manjili[‡,§], John R. Subjeck[¶],
Devanand Sarkar[*,†,‡], Paul B. Fisher[*,†,‡], Xiang-Yang Wang[*,†,‡,1]

[*]Department of Human and Molecular Genetics, Virginia Commonwealth University School of Medicine, Richmond, Virginia, USA
[†]VCU Institute of Molecular Medicine, Virginia Commonwealth University School of Medicine, Richmond, Virginia, USA
[‡]VCU Massey Cancer Center, Virginia Commonwealth University School of Medicine, Richmond, Virginia, USA
[§]Department of Microbiology and Immunology, Virginia Commonwealth University School of Medicine, Richmond, Virginia, USA
[¶]Department of Cell Stress Biology, Roswell Park Cancer Institute, Buffalo, New York, USA
[1]Corresponding author: e-mail address: xywang@vcu.edu

Contents

Advances in Cancer Research, Volume 119
ISSN 0065-230X
http://dx.doi.org/10.1016/B978-0-12-407190-2.00007-1

Abstract

Therapeutic vaccines represent a viable option for active immunotherapy of cancers that aim to treat late stage disease by using a patient's own immune system. The promising results from clinical trials recently led to the approval of the first therapeutic cancer vaccine by the U.S. Food and Drug Administration. This major breakthrough not only provides a new treatment modality for cancer management but also paves the way for rationally designing and optimizing future vaccines with improved anticancer efficacy. Numerous vaccine strategies are currently being evaluated both preclinically and clinically. This review discusses therapeutic cancer vaccines from diverse platforms or targets as well as the preclinical and clinical studies employing these therapeutic vaccines. We also consider tumor-induced immune suppression that hinders the potency of therapeutic vaccines, and potential strategies to counteract these mechanisms for generating more robust and durable antitumor immune responses.

1. INTRODUCTION

Unlike prophylactic vaccines that are generally administered to healthy individuals, therapeutic cancer vaccines are administered to cancer patients and are designed to eradicate cancer cells through strengthening the patient's own immune responses (Lollini, Cavallo, Nanni, & Forni, 2006). The various immune effector mechanisms mobilized by therapeutic vaccination specifically attack and destroy cancer cells and spare normal cells. Thus, therapeutic cancer vaccines, in principle, may be utilized to inhibit further growth of advanced cancers and/or relapsed tumors that are refractory to conventional therapies such as surgery, radiation therapy, and chemotherapy.

In 1891, Dr. William Coley made the first attempt to stimulate the immune system for improving a cancer patient's condition by intratumoral injections of inactivated *Streptococcus pyogenes* and *Serratia marcescens* (Coley's toxin) (McCarthy, 2006). The idea came from the observation of spontaneous remissions of sarcomas in rare-cancer patients who had developed erysipelas. Despite his reported effective responses in patients, his work was viewed with skepticism by the scientific community. Today, the field of immunology has developed into a highly sophisticated specialty, and the modern science of immunology has shown that Coley's principles were correct. Indeed, the bacillus calmette–guerin (BCG) which is similar to Coley's toxin is still being used intravesically to treat superficial bladder cancer (Lamm et al., 1991; Morales, Eidinger, & Bruce, 1976; van der Meijden, Sylvester, Oosterlinck, Hoeltl, & Bono, 2003).

Despite considerable efforts to develop cancer vaccines, for decades the clinical translation of cancer vaccines into efficacious therapies has been challenging. Nonetheless, the U.S. Food and Drug Administration (FDA) have approved two prophylactic vaccines, including one for hepatitis B virus that can cause liver cancer and another for human papillomavirus accounting for about 70% of cervical cancers. More encouragingly, recent advances in cancer immunology have achieved clinical proof-of-concept of therapeutic cancer vaccine. Sipuleucel-T, an immune cell-based vaccine, resulted in increased overall survival (OS) in hormone-refractory prostate cancer patients for the first time. This led to FDA approval of this vaccine with the brand name Provenge (Dendreon) in 2010 (Cheever & Higano, 2011).

Although the challenge of developing an effective cancer vaccine remains (Schreiber, Old, & Smyth, 2011; Zhou & Levitsky, 2012), many diverse therapeutic vaccination strategies are under development or being evaluated in clinical trials. Based on their format/content, they may be classified into several major categories, which include cell vaccines (tumor or immune cell), protein/peptide vaccines, and genetic (DNA, RNA, and viral) vaccines. In this review, we present a synopsis of the history of research in the field of therapeutic cancer vaccines, as well as the current state of vaccine therapeutics for the treatment of human cancers. In addition, the obstacles for effective cancer vaccine therapy are also discussed in order to provide future directions for improvement and optimization of cancer vaccines.

2. TUMOR CELL VACCINES

2.1. Autologous tumor cell vaccines

Autologous tumor vaccines prepared using patient-derived tumor cells represent one of the first types of cancer vaccines to be tested (Hanna & Peters, 1978). These tumor cells are typically irradiated, combined with an immunostimulatory adjuvant (e.g., BCG), and then administered to the individual from whom the tumor cells were isolated (Berger, Kreutz, Horst, Baldi, & Koff, 2007; Harris et al., 2000; Maver & McKneally, 1979; Schulof et al., 1988). Autologous tumor cell vaccines have been tested in various cancers, including lung cancer (Nemunaitis, 2003; Ruttinger et al., 2007; Schulof et al., 1988), colorectal cancer (de Weger et al., 2012; Hanna, Hoover, Vermorken, Harris, & Pinedo, 2001; Harris et al., 2000; Ockert et al., 1996), melanoma (Baars et al., 2002; Berd, Maguire, McCue, & Mastrangelo, 1990; Mendez et al., 2007), renal cell cancer (Antonia et al., 2002; Fishman et al., 2008; Kinoshita et al., 2001), and prostate cancer

(Berger et al., 2007). One major advantage of whole tumor cell vaccines is its potential to present the entire spectrum of tumor-associated antigens (TAAs) to the patient's immune system. However, preparation of autologous tumor cell vaccines requires sufficient tumor specimen, which limits this technology to only certain tumor types or stages.

Autologous tumor cells may be modified to confer higher immunostimulatory characteristics. Newcastle disease virus–infected autologous tumor cells were shown to induce tumor protective immunity in multiple animal tumor models, such as ESb lymphoma and B16 melanoma (Heicappell, Schirrmacher, von Hoegen, Ahlert, & Appelhans, 1986; Plaksin et al., 1994). Clinical trials demonstrated that these modified tumor cells were safe and had a positive effect on antitumor immune memory in cancer patients (Karcher et al., 2004; Ockert et al., 1996; Schirrmacher, 2005; Steiner et al., 2004). Immunization with tumor cells engineered to express IL-12, a key cytokine promoting Th1 immunity, also resulted in strong tumor suppression in mice, accompanied by high IFN-γ production and increased activation of cytotoxic T lymphocyte (CTL) and natural killer (NK) cells (Asada et al., 2002). In a recent phase II trial, treatment with renal cell carcinoma transduced with costimulatory molecule B7-1 showed promising antitumor effect, as indicated by 3% pathologic complete response, 5% partial response, 64% stable disease, and median survival of 21.8 months (Fishman et al., 2008).

Granulocyte–macrophage colony-stimulating factor (GM-CSF)-transduced autologous tumor cell vaccines (GVAX) have been extensively studied in preclinical and clinical studies (Armstrong et al., 1996; Dong, Yoneda, Kumar, & Fidler, 1998; Dranoff et al., 1993, 1997; Dunussi-Joannopoulos et al., 1998; Levitsky et al., 1996; Sampson et al., 1996; Soiffer et al., 2003, 1998; Wakimoto et al., 1996). Mechanistic studies showed that GVAX recruits dendritic cells (DCs) for the presentation of tumor antigens and priming of CD8+ T cells (Dranoff et al., 1993; Mach et al., 2000). GVAX also stimulates the maturation of DCs by upregulating B7-1 expression (Dranoff, 2002; Mach et al., 2000). Immunization with GVAX, when combined with blockade of CTL-associated antigen 4 (CTLA-4), an immune checkpoint inhibitor (Leach, Krummel, & Allison, 1996), promotes the rejection of established murine melanoma by altering the balance of T effector cells (Teff) and T regulatory cells (Treg) (Quezada, Peggs, Curran, & Allison, 2006; van Elsas, Hurwitz, & Allison, 1999). Enhanced antitumor efficacy was evident when tumor cell vaccine engineered to express Flt3 ligand (FVAX) was combined with blockade of CTLA-4 for the treatment of TRAMP prostate adenocarcinomas (Curran & Allison, 2009). In addition to CTLA-4, the programmed death-1 (PD-1) interaction with its ligand PD-L1/L2 or B7-1 also inhibits T-cell activation and cytokine production (Butte, Keir, Phamduy, Sharpe, & Freeman, 2007). Interestingly, combined blockade of PD-1 and CTLA-4 synergized with FVAX, but not GVAX, in controlling the outgrowth of pre-established B16 tumors (Curran, Montalvo, Yagita, & Allison, 2010), suggesting that blockade of negative costimulatory pathways favors the expansion of tumor-specific T cells and the maintenance of their effector functions, resulting in shifting the immunosuppressive tumor microenvironment (TME) to an inflammatory/immunostimulatory state. Other than targeting negative immunoregulatory pathways, GVAX has been formulated with lipopolysaccharide (LPS), a TLR4 agonist, for the treatment of several murine tumors (Davis et al., 2011). Intratumoral administration of LPS-absorbed GVAX markedly improved an antitumor response in comparison with GVAX alone. This enhanced antitumor effect correlated with increased tumor infiltration by activated DCs as well as $CD8^+$ and $CD4^+$ T cells.

2.2. Allogeneic tumor cell vaccines

Allogeneic whole tumor cell vaccines, which typically contain two or three established human tumor cell lines, may be used to overcome many limitations of autologous tumor cell vaccines. These include limitless sources of tumor antigens, standardized and large-scale vaccine production, reliable analysis of clinical outcomes, easy manipulation for expression of immunostimulatory molecules, and cost effectiveness.

Canvaxin™ is an allogeneic whole-cell vaccine consisting of three melanoma lines combined with BCG as an adjuvant (Morton et al., 1992). In a phase II trial, the median OS and 5-year rate of survival were significantly higher in stage III melanoma patients receiving Canvaxin™ as postoperative adjuvant therapy compared to control group (56.4 months and 49% vs. 31.9 months and 37%; $P < 0.001$) (Morton et al., 2002). In another phase II trial in patients with completely resected disseminated stage IV melanoma, treatment with Canvaxin™ resulted in a 39% 5-year OS compared to the control arm (20%) (Hsueh et al., 2002). However, two multi-institutional randomized phase III trials in patients with stages III and IV melanoma failed to achieve a determination of vaccine efficacy, and therefore these trials were discontinued (Sondak, Sabel, & Mule, 2006).

The clinical activity of allogeneic GVAX vaccine has been evaluated for the treatment of recurrent prostate cancer (Simons et al., 2006; Small et al., 2007), breast cancer (Emens et al., 2009), and pancreatic cancer (Lutz et al., 2011). Although the phase II results of allogeneic GVAX prostate cancer vaccine trials were encouraging, phase III clinical trials that were designed to examine GVAX or GVAX in combination with chemotherapies for the treatment of metastatic castrate-resistant prostate cancer (mCRPC) failed to achieve a survival benefit and were terminated (Antonarakis & Drake, 2010; Lassi & Dawson, 2010). Despite these disappointing results, other combination strategies involving GVAX prostate cancer vaccine and anti-CTLA-4 antibodies (i.e., ipilimumab), an immunomodulating agent recently approved by the FDA for the treatment of metastatic melanoma, are still being pursued (van den Eertwegh et al., 2012; Wang, Zuo, Sarkar, & Fisher, 2011).

Initial autologous tumor cell vaccines for non-small cell lung cancer (NSCLC) encountered manufacturing failures due to the nonavailability of tumor specimens, although the results from pilot studies were positive (Hege & Carbone, 2003; Nemunaitis et al., 2004; Salgia et al., 2003). An allogeneic tumor cell vaccine (belagenpumatucel-L) consisting of four NSCLC lines engineered to secrete antisense oligonucleotide to immuno-suppressive cytokine transforming growth factor $\beta2$ (TGF-$\beta2$) provides a promising strategy for the treatment of NSCLC. A dose-related survival difference was shown in a randomized phase II trial, in which stages II–IV NSCLC patients received intradermal injections of three dose levels of belagenpumatucel-L on a monthly or alternate month schedule to a maximum of 16 injections (Nemunaitis et al., 2006, 2009). The ongoing phase III investigation (STOP trial) involves the use of belagenpumatucel-L as a maintenance therapy in patients with unresectable stages III/IV NSCLC who have responded to or have stable disease after first-line platinum-based chemotherapy. The objective is to compare the OS of subjects treated with belagenpumatucel-L versus those treated with placebo. The study commenced in July 2008 and is expected to enroll 506 patients by October 2012 (Kelly & Giaccone, 2011).

3. DC VACCINES

3.1. The biology of DCs

DCs are the most potent professional antigen-presenting cells (APCs) (Banchereau & Steinman, 1998). They act as sentinels at peripheral tissues where they uptake, process, and present pathogen- or host-derived antigenic peptides to naive T lymphocytes at the lymphoid organs in the context of major histocompatibility (MHC) molecules (Banchereau et al., 2000; Timmerman & Levy, 1999). The significance of DCs in bridging innate and adaptive immunity is well established. Indeed, many cancer immunotherapeutic strategies target DCs directly or indirectly for the induction of antigen-specific immune responses. Earlier studies showed that different DC subsets direct development of distinct T-cell populations and regulate different classes of immune responses *in vivo* (Maldonado-Lopez et al., 1999; Pulendran et al., 1999). An animal study showed that $CD8^+CD205^+$ DCs present antigens through both MHC class I and MHC class II molecules, whereas $CD8^-33D1^+$ DCs utilize the MHC class II presentation pathway (Dudziak et al., 2007). In addition, targeting antigens to DCs does not always result in immune activation, because engagement of certain receptors on DCs may induce immune suppression (Li et al., 2012). It appears that DC maturation signals are critical for avoiding the induction of T-cell tolerance or augmentation of effective antitumor immunity (Bonifaz et al., 2004; Hawiger et al., 2001; Idoyaga et al., 2008; Wang et al., 2012; Wei et al., 2009). Extensive studies on the biology of DCs demonstrate that three interactive signals are generally required for functional activation of DCs and subsequent innate and adaptive immunity against cancers, including adequate loading of MHC–peptide complexes to DCs for T-cell priming, upregulation of costimulatory molecules such as CD40, CD80, and CD86, and production of cytokines capable of polarizing a Th1/Tc1 immune responses (Frankenberger & Schendel, 2012).

3.2. *Ex vivo* generated DCs as cancer vaccines

The pioneering work of Inaba, Steinman, and colleagues on culturing mouse DCs *ex vivo* from bone marrow precursors provided the basis of the development of DC vaccines two decades ago (Inaba et al., 1992). Similarly, human DCs can be generated in culture from $CD34^+$ hematopoietic progenitors or from peripheral blood–derived monocytes (Banchereau & Palucka, 2005). Preparation of DC vaccines can be achieved by loading TAAs to patients' autologous DCs that are simultaneously treated with adjuvants. These antigen-loaded, *ex vivo* matured DCs are administered back into patients to induce antitumor immunity. Antigens utilized for this purpose include tumor-derived proteins or peptides (Banchereau et al., 2001; Murphy, Tjoa, Ragde, Kenny, & Boynton, 1996; Schuler-Thurner et al., 2002), whole tumor cells (Berard et al., 2000; Geiger et al., 2001; Palucka et al., 2006; Salcedo et al., 2006), DNA/RNA/virus (Nair et al., 2002; Steele et al., 2011; Su et al., 2005), or fusion of tumor cells and DCs (Rosenblatt et al., 2011).

One of the first trials testing the immunogenicity of DCs was conducted in metastatic prostate cancer, in which patients received autologous DCs pulsed with HLA-A0201-restricted peptides derived from prostate-specific membrane antigen. Antigen-specific cellular responses and reduced PSA levels were observed in some patients, supporting the potential use of this vaccine therapy (Murphy et al., 1996). DC vaccines have been tested in clinical trials for the treatment of prostate cancer (Kantoff, Higano, et al., 2010; Small et al., 2000, 2006), melanoma (Lesterhuis et al., 2011; Nestle et al., 1998; Palucka et al., 2006; Romano et al., 2011; Thurner et al., 1999), renal cell carcinoma (Holtl et al., 1999), and glioma (Okada et al., 2011; Yu et al., 2001). However, this autologous vaccine regimen consists of leukaphereses to isolate peripheral blood mononuclear cells (PBMCs) from the patient and cell culture processing, which limits the number of vaccinations.

The Sipuleucel-T (Provenge™) was approved by the U.S. FDA in 2010 for the treatment of asymptomatic mCRPC (Longo, 2010). This autologous vaccine consists mainly of APCs from PBMCs that have been incubated with PA2024 that contains prostatic acid phosphatase (PAP, a prostate antigen) fused to GM-CSF. Although no difference in time to progression was observed, a survival advantage was achieved, with a statistically meaningful 4.1-month improvement in median survival in the active arm with respect to the placebo arm (25.8 months vs. 21.7 months). In view of its favorable toxicity profile and manageable route of administration, the success of Sipuleucel-T as the first therapeutic cancer vaccine opens exciting new paradigms for prostate cancer and other cancers.

3.3. Modification of DCs to improve vaccine potency

Despite the clinical success of the APC-based prostate cancer vaccine, the modest antitumor efficacy of Sipuleucel-T emphasizes the need for improvement and optimization of this approach. Considering that T-cell activation is finely controlled by costimulatory molecules expressed on DCs, modification of the expression levels of activating or inhibitory molecules could enhance the DC vaccine potency. CD40 stimulation on DCs provided by activated CD4$^+$ T cells is required for DC licensing and cross-priming of CD8$^+$ T-cell responses (Quezada, Jarvinen, Lind, & Noelle, 2004). CD40L overexpression in mouse DCs via virus transduction

(Feder-Mengus et al., 2005; Kikuchi, Moore, & Crystal, 2000; Koya et al., 2003) or messenger RNA (mRNA) electroporation (Tcherepanova et al., 2008) led to elevated expression of B7 molecules and enhanced production of IL-12p70, both of which are crucial for Th1-based antitumor immunity. Similarly, CD40L-expressing human DCs also resulted in increased activation of T-cell reactivity with the poorly immunogenic tumor antigens, such as glycoprotein 100 (gp100) and Melan-A (Bonehill et al., 2009; Knippertz et al., 2009). Modulation of other costimulatory molecules, such as CD70, GITRL, 4-1BBL (CD137L), and OX40L (Bonehill et al., 2008; Dannull et al., 2005; Grunebach et al., 2005; Tuyaerts et al., 2007), or proinflammatory factors, such as IL-12p70, IL-2, IL-18, CCR7, and CXCL10, also enhances DC functions by promoting its maturation/activation, migration, and its capacity to stimulate antigen-specific Th1 and CTL responses (Iinuma et al., 2006; Kang et al., 2009; Minkis et al., 2008; Ogawa, Iinuma, & Okinaga, 2004; Okada et al., 2005).

While activating molecules expressed on DCs are involved in a proinflammatory or antitumor T-cell response, certain suppressive molecules contribute to T tolerance or suppression. The ubiquitin-editing enzyme A20 negatively regulates both Toll-like receptor (TLR) and TNF receptor signaling-induced maturation of DCs and subsequent activation of $CD4^+$ T cells and CTLs in mouse models (Song et al., 2008). A20 silencing in human DCs also facilitates the development of IFN-γ producing Th1 cells and antigen-specific $CD8^+$ T cells (Breckpot et al., 2009). Suppressor of cytokine signaling 1 (SOCS1), an immunosuppressive molecule induced by cytokines, such as IFN-γ, IL-12, IL-2, IL-7, and GM-CSF, has also been shown to inhibit DC functions through signal transducer and activator of transcription signaling and impede antitumor immunity (Palmer & Restifo, 2009; Shen, Evel-Kabler, Strube, & Chen, 2004). Recently, our studies revealed that scavenger receptor SRA/CD204 represents a newly identified immune regulator. SRA/CD204 attenuates TLR4-engaged NF-κB-TRAF6 signaling pathways in DCs (Yu et al., 2011) and downregulates the immunogenicity of DCs and CTL-mediated antitumor immunity against several mouse tumors (Wang, Facciponte, Chen, Subjeck, & Repasky, 2007; Yi et al., 2009, 2012). The absence or genetic silencing of SRA/CD204 profoundly enhances the immunostimulating, antigen-presenting functions of DCs and consequent antitumor immune responses involving IFN-γ and CTLs (Guo, Yi, Yu, Hu, et al., 2012; Guo, Yi, Yu, Zuo, et al., 2012; Yi et al., 2011). These findings support the concept of targeting SRA/CD204 as a strategy to optimize the potency of current DC vaccines that may be used alone or in combination with conventional therapies such as radiotherapy.

4. PROTEIN/PEPTIDE-BASED CANCER VACCINES

4.1. TAAs as therapeutic targets

The availability of patient's samples or specimens and the complex procedure of preparing individualized vaccines greatly limit the broad use of autologous cancer vaccines, including whole tumor cells or DCs. Recombinant vaccines, which are based on peptides from defined TAAs, and usually administered together with an adjuvant or an immune modulator, clearly have advantages. *MAGE-1* is the first gene that was reported to encode a human tumor antigen recognized by T cells (van der Bruggen et al., 1991). The identification of TAAs has provided opportunities for the design of targeted therapeutic vaccines, and these antigens may be classified into several major categories. Cancer-testis antigens, such as MAGE, BAGE, NY-ESO-1, and SSX-2, are encoded by genes that are normally silenced in adult tissues but transcriptionally reactivated in tumor cells (De Smet et al., 1994; Gnjatic et al., 2010; Hofmann et al., 2008; Karbach et al., 2011). Tissue differentiation antigens are those of normal tissue origin and shared by both normal tissue and tumors, such as melanoma (gp100, Melan-A/Mart-1, and tyrosinase) (Bakker et al., 1994; Kawakami et al., 1994; Parkhurst et al., 1998), prostate cancer (PSA, PAP) (Correale et al., 1997; Kantoff, Higano, et al., 2010), and breast carcinomas (mammaglobin-A) (Jaramillo et al., 2002). Similar to these differentiation-associated antigens, several other tumor antigens, such as CEA (Tsang et al., 1995), MUC-1 (Finn et al., 2011; Kufe, 2009), HER-2/Neu (Disis et al., 2009), tumor suppressor genes (p53) (Azuma et al., 2003), hTERT (Vonderheide, Hahn, Schultze, & Nadler, 1999), and certain antiapoptotic proteins (i.e., livin and survivin) (Schmidt et al., 2003; Schmollinger et al., 2003), are also highly elevated in tumor tissues compared to normal counterparts. Unique tumor-specific antigens are often referred to as mutated oncogenes (RAS, BRAF) (Brichard & Lejeune, 2008; Parmiani, De Filippo, Novellino, & Castelli, 2007). Targeting these tumor-specific antigens involved in driving the neoplastic process has the advantage of resistance to immunoselection with potential to be more effective. Clinical trials are underway to test vaccines that target relatively few *RAS* mutations found in colorectal and pancreatic cancers. However, the tremendous effort required for the identification of such candidate mutations may hamper their broad clinical use (Fox, Salk, & Loeb, 2009). It is also difficult to target a wide array of frameshift mutations and unique mutations that occur in individual tumors. Other antigens for potential vaccine targets include molecules (SOX-2, OCT-4) that are associated with cancer "stem cells" (Dhodapkar & Dhodapkar, 2011; Dhodapkar et al., 2010; Spisek et al., 2007) and/or the epithelial–mesenchymal transition process (Polyak & Weinberg, 2009).

Protein/peptide-based vaccines are more cost-effective than autologous or individualized vaccines. However, they also have a potential drawback because they target only one epitope or a few epitopes of the TAA. It is generally believed that induction of both antigen-specific CTLs and antigen-specific CD4$^+$ helper T cells is necessary for a cancer vaccine to be optimally efficacious. Some polypeptide vaccines (e.g., Stimuvax) potentially contain both CD4 and CD8 epitopes. Other approaches to enhance immunogenicity of a self-antigen are to alter the peptide sequence of TAAs to introduce enhancer agonist epitopes, which increase peptide binding to the MHC molecule or the T-cell receptor, resulting in higher levels of T-cell responses and/or higher avidity T cells (Dzutsev, Belyakov, Isakov, Margulies, & Berzofsky, 2007; Hodge, Chakraborty, Kudo-Saito, Garnett, & Schlom, 2005; Hou, Kavanagh, & Fong, 2008; Jordan, McMahan, Kemmler, Kappler, & Slansky, 2010; Rosenberg et al., 1998).

Most peptide-based vaccines in clinical trials target cancer-testis antigens, differentiation-associated antigens, or certain oncofetal antigens (CEA, MUC-1). Although these vaccines were able to induce antigen-specific T-cell responses, clinical outcomes have been disappointing (Buonaguro, Petrizzo, Tornesello, & Buonaguro, 2011). In the phase III study that led to the approval of ipilimumab (Hodi et al., 2010), no difference in OS was observed in patients with unresectable stages III/IV melanoma between the ipilimumab group and ipilimumab plus gp100 group. However, the encouraging results came from a recent randomized phase III trial involving patients with stage IV or locally advanced stage III cutaneous melanoma (Schwartzentruber et al., 2011). The group treated with the gp100 (210M) peptide in Montanide ISA-51 adjuvant plus IL-2 demonstrated a statistically significant improvement in overall clinical response (16% vs. 6%, $P=0.03$) as well as longer progression-free survival (2.2 months vs. 1.6 months, $P=0.008$) compared with the IL-2 group. The median OS was also longer in the gp100 peptide vaccination plus IL-2 group than in the IL-2 group (OS = 17.8 months vs. 11.1 months; $P=0.06$)

(Schwartzentruber et al., 2011). Indeed, this was the first phase III trial to demonstrate a clinical benefit for a peptide vaccine in melanoma. The unique findings in this trial were not observed in three previous independent phase II clinical trials (Sosman et al., 2008).

4.2. Immunostimulatory adjuvants for protein/peptide-based vaccines

Given that TAAs are poorly immunogenic in nature, an immunostimulatory adjuvant is essential for the generation of an effective immune response. Aluminum salts (alum) have been used as adjuvants with great success for almost a century and have been particularly effective at promoting protective humoral immunity. However, alum is not optimally effective for diseases where cell-mediated immunity is required for protection. The recognition over the past two decades that activation of innate immunity is required to drive adaptive immune responses has radically altered theories as to how adjuvants promote adaptive immunity. In particular, the pioneering work of Charles Janeway demonstrated that adaptive immune responses are preceded by, and dependent on, innate immunity receptors triggered by microbial components (Janeway, 1992). Recognition of conserved moieties associated with pathogen or pathogen-associated molecular patterns (PAMPs) via pattern recognition receptors, for example, TLRs, engages coordinated innate and adaptive immunity against microbial pathogen or infected cells (Kawai & Akira, 2011). TLR-mediated activation of APCs, for example, DCs, is a crucial step in this process. Indeed, many established and experimental vaccines incorporate PAMPs not only to protect against infectious diseases, but also as part of therapeutic immunizations against cancer (Wille-Reece et al., 2006). The use of these molecularly and functionally defined molecules as adjuvants greatly facilitates the rational design of vaccines.

Supporting this view, long-used BCG for the treatment of bladder carcinoma has been relatively effective and shown to activate TLR2 and TLR4 (Heldwein et al., 2003; Uehori et al., 2003). LPS, a natural ligand of TLR4, was reported to possess anticancer properties as early as the 1960s (Mizuno, Yoshioka, Akamatu, & Kataoka, 1968; Prigal, 1961). Monophosphoryl lipid A (MPL) is a chemically modified derivative of *S. minnesota* endotoxin that exhibits greatly reduced toxicity but maintains most of the immunostimulatory properties of LPS (Mata-Haro et al., 2007). A plethora of studies have shown that MPL potently boosts a patient's immune response against viral and TAAs (Schwarz, 2009). FDA approved the Cervarix vaccine formulated with MPL and aluminum salt as a prophylactic vaccine against human papillomavirus (Schiffman & Wacholder, 2012). Imiquimod (a TLR7 agonist) was approved by FDA in 2004 for use in humans against actinic keratosis and superficial basal cell carcinoma (Hoffman, Smith, & Renaud, 2005). These TLR agonists have strong potential in promoting the immunogenicity of weakly immunogenic TAAs. Indeed, several peptide/protein-based cancer vaccines combined with TLR agonists are being tested in clinical trials; these include Ampligen targeting TLR3 (NCT01355393), Histonol targeting TLR3 (NCT00773097, NCT01585350, NCT01437605), MELITAC 12.1 targeting TLR4 (NCT01585350), and Resiquimod targeting TLR9 (NCT00960752). The family of PRRs has greatly expanded in recent years, so there is tremendous effort being expended to investigate the role of innate immune pathways in defining the mechanisms of adjuvant action as well as the roles of other PRRs (e.g., NLR, RLR) in the adjuvant activity of therapeutic cancer vaccines.

In addition to sensing pathogen-associated signals, PRRs also recognize endogenous "alarmins," such as stress/heat-shock proteins (HSPs) and HMGB-1 (Bianchi, 2007; Lotze et al., 2007; Todryk, Melcher, Dalgleish, & Vile, 2000). As intrinsic and highly conserved protein components of the cell, these damage-associated molecular patterns also communicate the nature and magnitude of cellular injury to the host immune system. Although HSPs are known to act as molecular chaperones that participate in intracellular protein quality control (Calderwood, Murshid, & Prince, 2009; Lindquist & Craig, 1988; Mayer & Bukau, 2005), studies in the last two decades have established the concept that certain HSPs are capable of integrating both innate and adaptive immune responses and can be utilized as immunostimulatory agents for cancer immunotherapy (Calderwood, Theriault, & Gong, 2005; Murshid, Gong, & Calderwood, 2008; Srivastava, 2002a, 2002b; Wang, Facciponte, & Subjeck, 2006).

Based on the early observations of Srivastava and his colleagues that HSPs isolated from cancer cells were able to induce tumor immunity (Srivastava, DeLeo, & Old, 1986; Udono, Levey, & Srivastava, 1994), it was proposed that the immunogenicity of HSPs was primarily attributed to their ability to bind antigenic peptides and transport these peptides to APCs for T-cell priming (Srivastava, 2002a, 2005). This is consistent with the well-recognized capacity of chaperones to bind polypeptide chains in response to physiological stress (Welch, 1993). To date, antitumor immunity elicited by HSPs, including the cytosolic HSPs Hsp70, Hsp90, Hsp110, or the endoplasmic reticulum (ER) resident Grp94/gp96, Grp170, and calreticulin

(CRT), has been shown against a variety of tumors of different histologic origins such as fibrosarcomas, lung carcinomas, melanomas, colon cancers, B-cell lymphoma, and prostate cancer (Graner et al., 2000; Janetzki et al., 2000; Srivastava et al., 1986; Tamura, Peng, Liu, Daou, & Srivastava, 1997; Vanaja, Grossmann, Celis, & Young, 2000; Wang, Kazim, Repasky, & Subjeck, 2001; Yedavelli et al., 1999). Interestingly, HSPs (e.g., Hsp70) prepared from DC–tumor fusion cells were shown to stimulate an enhanced T-cell response and antitumor immunity compared to tumor-derived HSPs (Enomoto et al., 2006; Gong et al., 2010). Purification of chaperones from a cancer is believed to copurify an antigenic peptide "fingerprint" of the cell of origin. Thus, vaccination with chaperone–peptide complexes derived from a tumor circumvents the need to identify CTL epitopes from individual cancers. This unique advantage extends the use of chaperone-based immunotherapy to cancers where specific tumor antigens have not yet been characterized.

The first autologous HSP vaccine, Oncophage (also known as HSP–peptide complex 96, HSPPC-96, Vitespen), has been examined in clinical trials of various types of malignancies, including metastatic colorectal carcinoma (Mazzaferro et al., 2003; Rivoltini et al., 2003), metastatic melanoma (Pilla et al., 2006; Testori et al., 2008), non-Hodgkin lymphoma (Younes, 2003), and RCC (Jonasch et al., 2008; Wood et al., 2008). Despite the positive results from early phase trials, the phase III trial conducted in stage IV melanoma patients failed to demonstrate survival benefits (Testori et al., 2008). However, introspective analysis revealed OS benefit within the early stage IV melanoma patients (M1a, distant skin, subcutaneous or nodal metastasis; M1b, lung metastasis) (Testori et al., 2008). Similarly, no difference in recurrence-free survival between vaccination group and observation (control) group was observed in a separate phase III trial of RCC, although stages I and II patients seemed to benefit from vaccination (Wood et al., 2008). Further analysis of the data showed that patients with stages I/II and T1/2/3a RCC had a recurrence-free survival of about 45% compared with the control group (Yang, 2008). As a result, Gp96-based vaccine (Oncophage/Vitespen) was approved in 2008 by the Russian Ministry of Health for adjuvant treatment of RCC (Carlson, 2008).

To overcome technical difficulties (tumor specimen requirement, time-consuming preparation, etc.) associated with the conventional autologous HSP vaccine approaches, we have developed a chaperoning technology to formulate recombinant HSP vaccines. This platform takes advantage of the exceptional protein-holding capability of large HSPs (Hsp110,

Grp170) (Easton, Kaneko, & Subjeck, 2000; Oh, Chen, & Subjeck, 1997; Park et al., 2003; Subjeck & Shyy, 1986) and generates chaperone complexes of the large HSP and clinically relevant tumor antigens (e.g., gp100, HER-2/Neu) *in vitro* (Manjili et al., 2002, 2003; Park et al., 2006; Wang, Arnouk, et al., 2006; Wang et al., 2003; Wang, Easton, & Subjeck, 2007; Wang et al., 2010). The whole protein antigen employed in this approach contains a large reservoir of potential peptides that allow the individual's own MHC alleles to select the appropriate epitope for presentation and increases the chance of polyepitope-directed T- and B-cell responses. This synthetic approach can serve as a model to develop many different antigen targets, either alone or in combination vaccines (Wang et al., 2010). The promising preclinical results have led to a phase I clinical trial of recombinant chaperone vaccine targeting melanoma that is to be launched soon.

The immunological function of chaperones that has received the most attention thus far is the ability to shuttle peptides into the endogenous presentation pathway of professional APCs. Several receptors, for example, CD91, LOX1, SRA/CD204, and SREC, have been identified as being involved in the HSP-facilitated cross-priming event (Basu, Binder, Ramalingam, & Srivastava, 2001; Berwin, Delneste, Lovingood, Post, & Pizzo, 2004; Berwin et al., 2003; Binder, Han, & Srivastava, 2000; Delneste et al., 2002; Facciponte, Wang, & Subjeck, 2007; Gong et al., 2009; Murshid, Gong, & Calderwood, 2010; Theriault, Adachi, & Calderwood, 2006). Intriguingly, our recent work revealed that SRA/CD204 absence markedly improved the therapeutic efficacy of the Hsp110/Grp170–gp100 vaccines in mice established with B16 melanoma (Qian et al., 2011; Wang, Facciponte, et al., 2007), suggesting the complex network of HSP-binding receptors and their potential distinct effects on HSP vaccine-induced immune responses.

5. GENETIC VACCINES

Another strategy to deliver antigen or antigen fragments *in vivo* is to utilize viral or plasmid DNA vectors carrying the expression cassettes. Upon administration, they transfect somatic cells (myocytes, keratinocytes) or DCs that infiltrate muscle or skin as a part of the inflammatory response to vaccination, resulting in a subsequent cross-priming or direct antigen presentation. One major advantage of genetic vaccines is the easy delivery of multiple antigens in one immunization and activation of various arms of immunity (Aurisicchio & Ciliberto, 2012).

5.1. DNA vaccines

DNA vaccines are bacterial plasmids that are constructed to function as a shuttle system to deliver and express tumor antigen (full-length, short peptides) for generating targeted cellular and humoral immunity (Liu, 2011). The transgene is usually driven by the cytomegalovirus immediate-early promoter and its adjacent intron A sequence to ensure transcription efficiency. Elevated expression of encoded antigen can be achieved by optimization of codon-usage, such as substitution of codons for rare tRNA (Stratford et al., 2000). The backbone of bacterial DNA itself acts as PAMPs to stimulate the activation of immune cells through TLRs or other innate pattern recognition molecules (Barber, 2011; Beutler et al., 2006; Spies et al., 2003).

The ability to incorporate multiple genes into the vector creates opportunities to modulate intracellular routing and modification of antigens as well as subsequent immune outcome. Addition of a leader sequence targeting antigens to the ER induced a humoral response (Walter & Johnson, 1994) and also facilitated generation of $CD8^+$ T-cell responses, probably due to retrograde transfer of antigen from the ER to cytosol and direct delivery of DNA to APC at the immunization site (Rice, Ottensmeier, & Stevenson, 2008). Fusing the single chain Fv of idiotypic immunoglobulin to fragment C derived from tetanus toxin in DNA vaccines results in the activation of fragment C-specific $CD4^+$ T helper cells, which facilitate anti-Id B cells to produce high levels of anti-Id antibodies for immune protection against lymphoma (King et al., 1998; Spellerberg et al., 1997).

In addition, DNA vaccines can be rationally combined with other immunostimulatory agents, such as TLR agonists, to optimize antibody responses. DNA cancer vaccine targeting HER-2/Neu or CEA, when used in conjunction with a novel TLR9 agonist IMO (Aurisicchio et al., 2009), or a TLR7 agonist SM360320 (Dharmapuri et al., 2009) resulted in greater antibody titers and antibody-dependent cellular cytotoxicity activity, which led to improved control of HER-2-positive mammary carcinoma or CEA-positive colon carcinoma in murine models. In a therapeutic setting, active immunization with HER-2/Neu DNA vaccine synergized with anti-HER-2/Neu monoclonal antibodies for enhanced inhibition of established mouse breast tumors (Orlandi et al., 2011).

Achieving an effective and durable CTL response remains the ultimate goal of cancer vaccines. Generation of $CD4^+$ T cell helps via a class II

MHC-dependent pathway is important for amplification of CD8$^+$ T-cell responses and maintenance of memory during DNA vaccination (Maecker, Umetsu, DeKruyff, & Levy, 1998). Given poor immunogenicity of self-TAAs, fusion of the TAAs to nonself-antigens or molecules, such as virus X coat protein (Savelyeva, Munday, Spellerberg, Lomonossoff, & Stevenson, 2001), GFP (Wolkers, Toebes, Okabe, Haanen, & Schumacher, 2002), a modified fragment C of tetanus toxin (Rice, Buchan, & Stevenson, 2002; Rice et al., 2006; Rice, Elliott, Buchan, & Stevenson, 2001) can provide T helper signals to CTLs, resulting in enhanced cross-presentation of TAAs and antitumor immunity against several murine tumors. DNA vaccines that were designed to target tumor antigens to costimulatory B7 molecules on APCs by fusing the extracellular domain of CTLA-4 to HER-2/Neu induced protective humoral and cellular immune responses, which delayed onset of HER-2/Neu-driven mammary carcinoma (Sloots et al., 2008).

DNA vaccines have also been tested for immune targeting of stable, proliferating endothelial cells in the tumor vasculature. A DNA vaccine with the expression cassette for vascular endothelial growth factor (VEGF) receptor 2 (FLK-1) promoted CTL-mediated killing of endothelial cells, resulting in potent therapeutic efficacy against several murine tumors (melanoma, colon carcinoma, and lung carcinoma) and reducing the dissemination of pulmonary metastases (Niethammer et al., 2002). Oral administration of a xenogenic DNA vaccine encoding human tumor endothelial marker 8 effectively suppressed tumor angiogenesis and protected mice from subsequent challenge with a lethal dose of tumor cells (Ruan et al., 2009). Mice immunized with a DNA vaccine encoding human papillomavirus type-16 E7 fused with CRT developed a strong tumor-specific CD8$^+$ T-cell response and also showed a dramatic reduction in microvessel density in lung tumor nodules, suggesting that the enhanced antitumor effect involves dual immune-mediated attack of both cancer cells and endothelial cells (Cheng et al., 2001). Other DNA vaccine approaches targeting the angiostatin receptor angiomotin also augmented immune-mediated blockade of angiogenesis and tumor inhibition. Interestingly, the increased tumor vessel permeability following DNA vaccination further enhanced the antitumor effect of a chemotherapeutic agent, doxorubicin (Arigoni et al., 2012; Holmgren et al., 2006).

Although DNA vaccine platforms have shown promise in preclinical studies (Xiang, Luo, Niethammer, & Reisfeld, 2008), they fail to translate from mice and rats to nonhuman primates and humans (Liu & Ulmer, 2005;

Rice et al., 2008). DNA vaccines are facing the obstacle of translation into the clinic due to problems of efficacy rather than toxicity. However, new constructs and methods of administration may enhance their utility. In addition to subcutaneous or intradermal injection, DNA vaccines can be injected directly into the lymph nodes to increase antigen uptake by APCs and promote local inflammatory signals. This is currently being tested in phase I/II trials for melanoma and other cancers (Ribas et al., 2011; Weber et al., 2011). Other approaches or carrier modalities, including gene gun, electroporation, ultrasound, laser, liposome, microparticles, and nanoparticles, have been used to enhance antigen expression and DNA vaccine efficacy (Bins et al., 2005; Buchan et al., 2005; Dupuis et al., 2000; Greenland & Letvin, 2007).

5.2. RNA vaccines

mRNA from autologous tumor tissues can also be used to induce a specific CTL response (Carralot et al., 2005; Scheel et al., 2005; Wolff et al., 1990). Administration of total RNA as a vaccine potentially generates immune responses against various tumor antigens to reduce the possibility of tumor escape. Unlike DNA vaccines, RNA vaccines are less likely to cause side effects or autoimmune diseases because of their rapid degradation and clearance. RNA vaccination is usually carried out together with other agents for stabilization or adjuvant effects, such as liposomes or protamines (Espuelas, Roth, Thumann, Frisch, & Schuber, 2005; Fotin-Mleczek et al., 2012; Qiu, Ziegelhoffer, Sun, & Yang, 1996; Scheel et al., 2005). Chemical modification of the phosphodiester backbone (phosphorothioate RNA) can also provide a "danger" signal for stimulating the DCs through the MyD88 pathway (Scheel et al., 2004). Other modifications of RNA vaccines by integrating an RNA replicase polyprotein derived from the Semliki forest virus to generate "self-replicating" RNA (Ying et al., 1999) or using β-globin UTR to stabilize the RNA vaccine (Carralot et al., 2004) also leads to enhanced antigen–specific immune responses. RNA-based cancer vaccines have only been clinically tested in phase I/II trials with patients with melanoma (Weide et al., 2008, 2009) or RCC (Oshiumi, Matsumoto, Funami, Akazawa, & Seya, 2003).

5.3. Viral-based vaccines

The rationale for using viruses as immunization vehicles is based on the phenomenon that viral infection often results in the presentation of

MHC class I/II restricted, virus-specific peptides on infected cells. The viral vectors with low disease-causing potential and low intrinsic immuno-genicity are engineered to encode TAAs or TAAs combined with immunomodulating molecules.

The first and most extensively evaluated viral-based vectors in cancer vaccine trials are from the Poxviridae family, such as vaccinia, modified vaccinia strain Ankara (MVA), and the avipoxviruses (fowlpox and canarypox; ALVAC) (Marshall et al., 1999, 2000; Walsh & Dolin, 2011). Poxviruses have the ability to accommodate large or several transgene inserts (Moss, 1996). Poxvirus replication and transcription are restricted to the cytoplasm, which minimizes risk to the host of insertional mutagenesis. It is believed that induction of a local inflammatory response by the host TLRs and other properties of vaccinia or MVA contribute to the enhanced immune response to inserted TAAs in preclinical studies.

One promising viral cancer vaccine is PROSTVAC, developed by Bavarian Nordic. This "off-the-shelf" platform consists of a replication-competent vaccinia priming vector and a replication-incompetent fowlpox-boosting vector. Each vector contains transgenes for PSA and three costimulatory molecules (CD80, CD54, and CD58) that are collectively designated TRICOM (Hodge et al., 2005). In a double-blinded, placebo-controlled phase II trial, PROSTVAC improved median OS relative to the control vector (25.1 months vs. 16.6 months, $P=0.006$) (Kantoff, Schuetz, et al., 2010). Similar improvement in the median OS was also observed in a second PROSTVAC single-arm phase II study (Gulley et al., 2010). The pivotal phase III trial following these encouraging data from phase II studies is ongoing (NCT01322490).

Trovax is an MVA vector-based cancer vaccine targeting renal cell carcinoma antigen 5T4. Phase III clinical trials of Trovax in metastatic renal cancer patients failed to meet the primary endpoint of OS (Amato et al., 2010). Another MVA vector-based vaccine TG4010 consists of expression cassettes encoding MUC-1 antigen and IL-2. In a phase II trial of renal cell carcinoma, TG4010 combined with interferon-α2a and IL-2 resulted in 22.4 months mean OS compared with 19.3 months for all patients. MUC-1-specific CD8$^+$ T cells were associated with the prolonged survival (Oudard et al., 2011). A separate phase II trial of TG4010 combined with first-line chemotherapy (cisplatin plus gemcitabine) in advanced NSCLC demonstrated a significant 6-month increase in median survival (17.1 months in the experimental arm vs. 11.3 months in the control arm). Activated NK cells were identified as predictive biomarkers for positive clinical outcome

(Quoix et al., 2011). A confirmatory phase IIb/III trial of TG4010 for the treatment of advanced stage (IV) NSCLC is ongoing (NCT01383148).

Recombinant adenovirus is another system that can be used as carrier for genetic vaccination. Adenoviruses are easy to engineer and propagate to high yields for clinical use. They also have the advantage of transducing both dividing and nondividing cells for high expression of transgenes. Indeed, adenoviruses are used extensively as cancer gene therapeutic agents (Das et al., 2012; Liu, Hwang, Bell, & Kirn, 2008; Raty, Pikkarainen, Wirth, & Yla-Herttuala, 2008). Although clinical evaluations of adenovirus platforms have been hindered by preexisting antiviral immunity, adenovirus vectors expressing various TAAs (PSA, HER-2/Neu) are currently being tested for their immunological and clinical efficacy (NCT00583024, NCT00197522). Newer, less immunogenic variants of adenoviruses and local delivery of adenovirus-based vaccines may circumvent this issue.

Herpes simplex virus type 1 (HSV-1) is an enveloped dsDNA virus with the ability to infect a wide variety of cell types and to incorporate single or multiple transgenes. An oncolytic HSV-1 encoding GM-CSF (Oncovex$^{\text{GM-CSF}}$) for direct injection into accessible melanoma lesions resulted in a 28% objective response rate in a phase II clinical trial (Senzer et al., 2009). Responding patients demonstrated regression of both injected and noninjected lesions, highlighting a dual mechanism of action of Oncovex$^{\text{GM-CSF}}$, which includes both a direct oncolytic activity in injected tumors and a secondary immune-mediated antitumor effect. The Oncovex$^{\text{GM-CSF}}$ Pivotal Trial in Melanoma (OPTIM), randomized phase III clinical trial, has been initiated to evaluate Oncovex$^{\text{GM-CSF}}$ in patients with unresectable, metastatic melanoma (Kaufman & Bines, 2010).

Like viral vectors, bacteria and yeasts have shown utility as vaccine vehicles in preclinical studies and may also be modified for immunizing cancer patients. Attenuated recombinant *Listeria monocytogenes* has been shown to induce both innate and adaptive antitumor immune responses (Singh & Paterson, 2006, 2007). *Saccharomyces cerevisiae* is inherently nonpathogenic and can be easily engineered and propagated for the preparation of a TAA-targeted vaccine (Remondo et al., 2009; Wansley et al., 2008).

6. CANCER VACCINE THERAPY COMBINED WITH OTHER TREATMENT MODALITIES

Given the existence of such diverse vaccine platforms that potentially engage the innate and adaptive immune components, it is feasible and attractive to use combinatorial cancer vaccine therapy. In addition to cancer vaccines, a wide range of other promising immunotherapeutic modalities is being tested or approved for cancer treatment. These include adoptive cell transfer of *ex vivo* expanded tumor infiltrating lymphocytes (Rosenberg et al., 1994), use of therapeutic antibodies (e.g., trastuzumab) for antagonizing oncogenic pathways and triggering antibody-dependent cytotoxicity and phagocytosis (Disis et al., 2009; Zhou & Levitsky, 2012), and administration of immune modulating antibodies targeting both coinhibitory and costimulatory receptors on activated T cells or the corresponding ligands on APCs, as well as tumor cells to enhance antitumor immune responses (Peggs, Segal, & Allison, 2007; Wolchok, Yang, & Weber, 2010). Therefore, a consensus view in cancer immunotherapy, that applications of rational combinations of multiple modalities targeting distinct aspects of tumor and immune pathways will achieve durable antitumor effects and more effective therapeutic outcomes, is developing.

A recent study demonstrated that recombinant CEA vaccines based on different poxviral and yeast platforms activated different T–cell repertoire and cytokine profiles, resulting in enhanced antitumor activity in mice (Boehm, Higgins, Franzusoff, Schlom, & Hodge, 2010). Preclinical studies also showed that cancer DNA vaccine targeting CEA in combination with multiple costimulatory molecules (B7-1, ICAM-1, LFA-3, and GM-CSF) amplified T–cell response and greatly enhanced antitumor responses (Grosenbach, Barrientos, Schlom, & Hodge, 2001). Cancer vaccines combined with the administration of cytokines, such as IL-7 (Pellegrini et al., 2009), IFN-α (Pace et al., 2010; Sikora et al., 2009), can synergize to induce immune stimulation of DCs and T cells, as well as antagonize Treg-mediated immune suppression, which leads to optimized and improved antitumor immune efficacy.

The recent FDA approval of anti-CTLA-4 antibodies (ipilimumab) for metastatic melanoma undoubtedly supports the rational combination of this immune checkpoint inhibitor with other vaccine therapies (Hodi et al., 2010; Lipson & Drake, 2011; Wang, Zuo, et al., 2011). Although no significant difference in the OS was seen in the recent phase III trial between the ipilimumab alone group and the ipilimumab plus gp100 vaccine group (Hodi et al., 2010), ipilimumab has been shown in several preclinical and clinical studies to enhance the avidity of T cells and to enhance antitumor effects in combination with vaccines (Brahmer et al., 2010; Chakraborty, Schlom, & Hodge, 2007; Hodi et al., 2003; van Elsas et al., 1999; Yuan et al., 2008). In addition, administration of ipilimumab after vaccination with GVAX generated clinically meaningful antitumor immunity in a majority of metastatic melanoma patients (Hodi et al., 2008). Clinical trials evaluating different combinations of ipilimumab with vaccines are planned or ongoing in the adjuvant and metastatic setting for the treatment of different types of cancer (ClinicalTrials.gov Identifier: NCT01302496, NCT00124670, NCT00836407).

Other promising candidates for immune modulation to enhance clinical vaccine efficacy include antibodies against PD-1 or PD-1L1 (Brahmer et al., 2010; Curran et al., 2010; Sakuishi et al., 2010), lymphocyte-activation gene-3 (LAG-3), T-cell immunoglobulin mucin-3 (Fourcade et al., 2010; Sakuishi et al., 2010), CD40 (Advani et al., 2009; Beatty et al., 2011), and inhibitors of TGF-β (Bogdahn et al., 2011; Bueno et al., 2008). The combination of PD-1 blockade with GM-CSF-secreting tumor cell immunotherapy leads to significantly improved antitumor responses in preclinical models (Li et al., 2009).

Emerging evidence from preclinical or clinical studies also supports the idea of combining cancer vaccines with conventional therapies (radiation, chemotherapy) to achieve additive or synergistic effects, even though the dose and scheduling of the combining agent require additional studies for optimization. Certain chemotherapeutic agents (e.g., doxorubicin) can induce immunogenic cancer cell death, resulting in enhanced cross-priming of TAA-specific T cells and subsequent antitumor immunity (Apetoh et al., 2007; Ghiringhelli et al., 2009; Kepp et al., 2011; Tesniere et al., 2010; Zitvogel et al., 2010). Low doses of cyclophosphamide and doxorubicin also enhance the therapeutic efficacy of GM-CSF-secreting whole tumor cell vaccines in tumor-bearing mice and cancer patients, probably due to their ability to diminish the number of Tregs (Emens et al., 2009; Machiels et al., 2001). Docetaxel has been reported to increase the expression of TAAs, peptide–MHC complexes, and the death receptors expressed on tumor cells, thus sensitizing tumors to vaccine-induced T cell killing (Garnett, Schlom, & Hodge, 2008). In addition, certain small molecule targeted therapeutics, such as BCL-2 inhibitor (Farsaci et al., 2010), BRAF inhibitor (Boni et al., 2010), and the tyrosine kinase inhibitor sunitinib (Farsaci, Higgins, & Hodge, 2012; Finke et al., 2008; Ko et al., 2009), demonstrate the ability to enhance T-cell functions and antitumor efficacy in preclinical studies. Recent studies also showed that the mTOR inhibitor rapamycin promotes production of IL-12 and development of memory $CD8^+$ T cells, leading to enhanced vaccine potency (Araki et al., 2009; Ohtani et al., 2008; Wang, Wang, Subjeck, Shrikant, & Kim, 2011).

Local radiation not only debulks the tumor but also generates an inflammatory microenvironment, thereby promoting presentation of dying tumor-released TAAs by DCs and subsequent T-cell priming (Guo, Yi, Yu, Zuo, et al., 2012; Hodge, Guha, Neefjes, & Gulley, 2008). In addition, radiation renders tumor cells more susceptible to attack by tumor-specific CTLs (Chakraborty et al., 2003; Garnett et al., 2004; Reits et al., 2006). Indeed, radiation therapy combined with a PSA-targeted vaccine displayed a favorable toxicity profile and generated significant T-cell responses in prostate cancer patients (Gulley et al., 2005; Lechleider et al., 2008). Moreover, the preclinical and clinical evidence indicates potential benefits of hormonal therapy in combination with vaccine therapy (Arredouani et al., 2010; Mercader et al., 2001). Randomized clinical trials of PROSTVAC vaccine also suggest that vaccination combined with nilutamide hormone therapy potentially results in improved survival in patients with nonmetastatic prostate cancer (Madan et al., 2008).

7. LESSONS LEARNED FROM CANCER VACCINE TRIALS

In contrast to other cytotoxic therapies, cancer vaccines have demonstrated minimal toxicity in all clinical trials that have been reported to date. Despite expression of many target TAAs in normal tissues, little evidence of autoimmunity has been observed, with the exception of vitiligo, which is seen in patients receiving some melanoma vaccines (Banchereau et al., 2001; Luiten et al., 2005). Therapeutic cancer vaccines of different forms are being actively evaluated in the clinic. Ongoing phase III trials are summarized in Table 7.1.

Clinical studies have now shown that patients who have received less prior chemotherapy are generally more responsive to vaccines (von Mehren et al., 2001). Thus, vaccine treatment of patients with a lower tumor burden may result in significantly improved outcomes (Gulley, Madan, & Schlom, 2011), highlighting the importance of selection of appropriate patient populations to be used in randomized vaccine trials. Strikingly, the mechanism of action and kinetics of clinical responses following vaccine therapy appear to differ significantly from that of chemotherapy (Stein et al., 2011). It may be explained by the time needed to establish the immune response, which is followed by continuing tumor cell destruction and cross-priming of Teff reactive with additional TAAs. Thus, antitumor activity of vaccine-induced immune activation over a long period results in a slower tumor growth rate and improved OS, even though patients fail to

Table 7.1 Ongoing phase III trials of therapeutic cancer vaccines

Vaccines		Description	Cancer type	NCI ID
DC/ APCs	AGS-003	Autologous DCs transfected with tumor and CD40L RNAs	RCC	NCT01582672
	DCVax®-L	Autologous DCs loaded with tumor lysate	GBM	NCT00045968
	Cvac	Autologous DCs pulsed with MUC-1-mannan fusion protein	EOC	NCT01521143
Peptides/ proteins	GV1001	hTERT peptide	NSCLC	NCT01579188
	GV1001	hTERT peptide	Pancreatic cancer	NCT00425360
	NeuVax™	HER-2/Neu peptide combined with GM-CSF	Breast cancer	NCT01479244
	N/A	MAGE-A3 and NY-ESO-1 peptides combined with GM-CSF	Multiple myeloma	NCT00090493
	Stimuvax	Liposome-encapsulated synthetic peptide derived from MUC-1	NSCLC	NCT01015443
	Rindopepimut	hEGFR variant III-specific peptide conjugated to KLH	GBM	NCT01480479
	POL-103A	Protein antigens from three melanoma cell lines with alum adjuvant	Melanoma	NCT01546571
Virus vectors	PROSTVAC	Recombinant fowlpox/ vaccinia virus encoding hPSA and TRICOM	Metastatic prostate cancer	NCT01322490
	CG0070	Oncolytic adenovirus encoding GM-CSF	Bladder cancer	NCT01438112
	INGN 201	Adenovirus encoding p53	SCCHN	NCT00041613
	INGN 201	Adenovirus encoding p53 combined with cisplatin and fluorouracil	SCCHN	NCT00041626

Table 7.1 Ongoing phase III trials of therapeutic cancer vaccines—cont'd

Vaccines	Description	Cancer type	NCI ID
TG4010	Modified vaccinia virus encoding human MUC-1 and IL-2	NSCLC	NCT01383148

EGFR, epidermal growth factor receptor; EOC, epithelial ovarian cancer; GBM, glioblastoma; GM-CSF, granulocyte–macrophage colony-stimulating factor; hPSA, human prostate-specific antigen; hTERT, human telomerase reverse transcriptase; KLH, keyhole limpet hemocyanin; MUC-1, mucin 1; NSCLC, nonsmall cell lung cancer; RCC, renal cell carcinoma; SCCHN, squamous cell cancer of the head and neck; TRICOM, recombinant vaccinia virus vaccine encoding three costimulatory molecule transgenes B7.1, ICAM-1, and LFA-3.

show substantial reductions in tumor burden and an improvement in relapse-free survival (Madan, Gulley, Fojo, & Dahut, 2010). Similar findings have been reported in the clinical trials evaluating ipilimumab treatment of metastatic melanoma, in which patients treated with ipilimumab showed a statistically significant advantage in OS without a statistically significant difference in time to progression (Hodi et al., 2010).

These findings indicate that traditional response criteria may not be adequate for evaluating clinical responses to vaccine therapy or immunotherapy. Classic Response Evaluation Criteria in Solid Tumors (RECIST criteria) were initially developed to monitor patients treated with cytotoxic chemotherapies (Therasse, Eisenhauer, & Verweij, 2006). Indeed, new guidelines or "immune response criteria" for the evaluation of immunotherapeutic activity in solid tumors have been developed to better classify and evaluate clinical activity (Wolchok et al., 2009).

Numerous studies have demonstrated that analysis of the immune infiltrates in cancer biopsies and "immune signature" can serve as independent prognostic predictors for survival (Ascierto et al., 2011, 2012; Camus et al., 2009; Galon et al., 2006). Future efforts should be focused on the identification and validation of diagnostic biomarkers in response to vaccine treatment. Obtaining information on the biomarkers of immune and clinical responsiveness to effective treatment will greatly facilitate the clinical development of therapeutic cancer vaccines.

8. TUMOR-INDUCED IMMUNE SUPPRESSION AND TME

Tumor-induced immunosuppressive mechanisms in the TME are one of the major reasons for the limited current success of therapeutic cancer vaccines. The original immunosurveillance concept proposed to interpret the cross-talk between the immune system and the tumor (Burnet, 1970) has been elaborated on by Schreiber et al. proposing the cancer "immunoediting" theory (Dunn, Bruce, Ikeda, Old, & Schreiber, 2002; Dunn, Old, & Schreiber, 2004). In this model, it is suggested that tumor cells that escape initial immunosurveillance may enter an equilibrium phase where they are kept in check by the immune system; as soon as the immune response is suppressed or epigenetic changes in the quiescent tumor cells result in antigen loss or HLA loss, tumor escape and recurrence will occur. It is also believed that tumor "immunoediting" occurs during vaccination therapy of established tumors and contributes to tumor progression or relapse (Schreiber et al., 2011).

These immunosuppression mechanisms constitute the principle obstacles for the development of effective therapeutic cancer vaccines (Fig. 7.1). Tumor cells can change themselves by alterations in the antigen processing–presenting machinery, loss of antigen, or induction of anti-apoptotic mechanisms (Dunn et al., 2002; Khong & Restifo, 2002; Racanelli et al., 2010; Respa et al., 2011; Seliger et al., 2010). The lack of T-cell costimulatory molecules on most solid tumors and chronic exposure to TAAs may enable activated T cells to become anergized during activation (Kim & Ahmed, 2010). The TME contains a range of immunosuppressive leukocyte populations, including myeloid-derived suppressor cells (MDSCs), tumor-associated macrophages (TAMs), and Tregs. Analysis of PBMCs from patients with different types of cancer has also shown increased levels of MDSCs and Tregs with increased suppressive functions (Cesana et al., 2006; Vergati et al., 2011). These suppressive cells, tumor cells, and TAMs residing in the TME also release a number of immunosuppressive soluble factors, including TGF-β, IL-10, indoleamine-pyrrole 2,3 dioxygenase (IDO), galectin, and VEGF, which promote and establish an immunosuppressive state at the tumor site (Vesely, Kershaw, Schreiber, & Smyth, 2011).

MDSCs are immature myeloid cells that express CD11b and Gr-1 markers in tumor-bearing mice (Peranzoni et al., 2010). These include monocytic MDSCs and polymorphonuclear MDSCs (granulocytic MDSCs). In cancer patients, MDSCs are characterized as LIN$^-$HLA-DR$^-$CD33$^+$CD11b$^+$ cells in blood. There is a positive correlation between the frequency of MDSCs and advanced stage tumors (Diaz-Montero et al., 2009; Kusmartsev et al., 2008; Raychaudhuri et al., 2011). MDSCs inhibit T-cell activation via arginase, inducible nitric oxide synthase, reactive oxygen species, or reactive nitrogen species (Movahedi et al., 2008; Youn,

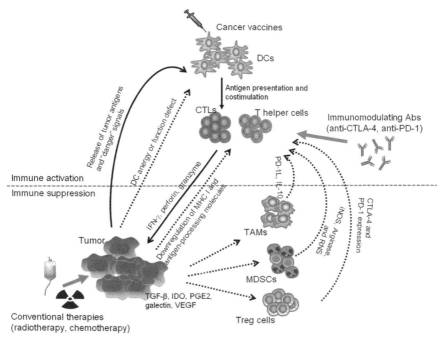

Figure 7.1 Counteracting tumor-induced immune suppression to achieve effective cancer vaccine therapy. Active immunization with therapeutic vaccines generally targets the host DCs for effective presentation of tumor-associated antigens and subsequent priming of CD8$^+$ CTLs and CD4$^+$ T helper cells. These tumor-specific T effector cells together with other innate immune cells can result in inhibition or destruction of cancer cells. In the tumor microenvironment, cancer cells produce immunosuppressive soluble factors (TGF-β, IL-10, IDO, galectin, and VEGF) and expand or recruit immune regulatory cells (MDSCs, Tregs, and TAMs), which establish an immunosuppressive state at the tumor site. This complex molecular and cellular network attenuates vaccine-induced antitumor immune responses and promotes tumor escape from immune attack. To overcome the immune suppressive mechanisms, novel immune modulators (anti-CTLA-4 and anti-PD1 antibodies) may be used to enhance vaccine potency and restore durable antitumor immunity. Cancer vaccines can also be combined with conventional cancer treatments, such as radiotherapy and chemotherapy, to engage multivalent antitumor effects for optimized therapeutic efficacy.

Nagaraj, Collazo, & Gabrilovich, 2008). Various mechanisms are involved in MDSC-mediated immune suppression, which include depletion of nutrients necessary for lymphocytes (Rodriguez et al., 2004; Srivastava, Sinha, Clements, Rodriguez, & Ostrand-Rosenberg, 2010), generation of oxidative stress to induce the loss of TCR ζ-chain expression on T cells (Schmielau & Finn, 2001) and disruption of IL-2 receptor signaling (Mazzoni et al., 2002), interference with lymphocyte trafficking (Hanson,

Clements, Sinha, Ilkovitch, & Ostrand-Rosenberg, 2009; Molon et al., 2011), and promoting activation of Tregs by CD40-CD40L ligation (Pan et al., 2010) and production of IL-10 or TGF-β (Huang et al., 2006). Contact-dependent mechanisms of T-cell suppression have also been reported in a mouse tumor model (Morales et al., 2009).

Macrophages are derived from circulating monocytes and terminally differentiate in various tissues. They express various surface markers and function differently in response to the local environmental cues (Mosser & Edwards, 2008). It is well known that TAM facilitates tumor progression and is associated with poor clinical outcomes (Mantovani & Sica, 2010; Qian & Pollard, 2010). TAMs are M2-like or alternatively activated macrophages that facilitate tumor angiogenesis and promote tumor invasion or metastasis (Lin et al., 2006; Qian et al., 2009). TAMs also promote tumor growth by producing IL-10 to drive the development of IL-4-expressing Th2 cells, which provides a positive feedback for stimulating TAM expansion (DeNardo et al., 2009). CCL22 produced by TAMs recruits Tregs to suppress CTL function (Curiel et al., 2004). Expression of PD1 ligand on monocytes/macrophages can induce apoptosis of activated T cells (Kuang et al., 2009).

Tregs not only suppress physiological and pathological immune responses against self-, nonself-, and quasi-self-tumor antigens, but are also able to attenuate antitumor functions of CD4$^+$ helper T cells, NK cells, NKT cells, and CD8$^+$ T cells (Sakaguchi, 2004; Shevach, 2002). A large number of Tregs can be recruited into the TME of tumor-bearing mice or cancer patients, due to the self-antigens released by dying tumor cells and inflammatory TME (Nishikawa et al., 2005; Pardoll, 2003). An increased presence of CD4$^+$CD25$^+$Foxp3$^+$ Tregs over CD8$^+$ T cells at the tumor site correlates with poor prognosis and therapeutic outcomes in cancer patients (Bates et al., 2006; Curiel et al., 2004; Sato et al., 2005). Although not clearly defined, expression of the inhibitory surface molecules CTLA-4 and PD-1, secretion of the immunosuppressive soluble factors TGF-β, IL-10, and IL-35 as well as certain cytolytic molecules may mediate immunosuppression by Tregs (Vignali, Collison, & Workman, 2008).

In addition to the immunosuppressive TME, the immune counteracting mechanisms engaged during cancer vaccine responses may also compromise antitumor responses and contribute to tumor escape. Tregs can be expanded in response to viral-based vaccination or multiple cycles of GVAX vaccines (LaCelle, Jensen, & Fox, 2009; Zhou, Drake, & Levitsky, 2006; Zhou &

Levitsky, 2007). Anti–CTLA-4 antibodies abrogate Treg-mediated suppression by decreasing Tregs in the TME, but expanding the overall Tregs (Kavanagh et al., 2008; Quezada et al., 2006). Widely used TLR agonists as vaccine adjuvants (Caramalho et al., 2003; Conroy, Marshall, & Mills, 2008; Crellin et al., 2005) and PD-1 blockade (Currie et al., 2009) also enhance the proliferation or amplify the suppressive function of Tregs. GM-CSF-based cancer vaccines could potentially attenuate antitumor responses by expanding MDSCs in animal models of cancer (Serafini et al., 2004) and in cancer patients (Filipazzi et al., 2007; Slingluff et al., 2009). Other factors induced by vaccination include IL-6, IL-17, and IL-1β, which drive the expansion of MDSCs due to their regulatory properties (Bunt et al., 2007; He et al., 2010; Rider et al., 2011). Therefore, innovative strategies are mandatory for overcoming these tumor-dependent and independent immunoregulatory or immunosuppressive mechanisms or pathways to achieve beneficial clinical outcomes in cancer vaccine therapy.

9. CONCLUDING REMARKS

Effective, safe, and enduring cancer treatments constitute major challenges of medical sciences, with therapeutic cancer vaccines emerging as attractive approaches for provoking long-lasting protective antitumor immunity. Recent approval of the first therapeutic cancer vaccine will pave the way for developing innovative, next generation vaccines with enhanced antitumor potency. Based on current data from clinical trials and the safety profiles of therapeutic vaccines, they will most probably be used in the adjuvant or neoadjuvant setting for the treatment of patients with minimal residual disease or more indolent metastatic disease, or those patients with a high risk of recurrence. Ultimate translation of cancer vaccines into clinically available medications with broad applications will require overcoming the immune tolerance/suppression pathways in the TME. A better understanding of host–tumor interactions and tumor immune escape mechanisms are required to develop effective cancer vaccines. Identification of unique tumor gene or protein products responsible for transformation of normal cells into tumor cells and promoting cancer progression will also uncover new potential targets for vaccine therapy. In addition, "immune signatures" will have to be established and exploited to define patient populations who will most likely respond to and benefit from vaccine therapies. Strategically combining vaccine strategies with other agents or approaches that synergistically enhance antitumor immunity and/or engage complementary antitumor responses should also lead to further improved clinical outcomes.

ACKNOWLEDGMENTS

The work was supported in part by American Cancer Society Scholarship RSG-08-187-01-LIB (X.-Y. W.), NIH CA129111 (X.-Y. W.), CA154708 (X.-Y. W.), CA097318 (P. B. F.), DOD Prostate Cancer Research Program W81XWH-10-PCRP-SIDA (P. B. F. and X.-Y. W.), and NCI Cancer Center Support Grants to Massey Cancer Center. X.-Y. W. and D. S. are Harrison Scholars and P. B. F. holds the Thelma Newmeyer Corman Chair in Cancer Research in the VCU Massey Cancer Center.
Conflict of interest statement: The authors declare no conflict of interest.

REFERENCES

Advani, R., Forero-Torres, A., Furman, R. R., Rosenblatt, J. D., Younes, A., Ren, H., et al. (2009). Phase I study of the humanized anti-CD40 monoclonal antibody dacetuzumab in refractory or recurrent non-Hodgkin's lymphoma. *Journal of Clinical Oncology: Official Journal of the American Society of Clinical Oncology, 27,* 4371–4377.

Amato, R. J., Hawkins, R. E., Kaufman, H. L., Thompson, J. A., Tomczak, P., Szczylik, C., et al. (2010). Vaccination of metastatic renal cancer patients with MVA-5T4: A randomized, double-blind, placebo-controlled phase III study. *Clinical Cancer Research: An Official Journal of the American Association for Cancer Research, 16,* 5539–5547.

Antonarakis, E. S., & Drake, C. G. (2010). Current status of immunological therapies for prostate cancer. *Current Opinion in Urology, 20,* 241–246.

Antonia, S. J., Seigne, J., Diaz, J., Muro-Cacho, C., Extermann, M., Farmelo, M. J., et al. (2002). Phase I trial of a B7-1 (CD80) gene modified autologous tumor cell vaccine in combination with systemic interleukin-2 in patients with metastatic renal cell carcinoma. *Journal of Urology, 167,* 1995–2000.

Apetoh, L., Ghiringhelli, F., Tesniere, A., Obeid, M., Ortiz, C., Criollo, A., et al. (2007). Toll-like receptor 4-dependent contribution of the immune system to anticancer chemotherapy and radiotherapy. *Nature Medicine, 13,* 1050–1059.

Araki, K., Turner, A. P., Shaffer, V. O., Gangappa, S., Keller, S. A., Bachmann, M. F., et al. (2009). mTOR regulates memory CD8 T-cell differentiation. *Nature, 460,* 108–112.

Arigoni, M., Barutello, G., Lanzardo, S., Longo, D., Aime, S., Curcio, C., et al. (2012). A vaccine targeting angiomotin induces an antibody response which alters tumor vessel permeability and hampers the growth of established tumors. *Angiogenesis, 15,* 305–316.

Armstrong, C. A., Botella, R., Galloway, T. H., Murray, N., Kramp, J. M., Song, I. S., et al. (1996). Antitumor effects of granulocyte-macrophage colony-stimulating factor production by melanoma cells. *Cancer Research, 56,* 2191–2198.

Arredouani, M. S., Tseng-Rogenski, S. S., Hollenbeck, B. K., Escara-Wilke, J., Leander, K. R., Defeo-Jones, D., et al. (2010). Androgen ablation augments human HLA2.1-restricted T cell responses to PSA self-antigen in transgenic mice. *Prostate, 70,* 1002–1011.

Asada, H., Kishida, T., Hirai, H., Satoh, E., Ohashi, S., Takeuchi, M., et al. (2002). Significant antitumor effects obtained by autologous tumor cell vaccine engineered to secrete interleukin (IL)-12 and IL-18 by means of the EBV/lipoplex. *Molecular Therapy, 5,* 609–616.

Ascierto, M. L., De Giorgi, V., Liu, Q., Bedognetti, D., Spivey, T. L., Murtas, D., et al. (2011). An immunologic portrait of cancer. *Journal of Translational Medicine, 9,* 146.

Ascierto, M. L., Kmieciak, M., Idowu, M. O., Manjili, R., Zhao, Y., Grimes, M., et al. (2012). A signature of immune function genes associated with recurrence-free survival in breast cancer patients. *Breast Cancer Research and Treatment*, *131*, 871–880.

Aurisicchio, L., & Ciliberto, G. (2012). Genetic cancer vaccines: Current status and perspectives. *Expert Opinion on Biological Therapy*, *12*, 1043–1058.

Aurisicchio, L., Peruzzi, D., Conforti, A., Dharmapuri, S., Biondo, A., Giampaoli, S., et al. (2009). Treatment of mammary carcinomas in HER-2 transgenic mice through combination of genetic vaccine and an agonist of Toll-like receptor 9. *Clinical Cancer Research: An Official Journal of the American Association for Cancer Research*, *15*, 1575–1584.

Azuma, K., Shichijo, S., Maeda, Y., Nakatsura, T., Nonaka, Y., Fujii, T., et al. (2003). Mutated p53 gene encodes a nonmutated epitope recognized by HLA-B* 4601-restricted and tumor cell-reactive CTLs at tumor site. *Cancer Research*, *63*, 854–858.

Baars, A., van Riel, J. M., Cuesta, M. A., Jaspars, E. H., Pinedo, H. M., & van den Eertwegh, A. J. (2002). Metastasectomy and active specific immunotherapy for a large single melanoma metastasis. *Hepato-Gastroenterology*, *49*, 691–693.

Bakker, A. B., Schreurs, M. W., de Boer, A. J., Kawakami, Y., Rosenberg, S. A., Adema, G. J., et al. (1994). Melanocyte lineage-specific antigen gp100 is recognized by melanoma-derived tumor-infiltrating lymphocytes. *The Journal of Experimental Medicine*, *179*, 1005–1009.

Banchereau, J., Briere, F., Caux, C., Davoust, J., Lebecque, S., Liu, Y. J., et al. (2000). Immunobiology of dendritic cells. *Annual Review of Immunology*, *18*, 767–811.

Banchereau, J., & Palucka, A. K. (2005). Dendritic cells as therapeutic vaccines against cancer. *Nature Reviews. Immunology*, *5*, 296–306.

Banchereau, J., Palucka, A. K., Dhodapkar, M., Burkeholder, S., Taquet, N., Rolland, A., et al. (2001). Immune and clinical responses in patients with metastatic melanoma to CD34(+) progenitor-derived dendritic cell vaccine. *Cancer Research*, *61*, 6451–6458.

Banchereau, J., & Steinman, R. M. (1998). Dendritic cells and the control of immunity. *Nature*, *392*, 245–252.

Barber, G. N. (2011). Cytoplasmic DNA innate immune pathways. *Immunological Reviews*, *243*, 99–108.

Basu, S., Binder, R. J., Ramalingam, T., & Srivastava, P. K. (2001). CD91 is a common receptor for heat shock proteins gp96, hsp90, hsp70, and calreticulin. *Immunity*, *14*, 303–313.

Bates, G. J., Fox, S. B., Han, C., Leek, R. D., Garcia, J. F., Harris, A. L., et al. (2006). Quantification of regulatory T cells enables the identification of high-risk breast cancer patients and those at risk of late relapse. *Journal of Clinical Oncology: Official Journal of the American Society of Clinical Oncology*, *24*, 5373–5380.

Beatty, G. L., Chiorean, E. G., Fishman, M. P., Saboury, B., Teitelbaum, U. R., Sun, W., et al. (2011). CD40 agonists alter tumor stroma and show efficacy against pancreatic carcinoma in mice and humans. *Science*, *331*, 1612–1616.

Berard, F., Blanco, P., Davoust, J., Neidhart-Berard, E. M., Nouri-Shirazi, M., Taquet, N., et al. (2000). Cross-priming of naive CD8 T cells against melanoma antigens using dendritic cells loaded with killed allogeneic melanoma cells. *The Journal of Experimental Medicine*, *192*, 1535–1544.

Berd, D., Maguire, H. C., Jr., McCue, P., & Mastrangelo, M. J. (1990). Treatment of metastatic melanoma with an autologous tumor-cell vaccine: Clinical and immunologic results in 64 patients. *Journal of Clinical Oncology: Official Journal of the American Society of Clinical Oncology*, *8*, 1858–1867.

Berger, M., Kreutz, F. T., Horst, J. L., Baldi, A. C., & Koff, W. J. (2007). Phase I study with an autologous tumor cell vaccine for locally advanced or metastatic prostate cancer. *Journal of Pharmacy and Pharmaceutical Sciences*, *10*, 144–152.

Berwin, B., Delneste, Y., Lovingood, R. V., Post, S. R., & Pizzo, S. V. (2004). SREC-I, a type F scavenger receptor, is an endocytic receptor for calreticulin. *Journal of Biological Chemistry, 279*, 51250–51257.

Berwin, B., Hart, J. P., Rice, S., Gass, C., Pizzo, S. V., Post, S. R., et al. (2003). Scavenger receptor-A mediates gp96/GRP94 and calreticulin internalization by antigen-presenting cells. *EMBO Journal, 22*, 6127–6136.

Beutler, B., Jiang, Z., Georgel, P., Crozat, K., Croker, B., Rutschmann, S., et al. (2006). Genetic analysis of host resistance: Toll-like receptor signaling and immunity at large. *Annual Review of Immunology, 24*, 353–389.

Bianchi, M. E. (2007). DAMPs, PAMPs and alarmins: All we need to know about danger. *Journal of Leukocyte Biology, 81*, 1–5.

Binder, R. J., Han, D. K., & Srivastava, P. K. (2000). CD91: A receptor for heat shock protein gp96. *Nature Immunology, 1*, 151–155.

Bins, A. D., Jorritsma, A., Wolkers, M. C., Hung, C. F., Wu, T. C., Schumacher, T. N., et al. (2005). A rapid and potent DNA vaccination strategy defined by in vivo monitoring of antigen expression. *Nature Medicine, 11*, 899–904.

Boehm, A. L., Higgins, J., Franzusoff, A., Schlom, J., & Hodge, J. W. (2010). Concurrent vaccination with two distinct vaccine platforms targeting the same antigen generates phenotypically and functionally distinct T-cell populations. *Cancer Immunology, Immunotherapy: CII, 59*, 397–408.

Bogdahn, U., Hau, P., Stockhammer, G., Venkataramana, N. K., Mahapatra, A. K., Suri, A., et al. (2011). Targeted therapy for high-grade glioma with the TGF-beta2 inhibitor trabedersen: Results of a randomized and controlled phase IIb study. *Neuro-Oncology, 13*, 132–142.

Bonehill, A., Tuyaerts, S., Van Nuffel, A. M., Heirman, C., Bos, T. J., Fostier, K., et al. (2008). Enhancing the T-cell stimulatory capacity of human dendritic cells by co-electroporation with CD40L, CD70 and constitutively active TLR4 encoding mRNA. *Molecular Therapy, 16*, 1170–1180.

Bonehill, A., Van Nuffel, A. M., Corthals, J., Tuyaerts, S., Heirman, C., Francois, V., et al. (2009). Single-step antigen loading and activation of dendritic cells by mRNA electroporation for the purpose of therapeutic vaccination in melanoma patients. *Clinical Cancer Research: An Official Journal of the American Association for Cancer Research, 15*, 3366–3375.

Boni, A., Cogdill, A. P., Dang, P., Udayakumar, D., Njauw, C. N., Sloss, C. M., et al. (2010). Selective BRAFV600E inhibition enhances T-cell recognition of melanoma without affecting lymphocyte function. *Cancer Research, 70*, 5213–5219.

Bonifaz, L. C., Bonnyay, D. P., Charalambous, A., Darguste, D. I., Fujii, S., Soares, H., et al. (2004). In vivo targeting of antigens to maturing dendritic cells via the DEC-205 receptor improves T cell vaccination. *The Journal of Experimental Medicine, 199*, 815–824.

Brahmer, J. R., Drake, C. G., Wollner, I., Powderly, J. D., Picus, J., Sharfman, W. H., et al. (2010). Phase I study of single-agent anti-programmed death-1 (MDX-1106) in refractory solid tumors: Safety, clinical activity, pharmacodynamics, and immunologic correlates. *Journal of Clinical Oncology: Official Journal of the American Society of Clinical Oncology, 28*, 3167–3175.

Breckpot, K., Aerts-Toegaert, C., Heirman, C., Peeters, U., Beyaert, R., Aerts, J. L., et al. (2009). Attenuated expression of A20 markedly increases the efficacy of double-stranded RNA-activated dendritic cells as an anti-cancer vaccine. *Journal of Immunology, 182*, 860–870.

Brichard, V. G., & Lejeune, D. (2008). Cancer immunotherapy targeting tumour-specific antigens: Towards a new therapy for minimal residual disease. *Expert Opinion on Biological Therapy, 8*, 951–968.

Buchan, S., Gronevik, E., Mathiesen, I., King, C. A., Stevenson, F. K., & Rice, J. (2005). Electroporation as a "prime/boost" strategy for naked DNA vaccination against a tumor antigen. *Journal of Immunology*, *174*, 6292–6298.

Bueno, L., de Alwis, D. P., Pitou, C., Yingling, J., Lahn, M., Glatt, S., et al. (2008). Semi-mechanistic modelling of the tumour growth inhibitory effects of LY2157299, a new type I receptor TGF-beta kinase antagonist, in mice. *European Journal of Cancer*, *44*, 142–150.

Bunt, S. K., Yang, L., Sinha, P., Clements, V. K., Leips, J., & Ostrand-Rosenberg, S. (2007). Reduced inflammation in the tumor microenvironment delays the accumulation of myeloid-derived suppressor cells and limits tumor progression. *Cancer Research*, *67*, 10019–10026.

Buonaguro, L., Petrizzo, A., Tornesello, M. L., & Buonaguro, F. M. (2011). Translating tumor antigens into cancer vaccines. *Clinical and Vaccine Immunology*, *18*, 23–34.

Burnet, F. M. (1970). The concept of immunological surveillance. *Progress in Experimental Tumor Research*, *13*, 1–27.

Butte, M. J., Keir, M. E., Phamduy, T. B., Sharpe, A. H., & Freeman, G. J. (2007). Programmed death-1 ligand 1 interacts specifically with the B7-1 costimulatory molecule to inhibit T cell responses. *Immunity*, *27*, 111–122.

Calderwood, S. K., Murshid, A., & Prince, T. (2009). The shock of aging: Molecular chaperones and the heat shock response in longevity and aging—A mini-review. *Gerontology*, *55*, 550–558.

Calderwood, S. K., Theriault, J. R., & Gong, J. (2005). Message in a bottle: Role of the 70-kDa heat shock protein family in anti-tumor immunity. *European Journal of Immunology*, *35*, 2518–2527.

Camus, M., Tosolini, M., Mlecnik, B., Pages, F., Kirilovsky, A., Berger, A., et al. (2009). Coordination of intratumoral immune reaction and human colorectal cancer recurrence. *Cancer Research*, *69*, 2685–2693.

Caramalho, I., Lopes-Carvalho, T., Ostler, D., Zelenay, S., Haury, M., & Demengeot, J. (2003). Regulatory T cells selectively express toll-like receptors and are activated by lipopolysaccharide. *The Journal of Experimental Medicine*, *197*, 403–411.

Carlson, B. (2008). Research, conferences, and FDA actions. *Biotechnology Healthcare*, *5*, 7–16.

Carralot, J. P., Probst, J., Hoerr, I., Scheel, B., Teufel, R., Jung, G., et al. (2004). Polarization of immunity induced by direct injection of naked sequence-stabilized mRNA vaccines. *Cellular and Molecular Life Sciences*, *61*, 2418–2424.

Carralot, J. P., Weide, B., Schoor, O., Probst, J., Scheel, B., Teufel, R., et al. (2005). Production and characterization of amplified tumor-derived cRNA libraries to be used as vaccines against metastatic melanomas. *Genetic Vaccines and Therapy*, *3*, 6.

Cesana, G. C., DeRaffele, G., Cohen, S., Moroziewicz, D., Mitcham, J., Stoutenburg, J., et al. (2006). Characterization of CD4+CD25+ regulatory T cells in patients treated with high-dose interleukin-2 for metastatic melanoma or renal cell carcinoma. *Journal of Clinical Oncology: Official Journal of the American Society of Clinical Oncology*, *24*, 1169–1177.

Chakraborty, M., Abrams, S. I., Camphausen, K., Liu, K., Scott, T., Coleman, C. N., et al. (2003). Irradiation of tumor cells up-regulates Fas and enhances CTL lytic activity and CTL adoptive immunotherapy. *Journal of Immunology*, *170*, 6338–6347.

Chakraborty, M., Schlom, J., & Hodge, J. W. (2007). The combined activation of positive costimulatory signals with modulation of a negative costimulatory signal for the enhancement of vaccine-mediated T-cell responses. *Cancer Immunology, Immunotherapy: CII*, *56*, 1471–1484.

Cheever, M. A., & Higano, C. S. (2011). PROVENGE (Sipuleucel-T) in prostate cancer: The first FDA-approved therapeutic cancer vaccine. *Clinical Cancer Research: An Official Journal of the American Association for Cancer Research*, *17*, 3520–3526.

Cheng, W. F., Hung, C. F., Chai, C. Y., Hsu, K. F., He, L., Ling, M., et al. (2001). Tumor-specific immunity and antiangiogenesis generated by a DNA vaccine encoding cal-reticulin linked to a tumor antigen. *The Journal of Clinical Investigation, 108*, 669–678.

Conroy, H., Marshall, N. A., & Mills, K. H. (2008). TLR ligand suppression or enhancement of Treg cells? A double-edged sword in immunity to tumours. *Oncogene, 27*, 168–180.

Correale, P., Walmsley, K., Nieroda, C., Zaremba, S., Zhu, M., Schlom, J., et al. (1997). In vitro generation of human cytotoxic T lymphocytes specific for peptides derived from prostate-specific antigen. *Journal of the National Cancer Institute, 89*, 293–300.

Crellin, N. K., Garcia, R. V., Hadisfar, O., Allan, S. E., Steiner, T. S., & Levings, M. K. (2005). Human CD4+ T cells express TLR5 and its ligand flagellin enhances the sup-pressive capacity and expression of FOXP3 in CD4+CD25+ T regulatory cells. *Journal of Immunology, 175*, 8051–8059.

Curiel, T. J., Coukos, G., Zou, L., Alvarez, X., Cheng, P., Mottram, P., et al. (2004). Spe-cific recruitment of regulatory T cells in ovarian carcinoma fosters immune privilege and predicts reduced survival. *Nature Medicine, 10*, 942–949.

Curran, M. A., & Allison, J. P. (2009). Tumor vaccines expressing flt3 ligand synergize with ctla-4 blockade to reject preimplanted tumors. *Cancer Research, 69*, 7747–7755.

Curran, M. A., Montalvo, W., Yagita, H., & Allison, J. P. (2010). PD-1 and CTLA-4 com-bination blockade expands infiltrating T cells and reduces regulatory T and myeloid cells within B16 melanoma tumors. *Proceedings of the National Academy of Sciences of the United States of America, 107*, 4275–4280.

Currie, A. J., Prosser, A., McDonnell, A., Cleaver, A. L., Robinson, B. W., Freeman, G. J., et al. (2009). Dual control of antitumor CD8 T cells through the programmed death-1/programmed death-ligand 1 pathway and immunosuppressive CD4 T cells: Regulation and counterregulation. *Journal of Immunology, 183*, 7898–7908.

Dannull, J., Nair, S., Su, Z., Boczkowski, D., DeBeck, C., Yang, B., et al. (2005). Enhancing the immunostimulatory function of dendritic cells by transfection with mRNA encoding OX40 ligand. *Blood, 105*, 3206–3213.

Das, S. K., Sarkar, S., Dash, R., Dent, P., Wang, X. Y., Sarkar, D., et al. (2012). Cancer terminator viruses and approaches for enhancing therapeutic outcomes. *Advances in Cancer Research, 115*, 1–38.

Davis, M. B., Vasquez-Dunddel, D., Fu, J., Albesiano, E., Pardoll, D., & Kim, Y. J. (2011). Intratumoral administration of TLR4 agonist absorbed into a cellular vector improves antitumor responses. *Clinical Cancer Research: An Official Journal of the American Association for Cancer Research, 17*, 3984–3992.

Delneste, Y., Magistrelli, G., Gauchat, J., Haeuw, J., Aubry, J., Nakamura, K., et al. (2002). Involvement of LOX-1 in dendritic cell-mediated antigen cross-presentation. *Immunity, 17*, 353–362.

DeNardo, D. G., Barreto, J. B., Andreu, P., Vasquez, L., Tawfik, D., Kolhatkar, N., et al. (2009). CD4(+) T cells regulate pulmonary metastasis of mammary carcinomas by enhancing protumor properties of macrophages. *Cancer Cell, 16*, 91–102.

De Smet, C., Lurquin, C., van der Bruggen, P., De Plaen, E., Brasseur, F., & Boon, T. (1994). Sequence and expression pattern of the human MAGE2 gene. *Immunogenetics, 39*, 121–129.

de Weger, V. A., Turksma, A. W., Voorham, Q. J., Euler, Z., Bril, H., van den Eertwegh, A. J., et al. (2012). Clinical effects of adjuvant active specific immunotherapy differ between patients with microsatellite-stable and microsatellite-instable colon cancer. *Clinical Cancer Research: An Official Journal of the American Association for Cancer Research, 18*, 882–889.

Dharmapuri, S., Aurisicchio, L., Neuner, P., Verdirame, M., Ciliberto, G., & La Monica, N. (2009). An oral TLR7 agonist is a potent adjuvant of DNA vaccination in transgenic mouse tumor models. *Cancer Gene Therapy, 16*, 462–472.

Dhodapkar, M. V., & Dhodapkar, K. M. (2011). Spontaneous and therapy-induced immunity to pluripotency genes in humans: Clinical implications, opportunities and challenges. *Cancer Immunology, Immunotherapy: CII, 60*, 413–418.

Dhodapkar, K. M., Feldman, D., Matthews, P., Radfar, S., Pickering, R., Turkula, S., et al. (2010). Natural immunity to pluripotency antigen OCT4 in humans. *Proceedings of the National Academy of Sciences of the United States of America, 107*, 8718–8723.

Diaz-Montero, C. M., Salem, M. L., Nishimura, M. I., Garrett-Mayer, E., Cole, D. J., & Montero, A. J. (2009). Increased circulating myeloid-derived suppressor cells correlate with clinical cancer stage, metastatic tumor burden, and doxorubicin-cyclophosphamide chemotherapy. *Cancer Immunology, Immunotherapy: CII, 58*, 49–59.

Disis, M. L., Wallace, D. R., Gooley, T. A., Dang, Y., Slota, M., Lu, H., et al. (2009). Concurrent trastuzumab and HER2/neu-specific vaccination in patients with metastatic breast cancer. *Journal of Clinical Oncology: Official Journal of the American Society of Clinical Oncology, 27*, 4685–4692.

Dong, Z., Yoneda, J., Kumar, R., & Fidler, I. J. (1998). Angiostatin-mediated suppression of cancer metastases by primary neoplasms engineered to produce granulocyte/macrophage colony-stimulating factor. *The Journal of Experimental Medicine, 188*, 755–763.

Dranoff, G. (2002). GM-CSF-based cancer vaccines. *Immunological Reviews, 188*, 147–154.

Dranoff, G., Jaffee, E., Lazenby, A., Golumbek, P., Levitsky, H., Brose, K., et al. (1993a). Vaccination with irradiated tumor cells engineered to secrete murine granulocyte-macrophage colony-stimulating factor stimulates potent, specific, and long-lasting anti-tumor immunity. *Proceedings of the National Academy of Sciences of the United States of America, 90*, 3539–3543.

Dranoff, G., Soiffer, R., Lynch, T., Mihm, M., Jung, K., Kolesar, K., et al. (1997). A phase I study of vaccination with autologous, irradiated melanoma cells engineered to secrete human granulocyte-macrophage colony stimulating factor. *Human Gene Therapy, 8*, 111–123.

Dudziak, D., Kamphorst, A. O., Heidkamp, G. F., Buchholz, V. R., Trumpfheller, C., Yamazaki, S., et al. (2007). Differential antigen processing by dendritic cell subsets in vivo. *Science, 315*, 107–111.

Dunn, G. P., Bruce, A. T., Ikeda, H., Old, L. J., & Schreiber, R. D. (2002). Cancer immunoediting: From immunosurveillance to tumor escape. *Nature Immunology, 3*, 991–998.

Dunn, G. P., Old, L. J., & Schreiber, R. D. (2004). The three Es of cancer immunoediting. *Annual Review of Immunology, 22*, 329–360.

Dunussi-Joannopoulos, K., Dranoff, G., Weinstein, H. J., Ferrara, J. L., Bierer, B. E., & Croop, J. M. (1998). Gene immunotherapy in murine acute myeloid leukemia: Granulocyte-macrophage colony-stimulating factor tumor cell vaccines elicit more potent antitumor immunity compared with B7 family and other cytokine vaccines. *Blood, 91*, 222–230.

Dupuis, M., Denis-Mize, K., Woo, C., Goldbeck, C., Selby, M. J., Chen, M., et al. (2000). Distribution of DNA vaccines determines their immunogenicity after intramuscular injection in mice. *Journal of Immunology, 165*, 2850–2858.

Dzutsev, A. H., Belyakov, I. M., Isakov, D. V., Margulies, D. H., & Berzofsky, J. A. (2007). Avidity of CD8 T cells sharpens immunodominance. *International Immunology, 19*, 497–507.

Easton, D. P., Kaneko, Y., & Subjeck, J. R. (2000). The hsp110 and Grp170 stress proteins: Newly recognized relatives of the Hsp70s. *Cell Stress & Chaperones, 5*, 276–290.

Emens, L. A., Asquith, J. M., Leatherman, J. M., Kobrin, B. J., Petrik, S., Laiko, M., et al. (2009). Timed sequential treatment with cyclophosphamide, doxorubicin, and an allogeneic granulocyte-macrophage colony-stimulating factor-secreting breast tumor vaccine: A chemotherapy dose-ranging factorial study of safety and immune activation.

Journal of Clinical Oncology: Official Journal of the American Society of Clinical Oncology, 27, 5911–5918.

Enomoto, Y., Bharti, A., Khaleque, A. A., Song, B., Liu, C., Apostolopoulos, V., et al. (2006). Enhanced immunogenicity of heat shock protein 70 peptide complexes from dendritic cell-tumor fusion cells. *Journal of Immunology, 177,* 5946–5955.

Espuelas, S., Roth, A., Thumann, C., Frisch, B., & Schuber, F. (2005). Effect of synthetic lipopeptides formulated in liposomes on the maturation of human dendritic cells. *Molecular Immunology, 42,* 721–729.

Facciponte, J. G., Wang, X. Y., & Subjeck, J. R. (2007). Hsp110 and Grp170, members of the Hsp70 superfamily, bind to scavenger receptor-A and scavenger receptor expressed by endothelial cells-I. *European Journal of Immunology, 37,* 2268–2279.

Farsaci, B., Higgins, J. P., & Hodge, J. W. (2012). Consequence of dose scheduling of sunitinib on host immune response elements and vaccine combination therapy. *International Journal of Cancer, 130,* 1948–1959.

Farsaci, B., Sabzevari, H., Higgins, J. P., Di Bari, M. G., Takai, S., Schlom, J., et al. (2010). Effect of a small molecule BCL-2 inhibitor on immune function and use with a recombinant vaccine. *International Journal of Cancer. Journal International du Cancer, 127,* 1603–1613.

Feder-Mengus, C., Schultz-Thater, E., Oertli, D., Marti, W. R., Heberer, M., Spagnoli, G. C., et al. (2005). Nonreplicating recombinant vaccinia virus expressing CD40 ligand enhances APC capacity to stimulate specific CD4+ and CD8+ T cell responses. *Human Gene Therapy, 16,* 348–360.

Filipazzi, P., Valenti, R., Huber, V., Pilla, L., Canese, P., Iero, M., et al. (2007). Identification of a new subset of myeloid suppressor cells in peripheral blood of melanoma patients with modulation by a granulocyte-macrophage colony-stimulation factor-based antitumor vaccine. *Journal of Clinical Oncology: Official Journal of the American Society of Clinical Oncology, 25,* 2546–2553.

Finke, J. H., Rini, B., Ireland, J., Rayman, P., Richmond, A., Golshayan, A., et al. (2008). Sunitinib reverses type-1 immune suppression and decreases T-regulatory cells in renal cell carcinoma patients. *Clinical Cancer Research: An Official Journal of the American Association for Cancer Research, 14,* 6674–6682.

Finn, O. J., Gantt, K. R., Lepisto, A. J., Pejawar-Gaddy, S., Xue, J., & Beatty, P. L. (2011). Importance of MUC1 and spontaneous mouse tumor models for understanding the immunobiology of human adenocarcinomas. *Immunologic Research, 50,* 261–268.

Fishman, M., Hunter, T. B., Soliman, H., Thompson, P., Dunn, M., Smilee, R., et al. (2008). Phase II trial of B7-1 (CD-86) transduced, cultured autologous tumor cell vaccine plus subcutaneous interleukin-2 for treatment of stage IV renal cell carcinoma. *Journal of Immunotherapy, 31,* 72–80.

Fotin-Mleczek, M., Zanzinger, K., Heidenreich, R., Lorenz, C., Thess, A., Duchardt, K. M., et al. (2012). Highly potent mRNA based cancer vaccines represent an attractive platform for combination therapies supporting an improved therapeutic effect. *The Journal of Gene Medicine, 14,* 428–439.

Fourcade, J., Sun, Z., Benallaoua, M., Guillaume, P., Luescher, I. F., Sander, C., et al. (2010). Upregulation of Tim-3 and PD-1 expression is associated with tumor antigen-specific CD8+ T cell dysfunction in melanoma patients. *The Journal of Experimental Medicine, 207,* 2175–2186.

Fox, E. J., Salk, J. J., & Loeb, L. A. (2009). Cancer genome sequencing—An interim analysis. *Cancer Research, 69,* 4948–4950.

Frankenberger, B., & Schendel, D. J. (2012). Third generation dendritic cell vaccines for tumor immunotherapy. *European Journal of Cell Biology, 91,* 53–58.

Galon, J., Costes, A., Sanchez-Cabo, F., Kirilovsky, A., Mlecnik, B., Lagorce-Pages, C., et al. (2006). Type, density, and location of immune cells within human colorectal tumors predict clinical outcome. *Science, 313,* 1960–1964.

Garnett, C. T., Palena, C., Chakraborty, M., Tsang, K. Y., Schlom, J., & Hodge, J. W. (2004). Sublethal irradiation of human tumor cells modulates phenotype resulting in enhanced killing by cytotoxic T lymphocytes. *Cancer Research, 64*, 7985–7994.

Garnett, C. T., Schlom, J., & Hodge, J. W. (2008). Combination of docetaxel and recombinant vaccine enhances T-cell responses and antitumor activity: Effects of docetaxel on immune enhancement. *Clinical Cancer Research: An Official Journal of the American Association for Cancer Research, 14*, 3536–3544.

Geiger, J. D., Hutchinson, R. J., Hohenkirk, L. F., McKenna, E. A., Yanik, G. A., Levine, J. E., et al. (2001). Vaccination of pediatric solid tumor patients with tumor lysate-pulsed dendritic cells can expand specific T cells and mediate tumor regression. *Cancer Research, 61*, 8513–8519.

Ghiringhelli, F., Apetoh, L., Tesniere, A., Aymeric, L., Ma, Y., Ortiz, C., et al. (2009). Activation of the NLRP3 inflammasome in dendritic cells induces IL-1beta-dependent adaptive immunity against tumors. *Nature Medicine, 15*, 1170–1178.

Gnjatic, S., Ritter, E., Buchler, M. W., Giese, N. A., Brors, B., Frei, C., et al. (2010). Seromic profiling of ovarian and pancreatic cancer. *Proceedings of the National Academy of Sciences of the United States of America, 107*, 5088–5093.

Gong, J., Zhang, Y., Durfee, J., Weng, D., Liu, C., Koido, S., et al. (2010). A heat shock protein 70-based vaccine with enhanced immunogenicity for clinical use. *Journal of Immunology, 184*, 488–496.

Gong, J., Zhu, B., Murshid, A., Adachi, H., Song, B., Lee, A., et al. (2009). T cell activation by heat shock protein 70 vaccine requires TLR signaling and scavenger receptor expressed by endothelial cells-1. *Journal of Immunology, 183*, 3092–3098.

Graner, M., Raymond, A., Romney, D., He, L., Whitesell, L., & Katsanis, E. (2000). Immunoprotective activities of multiple chaperone proteins isolated from murine B-cell leukemia/lymphoma. *Clinical Cancer Research, 6*, 909–915.

Greenland, J. R., & Letvin, N. L. (2007). Chemical adjuvants for plasmid DNA vaccines. *Vaccine, 25*, 3731–3741.

Grosenbach, D. W., Barrientos, J. C., Schlom, J., & Hodge, J. W. (2001). Synergy of vaccine strategies to amplify antigen-specific immune responses and antitumor effects. *Cancer Research, 61*, 4497–4505.

Grunebach, F., Kayser, K., Weck, M. M., Muller, M. R., Appel, S., & Brossart, P. (2005). Cotransfection of dendritic cells with RNA coding for HER-2/neu and 4-1BBL increases the induction of tumor antigen specific cytotoxic T lymphocytes. *Cancer Gene Therapy, 12*, 749–756.

Gulley, J. L., Arlen, P. M., Bastian, A., Morin, S., Marte, J., Beetham, P., et al. (2005). Combining a recombinant cancer vaccine with standard definitive radiotherapy in patients with localized prostate cancer. *Clinical Cancer Research, 11*, 3353–3362.

Gulley, J. L., Arlen, P. M., Madan, R. A., Tsang, K. Y., Pazdur, M. P., Skarupa, L., et al. (2010). Immunologic and prognostic factors associated with overall survival employing a poxviral-based PSA vaccine in metastatic castrate-resistant prostate cancer. *Cancer Immunology, Immunotherapy: CII, 59*, 663–674.

Gulley, J. L., Madan, R. A., & Schlom, J. (2011). Impact of tumour volume on the potential efficacy of therapeutic vaccines. *Current Oncology, 18*, e150–e157.

Guo, C., Yi, H., Yu, X., Hu, F., Zuo, D., Subjeck, J. R., et al. (2012). Absence of scavenger receptor A promotes dendritic cell-mediated cross-presentation of cell-associated antigen and antitumor immune response. *Immunology and Cell Biology, 90*, 101–108.

Guo, C., Yi, H., Yu, X., Zuo, D., Qian, J., Yang, G., et al. (2012). In situ vaccination with CD204 gene-silenced dendritic cell, not unmodified dendritic cell, enhances radiation therapy of prostate cancer. *Molecular Cancer Therapeutics, 11*, 2331–2341.

Hanna, M. G., Jr., Hoover, H. C., Jr., Vermorken, J. B., Harris, J. E., & Pinedo, H. M. (2001). Adjuvant active specific immunotherapy of stage II and stage III colon cancer with an autologous tumor cell vaccine: First randomized phase III trials show promise. *Vaccine, 19*, 2576–2582.

Hanna, M. G., Jr., & Peters, L. C. (1978). Specific immunotherapy of established visceral micrometastases by BCG-tumor cell vaccine alone or as an adjunct to surgery. *Cancer, 42*, 2613–2625.

Hanson, E. M., Clements, V. K., Sinha, P., Ilkovitch, D., & Ostrand-Rosenberg, S. (2009). Myeloid-derived suppressor cells down-regulate L-selectin expression on CD4+ and CD8+ T cells. *Journal of Immunology, 183*, 937–944.

Harris, J. E., Ryan, L., Hoover, H. C., Jr., Stuart, R. K., Oken, M. M., Benson, A. B., 3rd., et al. (2000). Adjuvant active specific immunotherapy for stage II and III colon cancer with an autologous tumor cell vaccine: Eastern Cooperative Oncology Group Study E5283. *Journal of Clinical Oncology: Official Journal of the American Society of Clinical Oncology, 18*, 148–157.

Hawiger, D., Inaba, K., Dorsett, Y., Guo, M., Mahnke, K., Rivera, M., et al. (2001). Dendritic cells induce peripheral T cell unresponsiveness under steady state conditions in vivo. *The Journal of Experimental Medicine, 194*, 769–779.

He, D., Li, H., Yusuf, N., Elmets, C. A., Li, J., Mountz, J. D., et al. (2010). IL-17 promotes tumor development through the induction of tumor promoting microenvironments at tumor sites and myeloid-derived suppressor cells. *Journal of Immunology, 184*, 2281–2288.

Hege, K. M., & Carbone, D. P. (2003). Lung cancer vaccines and gene therapy. *Lung Cancer, 41*(Suppl. 1), S103–S113.

Heicappell, R., Schirrmacher, V., von Hoegen, P., Ahlert, T., & Appelhans, B. (1986). Prevention of metastatic spread by postoperative immunotherapy with virally modified autologous tumor cells. I. Parameters for optimal therapeutic effects. *International Journal of Cancer, 37*, 569–577.

Heldwein, K. A., Liang, M. D., Andresen, T. K., Thomas, K. E., Marty, A. M., Cuesta, N., et al. (2003). TLR2 and TLR4 serve distinct roles in the host immune response against Mycobacterium bovis BCG. *Journal of Leukocyte Biology, 74*, 277–286.

Hodge, J. W., Chakraborty, M., Kudo-Saito, C., Garnett, C. T., & Schlom, J. (2005). Multiple costimulatory modalities enhance CTL avidity. *Journal of Immunology, 174*, 5994–6004.

Hodge, J. W., Guha, C., Neefjes, J., & Gulley, J. L. (2008). Synergizing radiation therapy and immunotherapy for curing incurable cancers. Opportunities and challenges. *Oncology (Williston Park, N.Y.), 22*, 1064–1070, discussion 1075, 1080–1081, 1084.

Hodi, F. S., Butler, M., Oble, D. A., Seiden, M. V., Haluska, F. G., Kruse, A., et al. (2008). Immunologic and clinical effects of antibody blockade of cytotoxic T lymphocyte-associated antigen 4 in previously vaccinated cancer patients. *Proceedings of the National Academy of Sciences of the United States of America, 105*, 3005–3010.

Hodi, F. S., Mihm, M. C., Soiffer, R. J., Haluska, F. G., Butler, M., Seiden, M. V., et al. (2003). Biologic activity of cytotoxic T lymphocyte-associated antigen 4 antibody blockade in previously vaccinated metastatic melanoma and ovarian carcinoma patients. *Proceedings of the National Academy of Sciences of the United States of America, 100*, 4712–4717.

Hodi, F. S., O'Day, S. J., McDermott, D. F., Weber, R. W., Sosman, J. A., Haanen, J. B., et al. (2010). Improved survival with ipilimumab in patients with metastatic melanoma. *The New England Journal of Medicine, 363*, 711–723.

Hoffman, E. S., Smith, R. E., & Renaud, R. C., Jr. (2005). From the analyst's couch: TLR-targeted therapeutics. *Nature Reviews. Drug Discovery, 4*, 879–880.

Hofmann, O., Caballero, O. L., Stevenson, B. J., Chen, Y. T., Cohen, T., Chua, R., et al. (2008). Genome-wide analysis of cancer/testis gene expression. *Proceedings of the National Academy of Sciences of the United States of America, 105*, 20422–20427.

Holmgren, L., Ambrosino, E., Birot, O., Tullus, C., Veitonmaki, N., Levchenko, T., et al. (2006). A DNA vaccine targeting angiomotin inhibits angiogenesis and suppresses tumor growth. *Proceedings of the National Academy of Sciences of the United States of America, 103,* 9208–9213.

Holtl, L., Rieser, C., Papesh, C., Ramoner, R., Herold, M., Klocker, H., et al. (1999). Cellular and humoral immune responses in patients with metastatic renal cell carcinoma after vaccination with antigen pulsed dendritic cells. *Journal of Urology, 161,* 777–782.

Hou, Y., Kavanagh, B., & Fong, L. (2008). Distinct CD8+ T cell repertoires primed with agonist and native peptides derived from a tumor-associated antigen. *Journal of Immunology, 180,* 1526–1534.

Hsueh, E. C., Essner, R., Foshag, L. J., Ollila, D. W., Gammon, G., O'Day, S. J., et al. (2002). Prolonged survival after complete resection of disseminated melanoma and active immunotherapy with a therapeutic cancer vaccine. *Journal of Clinical Oncology, 20,* 4549–4554.

Huang, B., Pan, P. Y., Li, Q., Sato, A. I., Levy, D. E., Bromberg, J., et al. (2006). Gr-1+CD115+ immature myeloid suppressor cells mediate the development of tumor-induced T regulatory cells and T-cell anergy in tumor-bearing host. *Cancer Research, 66,* 1123–1131.

Idoyaga, J., Cheong, C., Suda, K., Suda, N., Kim, J. Y., Lee, H., et al. (2008). Cutting edge: Langerin/CD207 receptor on dendritic cells mediates efficient antigen presentation on MHC I and II products in vivo. *Journal of Immunology, 180,* 3647–3650.

Iinuma, H., Okinaga, K., Fukushima, R., Inaba, T., Iwasaki, K., Okinaga, A., et al. (2006). Superior protective and therapeutic effects of IL-12 and IL-18 gene-transduced dendritic neuroblastoma fusion cells on liver metastasis of murine neuroblastoma. *Journal of Immunology, 176,* 3461–3469.

Inaba, K., Inaba, M., Romani, N., Aya, H., Deguchi, M., Ikehara, S., et al. (1992). Generation of large numbers of dendritic cells from mouse bone marrow cultures supplemented with granulocyte/macrophage colony-stimulating factor. *The Journal of Experimental Medicine, 176,* 1693–1702.

Janetzki, S., Palla, D., Rosenhauer, V., Lochs, H., Lewis, J. J., & Srivastava, P. K. (2000). Immunization of cancer patients with autologous cancer-derived heat shock protein gp96 preparations: A pilot study. *International Journal of Cancer, 88,* 232–238.

Janeway, C. A., Jr. (1992). The immune system evolved to discriminate infectious nonself from noninfectious self. *Immunology Today, 13,* 11–16.

Jaramillo, A., Majumder, K., Manna, P. P., Fleming, T. P., Doherty, G., Dipersio, J. F., et al. (2002). Identification of HLA-A3-restricted CD8+ T cell epitopes derived from mammaglobin-A, a tumor-associated antigen of human breast cancer. *International Journal of Cancer, 102,* 499–506.

Jonasch, E., Wood, C., Tamboli, P., Pagliaro, L. C., Tu, S. M., Kim, J., et al. (2008). Vaccination of metastatic renal cell carcinoma patients with autologous tumour-derived vitespen vaccine: Clinical findings. *British Journal of Cancer, 98,* 1336–1341.

Jordan, K. R., McMahan, R. H., Kemmler, C. B., Kappler, J. W., & Slansky, J. E. (2010). Peptide vaccines prevent tumor growth by activating T cells that respond to native tumor antigens. *Proceedings of the National Academy of Sciences of the United States of America, 107,* 4652–4657.

Kang, T. H., Bae, H. C., Kim, S. H., Seo, S. H., Son, S. W., Choi, E. Y., et al. (2009). Modification of dendritic cells with interferon-gamma-inducible protein-10 gene to enhance vaccine potency. *The Journal of Gene Medicine, 11,* 889–898.

Kantoff, P. W., Higano, C. S., Shore, N. D., Berger, E. R., Small, E. J., Penson, D. F., et al. (2010). Sipuleucel-T immunotherapy for castration-resistant prostate cancer. *The New England Journal of Medicine, 363,* 411–422.

Kantoff, P. W., Schuetz, T. J., Blumenstein, B. A., Glode, L. M., Bilhartz, D. L., Wyand, M., et al. (2010). Overall survival analysis of a phase II randomized controlled trial of a Poxviral-based PSA-targeted immunotherapy in metastatic castration-resistant prostate cancer. *Journal of Clinical Oncology: Official Journal of the American Society of Clinical Oncology*, *28*, 1099–1105.

Karbach, J., Neumann, A., Atmaca, A., Wahle, C., Brand, K., von Boehmer, L., et al. (2011). Efficient in vivo priming by vaccination with recombinant NY-ESO-1 protein and CpG in antigen naive prostate cancer patients. *Clinical Cancer Research: An Official Journal of the American Association for Cancer Research*, *17*, 861–870.

Karcher, J., Dyckhoff, G., Beckhove, P., Reisser, C., Brysch, M., Ziouta, Y., et al. (2004). Antitumor vaccination in patients with head and neck squamous cell carcinomas with autologous virus-modified tumor cells. *Cancer Research*, *64*, 8057–8061.

Kaufman, H. L., & Bines, S. D. (2010). OPTIM trial: A Phase III trial of an oncolytic herpes virus encoding GM-CSF for unresectable stage III or IV melanoma. *Future Oncology*, *6*, 941–949.

Kavanagh, B., O'Brien, S., Lee, D., Hou, Y., Weinberg, V., Rini, B., et al. (2008). CTLA4 blockade expands FoxP3+ regulatory and activated effector CD4+ T cells in a dose-dependent fashion. *Blood*, *112*, 1175–1183.

Kawai, T., & Akira, S. (2011). Toll-like receptors and their crosstalk with other innate receptors in infection and immunity. *Immunity*, *34*, 637–650.

Kawakami, Y., Eliyahu, S., Sakaguchi, K., Robbins, P. F., Rivoltini, L., Yannelli, J. R., et al. (1994). Identification of the immunodominant peptides of the MART-1 human melanoma antigen recognized by the majority of HLA-A2-restricted tumor infiltrating lymphocytes. *The Journal of Experimental Medicine*, *180*, 347–352.

Kelly, R. J., & Giaccone, G. (2011). Lung cancer vaccines. *Cancer Journal*, *17*, 302–308.

Kepp, O., Galluzzi, L., Martins, I., Schlemmer, F., Adjemian, S., Michaud, M., et al. (2011). Molecular determinants of immunogenic cell death elicited by anticancer chemotherapy. *Cancer Metastasis Reviews*, *30*, 61–69.

Khong, H. T., & Restifo, N. P. (2002). Natural selection of tumor variants in the generation of "tumor escape" phenotypes. *Nature Immunology*, *3*, 999–1005.

Kikuchi, T., Moore, M. A., & Crystal, R. G. (2000). Dendritic cells modified to express CD40 ligand elicit therapeutic immunity against preexisting murine tumors. *Blood*, *96*, 91–99.

Kim, P. S., & Ahmed, R. (2010). Features of responding T cells in cancer and chronic infection. *Current Opinion in Immunology*, *22*, 223–230.

King, C. A., Spellerberg, M. B., Zhu, D., Rice, J., Sahota, S. S., Thompsett, A. R., et al. (1998). DNA vaccines with single-chain Fv fused to fragment C of tetanus toxin induce protective immunity against lymphoma and myeloma. *Nature Medicine*, *4*, 1281–1286.

Kinoshita, Y., Kono, T., Yasumoto, R., Kishimoto, T., Wang, C. Y., Haas, G. P., et al. (2001). Antitumor effect on murine renal cell carcinoma by autologous tumor vaccines genetically modified with granulocyte-macrophage colony-stimulating factor and interleukin-6 cells. *Journal of Immunotherapy*, *24*, 205–211.

Knippertz, I., Hesse, A., Schunder, T., Kampgen, E., Brenner, M. K., Schuler, G., et al. (2009). Generation of human dendritic cells that simultaneously secrete IL-12 and have migratory capacity by adenoviral gene transfer of hCD40L in combination with IFN-gamma. *Journal of Immunotherapy*, *32*, 524–538.

Ko, J. S., Zea, A. H., Rini, B. I., Ireland, J. L., Elson, P., Cohen, P., et al. (2009). Sunitinib mediates reversal of myeloid-derived suppressor cell accumulation in renal cell carcinoma patients. *Clinical Cancer Research*, *15*, 2148–2157.

Koya, R. C., Kasahara, N., Favaro, P. M., Lau, R., Ta, H. Q., Weber, J. S., et al. (2003). Potent maturation of monocyte-derived dendritic cells after CD40L lentiviral gene delivery. *Journal of Immunotherapy*, *26*, 451–460.

Kuang, D. M., Zhao, Q., Peng, C., Xu, J., Zhang, J. P., Wu, C., et al. (2009). Activated monocytes in peritumoral stroma of hepatocellular carcinoma foster immune privilege and disease progression through PD-L1. *The Journal of Experimental Medicine, 206*, 1327–1337.

Kufe, D. W. (2009). Mucins in cancer: Function, prognosis and therapy. *Nature Reviews. Cancer, 9*, 874–885.

Kusmartsev, S., Su, Z., Heiser, A., Dannull, J., Eruslanov, E., Kubler, H., et al. (2008). Reversal of myeloid cell-mediated immunosuppression in patients with metastatic renal cell carcinoma. *Clinical Cancer Research, 14*, 8270–8278.

LaCelle, M. G., Jensen, S. M., & Fox, B. A. (2009). Partial CD4 depletion reduces regulatory T cells induced by multiple vaccinations and restores therapeutic efficacy. *Clinical Cancer Research: An Official Journal of the American Association for Cancer Research, 15*, 6881–6890.

Lamm, D. L., Blumenstein, B. A., Crawford, E. D., Montie, J. E., Scardino, P., Grossman, H. B., et al. (1991). A randomized trial of intravesical doxorubicin and immunotherapy with bacille Calmette-Guerin for transitional-cell carcinoma of the bladder. *The New England Journal of Medicine, 325*, 1205–1209.

Lassi, K., & Dawson, N. A. (2010). Update on castrate-resistant prostate cancer: 2010. *Current Opinion in Oncology, 22*, 263–267.

Leach, D. R., Krummel, M. F., & Allison, J. P. (1996). Enhancement of antitumor immunity by CTLA-4 blockade. *Science, 271*, 1734–1736.

Lechleider, R. J., Arlen, P. M., Tsang, K. Y., Steinberg, S. M., Yokokawa, J., Cereda, V., et al. (2008). Safety and immunologic response of a viral vaccine to prostate-specific antigen in combination with radiation therapy when metronomic-dose interleukin 2 is used as an adjuvant. *Clinical Cancer Research, 14*, 5284–5291.

Lesterhuis, W. J., de Vries, I. J., Schreibelt, G., Lambeck, A. J., Aarntzen, E. H., Jacobs, J. F., et al. (2011). Route of administration modulates the induction of dendritic cell vaccine-induced antigen-specific T cells in advanced melanoma patients. *Clinical Cancer Research, 17*, 5725–5735.

Levitsky, H. I., Montgomery, J., Ahmadzadeh, M., Staveley-O'Carroll, K., Guarnieri, F., Longo, D. L., et al. (1996). Immunization with granulocyte-macrophage colony-stimulating factor-transduced, but not B7-1-transduced, lymphoma cells primes idiotype-specific T cells and generates potent systemic antitumor immunity. *Journal of Immunology, 156*, 3858–3865.

Li, D., Romain, G., Flamar, A. L., Duluc, D., Dullaers, M., Li, X. H., et al. (2012). Targeting self- and foreign antigens to dendritic cells via DC-ASGPR generates IL-10-producing suppressive CD4+ T cells. *The Journal of Experimental Medicine, 209*, 109–121.

Li, B., VanRoey, M., Wang, C., Chen, T. H., Korman, A., & Jooss, K. (2009). Anti-programmed death-1 synergizes with granulocyte macrophage colony-stimulating factor-secreting tumor cell immunotherapy providing therapeutic benefit to mice with established tumors. *Clinical Cancer Research: An Official Journal of the American Association for Cancer Research, 15*, 1623–1634.

Lin, E. Y., Li, J. F., Gnatovskiy, L., Deng, Y., Zhu, L., Grzesik, D. A., et al. (2006). Macrophages regulate the angiogenic switch in a mouse model of breast cancer. *Cancer Research, 66*, 11238–11246.

Lindquist, S., & Craig, E. A. (1988). The heat-shock proteins. *Annual Review of Genetics, 22*, 631–677.

Lipson, E. J., & Drake, C. G. (2011). Ipilimumab: An anti-CTLA-4 antibody for metastatic melanoma. *Clinical Cancer Research: An Official Journal of the American Association for Cancer Research, 17*, 6958–6962.

Liu, M. A. (2011). DNA vaccines: An historical perspective and view to the future. *Immunological Reviews, 239*, 62–84.

Liu, T. C., Hwang, T. H., Bell, J. C., & Kirn, D. H. (2008). Translation of targeted oncolytic virotherapeutics from the lab into the clinic, and back again: A high-value iterative loop. *Molecular Therapy*, *16*, 1006–1008.

Liu, M. A., & Ulmer, J. B. (2005). Human clinical trials of plasmid DNA vaccines. *Advances in Genetics*, *55*, 25–40.

Lollini, P. L., Cavallo, F., Nanni, P., & Forni, G. (2006). Vaccines for tumour prevention. *Nature Reviews. Cancer*, *6*, 204–216.

Longo, D. L. (2010). New therapies for castration-resistant prostate cancer. *The New England Journal of Medicine*, *363*, 479–481.

Lotze, M. T., Zeh, H. J., Rubartelli, A., Sparvero, L. J., Amoscato, A. A., Washburn, N. R., et al. (2007). The grateful dead: Damage-associated molecular pattern molecules and reduction/oxidation regulate immunity. *Immunological Reviews*, *220*, 60–81.

Luiten, R. M., Kueter, E. W., Mooi, W., Gallee, M. P., Rankin, E. M., Gerritsen, W. R., et al. (2005). Immunogenicity, including vitiligo, and feasibility of vaccination with autologous GM-CSF-transduced tumor cells in metastatic melanoma patients. *Journal of Clinical Oncology: Official Journal of the American Society of Clinical Oncology*, *23*, 8978–8991.

Lutz, E., Yeo, C. J., Lillemoe, K. D., Biedrzycki, B., Kobrin, B., Herman, J., et al. (2011). A lethally irradiated allogeneic granulocyte-macrophage colony stimulating factor-secreting tumor vaccine for pancreatic adenocarcinoma. A Phase II trial of safety, efficacy, and immune activation. *Annals of Surgery*, *253*, 328–335.

Mach, N., Gillessen, S., Wilson, S. B., Sheehan, C., Mihm, M., & Dranoff, G. (2000). Differences in dendritic cells stimulated in vivo by tumors engineered to secrete granulocyte-macrophage colony-stimulating factor or Flt3-ligand. *Cancer Research*, *60*, 3239–3246.

Machiels, J. P., Reilly, R. T., Emens, L. A., Ercolini, A. M., Lei, R. Y., Weintraub, D., et al. (2001). Cyclophosphamide, doxorubicin, and paclitaxel enhance the antitumor immune response of granulocyte/macrophage-colony stimulating factor-secreting whole-cell vaccines in HER-2/neu tolerized mice. *Cancer Research*, *61*, 3689–3697.

Madan, R. A., Gulley, J. L., Fojo, T., & Dahut, W. L. (2010). Therapeutic cancer vaccines in prostate cancer: The paradox of improved survival without changes in time to progression. *The Oncologist*, *15*, 969–975.

Madan, R. A., Gulley, J. L., Schlom, J., Steinberg, S. M., Liewehr, D. J., Dahut, W. L., et al. (2008). Analysis of overall survival in patients with nonmetastatic castration-resistant prostate cancer treated with vaccine, nilutamide, and combination therapy. *Clinical Cancer Research: An Official Journal of the American Association for Cancer Research*, *14*, 4526–4531.

Maecker, H. T., Umetsu, D. T., DeKruyff, R. H., & Levy, S. (1998). Cytotoxic T cell responses to DNA vaccination: Dependence on antigen presentation via class II MHC. *Journal of Immunology*, *161*, 6532–6536.

Maldonado-Lopez, R., De Smedt, T., Michel, P., Godfroid, J., Pajak, B., Heirman, C., et al. (1999). CD8alpha+ and CD8alpha- subclasses of dendritic cells direct the development of distinct T helper cells in vivo. *The Journal of Experimental Medicine*, *189*, 587–592.

Manjili, M. H., Henderson, R., Wang, X. Y., Chen, X., Li, Y., Repasky, E., et al. (2002). Development of a recombinant HSP110-HER-2/neu vaccine using the chaperoning properties of HSP110. *Cancer Research*, *62*, 1737–1742.

Manjili, M. H., Wang, X. Y., Chen, X., Martin, T., Repasky, E. A., Henderson, R., et al. (2003). HSP110-HER2/neu chaperone complex vaccine induces protective immunity against spontaneous mammary tumors in HER-2/neu transgenic mice. *Journal of Immunology*, *171*, 4054–4061.

Mantovani, A., & Sica, A. (2010). Macrophages, innate immunity and cancer: Balance, tolerance, and diversity. *Current Opinion in Immunology*, *22*, 231–237.

Marshall, J. L., Hawkins, M. J., Tsang, K. Y., Richmond, E., Pedicano, J. E., Zhu, M. Z., et al. (1999). Phase I study in cancer patients of a replication-defective avipox recombinant vaccine that expresses human carcinoembryonic antigen. *Journal of Clinical Oncology: Official Journal of the American Society of Clinical Oncology, 17*, 332–337.

Marshall, J. L., Hoyer, R. J., Toomey, M. A., Faraguna, K., Chang, P., Richmond, E., et al. (2000). Phase I study in advanced cancer patients of a diversified prime-and-boost vaccination protocol using recombinant vaccinia virus and recombinant nonreplicating avipox virus to elicit anti-carcinoembryonic antigen immune responses. *Journal of Clinical Oncology: Official Journal of the American Society of Clinical Oncology, 18*, 3964–3973.

Mata-Haro, V., Cekic, C., Martin, M., Chilton, P. M., Casella, C. R., & Mitchell, T. C. (2007). The vaccine adjuvant monophosphoryl lipid A as a TRIF-biased agonist of TLR4. *Science, 316*, 1628–1632.

Maver, C., & McKneally, M. (1979). Preparation of autologous tumor cell vaccine from human lung cancer. *Cancer Research, 39*, 3276.

Mayer, M. P., & Bukau, B. (2005). Hsp70 chaperones: Cellular functions and molecular mechanism. *Cellular and Molecular Life Sciences, 62*, 670–684.

Mazzaferro, V., Coppa, J., Carrabba, M. G., Rivoltini, L., Schiavo, M., Regalia, E., et al. (2003). Vaccination with autologous tumor-derived heat-shock protein gp96 after liver resection for metastatic colorectal cancer. *Clinical Cancer Research, 9*, 3235–3245.

Mazzoni, A., Bronte, V., Visintin, A., Spitzer, J. H., Apolloni, E., Serafini, P., et al. (2002). Myeloid suppressor lines inhibit T cell responses by an NO-dependent mechanism. *Journal of Immunology, 168*, 689–695.

McCarthy, E. F. (2006). The toxins of William B. Coley and the treatment of bone and soft-tissue sarcomas. *The Iowa Orthopaedic Journal, 26*, 154–158.

Mendez, R., Ruiz-Cabello, F., Rodriguez, T., Del Campo, A., Paschen, A., Schadendorf, D., et al. (2007). Identification of different tumor escape mechanisms in several metastases from a melanoma patient undergoing immunotherapy. *Cancer Immunology, Immunotherapy: CII, 56*, 88–94.

Mercader, M., Bodner, B. K., Moser, M. T., Kwon, P. S., Park, E. S., Manecke, R. G., et al. (2001). T cell infiltration of the prostate induced by androgen withdrawal in patients with prostate cancer. *Proceedings of the National Academy of Sciences of the United States of America, 98*, 14565–14570.

Minkis, K., Kavanagh, D. G., Alter, G., Bogunovic, D., O'Neill, D., Adams, S., et al. (2008). Type 2 Bias of T cells expanded from the blood of melanoma patients switched to type 1 by IL-12p70 mRNA-transfected dendritic cells. *Cancer Research, 68*, 9441–9450.

Mizuno, D., Yoshioka, O., Akamatu, M., & Kataoka, T. (1968). Antitumor effect of intracutaneous injection of bacterial lipopolysaccharide. *Cancer Research, 28*, 1531–1537.

Molon, B., Ugel, S., Del Pozzo, F., Soldani, C., Zilio, S., Avella, D., et al. (2011). Chemokine nitration prevents intratumoral infiltration of antigen-specific T cells. *The Journal of Experimental Medicine, 208*, 1949–1962.

Morales, A., Eidinger, D., & Bruce, A. W. (1976). Intracavitary Bacillus Calmette-Guerin in the treatment of superficial bladder tumors. *Journal of Urology, 116*, 180–183.

Morales, J. K., Kmieciak, M., Graham, L., Feldmesser, M., Bear, H. D., & Manjili, M. H. (2009). Adoptive transfer of HER2/neu-specific T cells expanded with alternating gamma chain cytokines mediate tumor regression when combined with the depletion of myeloid-derived suppressor cells. *Cancer Immunology Immunotherapy, 58*, 941–953.

Morton, D. L., Foshag, L. J., Hoon, D. S., Nizze, J. A., Famatiga, E., Wanek, L. A., et al. (1992). Prolongation of survival in metastatic melanoma after active specific immunotherapy with a new polyvalent melanoma vaccine. *Annals of Surgery, 216*, 463–482.

Morton, D. L., Hsueh, E. C., Essner, R., Foshag, L. J., O'Day, S. J., Bilchik, A., et al. (2002). Prolonged survival of patients receiving active immunotherapy with Canvaxin therapeutic polyvalent vaccine after complete resection of melanoma metastatic to regional lymph nodes. *Annals of Surgery, 236,* 438–448, discussion 448–449.

Moss, B. (1996). Genetically engineered poxviruses for recombinant gene expression, vaccination, and safety. *Proceedings of the National Academy of Sciences of the United States of America, 93,* 11341–11348.

Mosser, D. M., & Edwards, J. P. (2008). Exploring the full spectrum of macrophage activation. *Nature Reviews. Immunology, 8,* 958–969.

Movahedi, K., Guilliams, M., Van den Bossche, J., Van den Bergh, R., Gysemans, C., Beschin, A., et al. (2008). Identification of discrete tumor-induced myeloid-derived suppressor cell subpopulations with distinct T cell-suppressive activity. *Blood, 111,* 4233–4244.

Murphy, G., Tjoa, B., Ragde, H., Kenny, G., & Boynton, A. (1996). Phase I clinical trial: T-cell therapy for prostate cancer using autologous dendritic cells pulsed with HLA-A0201-specific peptides from prostate-specific membrane antigen. *Prostate, 29,* 371–380.

Murshid, A., Gong, J., & Calderwood, S. K. (2008). Heat-shock proteins in cancer vaccines: Agents of antigen cross-presentation. *Expert Review of Vaccines, 7,* 1019–1030.

Murshid, A., Gong, J., & Calderwood, S. K. (2010). Heat shock protein 90 mediates efficient antigen cross presentation through the scavenger receptor expressed by endothelial cells-I. *Journal of Immunology, 185,* 2903–2917.

Nair, S. K., Morse, M., Boczkowski, D., Cumming, R. I., Vasovic, L., Gilboa, E., et al. (2002). Induction of tumor-specific cytotoxic T lymphocytes in cancer patients by autologous tumor RNA-transfected dendritic cells. *Annals of Surgery, 235,* 540–549.

Nemunaitis, J. (2003). Granulocyte-macrophage colony-stimulating factor gene-transfected autologous tumor cell vaccine: Focus[correction to fcous] on non-small-cell lung cancer. *Clinical Lung Cancer, 5,* 148–157.

Nemunaitis, J., Dillman, R. O., Schwarzenberger, P. O., Senzer, N., Cunningham, C., Cutler, J., et al. (2006). Phase II study of belagenpumatucel-L, a transforming growth factor beta-2 antisense gene-modified allogeneic tumor cell vaccine in non-small-cell lung cancer. *Journal of Clinical Oncology, 24,* 4721–4730.

Nemunaitis, J., Nemunaitis, M., Senzer, N., Snitz, P., Bedell, C., Kumar, P., et al. (2009). Phase II trial of Belagenpumatucel-L, a TGF-beta2 antisense gene modified allogeneic tumor vaccine in advanced non small cell lung cancer (NSCLC) patients. *Cancer Gene Therapy, 16,* 620–624.

Nemunaitis, J., Sterman, D., Jablons, D., Smith, J. W., 2nd., Fox, B., Maples, P., et al. (2004). Granulocyte-macrophage colony-stimulating factor gene-modified autologous tumor vaccines in non-small-cell lung cancer. *Journal of the National Cancer Institute, 96,* 326–331.

Nestle, F. O., Alijagic, S., Gilliet, M., Sun, Y., Grabbe, S., Dummer, R., et al. (1998). Vaccination of melanoma patients with peptide- or tumor lysate-pulsed dendritic cells. *Nature Medicine, 4,* 328–332.

Niethammer, A. G., Xiang, R., Becker, J. C., Wodrich, H., Pertl, U., Karsten, G., et al. (2002). A DNA vaccine against VEGF receptor 2 prevents effective angiogenesis and inhibits tumor growth. *Nature Medicine, 8,* 1369–1375.

Nishikawa, H., Kato, T., Tawara, I., Saito, K., Ikeda, H., Kuribayashi, K., et al. (2005). Definition of target antigens for naturally occurring CD4(+) CD25(+) regulatory T cells. *The Journal of Experimental Medicine, 201,* 681–686.

Ockert, D., Schirrmacher, V., Beck, N., Stoelben, E., Ahlert, T., Flechtenmacher, J., et al. (1996). Newcastle disease virus-infected intact autologous tumor cell vaccine for adjuvant active specific immunotherapy of resected colorectal carcinoma. *Clinical Cancer Research: An Official Journal of the American Association for Cancer Research, 2,* 21–28.

Ogawa, F., Iinuma, H., & Okinaga, K. (2004). Dendritic cell vaccine therapy by immunization with fusion cells of interleukin-2 gene-transduced, spleen-derived dendritic cells and tumour cells. *Scandinavian Journal of Immunology, 59,* 432–439.

Oh, H. J., Chen, X., & Subjeck, J. R. (1997). Hsp110 protects heat-denatured proteins and confers cellular thermoresistance. *Journal of Biological Chemistry*, *272*, 31636–31640.

Ohtani, M., Nagai, S., Kondo, S., Mizuno, S., Nakamura, K., Tanabe, M., et al. (2008). Mammalian target of rapamycin and glycogen synthase kinase 3 differentially regulate lipopolysaccharide-induced interleukin-12 production in dendritic cells. *Blood*, *112*, 635–643.

Okada, H., Kalinski, P., Ueda, R., Hoji, A., Kohanbash, G., Donegan, T. E., et al. (2011). Induction of CD8+ T-cell responses against novel glioma-associated antigen peptides and clinical activity by vaccinations with {alpha}-type 1 polarized dendritic cells and polyinosinic-polycytidylic acid stabilized by lysine and carboxymethylcellulose in patients with recurrent malignant glioma. *Journal of Clinical Oncology: Official Journal of the American Society of Clinical Oncology.*, *29*, 330–336.

Okada, N., Mori, N., Koretomo, R., Okada, Y., Nakayama, T., Yoshie, O., et al. (2005). Augmentation of the migratory ability of DC-based vaccine into regional lymph nodes by efficient CCR7 gene transduction. *Gene Therapy*, *12*, 129–139.

Orlandi, F., Guevara-Patino, J. A., Merghoub, T., Wolchok, J. D., Houghton, A. N., & Gregor, P. D. (2011). Combination of epitope-optimized DNA vaccination and passive infusion of monoclonal antibody against HER2/neu leads to breast tumor regression in mice. *Vaccine*, *29*, 3646–3654.

Oshiumi, H., Matsumoto, M., Funami, K., Akazawa, T., & Seya, T. (2003). TICAM-1, an adaptor molecule that participates in Toll-like receptor 3-mediated interferon-beta induction. *Nature Immunology*, *4*, 161–167.

Oudard, S., Rixe, O., Beuselinck, B., Linassier, C., Banu, E., Machiels, J. P., et al. (2011). A phase II study of the cancer vaccine TG4010 alone and in combination with cytokines in patients with metastatic renal clear-cell carcinoma: Clinical and immunological findings. *Cancer Immunology, Immunotherapy: CII*, *60*, 261–271.

Pace, L., Vitale, S., Dettori, B., Palombi, C., La Sorsa, V., Belardelli, F., et al. (2010). APC activation by IFN-alpha decreases regulatory T cell and enhances Th cell functions. *Journal of Immunology*, *184*, 5969–5979.

Palmer, D. C., & Restifo, N. P. (2009). Suppressors of cytokine signaling (SOCS) in T cell differentiation, maturation, and function. *Trends in Immunology*, *30*, 592–602.

Palucka, A. K., Ueno, H., Connolly, J., Kerneis-Norvell, F., Blanck, J. P., Johnston, D. A., et al. (2006). Dendritic cells loaded with killed allogeneic melanoma cells can induce objective clinical responses and MART-1 specific CD8+ T-cell immunity. *Journal of Immunotherapy*, *29*, 545–557.

Pan, P. Y., Ma, G., Weber, K. J., Ozao-Choy, J., Wang, G., Yin, B., et al. (2010). Immune stimulatory receptor CD40 is required for T-cell suppression and T regulatory cell activation mediated by myeloid-derived suppressor cells in cancer. *Cancer Research*, *70*, 99–108.

Pardoll, D. (2003). Does the immune system see tumors as foreign or self? *Annual Review of Immunology*, *21*, 807–839.

Park, J., Easton, D. P., Chen, X., MacDonald, I. J., Wang, X. Y., & Subjeck, J. R. (2003). The chaperoning properties of mouse grp170, a member of the third family of hsp70 related proteins. *Biochemistry*, *42*, 14893–14902.

Park, J., Facciponte, J. G., Chen, X., MacDonald, I. J., Repasky, E., Manjili, M. H., et al. (2006). Chaperoning function of stress protein grp170, a member of the hsp70 superfamily, is responsible for its immunoadjuvant activity. *Cancer Research*, *66*, 1161–1168.

Parkhurst, M. R., Fitzgerald, E. B., Southwood, S., Sette, A., Rosenberg, S. A., & Kawakami, Y. (1998). Identification of a shared HLA-A*0201-restricted T-cell epitope from the melanoma antigen tyrosinase-related protein 2 (TRP2). *Cancer Research*, *58*, 4895–4901.

Parmiani, G., De Filippo, A., Novellino, L., & Castelli, C. (2007). Unique human tumor antigens: Immunobiology and use in clinical trials. *Journal of Immunology*, *178*, 1975–1979.

Peggs, K. S., Segal, N. H., & Allison, J. P. (2007). Targeting immunosupportive cancer therapies: Accentuate the positive, eliminate the negative. *Cancer Cell*, *12*, 192–199.

Pellegrini, M., Calzascia, T., Elford, A. R., Shahinian, A., Lin, A. E., Dissanayake, D., et al. (2009). Adjuvant IL-7 antagonizes multiple cellular and molecular inhibitory networks to enhance immunotherapies. *Nature Medicine*, *15*, 528–536.

Peranzoni, E., Zilio, S., Marigo, I., Dolcetti, L., Zanovello, P., Mandruzzato, S., et al. (2010). Myeloid-derived suppressor cell heterogeneity and subset definition. *Current Opinion in Immunology*, *22*, 238–244.

Pilla, L., Patuzzo, R., Rivoltini, L., Maio, M., Pennacchioli, E., Lamaj, E., et al. (2006). A phase II trial of vaccination with autologous, tumor-derived heat-shock protein peptide complexes Gp96, in combination with GM-CSF and interferon-alpha in metastatic melanoma patients. *Cancer Immunology, Immunotherapy: CII*, *55*, 958–968.

Plaksin, D., Porgador, A., Vadai, E., Feldman, M., Schirrmacher, V., & Eisenbach, L. (1994). Effective anti-metastatic melanoma vaccination with tumor cells transfected with MHC genes and/or infected with Newcastle disease virus (NDV). *International Journal of Cancer*, *59*, 796–801.

Polyak, K., & Weinberg, R. A. (2009). Transitions between epithelial and mesenchymal states: Acquisition of malignant and stem cell traits. *Nature Reviews. Cancer*, *9*, 265–273.

Prigal, S. J. (1961). Development in mice of prolonged non-specific resistance to sarcoma implant and Staphylococcus infection following repository injection of lipopolysaccharide. *Nature*, *191*, 1111–1112.

Pulendran, B., Smith, J. L., Caspary, G., Brasel, K., Pettit, D., Maraskovsky, E., et al. (1999). Distinct dendritic cell subsets differentially regulate the class of immune response in vivo. *Proceedings of the National Academy of Sciences of the United States of America*, *96*, 1036–1041.

Qian, B., Deng, Y., Im, J. H., Muschel, R. J., Zou, Y., Li, J., et al. (2009). A distinct macrophage population mediates metastatic breast cancer cell extravasation, establishment and growth. *PLoS One*, *4*, e6562.

Qian, B. Z., & Pollard, J. W. (2010). Macrophage diversity enhances tumor progression and metastasis. *Cell*, *141*, 39–51.

Qian, J., Yi, H., Guo, C., Yu, X., Zuo, D., Chen, X., et al. (2011). CD204 suppresses large heat shock protein-facilitated priming of tumor antigen gp100-specific T cells and chaperone vaccine activity against mouse melanoma. *Journal of Immunology*, *187*, 2905–2914.

Qiu, P., Ziegelhoffer, P., Sun, J., & Yang, N. S. (1996). Gene gun delivery of mRNA in situ results in efficient transgene expression and genetic immunization. *Gene Therapy*, *3*, 262–268.

Quezada, S. A., Jarvinen, L. Z., Lind, E. F., & Noelle, R. J. (2004). CD40/CD154 interactions at the interface of tolerance and immunity. *Annual Review of Immunology*, *22*, 307–328.

Quezada, S. A., Peggs, K. S., Curran, M. A., & Allison, J. P. (2006). CTLA4 blockade and GM-CSF combination immunotherapy alters the intratumor balance of effector and regulatory T cells. *The Journal of Clinical Investigation*, *116*, 1935–1945.

Quoix, E., Ramlau, R., Westeel, V., Papai, Z., Madroszyk, A., Riviere, A., et al. (2011). Therapeutic vaccination with TG4010 and first-line chemotherapy in advanced non-small-cell lung cancer: A controlled phase 2B trial. *The Lancet Oncology*, *12*, 1125–1133.

Racanelli, V., Leone, P., Frassanito, M. A., Brunetti, C., Perosa, F., Ferrone, S., et al. (2010). Alterations in the antigen processing-presenting machinery of transformed plasma cells are associated with reduced recognition by CD8 + T cells and characterize the progression of MGUS to multiple myeloma. *Blood*, *115*, 1185–1193.

Raty, J. K., Pikkarainen, J. T., Wirth, T., & Yla-Herttuala, S. (2008). Gene therapy: The first approved gene-based medicines, molecular mechanisms and clinical indications. *Current Molecular Pharmacology, 1,* 13–23.

Raychaudhuri, B., Rayman, P., Ireland, J., Ko, J., Rini, B., Borden, E. C., et al. (2011). Myeloid-derived suppressor cell accumulation and function in patients with newly diagnosed glioblastoma. *Neuro-Oncology, 13,* 591–599.

Reits, E. A., Hodge, J. W., Herberts, C. A., Groothuis, T. A., Chakraborty, M., Wansley, E. K., et al. (2006). Radiation modulates the peptide repertoire, enhances MHC class I expression, and induces successful antitumor immunotherapy. *The Journal of Experimental Medicine, 203,* 1259–1271.

Remondo, C., Cereda, V., Mostbock, S., Sabzevari, H., Franzusoff, A., Schlom, J., et al. (2009). Human dendritic cell maturation and activation by a heat-killed recombinant yeast (Saccharomyces cerevisiae) vector encoding carcinoembryonic antigen. *Vaccine, 27,* 987–994.

Respa, A., Bukur, J., Ferrone, S., Pawelec, G., Zhao, Y., Wang, E., et al. (2011). Association of IFN-gamma signal transduction defects with impaired HLA class I antigen processing in melanoma cell lines. *Clinical Cancer Research: An Official Journal of the American Association for Cancer Research, 17,* 2668–2678.

Ribas, A., Weber, J. S., Chmielowski, B., Comin-Anduix, B., Lu, D., Douek, M., et al. (2011). Intra-lymph node prime-boost vaccination against Melan A and tyrosinase for the treatment of metastatic melanoma: Results of a phase 1 clinical trial. *Clinical Cancer Research: An Official Journal of the American Association for Cancer Research, 17,* 2987–2996.

Rice, J., Buchan, S., & Stevenson, F. K. (2002). Critical components of a DNA fusion vaccine able to induce protective cytotoxic T cells against a single epitope of a tumor antigen. *Journal of Immunology, 169,* 3908–3913.

Rice, J., Dunn, S., Piper, K., Buchan, S. L., Moss, P. A., & Stevenson, F. K. (2006). DNA fusion vaccines induce epitope-specific cytotoxic CD8(+) T cells against human leukemia-associated minor histocompatibility antigens. *Cancer Research, 66,* 5436–5442.

Rice, J., Elliott, T., Buchan, S., & Stevenson, F. K. (2001). DNA fusion vaccine designed to induce cytotoxic T cell responses against defined peptide motifs: Implications for cancer vaccines. *Journal of Immunology, 167,* 1558–1565.

Rice, J., Ottensmeier, C. H., & Stevenson, F. K. (2008). DNA vaccines: Precision tools for activating effective immunity against cancer. *Nature Reviews. Cancer, 8,* 108–120.

Rider, P., Carmi, Y., Guttman, O., Braiman, A., Cohen, I., Voronov, E., et al. (2011). IL-1alpha and IL-1beta recruit different myeloid cells and promote different stages of sterile inflammation. *Journal of Immunology, 187,* 4835–4843.

Rivoltini, L., Castelli, C., Carrabba, M., Mazzaferro, V., Pilla, L., Huber, V., et al. (2003). Human tumor-derived heat shock protein 96 mediates in vitro activation and in vivo expansion of melanoma- and colon carcinoma-specific T cells. *Journal of Immunology, 171,* 3467–3474.

Rodriguez, P. C., Quiceno, D. G., Zabaleta, J., Ortiz, B., Zea, A. H., Piazuelo, M. B., et al. (2004). Arginase I production in the tumor microenvironment by mature myeloid cells inhibits T-cell receptor expression and antigen-specific T-cell responses. *Cancer Research, 64,* 5839–5849.

Romano, E., Rossi, M., Ratzinger, G., de Cos, M. A., Chung, D. J., Panageas, K. S., et al. (2011). Peptide-loaded Langerhans cells, despite increased IL15 secretion and T-cell activation in vitro, elicit antitumor T-cell responses comparable to peptide-loaded monocyte-derived dendritic cells in vivo. *Clinical Cancer Research, 17,* 1984–1997.

Rosenberg, S. A., Yang, J. C., Schwartzentruber, D. J., Hwu, P., Marincola, F. M., Topalian, S. L., et al. (1998). Immunologic and therapeutic evaluation of a synthetic peptide vaccine for the treatment of patients with metastatic melanoma. *Nature Medicine, 4,* 321–327.

Rosenberg, S. A., Yannelli, J. R., Yang, J. C., Topalian, S. L., Schwartzentruber, D. J., Weber, J. S., et al. (1994). Treatment of patients with metastatic melanoma with autologous tumor-infiltrating lymphocytes and interleukin 2. *Journal of the National Cancer Institute*, *86*, 1159–1166.

Rosenblatt, J., Vasir, B., Uhl, L., Blotta, S., Macnamara, C., Somaiya, P., et al. (2011). Vaccination with dendritic cell/tumor fusion cells results in cellular and humoral antitumor immune responses in patients with multiple myeloma. *Blood*, *117*, 393–402.

Ruan, Z., Yang, Z., Wang, Y., Wang, H., Chen, Y., Shang, X., et al. (2009). DNA vaccine against tumor endothelial marker 8 inhibits tumor angiogenesis and growth. *Journal of Immunotherapy*, *32*, 486–491.

Ruttinger, D., van den Engel, N. K., Winter, H., Schlemmer, M., Pohla, H., Grutzner, S., et al. (2007). Adjuvant therapeutic vaccination in patients with non-small cell lung cancer made lymphopenic and reconstituted with autologous PBMC: First clinical experience and evidence of an immune response. *Journal of Translational Medicine*, *5*, 43.

Sakaguchi, S. (2004). Naturally arising CD4+ regulatory t cells for immunologic self-tolerance and negative control of immune responses. *Annual Review of Immunology*, *22*, 531–562.

Sakuishi, K., Apetoh, L., Sullivan, J. M., Blazar, B. R., Kuchroo, V. K., & Anderson, A. C. (2010). Targeting Tim-3 and PD-1 pathways to reverse T cell exhaustion and restore anti-tumor immunity. *The Journal of Experimental Medicine*, *207*, 2187–2194.

Salcedo, M., Bercovici, N., Taylor, R., Vereecken, P., Massicard, S., Duriau, D., et al. (2006). Vaccination of melanoma patients using dendritic cells loaded with an allogeneic tumor cell lysate. *Cancer Immunology, Immunotherapy: CII*, *55*, 819–829.

Salgia, R., Lynch, T., Skarin, A., Lucca, J., Lynch, C., Jung, K., et al. (2003). Vaccination with irradiated autologous tumor cells engineered to secrete granulocyte-macrophage colony-stimulating factor augments antitumor immunity in some patients with metastatic non-small-cell lung carcinoma. *Journal of Clinical Oncology: Official Journal of the American Society of Clinical Oncology*, *21*, 624–630.

Sampson, J. H., Archer, G. E., Ashley, D. M., Fuchs, H. E., Hale, L. P., Dranoff, G., et al. (1996). Subcutaneous vaccination with irradiated, cytokine-producing tumor cells stimulates CD8+ cell-mediated immunity against tumors located in the "immunologically privileged" central nervous system. *Proceedings of the National Academy of Sciences of the United States of America*, *93*, 10399–10404.

Sato, E., Olson, S. H., Ahn, J., Bundy, B., Nishikawa, H., Qian, F., et al. (2005). Intraepithelial CD8+ tumor-infiltrating lymphocytes and a high CD8+/regulatory T cell ratio are associated with favorable prognosis in ovarian cancer. *Proceedings of the National Academy of Sciences of the United States of America*, *102*, 18538–18543.

Savelyeva, N., Munday, R., Spellerberg, M. B., Lomonossoff, G. P., & Stevenson, F. K. (2001). Plant viral genes in DNA idiotypic vaccines activate linked CD4+ T-cell mediated immunity against B-cell malignancies. *Nature Biotechnology*, *19*, 760–764.

Scheel, B., Braedel, S., Probst, J., Carralot, J. P., Wagner, H., Schild, H., et al. (2004). Immunostimulating capacities of stabilized RNA molecules. *European Journal of Immunology*, *34*, 537–547.

Scheel, B., Teufel, R., Probst, J., Carralot, J. P., Geginat, J., Radsak, M., et al. (2005). Toll-like receptor-dependent activation of several human blood cell types by protamine-condensed mRNA. *European Journal of Immunology*, *35*, 1557–1566.

Schiffman, M., & Wacholder, S. (2012). Success of HPV vaccination is now a matter of coverage. *The Lancet Oncology*, *13*, 10–12.

Schirrmacher, V. (2005). Clinical trials of antitumor vaccination with an autologous tumor cell vaccine modified by virus infection: Improvement of patient survival based on improved antitumor immune memory. *Cancer Immunology, Immunotherapy: CII*, *54*, 587–598.

Schmidt, S. M., Schag, K., Muller, M. R., Weck, M. M., Appel, S., Kanz, L., et al. (2003). Survivin is a shared tumor-associated antigen expressed in a broad variety of malignancies and recognized by specific cytotoxic T cells. *Blood, 102*, 571–576.

Schmielau, J., & Finn, O. J. (2001). Activated granulocytes and granulocyte-derived hydrogen peroxide are the underlying mechanism of suppression of t-cell function in advanced cancer patients. *Cancer Research, 61*, 4756–4760.

Schmollinger, J. C., Vonderheide, R. H., Hoar, K. M., Maecker, B., Schultze, J. L., Hodi, F. S., et al. (2003). Melanoma inhibitor of apoptosis protein (ML-IAP) is a target for immune-mediated tumor destruction. *Proceedings of the National Academy of Sciences of the United States of America, 100*, 3398–3403.

Schreiber, R. D., Old, L. J., & Smyth, M. J. (2011). Cancer immunoediting: Integrating immunity's roles in cancer suppression and promotion. *Science, 331*, 1565–1570.

Schuler-Thurner, B., Schultz, E. S., Berger, T. G., Weinlich, G., Ebner, S., Woerl, P., et al. (2002). Rapid induction of tumor-specific type 1 T helper cells in metastatic melanoma patients by vaccination with mature, cryopreserved, peptide-loaded monocyte-derived dendritic cells. *The Journal of Experimental Medicine, 195*, 1279–1288.

Schulof, R. S., Mai, D., Nelson, M. A., Paxton, H. M., Cox, J. W., Jr., Turner, M. L., et al. (1988). Active specific immunotherapy with an autologous tumor cell vaccine in patients with resected non-small cell lung cancer. *Molecular Biotherapy, 1*, 30–36.

Schwartzentruber, D. J., Lawson, D. H., Richards, J. M., Conry, R. M., Miller, D. M., Treisman, J., et al. (2011). gp100 peptide vaccine and interleukin-2 in patients with advanced melanoma. *The New England Journal of Medicine, 364*, 2119–2127.

Schwarz, T. F. (2009). Clinical update of the AS04-adjuvanted human papillomavirus-16/18 cervical cancer vaccine, Cervarix. *Advances in Therapy, 26*, 983–998.

Seliger, B., Stoehr, R., Handke, D., Mueller, A., Ferrone, S., Wullich, B., et al. (2010). Association of HLA class I antigen abnormalities with disease progression and early recurrence in prostate cancer. *Cancer Immunology, Immunotherapy: CII, 59*, 529–540.

Senzer, N. N., Kaufman, H. L., Amatruda, T., Nemunaitis, M., Reid, T., Daniels, G., et al. (2009). Phase II clinical trial of a granulocyte-macrophage colony-stimulating factor-encoding, second-generation oncolytic herpesvirus in patients with unresectable metastatic melanoma. *Journal of Clinical Oncology: Official Journal of the American Society of Clinical Oncology, 27*, 5763–5771.

Serafini, P., Carbley, R., Noonan, K. A., Tan, G., Bronte, V., & Borrello, I. (2004). High-dose granulocyte-macrophage colony-stimulating factor-producing vaccines impair the immune response through the recruitment of myeloid suppressor cells. *Cancer Research, 64*, 6337–6343.

Shen, L., Evel-Kabler, K., Strube, R., & Chen, S. Y. (2004). Silencing of SOCS1 enhances antigen presentation by dendritic cells and antigen-specific anti-tumor immunity. *Nature Biotechnology, 22*, 1546–1553.

Shevach, E. M. (2002). CD4+ CD25+ suppressor T cells: More questions than answers. *Nature Reviews. Immunology, 2*, 389–400.

Sikora, A. G., Jaffarzad, N., Hailemichael, Y., Gelbard, A., Stonier, S. W., Schluns, K. S., et al. (2009). IFN-alpha enhances peptide vaccine-induced CD8+ T cell numbers, effector function, and antitumor activity. *Journal of Immunology, 182*, 7398–7407.

Simons, J. W., Carducci, M. A., Mikhak, B., Lim, M., Biedrzycki, B., Borellini, F., et al. (2006). Phase I/II trial of an allogeneic cellular immunotherapy in hormone-naive prostate cancer. *Clinical Cancer Research: An Official Journal of the American Association for Cancer Research, 12*, 3394–3401.

Singh, R., & Paterson, Y. (2006). Listeria monocytogenes as a vector for tumor-associated antigens for cancer immunotherapy. *Expert Review of Vaccines, 5*, 541–552.

Singh, R., & Paterson, Y. (2007). In the FVB/N HER-2/neu transgenic mouse both peripheral and central tolerance limit the immune response targeting HER-2/neu induced by

Listeria monocytogenes-based vaccines. *Cancer Immunology, Immunotherapy: CII, 56,* 927–938.

Slingluff, C. L., Jr., Petroni, G. R., Olson, W. C., Smolkin, M. E., Ross, M. I., Haas, N. B., et al. (2009). Effect of granulocyte/macrophage colony-stimulating factor on circulating CD8+ and CD4+ T-cell responses to a multipeptide melanoma vaccine: Outcome of a multicenter randomized trial. *Clinical Cancer Research: An Official Journal of the American Association for Cancer Research, 15,* 7036–7044.

Sloots, A., Mastini, C., Rohrbach, F., Weth, R., Curcio, C., Burkhardt, U., et al. (2008). DNA vaccines targeting tumor antigens to B7 molecules on antigen-presenting cells induce protective antitumor immunity and delay onset of HER-2/Neu-driven mammary carcinoma. *Clinical Cancer Research: An Official Journal of the American Association for Cancer Research, 14,* 6933–6943.

Small, E. J., Fratesi, P., Reese, D. M., Strang, G., Laus, R., Peshwa, M. V., et al. (2000). Immunotherapy of hormone-refractory prostate cancer with antigen-loaded dendritic cells. *Journal of Clinical Oncology, 18,* 3894–3903.

Small, E. J., Sacks, N., Nemunaitis, J., Urba, W. J., Dula, E., Centeno, A. S., et al. (2007). Granulocyte macrophage colony-stimulating factor–secreting allogeneic cellular immunotherapy for hormone-refractory prostate cancer. *Clinical Cancer Research: An Official Journal of the American Association for Cancer Research, 13,* 3883–3891.

Small, E. J., Schellhammer, P. F., Higano, C. S., Redfern, C. H., Nemunaitis, J. J., Valone, F. H., et al. (2006). Placebo-controlled phase III trial of immunologic therapy with sipuleucel-T (APC8015) in patients with metastatic, asymptomatic hormone refractory prostate cancer. *Journal of Clinical Oncology: Official Journal of the American Society of Clinical Oncology, 24,* 3089–3094.

Soiffer, R., Hodi, F. S., Haluska, F., Jung, K., Gillessen, S., Singer, S., et al. (2003). Vaccination with irradiated, autologous melanoma cells engineered to secrete granulocyte-macrophage colony-stimulating factor by adenoviral-mediated gene transfer augments antitumor immunity in patients with metastatic melanoma. *Journal of Clinical Oncology, 21,* 3343–3350.

Soiffer, R., Lynch, T., Mihm, M., Jung, K., Rhuda, C., Schmollinger, J. C., et al. (1998). Vaccination with irradiated autologous melanoma cells engineered to secrete human granulocyte-macrophage colony-stimulating factor generates potent antitumor immunity in patients with metastatic melanoma. *Proceedings of the National Academy of Sciences of the United States of America, 95,* 13141–13146.

Sondak, V. K., Sabel, M. S., & Mule, J. J. (2006). Allogeneic and autologous melanoma vaccines: Where have we been and where are we going? *Clinical Cancer Research, 12,* 2337s–2341s.

Song, X. T., Evel-Kabler, K., Shen, L., Rollins, L., Huang, X. F., & Chen, S. Y. (2008). A20 is an antigen presentation attenuator, and its inhibition overcomes regulatory T cell-mediated suppression. *Nature Medicine, 14,* 258–265.

Sosman, J. A., Carrillo, C., Urba, W. J., Flaherty, L., Atkins, M. B., Clark, J. I., et al. (2008). Three phase II cytokine working group trials of gp100 (210M) peptide plus high-dose interleukin-2 in patients with HLA-A2-positive advanced melanoma. *Journal of Clinical Oncology: Official Journal of the American Society of Clinical Oncology, 26,* 2292–2298.

Spellerberg, M. B., Zhu, D., Thompsett, A., King, C. A., Hamblin, T. J., & Stevenson, F. K. (1997). DNA vaccines against lymphoma: Promotion of anti-idiotypic antibody responses induced by single chain Fv genes by fusion to tetanus toxin fragment C. *Journal of Immunology, 159,* 1885–1892.

Spies, B., Hochrein, H., Vabulas, M., Huster, K., Busch, D. H., Schmitz, F., et al. (2003). Vaccination with plasmid DNA activates dendritic cells via Toll-like receptor 9 (TLR9) but functions in TLR9-deficient mice. *Journal of Immunology, 171,* 5908–5912.

Spisek, R., Kukreja, A., Chen, L. C., Matthews, P., Mazumder, A., Vesole, D., et al. (2007). Frequent and specific immunity to the embryonal stem cell-associated antigen SOX2 in patients with monoclonal gammopathy. *The Journal of Experimental Medicine, 204*, 831–840.

Srivastava, P. (2002a). Interaction of heat shock proteins with peptides and antigen presenting cells: Chaperoning of the innate and adaptive immune responses. *Annual Review of Immunology, 20*, 395–425.

Srivastava, P. (2002b). Roles of heat-shock proteins in innate and adaptive immunity. *Nature Reviews. Immunology, 2*, 185–194.

Srivastava, P. K. (2005). Immunotherapy for human cancer using heat shock protein-peptide complexes. *Current Oncology Reports, 7*, 104–108.

Srivastava, P. K., DeLeo, A. B., & Old, L. J. (1986). Tumor rejection antigens of chemically induced sarcomas of inbred mice. *Proceedings of the National Academy of Sciences of the United States of America, 83*, 3407–3411.

Srivastava, M. K., Sinha, P., Clements, V. K., Rodriguez, P., & Ostrand-Rosenberg, S. (2010). Myeloid-derived suppressor cells inhibit T-cell activation by depleting cystine and cysteine. *Cancer Research, 70*, 68–77.

Steele, J. C., Rao, A., Marsden, J. R., Armstrong, C. J., Berhane, S., Billingham, L. J., et al. (2011). Phase I/II trial of a dendritic cell vaccine transfected with DNA encoding melan A and gp100 for patients with metastatic melanoma. *Gene Therapy, 18*, 584–593.

Stein, W. D., Gulley, J. L., Schlom, J., Madan, R. A., Dahut, W., Figg, W. D., et al. (2011). Tumor regression and growth rates determined in five intramural NCI prostate cancer trials: The growth rate constant as an indicator of therapeutic efficacy. *Clinical Cancer Research: An Official Journal of the American Association for Cancer Research, 17*, 907–917.

Steiner, H. H., Bonsanto, M. M., Beckhove, P., Brysch, M., Geletneky, K., Ahmadi, R., et al. (2004). Antitumor vaccination of patients with glioblastoma multiforme: A pilot study to assess feasibility, safety, and clinical benefit. *Journal of Clinical Oncology: Official Journal of the American Society of Clinical Oncology, 22*, 4272–4281.

Stratford, R., Douce, G., Zhang-Barber, L., Fairweather, N., Eskola, J., & Dougan, G. (2000). Influence of codon usage on the immunogenicity of a DNA vaccine against tetanus. *Vaccine, 19*, 810–815.

Su, Z., Dannull, J., Yang, B. K., Dahm, P., Coleman, D., Yancey, D., et al. (2005). Telomerase mRNA-transfected dendritic cells stimulate antigen-specific CD8+ and CD4+ T cell responses in patients with metastatic prostate cancer. *Journal of Immunology, 174*, 3798–3807.

Subjeck, J. R., & Shyy, T. T. (1986). Stress protein systems of mammalian cells. *American Journal of Physiology, 250*, C1–C17.

Tamura, Y., Peng, P., Liu, K., Daou, M., & Srivastava, P. K. (1997). Immunotherapy of tumors with autologous tumor-derived heat shock protein preparations. *Science, 278*, 117–120.

Tcherepanova, I. Y., Adams, M. D., Feng, X., Hinohara, A., Horvatinovich, J., Calderhead, D., et al. (2008). Ectopic expression of a truncated CD40L protein from synthetic post-transcriptionally capped RNA in dendritic cells induces high levels of IL-12 secretion. *BMC Molecular Biology, 9*, 90.

Tesniere, A., Schlemmer, F., Boige, V., Kepp, O., Martins, I., Ghiringhelli, F., et al. (2010). Immunogenic death of colon cancer cells treated with oxaliplatin. *Oncogene, 29*, 482–491.

Testori, A., Richards, J., Whitman, E., Mann, G. B., Lutzky, J., Camacho, L., et al. (2008). Phase III comparison of vitespen, an autologous tumor-derived heat shock protein gp96 peptide complex vaccine, with physician's choice of treatment for stage IV melanoma: The C-100-21 Study Group. *Journal of Clinical Oncology, 26*, 955–962.

Therasse, P., Eisenhauer, E. A., & Verweij, J. (2006). RECIST revisited: A review of validation studies on tumour assessment. *European Journal of Cancer, 42,* 1031–1039.

Theriault, J. R., Adachi, H., & Calderwood, S. K. (2006). Role of scavenger receptors in the binding and internalization of heat shock protein 70. *Journal of Immunology, 177,* 8604–8611.

Thurner, B., Haendle, I., Roder, C., Dieckmann, D., Keikavoussi, P., Jonuleit, H., et al. (1999). Vaccination with mage-3A1 peptide-pulsed mature, monocyte-derived dendritic cells expands specific cytotoxic T cells and induces regression of some metastases in advanced stage IV melanoma. *The Journal of Experimental Medicine, 190,* 1669–1678.

Timmerman, J. M., & Levy, R. (1999). Dendritic cell vaccines for cancer immunotherapy. *Annual Review of Medicine, 50,* 507–529.

Todryk, S. M., Melcher, A. A., Dalgleish, A. G., & Vile, R. G. (2000). Heat shock proteins refine the danger theory. *Immunology, 99,* 334–337.

Tsang, K. Y., Zaremba, S., Nieroda, C. A., Zhu, M. Z., Hamilton, J. M., & Schlom, J. (1995). Generation of human cytotoxic T cells specific for human carcinoembryonic antigen epitopes from patients immunized with recombinant vaccinia-CEA vaccine. *Journal of the National Cancer Institute, 87,* 982–990.

Tuyaerts, S., Aerts, J. L., Corthals, J., Neyns, B., Heirman, C., Breckpot, K., et al. (2007). Current approaches in dendritic cell generation and future implications for cancer immunotherapy. *Cancer Immunology, Immunotherapy: CII, 56,* 1513–1537.

Udono, H., Levey, D. L., & Srivastava, P. K. (1994). Cellular requirements for tumor-specific immunity elicited by heat shock proteins: Tumor rejection antigen gp96 primes CD8 + T cells in vivo. *Proceedings of the National Academy of Sciences of the United States of America, 91,* 3077–3081.

Uehori, J., Matsumoto, M., Tsuji, S., Akazawa, T., Takeuchi, O., Akira, S., et al. (2003). Simultaneous blocking of human Toll-like receptors 2 and 4 suppresses myeloid dendritic cell activation induced by Mycobacterium bovis bacillus Calmette-Guerin peptidoglycan. *Infection and Immunity, 71,* 4238–4249.

Vanaja, D. K., Grossmann, M. E., Celis, E., & Young, C. Y. (2000). Tumor prevention and antitumor immunity with heat shock protein 70 induced by 15-deoxy-delta12,14-prostaglandin J2 in transgenic adenocarcinoma of mouse prostate cells. *Cancer Research, 60,* 4714–4718.

van den Eertwegh, A. J., Versluis, J., van den Berg, H. P., Santegoets, S. J., van Moorselaar, R. J., van der Sluis, T. M., et al. (2012). Combined immunotherapy with granulocyte-macrophage colony-stimulating factor-transduced allogeneic prostate cancer cells and ipilimumab in patients with metastatic castration-resistant prostate cancer: A phase 1 dose-escalation trial. *The Lancet Oncology, 13,* 509–517.

van der Bruggen, P., Traversari, C., Chomez, P., Lurquin, C., De Plaen, E., Van den Eynde, B., et al. (1991). A gene encoding an antigen recognized by cytolytic T lymphocytes on a human melanoma. *Science, 254,* 1643 1647.

van der Meijden, A. P., Sylvester, R. J., Oosterlinck, W., Hoeltl, W., & Bono, A. V. (2003). Maintenance Bacillus Calmette-Guerin for Ta T1 bladder tumors is not associated with increased toxicity: Results from a European Organisation for Research and Treatment of Cancer Genito-Urinary Group Phase III Trial. *European Urology, 44,* 429–434.

van Elsas, A., Hurwitz, A. A., & Allison, J. P. (1999). Combination immunotherapy of B16 melanoma using anti-cytotoxic T lymphocyte-associated antigen 4 (CTLA-4) and granulocyte/macrophage colony-stimulating factor (GM-CSF)-producing vaccines induces rejection of subcutaneous and metastatic tumors accompanied by autoimmune depigmentation. *The Journal of Experimental Medicine, 190,* 355–366.

Vergati, M., Cereda, V., Madan, R. A., Gulley, J. L., Huen, N. Y., Rogers, C. J., et al. (2011). Analysis of circulating regulatory T cells in patients with metastatic prostate cancer pre- versus post-vaccination. *Cancer Immunology, Immunotherapy: CII, 60,* 197–206.

Vesely, M. D., Kershaw, M. H., Schreiber, R. D., & Smyth, M. J. (2011). Natural innate and adaptive immunity to cancer. *Annual Review of Immunology, 29*, 235–271.

Vignali, D. A., Collison, L. W., & Workman, C. J. (2008). How regulatory T cells work. *Nature Reviews. Immunology, 8*, 523–532.

Vonderheide, R. H., Hahn, W. C., Schultze, J. L., & Nadler, L. M. (1999). The telomerase catalytic subunit is a widely expressed tumor-associated antigen recognized by cytotoxic T lymphocytes. *Immunity, 10*, 673–679.

von Mehren, M., Arlen, P., Gulley, J., Rogatko, A., Cooper, H. S., Meropol, N. J., et al. (2001). The influence of granulocyte macrophage colony-stimulating factor and prior chemotherapy on the immunological response to a vaccine (ALVAC-CEA B7.1) in patients with metastatic carcinoma. *Clinical Cancer Research: An Official Journal of the American Association for Cancer Research, 7*, 1181–1191.

Wakimoto, H., Abe, J., Tsunoda, R., Aoyagi, M., Hirakawa, K., & Hamada, H. (1996). Intensified antitumor immunity by a cancer vaccine that produces granulocyte-macrophage colony-stimulating factor plus interleukin 4. *Cancer Research, 56*, 1828–1833.

Walsh, S. R., & Dolin, R. (2011). Vaccinia viruses: Vaccines against smallpox and vectors against infectious diseases and tumors. *Expert Review of Vaccines, 10*, 1221–1240.

Walter, P., & Johnson, A. E. (1994). Signal sequence recognition and protein targeting to the endoplasmic reticulum membrane. *Annual Review of Cell Biology, 10*, 87–119.

Wang, X. Y., Arnouk, H., Chen, X., Kazim, L., Repasky, E. A., & Subjeck, J. R. (2006). Extracellular targeting of endoplasmic reticulum chaperone glucose-regulated protein 170 enhances tumor immunity to a poorly immunogenic melanoma. *Journal of Immunology, 177*, 1543–1551.

Wang, X. Y., Chen, X., Manjili, M. H., Repasky, E., Henderson, R., & Subjeck, J. R. (2003). Targeted immunotherapy using reconstituted chaperone complexes of heat shock protein 110 and melanoma-associated antigen gp100. *Cancer Research, 63*, 2553–2560.

Wang, X.-Y. , Easton, D. P., & Subjeck, J. R. (2007). *Large mammalian hsp70 family proteins, hsp110 and grp170, and their roles in biology and cancer therapy*. New York: Springer.

Wang, X. Y., Facciponte, J., Chen, X., Subjeck, J. R., & Repasky, E. A. (2007). Scavenger receptor-A negatively regulates antitumor immunity. *Cancer Research, 67*, 4996–5002.

Wang, X. Y., Facciponte, J. G., & Subjeck, J. R. (2006). Molecular chaperones and cancer immunotherapy. *Handbook of Experimental Pharmacology, 172*, 305–329.

Wang, X. Y., Kazim, L., Repasky, E. A., & Subjeck, J. R. (2001). Characterization of heat shock protein 110 and glucose-regulated protein 170 as cancer vaccines and the effect of fever-range hyperthermia on vaccine activity. *Journal of Immunology, 166*, 490–497.

Wang, X. Y., Sun, X., Chen, X., Facciponte, J., Repasky, E. A., Kane, J., et al. (2010). Superior antitumor response induced by large stress protein chaperoned protein antigen compared with peptide antigen. *Journal of Immunology, 184*, 6309–6319.

Wang, Y., Wang, X. Y., Subjeck, J. R., Shrikant, P. A., & Kim, H. L. (2011). Temsirolimus, an mTOR inhibitor, enhances anti-tumour effects of heat shock protein cancer vaccines. *British Journal of Cancer, 104*, 643–652.

Wang, B., Zaidi, N., He, L. Z., Zhang, L., Kuroiwa, J. M., Keler, T., et al. (2012). Targeting of the non-mutated tumor antigen HER2/neu to mature dendritic cells induces an integrated immune response that protects against breast cancer in mice. *Breast Cancer Research, 14*, R39.

Wang, X. Y., Zuo, D., Sarkar, D., & Fisher, P. B. (2011). Blockade of cytotoxic T-lymphocyte antigen-4 as a new therapeutic approach for advanced melanoma. *Expert Opinion on Pharmacotherapy, 12*, 2695–2706.

Wansley, E. K., Chakraborty, M., Hance, K. W., Bernstein, M. B., Boehm, A. L., Guo, Z., et al. (2008). Vaccination with a recombinant Saccharomyces cerevisiae expressing a tumor antigen breaks immune tolerance and elicits therapeutic antitumor responses. *Clinical Cancer Research: An Official Journal of the American Association for Cancer Research*, *14*, 4316–4325.

Weber, J. S., Vogelzang, N. J., Ernstoff, M. S., Goodman, O. B., Cranmer, L. D., Marshall, J. L., et al. (2011). A phase 1 study of a vaccine targeting preferentially expressed antigen in melanoma and prostate-specific membrane antigen in patients with advanced solid tumors. *Journal of Immunotherapy*, *34*, 556–567.

Wei, H., Wang, S., Zhang, D., Hou, S., Qian, W., Li, B., et al. (2009). Targeted delivery of tumor antigens to activated dendritic cells via CD11c molecules induces potent antitumor immunity in mice. *Clinical Cancer Research*, *15*, 4612–4621.

Weide, B., Carralot, J. P., Reese, A., Scheel, B., Eigentler, T. K., Hoerr, I., et al. (2008). Results of the first phase I/II clinical vaccination trial with direct injection of mRNA. *Journal of Immunotherapy*, *31*, 180–188.

Weide, B., Pascolo, S., Scheel, B., Derhovanessian, E., Pflugfelder, A., Eigentler, T. K., et al. (2009). Direct injection of protamine-protected mRNA: Results of a phase 1/2 vaccination trial in metastatic melanoma patients. *Journal of Immunotherapy*, *32*, 498–507.

Welch, W. J. (1993). Heat shock proteins functioning as molecular chaperones: Their roles in normal and stressed cells. *Philosophical Transactions of the Royal Society of London. Series B, Biological Sciences*, *339*, 327–333.

Wille-Reece, U., Flynn, B. J., Lore, K., Koup, R. A., Miles, A. P., Saul, A., et al. (2006). Toll-like receptor agonists influence the magnitude and quality of memory T cell responses after prime-boost immunization in nonhuman primates. *The Journal of Experimental Medicine*, *203*, 1249–1258.

Wolchok, J. D., Hoos, A., O'Day, S., Weber, J. S., Hamid, O., Lebbe, C., et al. (2009). Guidelines for the evaluation of immune therapy activity in solid tumors: Immune-related response criteria. *Clinical Cancer Research*, *15*, 7412–7420.

Wolchok, J. D., Yang, A. S., & Weber, J. S. (2010). Immune regulatory antibodies: Are they the next advance? *Cancer Journal*, *16*, 311–317.

Wolff, J. A., Malone, R. W., Williams, P., Chong, W., Acsadi, G., Jani, A., et al. (1990). Direct gene transfer into mouse muscle in vivo. *Science*, *247*, 1465–1468.

Wolkers, M. C., Toebes, M., Okabe, M., Haanen, J. B., & Schumacher, T. N. (2002). Optimizing the efficacy of epitope-directed DNA vaccination. *Journal of Immunology*, *168*, 4998–5004.

Wood, C., Srivastava, P., Bukowski, R., Lacombe, L., Gorelov, A. I., Gorelov, S., et al. (2008). An adjuvant autologous therapeutic vaccine (HSPPC-96; vitespen) versus observation alone for patients at high risk of recurrence after nephrectomy for renal cell carcinoma: A multicentre, open-label, randomised phase III trial. *Lancet*, *372*, 145–154.

Xiang, R., Luo, Y., Niethammer, A. G., & Reisfeld, R. A. (2008). Oral DNA vaccines target the tumor vasculature and microenvironment and suppress tumor growth and metastasis. *Immunological Reviews*, *222*, 117–128.

Yang, J. C. (2008). Vitespen: A vaccine for renal cancer? *Lancet*, *372*, 92–93.

Yedavelli, S. P., Guo, L., Daou, M. E., Srivastava, P. K., Mittelman, A., & Tiwari, R. K. (1999). Preventive and therapeutic effect of tumor derived heat shock protein, gp96, in an experimental prostate cancer model. *International Journal of Molecular Medicine*, *4*, 243–248.

Yi, H., Guo, C., Yu, X., Gao, P., Qian, J., Zuo, D., et al. (2011). Targeting the immunoregulator SRA/CD204 potentiates specific dendritic cell vaccine–induced T cell response and antitumor immunity. *Cancer Research*, *71*, 6611–6620.

Yi, H., Yu, X., Gao, P., Wang, Y., Baek, S. H., Chen, X., et al. (2009). Pattern recognition scavenger receptor SRA/CD204 down-regulates Toll-like receptor 4 signaling-dependent CD8 T-cell activation. *Blood*, *113*, 5819–5828.

Yi, H., Zuo, D., Yu, X., Hu, F., Manjili, M. H., Chen, Z., et al. (2012). Suppression of antigen-specific CD4(+) T cell activation by SRA/CD204 through reducing the immunostimulatory capability of antigen-presenting cell. *Journal of Molecular Medicine, 90*, 413–426.

Ying, H., Zaks, T. Z., Wang, R. F., Irvine, K. R., Kammula, U. S., Marincola, F. M., et al. (1999). Cancer therapy using a self-replicating RNA vaccine. *Nature Medicine, 5*, 823–827.

Youn, J. I., Nagaraj, S., Collazo, M., & Gabrilovich, D. I. (2008). Subsets of myeloid-derived suppressor cells in tumor-bearing mice. *Journal of Immunology, 181*, 5791–5802.

Younes, A. (2003). A phase II study of heat shock protein-peptide complex-96 vaccine therapy in patients with indolent non-Hodgkin's lymphoma. *Clinical Lymphoma, 4*, 183–185.

Yu, J. S., Wheeler, C. J., Zeltzer, P. M., Ying, H., Finger, D. N., Lee, P. K., et al. (2001). Vaccination of malignant glioma patients with peptide-pulsed dendritic cells elicits systemic cytotoxicity and intracranial T-cell infiltration. *Cancer Research, 61*, 842–847.

Yu, X., Yi, H., Guo, C., Zuo, D., Wang, Y., Kim, H. L., et al. (2011). Pattern recognition scavenger receptor CD204 attenuates Toll-like receptor 4-induced NF-kappaB activation by directly inhibiting ubiquitination of tumor necrosis factor (TNF) receptor-associated factor 6. *Journal of Biological Chemistry, 286*, 18795–18806.

Yuan, J., Gnjatic, S., Li, H., Powel, S., Gallardo, H. F., Ritter, E., et al. (2008). CTLA-4 blockade enhances polyfunctional NY-ESO-1 specific T cell responses in metastatic melanoma patients with clinical benefit. *Proceedings of the National Academy of Sciences of the United States of America, 105*, 20410–20415.

Zhou, G., Drake, C. G., & Levitsky, H. I. (2006). Amplification of tumor-specific regulatory T cells following therapeutic cancer vaccines. *Blood, 107*, 628–636.

Zhou, G., & Levitsky, H. I. (2007). Natural regulatory T cells and de novo-induced regulatory T cells contribute independently to tumor-specific tolerance. *Journal of Immunology, 178*, 2155–2162.

Zhou, G., & Levitsky, H. (2012). Towards curative cancer immunotherapy: Overcoming posttherapy tumor escape. *Clinical and Developmental Immunology, 2012*, 124187.

Zitvogel, L., Kepp, O., Senovilla, L., Menger, L., Chaput, N., & Kroemer, G. (2010). Immunogenic tumor cell death for optimal anticancer therapy: The calreticulin exposure pathway. *Clinical Cancer Research, 16*, 3100–3104.

INDEX

Note: Page numbers followed by "*f*" indicate figures, and "*t*" indicate tables.

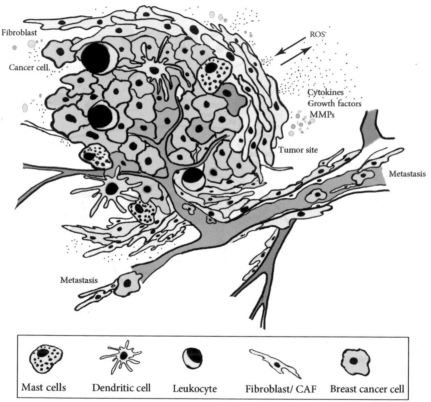

Fibroblast

Cancer cell

ROS

Cytokines
Growth factors
MMPs

Tumor site

Metastasis

Metastasis

| Mast cells | Dendritic cell | Leukocyte | Fibroblast/ CAF | Breast cancer cell |

Figure 3.1, Agnieszka Jezierska-Drutel *et al.* (See Page 109 of this volume.)

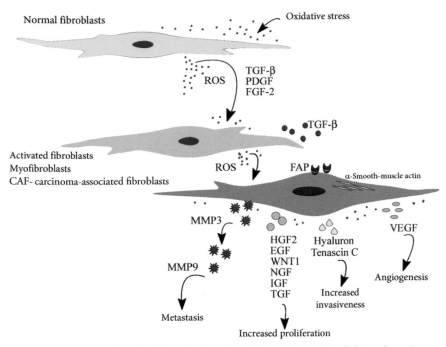

Figure 3.2, Agnieszka Jezierska-Drutel *et al.* (See Page 110 of this volume.)

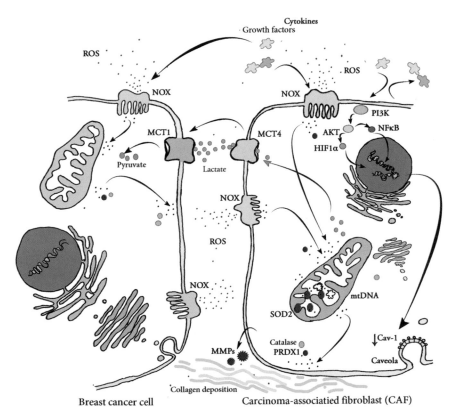

Figure 3.3, Agnieszka Jezierska-Drutel *et al.* (See Page 119 of this volume.)

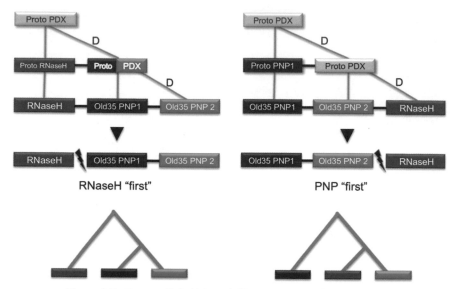

RNaseH "first"

PNP "first"

Figure 5.5, Upneet K. Sokhi *et al.* (See Page 173 of this volume.)

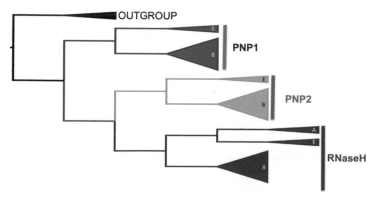

Figure 5.6, Upneet K. Sokhi *et al.* (See Page 174 of this volume.)